Further Mathematics

$$(+, -, \circ, \div)$$

ARSAL
———————
2018

Further Mathematics

George N. Frempong

Copyright © 2017 by George N. Frempong

ISBN – 13:978 – 1546790020

ISBN – 10:1546790020

All rights reserved. No part of this book may be reproduced or transmitted in any form or by any means, electronic or mechanical, including photocopying, recording, or by any information storage and retrieval system, without permission in writing from the copyright owner

Printed by CreateSpace

www.CreateSpace.com/TITLEID

Available from Amazon.com and other retail outlets

Contents

Preface

1. Topics in Algebra	1
2. Logic	39
3. Sets	59
4. Surds	73
5. Binary Operations	85
6. Functions	93
7. Linear Functions	113
8. Linear Inequalities in Two Variables	137
9. Quadratic Functions	151
10. Polynomials	177
11. Rational Functions	187
Test covering Chapter 1 – 11	195
12. Conic Sections	199
13. Binomial Expansions	219
14. Indices and Logarithms	229
15. Sequences	255
16. Trigonometry	277
17. Matrices and Transformations	325
Test covering Chapter 12 – 17	347
18. Differentiation and Integration	351

19. Vectors	431
20. Mechanics	461
21. Permutations and Combinations	513
22. Probabilities	525
23. Statistics	547
Test covering Chapter 18 – 23	599
Answer to exercises	605

PREFACE

This book has been developed with the student in mind, and written to meet varied curricular requirements.

The purpose of this book is to provide you with the tools you need to be successful in your mathematics course. The exercises at the end of each section are varied, providing basic skills and the understanding of the concepts. During more than 30 years of teaching, I have learned that students are most successful when they learn and retain mathematical concepts.

Every chapter ends with a review exercise. The review exercises give you the opportunity to see where you need help and provide additional practice with concepts in the chapter or previous chapters. The test offer you opportunities for self-assessment, help you track your progress and also help you retain mastery of earlier topics.

The book is easy to follow with little help and its clarity makes this book excellent for self-study.

Over the years I have accumulated and absorbed ideas from textbooks, my colleagues and students. It is impossible to look back and recall the origin of most of the material in this book. I am very grateful to all whose work I have benefitted from.

1.1

1) $2(x+3)$
   $= 2x+6$

2) $4(x-y)$
   $= 4x-4y$

3) $-3(a-2b)$
   $= -3a+6b$

4) $3(2x+y)$
   $= 6x+3y$

5) $-4y(y-2x)$
   $= -4y^2 + 8xy$

6) $3x^2(x-y)$
   $= 3x^3 - 3x^2y$

7) $2(3x-y)+4y$
   $= 6x - 2y + 4y$
   $= 6x + 2y$

8) $3x \cdot 2(3x-y)$
   $= 9x^2 - 3xy - 6x + 2y$

9) $-a(a^2-b) + 2a^3$
   $= -a^3 + ab + 2a^3$
   $= a^3 + ab$

10) $5(x+2) + 3(2x-1)$
    $= 5x + 10 + 6x - 3$
    $= 11x + 7$

11) $3(2x-1) - 7(x-3)$
    $= 6x - 3 - 7x + 21$
    $= 18 - x$

12) $3(x+2y) + 2(x-3y)$
    $= 3x + 6y + 2x - 6y$
    $= 5x$

13) $x(2x+y) - 2x(3x-y)$
    $= 2x^2 + xy - 6x^2 + 2xy$
    $= -4x^2 + 3xy$

14) $y(y-3z) - (3y^2 - 5yz)$
    $= y^2 - 3yz - 3y^2 + 5yz$
    $= -2y^2 + 2yz$
    $= 2y(-y + z)$

15) $y(y^2 - 2y + 1) - 2y(y^2 - 3y - 2)$
    $= y^3 - 2y^2 + y - 2y^3 + 6y^2 + 4y$
    $= -y^3 + 4y^2 + 5y$
    $= y(-y^2 + 4y + 5)$

1 Topics in Algebra

1.1 Expanding Brackets

To expand an expression such as $x(3x + 2)$ we use the Distributive Property;

$$a(b + c) = ab + ac$$

Multiplying each term inside the bracket by x gives

$$x(3x + 2) = x \cdot 3x + x \cdot 2$$
$$= 3x^2 + 2x$$

With practice you can do the first step mentally.

Example 1

(a) Expand $-3a^2(2a^3 - 4a)$

Solution $-3a^2(2a^3 - 4a) = -6a^5 + 12a^3$

Notice that if the term outside the bracket is negative then each term inside the bracket changes sign.

(b) Simplify $2(4x + 3y) - 3(2x - y)$

Solution $2(4x + 3y) - 3(2x - y) = 8x + 6y - 6x + 3y$
$$= 2x + 9y$$

Exercise 1.1

Expand and simplify each of the following expressions if possible:

1. $2(x + 3)$
2. $4(x - y)$
3. $-3(a - 2b)$
4. $3(2x + y)$
5. $-4y(y - 2x)$
6. $3x^2(x - y)$
7. $2(3x - y) + 4y$
8. $3x - 2(3x - y)$
9. $-a(a^2 - b) + 2a^3$
10. $5(x + 2) + 3(2x - 1)$
11. $3(2x - 1) - 7(x - 3)$
12. $3(x + 2y) + 2(x - 3y)$
13. $x(2x + y) - 2x(3x - y)$
14. $y(y - 3z) - (3y^2 - 5yz)$
15. $y(y^2 - 2y + 1) - 2y(y^2 - 3y - 2)$

1.2 Factorising Binomials

In the preceding section we learned to expand some expressions such as $3(2x + 4)$ by multiplying each term inside a bracket by the term outside the bracket. Using the Distributive Property backward we can write an algebraic expression as a product of factors. This process is called factorisation.

Consider the following expression $8x^3 + 20x^2$.

To factorise this expression we first identify the largest common factor of $8x^3$ and $20x^2$. Then remove this factor from each of the terms and place it outside the bracket.

You can find the largest common factor by looking for the largest common numerical coefficient and the largest common power of each common variable. In this case, the largest common factor of $8x^3$ and $20x^2$ is $4x^2$. Factorising out $4x^2$ gives

$$8x^3 + 20x^2 = 4x^2(2x + 5)$$

Example 2

Factorise $12x^2y + 18xy^2$

Solution The largest common factor is $6xy$. We write

$$12x^2y + 18xy^2 = 6xy(2x + 3y)$$

It is also true that $12x^2y + 18xy^2 = 3xy(4x + 6y)$. However this is not completely factorised. Remember that the common factor must be the largest common factor.

Exercise 1.2(a)

Factorise each of the following expressions:

1. $2x + 10$
2. $3x + 6y$
3. $3y + 15$
4. $5a - 20ab$
5. $3a^2x + 2ax^2$
6. $6xy^2 - 3x^2y$
7. $9x^3 + 6x^2y$
8. $9x^3y - 36x^2y^2$
9. $a^3b + a^2b^2$
10. $a^2b - ab^2$
11. $6p^3 + 18p^2$
12. $ax^2 - ax$

Factorising by Grouping

An algebraic expression with four or more terms can be factorised by splitting them into groups of terms with a common factor. This method is known as factorisation by grouping.

Example 3 Factorise each expression

(a) $a^2 + ab + ac + bc$

Solution Group the first and second terms and the third and fourth terms, and then factorise each group.

$$a^2 + ab + ac + bc = (a^2 + ab) + (ac + bc) \quad \text{Group terms}$$
$$= a(a + b) + c(a + b) \quad \text{Factorise grouped terms}$$
$$= (a + b)(a + c) \quad \text{Factorise out the binomial factor}$$

(b) $3ax - 4by + 6bx - 2ay$

Solution The first and second terms and the third and fourth terms do not share a common factor so we rewrite the expression before we group the terms.

$$3ax - 4by + 6bx - 2ay = 3ax + 6bx - 2ay - 4by \quad \text{Rearrange terms}$$
$$= (3ax + 6bx) - (2ay + 4by) \quad \text{Group terms}$$
$$= 3x(a + 2b) - 2y(a + 2b) \quad \text{Factorise}$$
$$= (a + 2b)(3x - 2y) \quad \text{Factorise}$$

Exercise 1.2(b)

Factorise each of the following expressions:

1. $ax + bx + ay + by$ $\;ax+ay+bx+by$
   $a(x+y)+b(x+y)$
2. $xy - 2y + 3x - 6$ $\;y(x-2)+3(x-2)$
3. $y^3 + y^2 + y + 1$
4. $x^2 - 2xy - 3x + 6y$
5. $ab - 6by + 3ay - 2b^2$
6. $x^2y + xy^2 - yz - xz$
7. $3ax - a - 6bx + 2b$
8. $21ab - xy - 3bx + 7ay$

1.3 The Product of Two Binomials

Another use of the distributive property is to multiply two binomials. For example, to find the product of $(x + 4)$ and $(x + 5)$ we treat $(x + 4)$ as a single quantity and multiply as follows:

$$(x + 4)(x + 5) = (x + 4) \cdot x + (x + 4) \cdot 5$$
$$= x \cdot x + 4 \cdot x + x \cdot 5 + 4 \cdot 5$$

$$= x^2 + 4x + 5x + 20$$

$$= x^2 + 5x + 4x + 20 \qquad \text{Using Commutative Property}$$

$$= x^2 + 9x + 20$$

With practice we could write down the product without writing out all of the steps. The fourth step can be obtained by multiplying each term of the second bracket by the first term and then by the second term of the first bracket.

Example 4 Expand each expression

(a) $(x - 3)(x + 2)$

Solution $(x - 3)(x + 2) = x^2 + 2x - 3x - 6$

$$= x^2 - x - 6$$

(b) $(x - 7)(x - 3)$

Solution $(x - 7)(x - 3) = x^2 - 3x - 7x + 21$

$$= x^2 - 10x + 21$$

Exercise 1.3(a)

Expand each of the following expression:

1. $(x + 1)(x + 2)$
2. $(x - 6)(x - 2)$
3. $(x - 2)(x + 5)$
4. $(2x + 3)(x - 2)$
5. $(3x + 2)(2x + 3)$
6. $(2x - 3)(2x - 5)$
7. $(x - 3)(x + 7) + (x - 8)(x + 3)$
8. $(x - 3)(x - 5) - (x - 1)(x - 4)$

Squares of Binomials

Consider the square of the binomial $(a + b)$. Note that $(a + b)^2$ can be expressed as $(a + b)(a + b)$. This is a product of two binomials and is multiplied as follows:

$$(a + b)^2 = (a + b)(a + b)$$

$$= a^2 + ab + ba + b^2$$

$$= a^2 + 2ab + b^2$$

In a similar manner we obtain $(a - b)^2 = a^2 - 2ab + b^2$

The two expressions $a^2 + 2ab + b^2$ and $a^2 - 2ab + b^2$ are called perfect square trinomials.

Notice that:

1. the first term of the square of a binomial is the square of the first term.

2. the middle term of the square of a binomial is twice the product of the two terms.

3. the last term of the square of a binomial is the square of the last term.

Example 5 Expand each expression

(a) $(2x + 3)^2$

Solution $(2x + 3)^2 = (2x)^2 + 2(2x)(3) + 3^2$

$$= 4x^2 + 12x + 9$$

Note that $(2x + 3)^2 \neq 4x^2 + 9$. A common mistake in squaring binomials is to forget the middle term.

(b) $(4x - y)^2$

Solution $(4x - y)^2 = 16x^2 - 8xy + y^2$

Exercise 1.3(b)

Expand each of the following expressions and simplify if possible:

1. $(x + 3)^2$
2. $(x - 1)^2$
3. $(x + 5)^2$
4. $(x - 6)^2$
5. $(2x + 3)^2$
6. $(3x - 5)^2$
7. $(2x - 5y)^2$
8. $(x + 3y)^2$
9. $(x + 1)^2 + (x - 1)^2$
10. $(x - 3)^2 + (2x + 1)^2$
11. $(3x - y)^2 - (2x + y)^2$
12. $2(x + 1)^2 + 3(x - 2)^2$

Product of Two Binomials that differ only in sign

Consider the product of the sum and difference of the same two terms such as $(a - b)(a + b)$

Since this is the product of two binomials we have

$(a - b)(a + b) = a^2 + ab - ba - b^2$

$$= a^2 - b^2$$

6 Further Mathematics

The binomial expression $a^2 - b^2$ is called difference of two squares.

Notice that the product of two binomials that differ only in the sign between the terms is the square of the first term minus the square of the second term.

Example 6 Expand each expression

(a) $(x + 3)(x - 3)$

Solution $(x + 3)(x - 3) = x^2 - 3^2$
$$= x^2 - 9$$

(b) $(3x - 2y)(3x + 2y)$

Solution $(3x - 2y)(3x + 2y) = 9x^2 - 4y^2$

Exercise 1.3(c)

Expand:

1. $(x + 2)(x - 2)$
2. $(x + 6)(x - 6)$
3. $(4 - 3x)(4 + 3x)$
4. $(3x + 2)(3x - 2)$
5. $(x^2 + y^2)(x^2 - y^2)$
6. $(2x^2 - 3)(2x^2 + 3)$
7. $(7x - 5y)(7x + 5y)$
8. $(a^2x^3 - 4y^2)(a^2x^3 + 4y^2)$

1.4 Factorising Trinomials

In Section 1.3 you may have noticed that the products of two binomials often have three terms. It follows that some trinomials can be factorised into the product of two binomial factors. In this section we will learn to factorise trinomials of the form $ax^2 + bx + c$.

Factorising $ax^2 + bx + c$ where $a = 1$

To factorise $x^2 + bx + c$ into a product of two binomials we look for two factors of c whose sum is b. It may be helpful to list all of the distinct pair of factors, and then choose the appropriate pair from the list.

Example 7 Factorise each expression

(a) $x^2 + 6x + 8$

Solution The numbers 2 and 4 have a product of 8 and a sum of 6. Thus
$$x^2 + 6x + 8 = (x + 2)(x + 4)$$

We could also write the answer as $(x + 4)(x + 2)$.

(b) $x^2 - 9x + 14$

Solution The number -7 and -2 have a product of 14 and a sum -9.

$$\text{So } x^2 - 9x + 14 = (x - 2)(x - 7)$$

(c) $x^2 - 2x - 15$

Solution The numbers -5 and 3 have a product -15 and a sum -2.

$$\text{So } x^2 - 2x - 15 = (x - 5)(x + 3)$$

Note that:

1. If b and c are both positive the two factors of c are both positive.

2. If b is negative and c is positive the two factors of c are both negative.

3. If b is positive and c is negative the two factors of c are of opposite signs (the value with the larger absolute value is positive).

4. If b is negative and c is negative the two factors of c are of opposite signs (the value with the larger absolute value is negative).

Exercise 1.4(a)

Factorise:

1. $x^2 + 5x + 6$
2. $x^2 + 7x + 10$
3. $x^2 - 6x + 5$
4. $x^2 - 7x + 12$
5. $x^2 + 5x - 14$
6. $x^2 + 2x - 3$
7. $x^2 - x - 20$
8. $x^2 - 7x - 18$
9. $x^2 - 12x + 20$
10. $x^2 - 2x - 48$
11. $x^2 - 11x + 24$
12. $x^2 + 11x - 60$
13. $x^2 - 2x - 80$
14. $x^2 - 10x - 24$
15. $x^2 - 17x + 60$

Factorising $ax^2 + bx + c$ where $a > 1$

A trinomial of the form $ax^2 + bx + c$, where $a > 1$, may be factorised by using the method of grouping.

To factorise $ax^2 + bx + c$ we multiply the coefficient of x^2 and the constant term and then find two factors of the product whose sum is equal to the coefficient of x. Next, we write the middle term as the sum of the factors and then factorise the resulting expression by grouping.

8   Further Mathematics

Example 8   Factorise each expression

(a) $2x^2 + 7x + 6$

Solution   Here $a = 2$ and $c = 6$ so $ac = 12$. The numbers whose product is 12 and whose sum is 7 are 3 and 4. Writing the middle term as $4x + 3x$, we have

$$2x^2 + 7x + 6 = 2x^2 + 4x + 3x + 6$$

$$= 2x(x + 2) + 3(x + 2)$$

$$= (x + 2)(2x + 3)$$

(b) $6x^2 - 17x + 7$

Solution   Here the product $ac = 42$. The numbers -14 and -3 have a product 42 and a sum -17. Thus

$$6x^2 - 17x + 7 = 6x^2 - 3x - 14x + 7$$

$$= 3x(2x - 1) - 7(2x - 1)$$

$$= (2x - 1)(3x - 7)$$

(c) $3x^2 + 13x - 10$

Solution   Here the product $ac = -30$. The numbers 15 and -2 have a product of -30 and a sum 13. Thus

$$3x^2 + 13x - 10 = 3x^2 + 15x - 2x - 10$$

$$= 3x(x + 5) - 2(x + 5)$$

$$= (x + 5)(3x - 2)$$

Trial and Error

So far we have use the method of grouping to factorise a trinomial of the form $ax^2 + bx + c$, where $a \neq 1$. An alternative method is to try combinations of possible factors until one works. This technique is called the trial and error method. We will not discuss this method.

Exercise 1.4(b)

Factorise each of the following expressions:

1. $2x^2 + 7x + 3$ 2. $6x^2 - 7x - 3$ 3. $5x^2 + 9x - 2$

4. $3x^2 + 5x + 2$
5. $4x^2 + 9x + 5$
6. $3x^2 - 20x - 7$

7. $8x^2 + 6x - 9$
8. $12x^2 - 7x + 1$
9. $3x^2 - 10x + 8$

10. $15 - 13x - 6x^2$
11. $3x^2 - 5x + 2$
12. $15x^2 - x - 6$

Difference of Two Squares

Recall that the product $(a - b)(a + b) = a^2 - b^2$. To factorise the expression on the right hand side we use the expression backwards.

Example 9 Factorise each expression

(a) $x^2 - 64$

Solution $x^2 - 64 = x^2 - 8^2$

$$= (x - 8)(x + 8)$$

(b) $9x^2 - 25y^2$

Solution $9x^2 - 25y^2 = (3x)^2 - (5y)^2$

$$= (3x - 5y)(3x + 5y)$$

(c) $27x^5y - 12xy^7$

Solution Begin by factorising $3xy$ from the expression

$$27x^5y - 12xy^2 = 3xy(9x^4 - 4y^6)$$

$$= 3xy(3x^2 - 2y^3)(3x^2 + 2y^3)$$

Exercise 1.4(c)

Factorise each of the following expressions:

1. $1 - x^2$
2. $4x^2 - y^2$
3. $x^2 - 16y^2$
4. $4x^2 - 9y^2$

5. $3x^2 - 3$
6. $9 - 4x^2$
7. $4x^2 - 25$
8. $16x^2 - 81y^2$

9. $4x^2 - 49y^2$
10. $5x^2 - 45y^2$
11. $81x^2 - 100y^6$
12. $75x^5 - 27x^3$

Perfect Square Trinomial

Remember that $a^2 + 2ab + b^2$ and $a^2 - 2ab + b^2$, called the perfect square trinomials, are the products of the square of $(a + b)$ and $(a - b)$ respectively. Notice that a perfect square trinomial has the following properties:

1. the first term and the last term are the square of the first term and the second term of the binomial respectively.

2. the middle term is twice the product of the two terms of the binomial. (The middle term can be positive or negative)

A perfect square trinomial can be factorised as follows; find the square roots of the first and the last terms and add if the middle term is positive or subtract if the middle term is negative.

Example 10 Factorise each expression

(a) $x^2 + 6x + 9$

Solution $x^2 + 6x + 9 = (x + 3)^2$

(b) $4x^2 - 20xy + 25y^2$

Solution $4x^2 - 20xy + 25y^2 = (2x - 5y)^2$

(c) $2x^3y + 20x^2y^2 + 50xy^3$

Solution First factorise out the largest common factor. This is $2xy$

$$2x^3y + 20x^2y^2 + 50xy^3 = 2xy(x^2 + 10xy + 25y^2)$$
$$= 2xy(x + 5y)^2$$

Exercise 1.4(d)

Factorise each of the following expression

1. $x^2 + 12x + 36$
2. $x^2 - 6x + 9$
3. $x^2 + 10x + 25$
4. $x^2 - 4x + 4$
5. $1 - 2x + x^2$
6. $4 + 12x + 9x^2$
7. $4x^2 + 4xy + y^2$
8. $9x^2 - 30xy + 25y^2$
9. $x^2 + 14xy + 49y^2$
10. $36x^2 - 12xy + y^2$
11. $45x^3 + 60x^2 + 20x$
12. $50y - 40y^2 + 8y^3$

1.5 Algebraic Fractions

An algebraic fraction is an expression of the form $\frac{P}{Q}$, where P and Q are polynomials and Q cannot have the value 0. P is called the numerator and Q is called the denominator. The following are examples of algebraic fractions:

$$\frac{2}{x}, \quad \frac{4}{3-x}, \quad \frac{2x+5}{x+4}, \quad \frac{2x^2+3x+5}{x^2-1}$$

Simplifying Algebraic Fraction

An algebraic fraction is said to be in its simplest form if its numerator and denominator have no common factor other than 1. To write a fraction in its simplest form:

1. factorise both the numerator and denominator completely if possible

2. divide out the common factors.

Example 11 Simplify

(a) $\dfrac{6a^2b}{8ab^2}$

Solution $\dfrac{6a^2b}{8ab^2} = \dfrac{3a \times 2ab}{4b \times 2ab} = \dfrac{3a}{4b}$

(b) $\dfrac{6x^2y}{-3xy^2}$

Solution $\dfrac{6x^2y}{-3xy^2} = \dfrac{2x \times 3xy}{-y \times 3xy} = \dfrac{-2x}{y}$

Note: There are three equivalent ways to write the answer.

$\dfrac{2x}{-y} = \dfrac{-2x}{y} = -\dfrac{2x}{y}$

The most common way to write the fraction is to have the minus sign in the numerator.

(c) $\dfrac{4a+8}{6a^2+12a}$

Solution $\dfrac{4a+8}{6a^2+12a} = \dfrac{2 \times 2(a+2)}{3a \times 2(a+2)} = \dfrac{2}{3a}$

(d) $\dfrac{25-x^2}{x^2-2x-15}$

Solution $\dfrac{25-x^2}{x^2-2x-15} = \dfrac{(5-x)(5+x)}{(x-5)(x+3)}$

$= \dfrac{-(x-5)(5+x)}{(x-5)(x+3)}$

$= -\dfrac{5+x}{x+3}$

Exercise 1.5(a)

Simplify each of the following fractions:

1. $\dfrac{9x^2y}{12xy}$

2. $\dfrac{-4x^3y}{6xy^2}$

3. $\dfrac{-15m^3n^3}{-20mn^4}$

4. $\dfrac{-10x^3y^2z^3}{15xy^4z}$

5. $\dfrac{a^2-b^2}{a^2-ab}$ 6. $\dfrac{x^2-9}{x^2+3x}$ 7. $\dfrac{9x^2-1}{3x^2+x}$ 8. $\dfrac{x^2-2x-15}{xy-5y}$

9. $\dfrac{x^2+5x+6}{x^2+x-2}$ 10. $\dfrac{2x^2+7x+6}{x^2+5x+6}$ 11. $\dfrac{x+3y}{9y^2-x^2}$ 12. $\dfrac{9x^2+12x+4}{9x^2-4}$

Multiplication and Division of Algebraic Fractions

Multiplication

To multiply two or more algebraic fractions multiply their numerators and multiply their denominators and write the new fraction in its simplest form where possible.

You may begin by writing the fractions in their simplest form where possible or dividing out the common factors. Then multiply the numerators and the denominators.

Example 12   Simplify

(a) $\dfrac{5x}{3y^2} \times \dfrac{9y^3}{10x^2}$

Solution $\dfrac{5x}{3y^2} \times \dfrac{9y^3}{10x^2} = \dfrac{45xy^3}{30x^2y^2} = \dfrac{3y}{2x}$

Alternatively we have

$\dfrac{5x}{3y^2} \times \dfrac{9y^3}{10x^2} = \dfrac{1}{1} \times \dfrac{3y}{2x} = \dfrac{3y}{2x}$

(b) $\dfrac{8a^2}{4a+4b} \times \dfrac{ab+b^2}{3a}$

Solution $\dfrac{8a^2}{4a+4b} \times \dfrac{ab+b^2}{3a} = \dfrac{8a^2}{4(a+b)} \times \dfrac{b(a+b)}{3a}$

$= \dfrac{2ab}{3}$

(c) $\dfrac{x^2-x-2}{x^2+4x+3} \times \dfrac{x^2+5x+6}{x^2-4}$

Solution $\dfrac{x^2-x-2}{x^2+4x+3} \times \dfrac{x^2+5x+6}{x^2-4} = \dfrac{(x+1)(x-2)}{(x+1)(x+3)} \times \dfrac{(x+2)(x+3)}{(x-2)(x+2)}$

$= 1$

Exercise 1.5(b)

Simplify each of the following expressions:

1. $\dfrac{8x^3}{25x^2y} \times \dfrac{15xy^2}{6y}$ 2. $\dfrac{a-b}{3a} \times \dfrac{12ab}{2a-2b}$ 3. $\dfrac{x+a}{2ax} \times \dfrac{4ax}{(x+a)^2}$

Topics in Algebra 13

4. $\dfrac{x-2}{x+3} \times \dfrac{xy+3y}{x^2-2x}$

5. $\dfrac{15x^2y}{x-y} \times \dfrac{x^2-y^2}{5xy^2}$

6. $\dfrac{x^2-25}{x^2+4x+3} \times \dfrac{2x+6}{3x-15}$

7. $\dfrac{x^2-2x}{2x+6} \times \dfrac{x^2+5x+6}{xy-2y}$

8. $\dfrac{x^2-3x+2}{x^2+3x-4} \times \dfrac{2x^2+8x}{3x-6}$

9. $\dfrac{x^2+3x}{x^2-9} \times \dfrac{x^2-3x}{x^2}$

10. $\dfrac{x^2+3x-4}{x+2} \times \dfrac{x^2+4x+4}{x^2+5x+4}$

11. $\dfrac{a^2+b^2}{x+2y} \times \dfrac{x^2-4y^2}{a^3+ab^2}$

12. $\dfrac{x^2-y^2}{x^2+xy} \times \dfrac{2y}{x-y}$

Division

To divide two algebraic fractions multiply the first fraction by the reciprocal of the second fraction (i.e. invert the divisor and multiply). The reciprocal of algebraic fractions are found by interchanging the numerator and the denominator. For example the reciprocal of $\dfrac{2x+3}{x-2}$ is $\dfrac{x-2}{2x+3}$.

Note that the fractions could only be simplified after the divisor is inverted.

Example 13 Simplify

(a) $\dfrac{9x^2}{y^2} \div \dfrac{3x}{y}$

Solution $\dfrac{9x^2}{y^2} \div \dfrac{3x}{y} = \dfrac{9x^2}{y^2} \times \dfrac{y}{3x} = \dfrac{3x}{y}$

(b) $\dfrac{x-1}{x^2} \div \dfrac{x^2-1}{x^3}$

Solution $\dfrac{x-1}{x^2} \div \dfrac{x^2-1}{x^3} = \dfrac{x-1}{x^2} \times \dfrac{x^3}{x^2-1}$

$= \dfrac{x-1}{x^2} \times \dfrac{x^3}{(x-1)(x+1)}$

$= \dfrac{x}{x+1}$

(c) $\dfrac{a^2+3a}{a^2-9} \div \dfrac{a+3}{a^2+6a+9}$

Solution $\dfrac{a^2+3a}{a^2-9} \div \dfrac{a+3}{a^2+6a+9} = \dfrac{a^2-3a}{a^2-9} \times \dfrac{a^2+6a+9}{a+3}$

$= \dfrac{a(a-3)}{(a-3)(a+3)} \times \dfrac{(a+3)^2}{a+3}$

$= a$

Exercise 1.5(c)

Simplify each of the following expressions:

1. $\dfrac{3x^2}{5y} \div \dfrac{9x^3}{15y}$

2. $\dfrac{16x^2}{5y^3} \div \dfrac{8x}{25y^2}$

3. $\dfrac{4y}{5} \div \dfrac{8y^2}{10}$

4. $\dfrac{12x^2y^2}{35ab^2} \div \dfrac{24xy^2}{7b^2}$

5. $\dfrac{6}{x^2-y^2} \div \dfrac{18}{x-y}$

6. $\dfrac{a+b}{a-2} \div \dfrac{a^2-b^2}{a^2-4}$

7. $\dfrac{x^2+3x}{4x^2-1} \div \dfrac{x+3}{2x+1}$

8. $\dfrac{x^2-9}{a^2-b^2} \div \dfrac{3x-9}{2a+2b}$

9. $\dfrac{x^2+2x-15}{x^2+7x+10} \div \dfrac{xy-3y}{x^2+2x}$

10. $\dfrac{2x^2-3x+1}{x^2-4x+3} \div \dfrac{2x^2+3x-2}{x^2-x-6}$

11. $\dfrac{x^2-y^2}{(x+y)^2} \div \dfrac{x-y}{5x+5y}$

12. $\dfrac{x^2+4x-12}{x^2+9x+18} \div \dfrac{3x-6}{6x+6}$

Addition and Subtraction of Algebraic Fractions

The procedure used to add or subtract two algebraic fractions depends on whether the fractions have the same denominators or different denominators.

Adding or subtracting algebraic fractions with the same denominators

To add or subtract two algebraic fractions with the same denominator simply add or subtract the numerators and place the result over the common denominator.

For example $\dfrac{2x}{x+3} + \dfrac{5}{x+3} = \dfrac{2x+5}{x+3}$

Adding or subtracting algebraic fractions with different denominators

To add or subtract algebraic fractions that do not have a common denominator you must first find the least common multiple (LCM) of the denominators of the fractions. Next rewrite the original fractions in an equivalent form with the same denominator and then add or subtract the resulting fractions using the rule for adding or subtracting fractions with the same denominators.

Example 14 Work out

(a) $\dfrac{4}{x+2} + \dfrac{3}{x-1}$

Solution The only factors of the denominators are $x+2$ and $x-1$ so the LCM is $(x+2)(x-1)$. Rewrite each fraction with a denominator $(x+2)(x-1)$.

$$\dfrac{4}{x+2} + \dfrac{3}{x-1} = \dfrac{4(x-1)}{(x+2)(x-1)} + \dfrac{3(x+2)}{(x+2)(x-1)}$$

$$= \dfrac{4(x-1)+3(x+2)}{(x+2)(x-1)}$$

$$= \frac{7x+2}{(x+2)(x-1)}$$

(b) $\frac{5}{2x+6} - \frac{4}{x^2+5x+6}$

Solution Factorise the denominators:

$$2x + 6 = 2(x + 3)$$

$$x^2 + 5x + 6 = (x + 2)(x + 3)$$

The LCM is $2(x + 2)(x + 3)$

So $\quad \frac{5}{2x+6} - \frac{4}{x^2+5x+6} = \frac{5(x+2)}{2(x+2)(x+3)} - \frac{8}{2(x+2)(x+3)}$

$$= \frac{5(x+2)-8}{2(x+2)(x+3)}$$

$$= \frac{5x+2}{2(x+2)(x+3)}$$

Exercise 1.5(d)

Work out each of the following:

1. $\frac{2x+3y}{10} + \frac{3x-8y}{10}$
2. $\frac{3x+y}{6} - \frac{x-3y}{6}$
3. $\frac{x^2}{x-3} - \frac{9}{x-3}$
4. $\frac{x^2}{x^2-25} + \frac{10x+25}{x^2-25}$
5. $\frac{6x}{3x-4} + \frac{8}{4-3x}$
6. $\frac{7}{x-3} + \frac{6}{x-2}$
7. $\frac{3}{x-1} - \frac{6}{x^2-1}$
8. $\frac{2x}{x-2} - \frac{2x-1}{x}$
9. $\frac{x}{x-y} - \frac{y}{x+y}$
10. $\frac{4y}{y^2+6y+5} + \frac{2y}{y^2-1}$
11. $\frac{6x}{x^2-9} - \frac{5x}{x^2+x-6}$
12. $\frac{2}{y^2+y-6} + \frac{3y}{y^2-2y-15}$

Complex Fractions

A fraction that has a fraction in its numerator or denominator or in both is called a complex fraction. Examples include

$$\frac{\frac{3}{4}}{\frac{5}{6}}, \quad \frac{\frac{3}{x^2}}{\frac{4}{x}} \quad \text{and} \quad \frac{a+2}{\frac{3}{a-3}}.$$

Simplifying Complex Algebraic Fractions

We will consider two methods used in simplifying complex fractions.

16 Further Mathematics

Method 1

Multiply the numerator and the denominator by the LCD of all fractions in the numerator and denominator.

Example 15 Simplify

(a) $\dfrac{\frac{x}{y}+2}{\frac{x}{y}-3}$

Solution You can multiply the numerator and denominator by y.

$$\dfrac{\frac{x}{y}+2}{\frac{x}{y}-3} = \dfrac{\left(\frac{x}{y}+2\right)\times y}{\left(\frac{x}{y}-3\right)\times y}$$

$$= \dfrac{x+2y}{x-3y}$$

(b) $\dfrac{1-\frac{2}{x-1}}{\frac{3}{x-1}-1}$

Solution $\dfrac{1-\frac{2}{x-1}}{\frac{3}{x-1}-1} = \dfrac{\left(1-\frac{2}{x-1}\right)(x-1)}{\left(\frac{3}{x-1}-1\right)(x-1)}$

$$= \dfrac{(x-1)-2}{3-(x-1)}$$

$$= \dfrac{x-3}{4-x}$$

Method 2

Write the numerator and denominator of the complex fraction as single fractions and then multiply by inverting the denominator as illustrated in Example 16.

Example 16 Simplify

(a) $\dfrac{\frac{y}{y+2}}{\frac{5}{3y+6}}$

Solution The numerator and denominator are already single fractions. Inverting the denominator and multiplying gives

$$\dfrac{\frac{y}{y+2}}{\frac{5}{3y+6}} = \dfrac{y}{y+2} \times \dfrac{3y+6}{5}$$

$$= \dfrac{y}{y+2} \times \dfrac{3(y+2)}{5}$$

$$= \frac{3y}{5}$$

(b) $\dfrac{1-\frac{a^2}{b^2}}{1+\frac{a}{b}}$

Solution $\dfrac{1-\frac{a^2}{b^2}}{1+\frac{a}{b}} = \dfrac{\frac{b^2-a^2}{b^2}}{\frac{b+a}{b}}$

$$= \frac{b^2-a^2}{b^2} \times \frac{b}{b+a}$$

$$= \frac{(b-a)(b+a)}{b(b+a)}$$

$$= \frac{b-a}{b}$$

Exercise 1.5(e)

Work out each of the following:

1. $\dfrac{1+\frac{1}{y}}{\frac{1}{2}-\frac{1}{2y}}$

2. $\dfrac{\frac{1}{4}+\frac{3}{4x}}{\frac{1}{2}-\frac{3}{2x}}$

3. $\dfrac{\frac{1}{x}+1}{1-\frac{1}{x^2}}$

4. $\dfrac{\frac{x^2}{y^2}-1}{\frac{x}{y}+1}$

5. $\dfrac{\frac{4}{x^2}-\frac{1}{y^2}}{\frac{2}{x}+\frac{1}{y}}$

6. $\dfrac{\frac{1}{x^2y}+\frac{2}{x}}{\frac{1}{xy^2}+\frac{2}{y}}$

7. $\dfrac{3+\frac{2}{x-1}}{\frac{1}{x-1}-2}$

8. $\dfrac{\frac{5}{x-2}+1}{1-\frac{1}{x-2}}$

9. $\dfrac{\frac{1}{y^2}-\frac{9}{x^2}}{\frac{1}{y^2}+\frac{5}{xy}+\frac{6}{x^2}}$

10. $\dfrac{\frac{2}{xy}+\frac{1}{y^2}}{\frac{4}{x}-\frac{x}{y^2}}$

11. $\dfrac{1+\frac{1}{x}-\frac{12}{x^2}}{1+\frac{2}{x}-\frac{8}{x^2}}$

12. $\dfrac{1-\frac{1}{x}-\frac{6}{x^2}}{1-\frac{3}{x}-\frac{10}{x^2}}$

1.6 Equations

An equation is a statement that two algebraic expressions are equal. Examples include $3x + 2 = 5$, $x^2 + 2x - 8 = 0$ and $|x - 2| = 5$.

Linear Equations

Equations that can be written in the form $ax + b = 0$, where a and b are real numbers with $a \neq 0$, are called linear equations. Examples include $3x + 5 = -2$, $5x - 6 = 3x$ and $2x + 7 = 8 - 3x$.

Solving Linear Equations

To solve a linear equation in x, we transform the equation into an equivalent form $x = a$, where a is a real number. The solution of this equation is the same as the original equation.

18 Further Mathematics

Example 17 shows how you can use the following properties to solve linear equations.

1. The same number may be added to both sides of an equation.

2. The same number may be subtracted from both sides of an equation.

3. Both sides of an equation may be multiplied by the same number.

4. Both sides of an equation may be divided by the same number.

Example 17 Solve each equation.

(a) $4x + 3 = 11$

Solution $\quad 4x + 3 = 11$

$\qquad 4x + 3 - 3 = 11 - 3 \qquad$ Subtract 3 from each side

$\qquad 4x = 8 \qquad$ Simplify

$\qquad x = 2 \qquad$ Divide each side by 4

(b) $\frac{1}{3}x - 9 = -5$

Solution $\quad \frac{1}{3}x - 9 = -5$

$\qquad \frac{1}{3}x - 9 + 9 = -5 + 9 \qquad$ Add 9 to both sides

$\qquad \frac{1}{3}x = 4 \qquad$ Simplify

$\qquad x = 12 \qquad$ Multiply both sides by 3

Equations having variables on both sides

Example 18

Solve the equation $3x + 2 = 12 - 2x$

Solution $\quad 3x + 2 = 12 - 2x$

$\qquad 3x + 2x = 12 - 2 \qquad$ Grouping like terms

$$5x = 10 \qquad \text{Combine like terms}$$

$$x = 2 \qquad \text{Divide both sides by 5}$$

Exercise 1.6(a)

Solve each of the following equations:

1. $5x + 12 = 37$
2. $6x - 5 = 7$
3. $5x - 16 = x$

4. $3x = 2x - 4$
5. $5x - 4 = 8 + 3x$
6. $5 - 2x = 20 - 7x$

7. $0.2x - 5.4 = 8.6 - 1.8x$
8. $0.2 - 0.5x = 0.9 - 0.7x$

9. $7x - 19 = 2 + 3x - 5$
10. $3 + 7x - 8 = 2x - 5$

11. $11 - 2x - 8 = 29 - 4x$
12. $9x + 2 - 6x = 13 - 8x$

Equations containing brackets

Example 19

Solve the equation $3(x + 2) - 2(x - 4) = 4(x + 5)$

Solution $\quad 3(x + 2) - 2(x - 4) = 4(x + 5)$

$$3x + 6 - 2x + 8 = 4x + 20 \qquad \text{Remove bracket}$$

$$x + 14 = 4x + 20 \qquad \text{Combine like terms}$$

$$x - 4x = 20 - 14 \qquad \text{Group like terms}$$

$$-3x = 6 \qquad \text{Simplify}$$

$$x = -2 \qquad \text{Divide both sides by } -3$$

Exercise 1.6(b)

Solve each of the following equations:

1. $3(x - 2) = x + 8$
2. $5(x + 2) = 3x - 4$

3. $3 - (5x - 7) = 3x - 2$
4. $3y - 2(5y - 6) = -30$

5. $2(x - 4) + 3 = 3x$
6. $5x - (3x + 7) = 3$

7. $7x - 4(3x + 4) = 9$
8. $4(x + 3) - 3(x - 2) = 5(x - 4)$

9. $3(4x - 1) - 2(x + 3) = 6x$
10. $3(3y - 2) - 2(5y - 3) + 9 = 0$

Equations Involving Fractions

To solve a fractional equation we first clear the equation of all fractions by multiplying each side of the equation by the LCD of the fractions.

Example 20 Solve each equation.

(a) $\frac{1}{2}(x-2) - 3 = \frac{1}{3}(x+3)$

Solution $\quad \frac{1}{2}(x-2) - 3 = \frac{1}{3}(x+3)$

$\qquad\qquad 3(x-2) - 18 = 2(x+3)$ Multiply each term by 6

$\qquad\qquad 3x - 6 - 18 = 2x + 6$ Remove bracket

$\qquad\qquad 3x - 24 = 2x + 6$ Combine like terms

$\qquad\qquad x = 30$

(b) $\frac{x}{3x+12} + \frac{x-1}{x+4} = \frac{2}{3}$

Solution $\quad \frac{x}{3x+12} + \frac{x-1}{x+4} = \frac{2}{3}$

$\qquad\qquad x + 3(x-1) = 2(x+4)$ Multiply both sides by the LCD

$\qquad\qquad x + 3x - 3 = 2x + 8$ Remove brackets

$\qquad\qquad 4x - 3 = 2x + 8$ Combine like terms

$\qquad\qquad 4x - 2x = 8 + 3$ Group like terms

$\qquad\qquad 2x = 11$

$\qquad\qquad x = \frac{11}{2}$ Divide both sides by 2

$\qquad\qquad x = 5\frac{1}{2}$

Exercise 1.6(c)

Solve each of the following equations:

1. $\frac{1}{3}x = \frac{1}{2}x + 1$

2. $\frac{x}{4} - \frac{x}{5} = 1\frac{1}{2}$

3. $\frac{3}{4}x = x + \frac{1}{3}$

4. $\frac{x+1}{3} = \frac{x-2}{4}$

5. $\frac{5}{x} - \frac{1}{2x} = \frac{1}{3}$

6. $\frac{3}{x} - 2 = \frac{9}{4x}$

7. $\dfrac{1-5y}{y} = \dfrac{8}{3y}$ 8. $\dfrac{2x+3}{x+3} = \dfrac{2x-3}{x-2}$ 9. $4 - \dfrac{1}{3}(x+3) = \dfrac{1}{4}(x-2)$

10. $\dfrac{7x-4}{5} - 4 = \dfrac{2x+1}{3}$ 11. $\dfrac{2}{3}(x+2) - \dfrac{3}{5}(1+x) = 1$ 12. $\dfrac{2}{x} - \dfrac{1}{x+5} = \dfrac{2}{2x-5}$

Changing the Subject of a Formula

Formulas are extremely useful tools in any field in which mathematics is applied. A formula is an equation that expresses a relationship between more than one letters (or variables).

The formula $C = 2\pi r$ gives the circumference C of a circle with radius r. The letter C is called the subject of the formula. Sometimes we may find it helpful to rearrange a formula to make a new subject in order to solve a particular problem. The process of rearranging a formula is called changing the subject of a formula. The rules used in this process are the same as those used in solving linear equations.

We can rearrange $C = 2\pi r$ to make r the subject as shown below.

$\dfrac{C}{2\pi} = \dfrac{2\pi r}{2\pi}$ Divide both sides by 2π

$\dfrac{C}{2\pi} = r$

or $r = \dfrac{C}{2\pi}$

Given a circle's circumference C you can use the equation $r = \dfrac{C}{2\pi}$ to find the radius.

Example 21

(a) Make h the subject of the formula $V = 2\pi r(r + h)$

Solution $V = 2\pi r(r + h)$ Original formula

$\dfrac{V}{2\pi r} = r + h$ Divide both sides by $2\pi r$

$\dfrac{V}{2\pi r} - r = h$ Subtract r from both sides

(b) Make R_1 the subject of the formula $\dfrac{1}{R} = \dfrac{1}{R_1} + \dfrac{1}{R_2}$

Solution $\dfrac{1}{R} = \dfrac{1}{R_1} + \dfrac{1}{R_2}$

$R_1 R_2 = RR_2 + RR_1$

$R_1 R_2 - RR_1 = RR_2$

$$R_1(R_2 - R) = RR_2$$

$$R_1 = \frac{RR_2}{R_2 - R}$$

This result can also be obtained as follows:

$$\frac{1}{R} = \frac{1}{R_1} + \frac{1}{R_2}$$

$$\frac{1}{R} - \frac{1}{R_2} = \frac{1}{R_1}$$

$$\frac{R_2 - R}{RR_2} = \frac{1}{R_1}$$

$$R_1 = \frac{RR_2}{R_2 - R}$$

(c) Mark r the subject of the formula $V = \frac{1}{3}\pi r^2 h$

Solution $V = \frac{1}{3}\pi r^2 h$

$$3V = \pi r^2 h \qquad \text{Multiply each side by 3}$$

$$\frac{3V}{\pi h} = r^2 \qquad \text{Divide each side by } \pi h$$

$$\sqrt{\frac{3V}{\pi h}} = r \qquad \text{Taking the positive square root}$$

(d) Rewrite the equation $y = \sqrt{\frac{x+a}{x}}$ with x as subject

Solution $y = \sqrt{\frac{x+a}{x}}$

$$y^2 = \frac{x+a}{x} \qquad \text{Square each side}$$

$$xy^2 = x + a \qquad \text{Multiply each side by } x$$

$$xy^2 - x = a \qquad \text{Subtract } x \text{ from each side}$$

$$x(y^2 - 1) = a \qquad \text{Factorise out } x$$

$$x = \frac{a}{y^2 - 1} \qquad \text{Divide each side by } (y^2 - 1)$$

Exercise 1.6(d)

Make the letter printed after each formula the subject.

1. $V = \frac{4}{3}\pi r^3$ r
2. $E = \frac{1}{2}mv^2$ v
3. $A = \frac{1}{2}h(a+b)$ a
4. $V = \pi r^2 h$ r
5. $s = \frac{1}{2}gt^2$ g
6. $\frac{1}{u} + \frac{1}{v} = \frac{1}{f}$ v
7. $T = 2\pi\sqrt{\frac{l}{g}}$ g
8. $s = ut + \frac{1}{2}gt^2$ g
9. $v^2 = u^2 + 2as$ s
10. $I = \frac{PRT}{100}$ R
11. $s = \frac{1}{2}(u+v)t$ u
12. $E = \frac{1}{2}m(v^2 - u^2)$ v

Simultaneous Linear Equations

Two linear equations with two unknowns which can be solved at the same time, and are satisfied by the same pair of values are called simultaneous linear equations.

Solving Simultaneous Linear Equations

Simultaneous linear equations in two unknowns can be solved by one of these methods:

1. Method of elimination

2. Method of substitution

3. Graphical method

Method of Elimination

To solve simultaneous equations by the method of elimination we eliminate one of the unknowns by either adding or subtracting the equations. The following examples will illustrate this method.

Example 22   Solve the simultaneous equations:

(a) $3x + 2y = 13$

   $-3x + y = -7$

Solution   In this case we can eliminate x by adding the two equations. This gives an equation in y which you can solve in the usual way.

$$3y = 6$$

$$y = 2$$

Next we find the value of x by substituting 2 for y in either of the original equations. Using the first equation we get

$$3x + 4 = 13$$
$$3x = 9$$
$$x = 3$$

The answers are $x = 3$ and $y = 2$

(b) $x + 3y = 17$

$2x - 3y = -2$

Solution The working can be presented as shown below.

Write out the two equations and label them

$$x + 3y = 17 \qquad (1)$$
$$2x + 3y = 22 \qquad (2)$$

$(1) - (2) \qquad -x = -5$

$$x = 5$$

Substitute $x = 5$ into (1)

$$5 + 3y = 17$$
$$y = 4$$

The answers are $x = 5$ and $y = 4$

You may have noticed that in both examples we eliminated one of the unknowns either by adding or subtracting the equations. We could do this because the coefficients of one of the unknowns are identical.

Sometimes you will need to multiply one or both equations by a number that will give identical terms before you add or subtract.

Example 23 Solve the simultaneous equations:

(a) $-2x + y = -7$

$3x + 4y = -6$

Solution One approach is to multiply the first equation by 4. The coefficient of y becomes 4.

$$-2x + y = -7 \quad (1)$$

$$3x + 4y = -6 \quad (2)$$

$(1) \times 4 \quad -8x + 4y = -28 \quad (3)$

$(2) - (3) \quad 11x = 22$

$$x = 2$$

Substitute $x = 2$ into (2)

$$6 + 4y = -6$$

$$y = -3$$

The answers are $x = 2$ and $y = -3$

(b) $3x + 4y = 18$

$2x + 3y = 12$

Solution In this case we will multiply both equations by numbers that will make the coefficients of one unknown identical in both equations.

$$3x + 4y = 17 \quad (1)$$

$$2x + 3y = 12 \quad (2)$$

$(1) \times 2 \quad 6x + 8y = 34 \quad (3)$

$(2) \times 3 \quad 6x + 9y = 36 \quad (4)$

$(3) - (4) \quad -y = -2$

$$y = 2$$

Substitute $y = 2$ into (2)

$$2x + 6 = 12$$

$$x = 3$$

Note that you can eliminate y by multiplying equation (1) by 3 and equation (2) by 4.

Exercise 1.6(e)

Solve the following simultaneous equations

1. $2x + y = 5$
   $x - y = 1$

2. $3x + y = 11$
   $2x + y = 8$

3. $x + 2y = 7$
   $3x - 2y = 9$

4. $3x + 2y = 8$
   $3x + 4y = 14$

5. $3x - 2y = 2$
   $3x + 4y = 14$

6. $3x + 2y = 4$
   $5x + 4y = 3$

7. $5x - 3y = -5$
   $3x + 11y = -3$

8. $x + 2y = 5$
   $3x - y = 1$

9. $2x - 5y = -9$
   $3x + 2y = 15$

10. $4x - 3y = 0$
    $5x + 2y = 23$

11. $7x - 2y = -17$
    $x + 4y = 4$

12. $x + y = 2$
    $4x + 7y = -1$

13. $4x - 5y = 5$
    $2x - 3y = 2$

14. $3x - 2y = 1$
    $5x - 3y = 3$

15. $-5x + 6y = -7$
    $4x + 3y = 16$

16. $-5x + 4y = 5$
    $-3x + 2y = 2$

17. $3x - y = 17$
    $5x + 3y = 5$

18. $4x - 3y = 1$
    $x - 2y = 4$

19. $5x + 2y = 2$
    $2x + 3y = -8$

20. $3x + 2y = 10$
    $4x - y = 6$

21. $x - 2y = -7$
    $4x + 3y = -6$

Method of Substitution

To solve simultaneous equations by substitution we take one of the equations and express one of the unknowns in terms of the other. Then we substitute this into the other equation. This gives an equation with one unknown.

Example 24 Solve the simultaneous equations:

(a) $2x + 3y = 12$

$x = 2y - 1$

Solution The second equation is already solved for x

Substitute $x = 2y - 1$ into the first equation

$$2(2y - 1) + 3y = 12$$

We solve this equation to get

$$7y = 14$$

$$y = 2$$

Now, we substitute $y = 2$ into the second equation to get

$$x = 4 - 1 = 3$$

The answers are $x = 3$ and $y = 2$

(b) $3x - 2y = 8$

$2x + y = 10$

Solution Begin by rewriting the second equation as $y = 10 - 2x$

Then substitute $y = 10 - 2x$ into the first equation

$$3x - 2(10 - 2x) = 8$$

Solving this equation we get

$$7x = 28$$

$$x = 4$$

Substituting $x = 4$ into the second equation gives

$$8 + y = 10$$

$$y = 2$$

The answers are $x = 4$ and $y = 2$

The choice of which algebraic method you use depends largely on the given simultaneous equations. When one of the equations expresses one of the unknowns in terms of the other then substitution is the preferred method, otherwise use the method of elimination.

Exercise 1.6(f)

Solve each of the following simultaneous equations by substitution

1. $y = x + 1$
   $x + y = 3$

2. $y = 2x - 4$
   $3x + y = 11$

3. $x + y = 4$
   $2x - y = 5$

4. $y - 2x = 1$
   $3x - 4y = 1$

5. $3x + 2y = 10$
   $4x - y = 6$

6. $4x + 3y = 9$
   $2x + 5y = 15$

7. $3x - 2y = 4$
   $2x + 3y = -6$

8. $5x + y = 1$
   $x + y = 5$

9. $x - y = 4$
   $-2x + 5y = 1$

10. $x + 2y = 4$
    $x - 2y = 4$

11. $4x - 3y = 11$
    $4x - y = 9$

12. $3x + 2y = -5$
    $x - 3y = -9$

Simultaneous Equations with Three Variables

Example 25   Solve the equation

$2x + y - z = 1$ \hfill (1)

$3x - y + 2z = 10$ \hfill (2)

$4x - 2y - 3z = 9$ \hfill (3)

Solution   First we choose two of the equations and eliminate one of the variables. Let us choose equation (1) and equation (2). Add equation (1) and equation (2) to eliminate y.

$5x + z = 11$ \hfill (4)

Next choose a different pair of equations to eliminate y. Multiplying equation (1) by 2 and then adding the result to equation (3) gives

$8x - 5z = 11$ \hfill (5)

We now have two equations (4) and (5) in the variables x and z. We now solve the equations in two variables using any of the methods discussed above. We multiply (4) by 5 to gives

$25x + 5z = 55$ \hfill (6)

Now we add equation (5) and equation (6) to give

$33x = 66$

$$x = 2$$

Substituting $x = 2$ into equation (4) gives

$$5(2) + z = 11$$
$$z = 1$$

Finally substituting $x = 2$ and $z = 1$ in equation (1) gives

$$2(2) + y - 1 = 1$$
$$y = -2$$

Exercise 1.6(g)

Solve each of the following simultaneous equations:

1. $x + 2y - z = 2$
   $2x - 3y + z = -1$
   $4x + y + 2z = 12$

2. $x + 2y + 3z = -4$
   $x - y - 3z = 8$
   $2x + y + 6z = -14$

3. $4x + 3y - z = 12$
   $3x - y = 5$
   $x + 2y + 2z = 2$

4. $x + y + z = 6$
   $2x + 3y + z = 11$
   $3x + 2y + 2z = 13$

5. $2x + 3y - 5z = 0$
   $3x + 2y - 4z = -2$
   $4x + y - z = 2$

6. $x + 2y + 3z = -1$
   $4x - 3y + 2z = 2$
   $3x - 8y - 5z = 11$

Simultaneous Equations with One Linear Equation

The method of substitution already explained can be used to solve simultaneous equations in two unknowns with one linear equation.

Example 26 Solve the simultaneous equations:

(a) $3x + 2y = 12$

$$xy = 6$$

Solution From the first equation we get $y = \dfrac{12 - 3x}{2}$

30 Further Mathematics

Substituting $y = \frac{12-3x}{2}$ into the second equation gives

$$x\left(\frac{12-3x}{2}\right) = 6$$

$$12x - 3x^2 = 12$$

Simplifying the equation we get

$$x^2 - 4x + 4 = 0$$

Factorising the quadratic expression we get

$$(x-2)^2 = 0$$

$$x = 2 \text{ twice}$$

Now substitute $x = 2$ into (2)

$$2y = 6$$

$$y = 3$$

The answers are $x = 2$ and $y = 3$

(b) $\quad x^2 + 3xy + y^2 = -5$

$$2x + y = 1$$

Solution Solving the second equation for y we get $y = 1 - 2x$

Substitute $y = 1 - 2x$ into the first equation

$$x^2 + 3x(1 - 2x) + (1 - 2x)^2 = -5$$

Simplifying this equation gives

$$x^2 + x - 6 = 0$$

Factorising the quadratic expression we have

$$(x-2)(x+3) = 0$$

$$x - 2 = 0 \quad \text{or} \quad x + 3 = 0$$

$$x = 2 \qquad x = -3$$

Substitute $x = 2$ into (2)

$$4 + y = 1$$

$$y = -3$$

Also substitute $x = -3$ into (2)

$$-6 + y = 1$$

$$y = 7$$

Check each pair of values in (1)

When $x = 2$ and $y = -3$,

we have $2^2 + 3(2)(-3) + (-3)^2 = 4 - 18 + 9 = -5$

When $x = -3$ and $y = 7$,

we have $(-3)^2 + 3(-3)(7) + 7^2 = 9 - 63 + 49 = -5$

The answers are $x = 2, y = -3$ and $x = -3, y = 7$

Exercise 1.6(h)

Solve the following simultaneous equations:

1. $3x + 2y = 18$
   $xy = 12$

2. $2x - 3y = 4$
   $xy = 10$

3. $x^2 + 2xy + y^2 = 9$
   $2x + y = 5$

4. $2x^2 + y^2 = 19$
   $3x + y = 10$

5. $2x - y = 2$
   $y^2 - 2x^2 - 14 = 0$

6. $3y - x = -1$
   $x^2 - 2xy - y^2 = 7$

7. $2x - y = 1$
   $3x^2 - xy + y^2 = 15$

8. $x^2 + 2xy - 4y^2 = -4$
   $3x - 2y = 8$

32 Further Mathematics

9. $2x + y = 3$

   $2x^2 + 3xy + 2y^2 = 9$

10. $3x - 2y = 1$

    $9x^2 - 4y^2 = 17$

11. $x^2 + 2xy = 8$

    $x + 2y = 2$

12. $x^2 + y^2 + 2xy + 6x = -2$

    $2x - y = -5$

Quadratic Equations

A quadratic equation is an equation that can be written in the general form $ax^2 + bx + c = 0$, where a, b and c are real numbers with $a \neq 0$. Examples are $x^2 + 4x = 3$ and $2x^2 = 3x$. Notice that the highest exponent of the variable of each equation is 2.

Solving Quadratic Equation by Factorisation

To solve a quadratic equation by factorisation write the equation in the general form and factorise the quadratic expression. Then set each factor to zero and solve the resulting linear equation.

Example 27   Solve each quadratic equation

(a) $x^2 - 5x - 24 = 0$

Solution $x^2 - 5x - 24 = 0$

$(x - 8)(x + 3) = 0$ Factorise the quadratic expression

$x - 8 = 0$ or $x + 3 = 0$ Set each factor equal to 0

$x = 8$ $x = -3$ Solve for x

The solutions of the equation are $x = -3$ and $x = 8$

(b) $2x^2 = 12 - 5x$

Solution $2x^2 = 12 - 5x$

$2x^2 + 5x - 12 = 0$ Write the equation in the general form

$(2x - 3)(x + 4) = 0$ Factorise the quadratic expression

$2x - 3 = 0$ or $x + 4 = 0$ Set each factor equal to 0

$x = 1\frac{1}{2}$ $x = -4$ Solve for x

The solutions of the equation are $x = -4$ and $x = 1\frac{1}{2}$

(c) $3x^2 - 4x = 0$

Solution $3x^2 - 4x = 0$

$\quad\quad\quad x(3x - 4) = 0$ $\quad\quad$ Factorise the quadratic expression

$\quad\quad\quad x = 0$ or $3x - 4 = 0$ $\quad\quad$ Set each factor equal to 0

$\quad\quad\quad\quad\quad\quad x = 1\frac{1}{3}$ $\quad\quad$ Solve for x

The solutions of the equation are $x = 0$ and $x = 1\frac{1}{3}$

(d) $9x^2 - 4 = 0$

Solution $(3x - 2)(3x + 2) = 0$ $\quad\quad$ Factorise the quadratic expression

$\quad\quad\quad 3x - 2 = 0$ and $3x + 2 = 0$ $\quad\quad$ Set each factor equal to 0

$\quad\quad\quad x = \frac{2}{3} \quad\quad\quad x = -\frac{2}{3}$

The solutions are $x = -\frac{2}{3}$ and $x = \frac{2}{3}$

An alternative solution is shown below

$\quad\quad\quad 9x^2 = 4$ $\quad\quad$ Isolate x^2 term

$\quad\quad\quad x^2 = \frac{4}{9}$ $\quad\quad$ Divide both sides by 9

$\quad\quad\quad x = \pm\frac{2}{3}$ $\quad\quad$ Take square root of both sides

So $x = -\frac{2}{3}$ and $x = \frac{2}{3}$

Exercise 1.6(i)

Solve each of the following quadratic equations:

1. $x^2 + 7x + 10 = 0$ $\quad\quad\quad\quad\quad\quad$ 2. $x^2 - 7x + 12 = 0$

3. $x^2 - 3x - 40 = 0$ $\quad\quad\quad\quad\quad\quad$ 4. $x^2 + 3x - 10 = 0$

5. $x^2 + 8x + 15 = 0$ $\quad\quad\quad\quad\quad\quad$ 6. $x^2 - 3x - 18 = 0$

34 Further Mathematics

7. $x^2 - 12x + 32 = 0$

8. $x^2 + 4x - 21 = 0$

9. $x^2 - 7x + 10 = 0$

10. $8 - 2x - x^2 = 0$

11. $42 + x - x^2 = 0$

12. $x^2 - 6x + 9 = 0$

13. $x^2 - 4x = 12$

14. $x^2 + 8x = -15$

15. $3x^2 - 4x - 20 = 0$

16. $2x^2 - 5x + 3 = 0$

17. $6x^2 = x + 1$

18. $4x^2 - 11x = 3$

19. $5x - 20x^2 = 0$

20. $3x^2 - 75 = 0$

21. $(x + 2)^2 = 16$

22. $(3x - 2)^2 = 25$

23. $(2x + 3)^2 = 16x^2$

24. $(x - 1)(x - 2) = 42$

25. $(x - 5)(x + 2) = 18$

26. $(x - 3)(x + 2) = 24$

27. $x(x - 10) = -24$

28. $8x(x + 1) = 30$

29. $(2x + 1)^2 = 3x^2 + 13$

30. $(3x - 2)^2 = x(6x - 4)$

Review Exercise 1

1. Expand:

(a) $7(2x + y)$

(b) $-2(3x - 5y)$

(c) $2x(x + 3y)$

(d) $-3x^2(x - y)$

(e) $2xy^2(3x^2 + 5xy)$

(f) $-3a^2b(2a - 3b)$

2. Simplify:

(a) $2x + 3(x + 5)$

(b) $7a^2 - 5a(a + 1)$

(c) $3x + 5y + 7(5x - 2y)$

(d) $2(a + b) + 3(2a - 3b)$

(e) $3(5x - 2y) - 2(2x - 3y)$

(f) $3x(x + y) + 4y(x - y)$

3. Factorise:

(a) $8xy + 12y$

(b) $9xy - 3zy$

(c) $6xy + 9xz$

(d) $a^2 + ab$

(e) $21y^2 - 14y$

(f) $3d + 9cd^2$

4. Factorise the following by grouping:

(a) $2pr + ps - 6qr - 3qs$

(b) $6x^2 + 12y - 9xy - 8x$

(c) $6wy - xz - 2xy + 3wz$

(d) $2am - 3an + 2bm - 3bn$

(e) $15ac - 20ad + 3bc - 4bd$

(f) $x^3 - 3x^2 - 9x + 27$

5. Expand:

(a) $(3x + 7)(2x - 3)$

(b) $(4x - 3)(3x - 5)$

(c) $(2x + 1)(x + 4)$

(d) $(x - 4)(x + 4)$

(e) $(2x + y)(2x - y)$

(f) $(3x - 4)(3x + 4)$

(g) $(x + 2y)^2$

(h) $(3a - 2b)^2$

(i) $(2x - 3)^2$

6. Factorise each of the following:

(a) $x^2 - 4x - 21$

(b) $x^2 - 7x + 10$

(c) $x^2 + 13x + 30$

(d) $x^2 - 11x + 30$

(e) $x^2 - 8x + 16$

(f) $x^2 - 10x + 25$

(g) $3x^2 - 12y^2$

(h) $16x^6 - 81y^4$

(i) $12a^3 - 27ab^4$

7. Factorise each of the following:

(a) $2x^2 + 7x + 5$

(b) $5y^2 - 2y - 7$

(c) $2x^2 - 3x - 14$

(d) $2y^2 + 7y + 6$

(e) $3a^2 + 2ab - 5b^2$

(f) $3 + 4x - 15x^2$

(g) $3x^2 + 4xy + y^2$

(h) $x^3 + 2x^2 - 24x$

(i) $y^3 - 10y^2 + 25y$

8. Factorise each of the following:

(a) $(3x - 5y)^2 - 9x^2$

(b) $(x + 2y)^2 - 4y^2$

(c) $(2a + 3)^2 - (6a - 5)^2$

(d) $(3x + 1)^2 - (2x - 1)^2$

(e) $4(a + 1)^2 - 9(a - 1)^2$

(f) $(3a - b)^2 - (2a + b)^2$

(g) $32(x + 1)^2 - 8(x - 2)^2$

(h) $3(2x + 1)^2 - 12(x - 2)^2$

9. Simplify:

(a) $\dfrac{5x^2}{3y} \times \dfrac{6y^2}{20x^3}$

(b) $\dfrac{x+a}{2ax} \times \dfrac{4ax}{(x+a)^2}$

c) $\dfrac{6a^2}{3a+3b} \times \dfrac{ab+b^2}{4a}$

(d) $\dfrac{(a+b)^2}{3ab} \times \dfrac{6ab}{a^2-b^2}$

(e) $\dfrac{x^2+3x-4}{x+2} \times \dfrac{x^2+4x+4}{x^2+5x+4}$ 　　(f) $\dfrac{x^2-16}{x^2+8x+16} \times \dfrac{3x^2-5x-2}{x^2-6x+8}$

10. Simplify:

(a) $\dfrac{3c^2b}{11cd^2} \div \dfrac{4ab^2}{33cd^3}$ 　　(b) $\dfrac{2x^2y}{3z^2} \div \dfrac{4x^3y}{9z}$

(c) $\dfrac{3a^2b-3ab}{2ab-7b^2} \div \dfrac{3a}{7b}$ 　　(d) $\dfrac{x^2-y^2}{(x+y)^2} \div \dfrac{x-y}{5x+5y}$

(e) $\dfrac{x^2+4x-12}{x^2+9x+18} \div \dfrac{3x+12}{6x+6}$ 　　(f) $\dfrac{y^2+5y+6}{y^2+4y-21} \div \dfrac{y^2-4}{y^2+7y}$

11. Simplify:

(a) $\dfrac{7}{x-3} + \dfrac{6}{x-2}$ 　　(b) $\dfrac{x}{x-y} - \dfrac{y}{x+y}$

(c) $\dfrac{3(x+1)}{x^2+5x+4} - \dfrac{1}{x-2}$ 　　(d) $\dfrac{x+4}{x-5} + \dfrac{x-5}{x+4}$

(e) $\dfrac{a}{a^2-5a+6} - \dfrac{1}{a^2-a-2}$ 　　(f) $\dfrac{1}{y^2-9y+20} + \dfrac{1}{y^2-11y+30}$

12. Simplify:

(a) $\dfrac{y - \dfrac{y^2}{y-x}}{1 + \dfrac{x^2}{y^2-x^2}}$ 　　(b) $\dfrac{\dfrac{a^2}{a-b} - a}{\dfrac{b^2}{a-b} + b}$

(c) $\dfrac{\dfrac{x}{y} - 2 + \dfrac{y}{x}}{\dfrac{x}{y} - \dfrac{y}{x}}$ 　　(d) $\dfrac{1 + \dfrac{2}{x} - \dfrac{15}{x^2}}{1 + \dfrac{4}{x} - \dfrac{5}{x^2}}$

(e) $2 - \dfrac{1}{1 - \dfrac{2}{a+2}}$ 　　(f) $1 - \dfrac{1}{1 - \dfrac{1}{1 - \dfrac{1}{x}}}$

13. Solve each of the following equations:

(a) $5(x-2) = 3(x+4)$ 　　(b) $3y - 3(5y-6) = -30$

(c) $5x - (4x+4) = 3$ 　　(d) $3(2x-5) = 3 - 4(x+2)$

(e) $\dfrac{1}{2}(x-3) + \dfrac{1}{3}(x-3) = 3$ 　　(f) $\dfrac{1}{4}(x+1) - \dfrac{1}{3}(x-1) = 1$

(g) $\dfrac{2}{x+3} = \dfrac{1}{3-x}$ 　　(h) $\dfrac{2}{x-1} - \dfrac{1}{x+1} = \dfrac{2}{1-x^2}$

14. Make the letter printed after each formula the subject:

(a) $v = u + at$, t

(b) $S = 4\pi r^2$, r

(c) $s = \frac{n}{2}(a + l)$, l

(d) $F = \frac{9}{5}C + 32$, C

(e) $l = \sqrt{h^2 + 4r^2}$, h

(f) $I = \frac{nE}{R+nr}$, n

15. Solve each of the following simultaneous equations:

(a) $2x + 3y = 13$
    $7x - 5y = -1$

(b) $6x - 7y = -1$
    $5x - 4y = 12$

(c) $3x + 2y = 20$
    $-5x + 3y = 11$

(d) $\frac{x}{6} + \frac{y}{3} = 8$
    $\frac{x}{4} - \frac{y}{9} = 1$

(e) $\frac{a}{2} - \frac{b}{3} = 1$
    $\frac{a}{4} - \frac{b}{9} = \frac{2}{3}$

(f) $\frac{2x}{3} - \frac{3y}{4} = -1$
    $\frac{x}{3} - \frac{y}{2} = -1$

16. Solve each of the following simultaneous equations:

(a) $x - y + 2z = 7$
    $2x + y - z = 3$
    $x + y + z = 9$

(b) $x - 2y + 3z = 7$
    $2x + y + z = 4$
    $-3x + 2y - 2z = -10$

(c) $x + y - z = 6$
    $3x - 2y + z = -5$
    $x + 3y - 2z = 14$

(d) $x + 2y - z = -3$
    $2x - 4y + z = -7$
    $-2x + 2y - 3z = 4$

17. Solve each of the following quadratic equations:

(a) $x^2 = 10x - 21$

(b) $x^2 + 10 = -7x$

(c) $x^2 - 1 = 12x + 12$

(d) $2x^2 - x - 3 = 0$

(e) $5x^2 - 13x + 6 = 0$

(f) $3x^2 + 11x = -6$

(g) $y^2 - 3y - 4 = 10 - y$

(h) $(2x + 1)^2 = 49$

18. Solve each of the following simultaneous equations:

(a) $x^2 + y^2 = 10$
    $y - x = 4$

(b) $x^2 + y^2 - 4 = 0$
    $x + y = 2$

(c) $x^2 - 2xy + y^2 = 36$
    $x + y = 2$

(d) $2y = x + 3$
    $x^2 + y^2 - 2x + 6y = 15$

2 Logic

In this Chapter we consider some fundamental concepts in logic which will help you develop patterns of reasoning. There are several reasons for studying logic. Understanding logic enable you to think clearly, communicate effectively and make more convincing arguments. In addition logic will enable you gain proficiency in correct mathematical reasoning.

2.1 Statements

In our daily interaction we communicate ideas in writing or in speech by using sentences. Consider the following sentences:

1. Dr Kwame Nkrumah was the first president of Ghana.

2. The largest city in Togo is Accra

3. Ama celebrated her fifth birthday yesterday.

These sentences express ideas that can be true or false but cannot be both true and false at the same time. A sentence that is either true or false is called a statement. Statement 1 is true and Statement 2 is false. We cannot determine whether statement 3 is true or false but definitely it may be one or the other. The three sentences are examples of simple statements because they convey only one idea.

Compound Statements

Consider the statement "Esi is a student and Esi likes reading". This statement is obtained by joining the two simple statements "Esi is a student," "Esi likes reading" by the word "and". A statement formed by joining two or more simple statements or by negating a statement is called a compound statement.

Connectives

We use words such as: and, or, if - then, and if and only if, to join simple statements. In logic we call these words connectives. Although "not" modifies a simple statement and does not join two statements it is common practice in mathematics to call it a connective.

It is customary to use lower-case letters such as p, q and r to represent simple statements and to represent the connectives with symbols. Table 1 shows the connectives and their symbols.

Table 1

Connective	Symbol
Not	~
Or	∨
And	∧
If - then	→
If and only if	↔

Each connective is a very important part of a compound statement because the truth value of a compound statement depends not only on the truth value of its components but also on the meaning we give to the connectives.

Types of Statement

Compound statements are given specific names depending on the connective used.

Negation

Sometimes you would be required to change a statement to its opposite meaning. To do so you can use the negation of the statement. For example the negation of the statement "Ama walked home after school" is "Ama did not walk home after school".

If p represents the statement "Ama walked home after school" then ~p (read not p) represents the statement "Ama did not walk home after school". To indicate the negation of a statement we can prefix it with the phrase "It is false that" or use the connective "not". Note that the negation of "It is hot" is not "It is cold". If the weather is not hot it is not necessarily cold. The correct negation is " The weather is not hot".

Consider the statement "Kofi is not a student". This statement contains the word not which indicates that it is a negation. To write this statement in symbol we let p represent "Kofi is a student". Then ~p would be "Kofi is not a student". By convention p, q and r are used to represent statements that are not negation.

An important point to remember is that a negation has the effect of negating only the statement that directly follows it. To negate a compound statement we must use parentheses. When a negation symbol is placed in front of a statement in parentheses it negates the entire statement in the parentheses.

Quantifiers

A word or phrase that conveys the idea of quantity is called a quantifier. Such words include all, none (or no), some and not all. The quantifiers all , every and each are interchangeable. The quantifiers some, there exit(s) and at least one are also interchangeable.

Negating Quantifiers

Care must be taken when negating statements containing quantifiers. Consider the statement "All birds can fly". This statement is false since at least one bird the chicken cannot fly. Therefore the negation of this statement must be true. No bird can fly is not the negation of the statement because it is false. The correct negation of the statement is "Some birds cannot fly" or "At least one bird cannot fly" which are true statements.

Table 2 shows some statements that contain quantifiers and their negation.

Table 2

Statement	Negation
1. All boys are intelligent	Some boys are not intelligent
2. No student has a driver's licence	Some students have driver's licence
3. Some students do not study hard	All students study hard
4. Some of the students had grade A	None of the student had grade A

Conjunction

A conjunction is a compound statement formed by joining two or more statements with the word " and".

If p and q are two statements then the conjunction p and q is denoted by the symbol p ∧ q, read p and q.

Let p and q represent the statements:

p: Ama is clever

q: Ama is in the science class

Then the conjunction p ∧ q represents the statement

"Ama is clever and Ama is in the science class".

The word and in this statement indicates that both events will take place. Other words used to express conjunction are but, however or nevertheless.

Disjunction

A disjunction is a compound statement formed by joining two or more statements with the word "or".

If p and q are two statements the disjunction p or q is denoted by the symbol p ∨ q, read p or q.

Let p and q represent the statements:

p: Kofi passed the exam

q: Kofi failed the exam

Then the disjunction p ∨ q represents the statement "Kofi passed the exam or Kofi failed the exam".

The word "or" can be used in two different ways as shown in the following statements.

Ama obtained a grade A or grade B in Mathematics.

We interpret the connective or as meaning either Ama obtained a grade A or Ama obtained a grade B but not both. The word "or" is used as exclusive disjunction.

The statement "Kofi passed the exam or Kofi received a scholarship" offers three possibilities. Kofi passed the exam, Kofi received a scholarship or Kofi passed the exam and received a scholarship. The word "or" is used as an inclusive disjunction.

Conditional

A conditional statement is a statement that can be written in the form "If -then".

If p and q are two statements then the conditional statement is denoted by the symbol $p \rightarrow q$, read if p then q.

The statement immediately following the word if is called the hypothesis (or premise) and the statement immediately following the word then is called the conclusion.

Let p and q represent the statement

p: The sun is shining

q: Asare is playing outside

The conditional statement is "If the sun is shining then Asare is playing outside".

Biconditional

If p and q are two statements then the statement "p if and only if q" is called a conditional statement, denoted by the symbol $p \leftrightarrow q$.

Let p and q represents the simple statements.

p: Mensah will be considered for promotion

q: Mensah works hard

The Biconditional statement p ↔ q is "Mensah will be considered for promotion if and only if Mensah works hard".

Dominance of Connectives

In symbolic logic statements are grouped by parentheses. In written logic statement, commas are used to indicate which simple statements are to be grouped together. When you write compound statement in symbols, group simple statements on the side of a comma within parentheses.

When there are no parentheses indicating groupings in a symbolic statement or by a comma in written statement the least dominant connective is applied first and the most dominant connective last. The level of the dominance of the connectives from the least dominant to the most dominant is as shown below.

1. Negation: ~

2. Conjunction: ∧   disjunction: ∨

3. Conditional: →

4. Biconditional: ↔

The conjunction and disjunction have the same level of dominance. Thus to indicate which one of these two connectives should be applied first we use parentheses. For the symbolic statement (p ∨ q) ∧ r, the connective ∨ is applied first.

Example 1

(a) Add parentheses to the statement p ∨ q ↔ q → ~p

Solution   The Biconditional has the greatest dominance so we place parentheses as follows:

(p ∨ q) ↔ (q → ~p)

This is a Biconditional statement since the Biconditional symbol is outside the parentheses.

(b) Write the following statement in symbolic form

Ama will read a book, or Ama will not read a book and Ama will watch TV

Solution   Let p and q represent the statements

    p: Ama will read a book.

    q: Ama will watch TV.

The placement of the comma indicates that the statement "Ama will not read a book" and "Ama will watch TV" are to be grouped together. The statement can be represented by the symbol p ∨ (∼ p ∧ q).

The statement is a disjunction since the disjunction symbol is outside the parenthesis.

(c) Write the following statement in symbolic form

   If I walk home or leave school after 3 pm then I will not attend the party this evening.

Solution   Let p: I walk home.

   q: I leave school after 3 pm.

   r: I will attend the party this evening.

Since no commas appear in the sentence we will evaluate it by using the dominance of connectives. Because the conditional has higher dominance than the disjunction, the conditional statement will be evaluated last. Thus the statement; I walk home and I leave school after 3 pm are to be grouped together. The statement written in symbolic form is (p ∨ q) → ∼ r. The statement is a conditional statement since the conditional symbol is outside the parenthesis.

Exercise 2.1

1. Determine which of the following sentences are statements:

(a) Ama went to the market.

(b) All the students in the class did the punishment.

(c) Pick the bowl up.

(d) He wrote a letter to his father

(e) Esi plays football

(f) Is Kwesi studying?

(g) Kofi got his passport today

(h) Will you visit your friend?

2. Classify each of the following statements as simple or compound. Write the connective of each compound statement in symbol.

(a) Not all birds can fly

(b) The dog barked or the man walked away.

(c) Adjoa will study if and only if the bulb is fixed.

(d) Dela slipped this afternoon

(e) If the sun shines today then Mensah will play outside

(f) Akosua was the guest speaker

(g) Dzifa went home and Esi studied mathematics

(h) Afua wrote a hit song

3. Indicate whether each of the following compound statement is a negation, conjunction, disjunction, conditional or biconditional.

(a) The colour of the car is blue and the colour of the door is red.

(b) The boy is a student or the man works at the factory.

(c) The girl is not intelligent.

(d) The girl will read a book if and only if she stays at home.

(e) If the car breaks down Esi will call the mechanics.

(f) Ama is playing football or Kwame is reading a book.

(g) Kofi will answer the phone if and only if the phone rings.

(h) Not everybody liked him.

4. Write the negation of each of the following statements:

(a) Some men have no car

(b) No shop opens on Sunday.

(c) All babies walk after two years.

(d) Some dogs do not bark.

(e) No one wants to buy my mobile phone

(f) None of the men like music.

(g) Some people who work hard do not own a house.

5. Let p and q represent respectively the two statements connected by the connectives; and, or, if-then and if and only if. Write each statement in symbolic form.

(a) If Kofi comes home then his sister will go out.

(b) Ama will attend a university or Kojo will attend a polytechnic

(c) Esi went to the market and Kwesi worked in the garden.

(d) If it rains today Kofi will stay indoors

(e) The football match will be postponed if and only if it rains today.

(f) Esi will go to the movie or Afua will attend the party.

(g) Kofi will qualify to be a minister of state if and only if he is a Ghanaian.

(h) If Adjoa is 12 years old today then she was born on a Sunday.

6. Let p denotes "Ofori went home" and q denotes "Kwebena studied French". Write each of the following compound statements in words:

(a) p ∨ q (b) p ∧ q (c) ~p (d) p → q (e) p ↔ q

7. Let p: He works hard and q: He has two cars.

Write each of the following statements in words:

(a) ~p ∨ ~q (b) ~p → ~q (c) ~q ↔ ~p

(d) p ∨ ~q (e) ~p ∧ q (e) ~q ∧ p

8. Add parentheses by using the dominance of connectives and state whether the statement is a negation, conjunction, disjunction, conditional or biconditional.

(a) p ∧ ~q ↔ ~p (b) p ∨ q → r (c) p ↔ q → ~r

(d) q → ~r ∨ p (e) p ∧ ~q → ~p ∧ q (f) p → q ∨ ~r

9. Represent the simple statements in each statement by the letters p, q or r and write each statement in symbol using parentheses. State whether the statement is a negation, conjunction, disjunction, conditional or biconditional.

(a) If the moon is out then Ama is in the garden, or she is jogging.

(b) If he stays at home then it is raining or his father had travelled.

(c) It is false that if you are a Ghanaian then you will not be eligible to contest the election, and you can be a minister of state.

(d) If you are intelligent or you study hard then you will be promoted.

(e) The office is open if and only if today is not Sunday or it is before 5 pm.

(f) If he works hard and attend lectures regularly, then he will pass the final exam.

(g) If today is Monday then tomorrow is Tuesday if and only if today is not Monday.

2.2 Truth Tables

The label true or false given to a statement is called its truth value. True or false are denoted by T and F respectively. A truth table is a table that shows the truth value of a compound statement for all possible cases. We will discuss five basic truth tables.

Negation

Table 3

p	~p
T	F
F	T

Conjunction

In a compound statement with two components, such as p ∧ q, there are at most four possibilities, called the logical possibilities, to be considered. The first two column of Table 4 list these four possibilities. The last column gives the truth values of p ∧ q.

Table 4

p	q	p ∧ q
T	T	T
T	F	F
F	T	F
F	F	F

The conjunction p ∧ q is true only when both statements in it are true.

Disjunction

The truth values of p ∨ q are given in Table 5. The symbol ∨ means the inclusive or.

Table 5

p	q	p ∨ q
T	T	T
T	F	T
F	T	T
F	F	F

The disjunction p ∨ q is true if at least one of the statements is true.

Conditional

The truth table for the conditional p → q is given in Table 6.

Table 6

p	q	p → q
T	T	T
T	F	F
F	T	T
F	F	T

Notice that a conditional statement is true in all cases excepts where the hypothesis is true and the conclusion is false.

Biconditional

The truth table for the Biconditional p ↔ q is given Table 7.

Table 7

p	q	p ↔ q
T	T	T
T	F	F
F	T	F
F	F	T

Notice that the biconditional statement is true where both statements are true and where both statements are false.

Derived Truth Tables

We can derive other truth tables from the five basic truth tables discussed above. For example we can use the truth tables for the negation and disjunction to write the truth values for ∼ (∼ p ∨ q). Notice that the statement ∼ (∼ p ∨ q) is true only when p is true and q is false.

Table 8

p	q	∼ p	∼ p ∨ q	∼ (∼p ∨ q)
T	T	F	T	F
T	F	F	F	T
F	T	T	T	F
F	F	T	T	F

The method used in constructing Table 8 is explained below.

The first two column list all the possible truth values of p and q. We then use the truth values of p and Table 3 to determine the entries in the third column. Next we use the entries in the third and second column, and Table 5 to determine the entries in the fourth column. Finally using the entries in the fourth column and Table 3 gives the truth values in the fifth column.

The Biconditional statement p ↔ q is the conjunction of p → q and q → p, written (p →q) ∧ (q → p). The truth values of p ↔ q can be determined by constructing the truth table for (p → q) ∧ (q →p). This is left as an exercise.

Converse, Inverse and Contrapositive

There are three other conditionals associated closely to a given conditional p → q. These related conditionals are converse, inverse and contrapositive.

Converse

Given the conditional p → q the conditional q → p is called the converse of p → q. Notice that the inverse of a conditional interchanges the hypothesis and conclusion. Table 9 shows the truth values of the conditional p → q and its converse q → p.

Table 9

p	q	p → q	q → p
T	T	T	T
T	F	F	T
F	T	T	F
F	F	T	T

Notice that the true values of p → q and q → p are not always the same. If a given conditional is true the converse is not necessary true. For example the statement "If ABCD is a square than ABCD has four equal sides" is true but its converse "If ABCD has four equal sides than ABCD is a square is not necessary true. Remember that a rhombus has four equal sides.

Inverse

The conditional ~ p → ~ q is called the inverse of the conditional p → q. Notice that the inverse of conditional negates both the hypothesis and the conclusion.

The truth values for p → q and its inverse ~ p → ~ q are given in Table 10.

Table 10

p	q	p → q	~ p → ~ q
T	T	T	T
T	F	F	T
F	T	T	F
F	F	T	T

If a given conditional is true, the inverse is not necessarily true. For example the inverse of the statement "If ABCD is a square than ABCD is a quadrilateral" is "If ABCD is not a square than ABCD is not a quadrilateral." The original statement is true but the inverse is not necessary true.

Contrapositive

The conditional ~ q → ~ p is called the contrapositive of the conditional p → q. Notice that the contrapositive of a conditional interchanges and negates both the hypothesis and the conclusion. The truth values for p → q and ~ q → ~ p are given in Table 11.

Table 11

p	q	~p	~q	p → q	~q → ~p
T	T	F	F	T	T
T	F	F	T	F	F
F	T	T	F	T	T
F	F	T	T	T	T

The contrapositive of a true conditional is always true and the contrapositive of a false conditional is always false. That is the conditional and its contrapositive always have the same truth values. For example the contrapositive of the statement "If a number is divisible by 2 than it is even" is "If a number is not even than the number is not divisible by 2." The conditional and the contrapositive are both true.

Exercise 2.2

1. Construct a truth table for each of the following compound statements.

(a) $p \vee \sim q$

(b) $\sim p \vee \sim q$

(c) $\sim p \wedge \sim q$

(d) $\sim p \wedge q$

(e) $(p \vee q) \rightarrow p$

(f) $(p \vee \sim q) \wedge \sim p$

(g) $\sim(\sim p \vee \sim q)$

(h) $(p \rightarrow q) \leftrightarrow (q \rightarrow p)$

(i) $(p \vee \sim q) \wedge (q \wedge \sim p)$

(j) $(\sim p \vee q) \leftrightarrow \sim(p \wedge \sim q)$

(k) $(\sim p \rightarrow q) \leftrightarrow (p \vee q)$

(m) $(p \wedge \sim q) \vee (q \wedge \sim p)$

(n) $(p \wedge q) \vee (p \wedge r)$

(o) $(p \wedge \sim q) \vee r$

(p) $(\sim p \vee q) \wedge \sim r$

2. Write the converse, inverse and contrapositive of each of the following conditional statements.

(a) If the car is new then the car has air conditioning.

(b) If the electric bill is too high then I will not able to pay the bill.

(c) If the light goes off then we will go for a walk.

(d) If the pay is not good then I will not take the job.

(e) If today is a holiday then the library is not open.

3. State whether or not the converse, inverse and contrapositive of each of the following statements are true.

(a) If the triangle is isosceles then two angles are equal.

(b) If two lines intersect in at least one point then the two lines are not parallel.

(c) If the quadrilateral is a rectangle then the quadrilateral is a parallelogram.

(d) If the sum of the interior angles of a polygon is 360° then the polygon is a quadrilateral.

(e) If the quadrilateral is a square then the quadrilateral is a rectangle.

(f) If 2 divides a natural number then 2 divides the units digit of the natural number.

4. Write the converse, inverse and contrapositive of each of the following statements.

(a) $\sim p \to q$ (b) $\sim p \to \sim q$ (c) $p \to \sim q$

(d) $(p \wedge q) \to r$ (e) $(p \wedge \sim q) \to \sim p$ (f) $(p \vee q) \to (\sim q \vee r)$

2.3 Logically Equivalent and Tautology

Statements with the same truth values are said to be logically equivalent.

From Table 11 it can be seen that a conditional statement and its contrapositive are equivalent. You can also see from Table 9 and Table 10 that the converse and inverse of a conditional statement are equivalent. A glance at Table 12 shows that the compound statements $p \vee q$ and $\sim (\sim p \wedge \sim q)$ are equivalent.

Table 12

p	q	p ∨ q	~p	~q	~p ∧ ~q	~(~p ∧ ~q)
T	T	T	F	F	F	T
T	F	T	F	T	F	T
F	T	T	T	F	F	T
F	F	F	T	T	T	F

The two compound statements in each pair of the statements below are equivalent.

$\sim (\sim p)$ and p $\sim (p \wedge q)$ and $(\sim p \vee \sim q)$

$\sim (p \to q)$ and $(p \wedge \sim q)$ $[(p \wedge q) \to r]$ and $[p \to (q \to r)]$

Example 2

(a) Use truth tables to show that the compound statements $\sim (p \wedge q)$ and $\sim p \vee \sim q$ are equivalent.

Solution

Table 13

1	2	3	4	5	6	7
p	q	p ∧ q	~(p ∧ q)	~p	~q	~p ∨ ~q
T	T	T	F	F	F	F
T	F	F	T	F	T	T
F	T	F	T	T	F	T
F	F	F	T	T	T	T

Since the entries in columns 4 and 7 of the truth table are the same the two compound statements are logically equivalent.

(b) Use truth tables to show that the compound statements p ∨ (q ∧ r) and (p ∨ q) ∧ (p ∨ r) are equivalent.

Solution

Table 14

p	q	r	q ∧ r	p ∨ q	p ∨ r	p ∨ (q ∧ r)	(p ∨ q) ∧ (p ∨ r)
T	T	T	T	T	T	T	T
T	T	F	F	T	T	T	T
T	F	T	F	T	T	T	T
T	F	F	F	T	T	T	T
F	T	T	T	T	T	T	T
F	T	F	F	T	F	F	F
F	F	T	F	F	T	F	F
F	F	F	F	F	F	F	F

The truth table shows that p ∨ (q ∧ r) and (p ∨ q) ∧ (p ∨ r) have the same truth values in each of all eight logical possibilities. Therefore p ∨ (q ∧ r) and (p ∨ q) ∧ (p ∨ r) are equivalent.

Tautology

There are some statement patterns that have no false instances. Table 15 is the truth table for the compound statement p ∨ ~ p. You may have observed from the table that the statement p ∨ ~ p is true in all logical possibilities.

Table 15

p	~p	p ∨ ~p
T	F	T
F	T	T

A compound statement that is true in each of all logical possibilities is called a tautology. You may have observed that every entry in the last column is " T " regardless of the truth value of p.

Table 16 shows that the statement p → (p ∨ q) is a tautology.

Table 16

p	q	p ∨ q	p → (p ∨ q)
T	T	T	T
T	F	T	T
F	T	T	T
F	F	F	T

Conditional and Biconditional statements which are tautology are of particular importance in mathematics. Some other examples of tautology are expressed below:

1. $[(p \to q) \wedge (q \to r)] \to (p \to r)$
2. $[p \wedge (p \to q)] \to q$
3. $(p \to q) \leftrightarrow (\sim q \to \sim p)$
4. $(p \leftrightarrow q) \leftrightarrow [(p \to q) \wedge (q \to p)]$

We leave it to the student to confirm that these statements are tautology.

Exercise 2.3

1. Show that the given statements are logically equivalent

(a) $p \wedge q, \sim (\sim p \vee \sim q)$
(b) $(p \to q) \wedge p,\ q$
(c) $p \to q, \sim (p \wedge \sim q)$
(d) $p \leftrightarrow q, (p \to q) \wedge (q \to p)$
(e) $(p \to q) \wedge \sim q, \sim p$
(f) $p \to q, (p \wedge \sim q) \to (q \wedge \sim q)$
(g) $\sim (q \to p) \vee r, (p \vee q) \wedge \sim r$
(h) $p \vee (q \wedge r), \sim p \to (q \wedge r)$

2. Prove that each statement is a tautology:

(a) $p \wedge (p \vee q) \leftrightarrow p$
(b) $p \vee (p \wedge q) \leftrightarrow p$
(c) $[(p \to q) \wedge p] \to q$
(d) $\sim (p \to q) \leftrightarrow (p \wedge \sim q)$
(e) $p \vee (q \wedge r) \leftrightarrow (p \vee q) \wedge (p \vee r)$
(f) $[(p \vee q) \wedge \sim q] \to p$

2.4 Arguments

A person presents an argument when he makes a sequence of statements and draws some conclusion from them. An argument consists of a set of statements, called premises (or hypotheses) and a conclusion. An argument is said to be valid if the conclusion of the argument follows logically from the given premises. Notice that a valid argument does not mean that the conclusion is true but merely that the conclusion follows from the premises. When the conclusion does not follow from the given set of premises we say that the argument is not valid. An argument that is not valid is called a fallacy.

Use of Truth Tables

The premises in a given logical argument may consist of two or more interrelated statements. By writing the premises in the form of a conjunction we can form a single conditional statement that represents the entire argument. If the conditional statement is a tautology then the argument is valid. If the statement is not a tautology, then the argument is not valid.

Consider the following argument

" Esi is at lunch or she is in a meeting. Esi is not in a meeting therefore she is at lunch."

First we number the premises and separate them from the conclusion with a line.

Statement 1: Esi is at lunch or she is in a meeting.

Statement 2: Esi is not in a meeting
_____

Conclusion: Esi is not at a lunch.

Next write the argument in symbolic form.

Let p and q represent the following statements.

p: Esi is at lunch.

q: Esi is in a meeting.

Symbolically the argument is written

Premises: p ∨ q

$\qquad$ ~q

Conclusion: p

The argument is represented by the conditional statement:

[(p ∨ q) ∧ ~ q] → p

Finally, construct a truth table for this statement.

Table 17 is the truth table for the statement [(p ∨ q) ∧ ~ q] → p.

Table 17

p	q	p ∨ q	~ q	(p ∨ q) ∧ ~ q	[(p ∨ q) ∧ ~ q] → p
T	T	T	F	F	T
T	F	T	T	T	T
F	T	T	F	F	T
F	F	F	T	F	T

The truth table for the statement [(p ∨ q) ∧ ~ q] → p is true in every case (see the last column) so the statement is a tautology. Hence the conclusion necessarily follows from the premises and the argument is valid.

Remember that when an argument is valid the conclusion necessarily follows from the premises, and it is not necessary for the hypotheses or the conclusion to be true statements.

Example 3

(a) Determine whether the conclusion of the argument is valid or not valid.

"If Kwame is jogging then it is raining. It is not raining therefore Kwame is not jogging."

Solution   Statement 1: If Kwame is jogging then it is raining.

Statement 2: It is not raining.

Conclusion: Kwame is not jogging.

Let p and q represent the following statements.

p: Kwame is jogging.

q: It is raining.

In symbols the argument becomes.

Premises: $p \rightarrow q$

$\sim q$

Conclusion: $\sim p$

The argument is represented by the conditional statement:

$[(p \rightarrow q) \wedge \sim q] \rightarrow \sim p$

Table 18 is the truth table for the conditional statement $[(p \rightarrow q) \wedge \sim q] \rightarrow \sim p$. Since the statement is a tautology, the conclusion necessary follows from the premises and the argument is valid.

Table 18

p	q	$p \rightarrow q$	$\sim q$	$(p \rightarrow q) \wedge \sim q$	$\sim p$	$[(p \rightarrow q) \wedge \sim q] \rightarrow \sim p$
T	T	T	F	F	F	T
T	F	F	T	F	F	T
F	T	T	F	F	T	T
F	F	T	T	T	T	T

(b) Determine whether the following argument is valid or not valid.

The shirt is blue or the shirt is large. The shirt is large therefore the shirt is blue.

Solution   Statement 1: The shirt is blue

Statement 2: The shirt is large
_____
Conclusion: The shirt is blue

Let p and q represent the following statements.

p: The shirt is blue

q: The shirt is large

In symbols the argument becomes.

Premises: p ∨ q

   q

Conclusion: p

This statement is represented by the conditional statement: $[(p \vee q) \wedge q] \rightarrow p$.

Table 19 shows the truth values for the conditional statement $[(p \vee q) \wedge q] \rightarrow p$. The last column is not true in every case. Therefore, the statement is not a tautology, and the argument is not valid.

Table 19

p	q	p ∨ q	(p ∨ q) ∧ q	[(p ∨ q) ∧ q] → p
T	T	T	T	T
T	F	T	F	T
F	T	T	T	F
F	F	F	F	F

An argument is valid if the conclusion is true whenever the premises are all true. If a truth table has a column for each premise and a column for the conclusion, and has a row where the conclusion is true while every premise is true then the argument is valid. If the conclusion is false while every premise is true then the argument is not valid.

Note that if an argument in a particular form is valid, all argument with exactly the same form will also be valid. Such valid argument forms have specific names. For example, the argument form $[(p \rightarrow q) \wedge \sim q] \rightarrow \sim p$ is called Law of Contraposition.

Logic   57

Exercise 2.4

1. Use truth table analysis to prove that each of the following statements is logically valid:

(a) $(p \vee q) \leftrightarrow \sim(\sim p \wedge \sim q)$ (b) $(p \rightarrow q) \leftrightarrow \sim(p \wedge \sim q)$ (c) $[(p \rightarrow q) \wedge \sim q] \rightarrow \sim q$

(d) $[(p \vee q) \wedge \sim q] \rightarrow p$ (e) $[p \wedge (p \rightarrow q)] \rightarrow q$ (f) $(p \leftrightarrow q) \leftrightarrow [(p \rightarrow q) \wedge (q \rightarrow p)]$

2. Determine whether the following arguments are valid by using truth tables.

(a) Hypotheses: I will play outside or I will go for a walk

   I will not play outside

   Conclusion: I will go for a walk.

(b) Hypotheses: If he receives good grades he will study science

   He will not study science

   Conclusion: He does not receive good grades

(c) Hypotheses: I will win the race if and only if I train regularly

   I train regularly

   Conclusion: I will win the race

(d) Hypotheses: It is not cloudy or it is not raining

   It is cloudy

   Conclusion: It is not raining

Review Exercise 2

1. Determine which of the following sentences are statements:

(a) Esi will sell her car   (b) When will he come?   (c) The boy is intelligent

(d) Akosua bought a book   (e) Take the child home

2. Classify each of the following statements as simple or compound. Name the connective of each compound statement.

(a) Mensah will sell his old Television set and buy a new one   (b) He is handsome

(c) The boy is intelligent or he is talented   (d) If Ama saves she will buy a car.

(e) The baby is crying. (f) I will stay home if and only if it rains.

3. Negate each statement

(a) I will study hard.

(b) Esi is talented.

(c) I will wash my car.

(d) Afua will do her homework.

4. Construct a truth table for each of the following statements:

(a) $p \wedge (q \vee \sim q)$

(b) $(p \vee q) \wedge (p \wedge \sim q)$

(c) $[(p \to q) \wedge \sim q] \to \sim p$

(d) $(p \wedge q) \vee (\sim p \wedge \sim q)$

5. Put each of the following statements into symbolic form:

(a) I will study or I will go fishing.

(b) The boy is intelligent and he will make a good grade.

(c) I will dance if and only if she sings.

(d) If I study hard I will pass the test.

6. Write the converse, contrapositive and inverse of each of the following statements.

(a) If it is cloudy then it is raining.

(b) If I walk home I will miss the program.

(c) If Esi does not obey school rules she will be expel.

(d) If I go to school I will not go fishing

7. Determine whether the following are logically equivalent.

(a) $\sim p \vee q$, $p \to q$

(b) $\sim(p \wedge q)$, $\sim p \wedge q$

(c) $(p \wedge q) \to r$, $p \to (q \to r)$

(d) $(p \to q) \wedge (\sim q \vee p)$, $p \leftrightarrow q$

(e) $q \leftrightarrow (p \wedge \sim r)$, $q \to (p \vee r)$

(f) $(p \vee q) \vee r$, $p \vee (q \vee r)$

8. Use truth tables to determine whether the following arguments are valid.

(a) All mathematics teachers are men and my mother is a mathematics teacher, therefore my mother is a man.

(b) If Kojo is innocent he will not appear before the disciplinary committee. Kojo appears before the disciplinary committee, therefore Kojo is guilty.

(c) If today is a holiday Kofi will not go to school. Kofi will not go to school, therefore today is a holiday.

3 Sets

3.1 Set and Set Notation

The word set is used to describe a collection of objects such as a collection of books, a group of students, a list of countries.

Definition of Sets

A set is a well - defined collection of objects. Hence you can easily tell if whether any given object is or is not in the set.

Each object in a set is called an element or a member of the set.

Set Notation

A set can be described by listing all its elements inside curly brackets. For example, the set containing the elements 2, 4, 6 and 8 can be written as {2, 4, 6, 8}. Notice that each element is separated by commas. The curly bracket { } is read " the set of".

Naming Sets

Capital letters such as A, B and C are often used to represent sets and lower case letters such as a, b and c are used to represent elements of sets. For example, we may name the set $\{a, b, c, d, e\}$ by the letter A, and write $A = \{a, b, c, d, e\}$.

Elements of a Set

The set of numbers 1, 3, 2, 2, 4, 2, 4, 5 and 6 can be listed as {1, 2, 3, 4, 5, 6}. Identical objects cannot be listed in the set more than once. The elements in a set can be listed in any order. For example, {1, 2, 3}, {2, 1, 3} and {3, 1, 2} are different listing of the same set.

Indicating Membership of a Set

The symbol " $\in$ " means "belongs to" or " is a member of" or " is an element of" For example, if $A = \{2, 4, 5, 6, 7, 8\}$, then $6 \in A$.

The symbol " $\notin$ " means " does not belong to " or " is not a member of " or " is not an element of ". For example $9 \notin A$.

Set-Builder Notation

Sometimes we describe a set by stating a property of its members. For example, the set of real numbers between 1 and 2 cannot be named by listing all its members because there

is infinite number of them. This set can be written in the set-builder notation as follows
$\{x : x \in R, 1 < x < 2\}$

This is read " the set of all x such that x is a real number and x is greater than 1 and less than 2". The colon " : " means such that. You can use the vertical bar | instead of :.

Finite and Infinite Sets

A set is said to be finite if it is either empty or its elements can be counted.

For example, $\{-2, -1, 0, 1, 2, 3\}$ is a finite set

A set is infinite if there is no end in counting its elements. For example, the set of natural numbers is an infinite set. An infinite set may be described by listing few elements followed by three dots. The set A = {natural numbers} can be listed as $\{1, 2, 3, 4, 5, \cdots\}$. The three dots indicate that the elements in the set continue in the same pattern. For instance the set also contains the numbers, 99, 100, 101 and so on.

The Number of Elements in a Set

The number of elements in a set A is denoted by $n(A)$. For example, if $A = \{10, 11, 12, 13, 14, 15\}$, then $n(A) = 6$

Equal Sets

Two sets A and B are equal, written $A = B$ if they contain exactly the same elements. For example the set $A = \{x : x \in N, x < 4\}$ and $B = \{1, 2, 3\}$ are equal, so we write $A = B$.

Equivalent Sets

Two sets A and B are said to be equivalent if they contain the same number of elements. For example, the set $A = \{0, 1, 2\}$ and set $B = \{13, 14, 15\}$ are equivalent sets.

Empty Set

The set that has no element is called the empty set or null set. The empty set is denoted by $\emptyset$ or { }. It is important to note that the empty set is not written as $\{\emptyset\}$. The set $\{\emptyset\}$ contains the element $\emptyset$.

Universal Sets

Often we use a set whose elements are contained in a larger set, called the universal set. For example, the set of integers include whole numbers. We can consider the set of integers as the universal set for the set of whole numbers. In general the universal set, denoted by ξ or

$\cup$, is the set which contains the entire elements for specific discussion.

Subsets

If every element of set A is also an element of set B, then A is a subset of B. This is denoted by $A \subseteq B$. For example the set of natural numbers is a subset of the set of integers.

Proper Subsets

Set A is a proper subset of Set B if every element of A is in B, and there is at least one element in B which is not in A, denoted by $A \subset B$.

For example $\{2, 5, 6\}$ is a proper subset of $\{1, 2, 5, 6, 8\}$ because 1 and 8 are not in the first set.

Note that:

1. for any set A $A \subseteq A$, that is every set is a subset of itself

2. the empty set is a subset of every set

3. $A = B$ if and only if $A \subset B$ and $B \subset A$

Number of Subsets

A set may have several subsets. For example, $\emptyset$, $\{0\}$, $\{1\}$, $\{0,1\}$ are subsets of $\{0, 1\}$. A set with n elements has 2^n subsets. For instance, a set containing 5 elements has 2^5 i.e. 32 subsets

The Complement of a Set

The complement of Set A, written A', is the set of all the elements in the universal set that are not in A. For example given that

U= $\{1, 2, 3, 4, 5, 6, 7, 8, 9, 10\}$ and A = $\{1, 3, 5, 7, 9\}$ then $A' = \{2,4,6,8,10\}$.

Exercise 3.1

1. Determine whether each of the following is true or false.

(a) $6 \in \{5, 6, 7\}$ (b) $7 \notin \{2,4,5\}$ (c) $2 \in \{3,4,5\}$

(d) $8 \notin \{7,8,9\}$ (e) $2 \in \{prime\ numbers\}$ (f) $3 \in \{even\ numbers\}$

2. List the elements of the following sets:

(a) $\{x: x \text{ is an integer between } 10 \text{ and } 12\}$

(b) $\{x: x \text{ is a month with } 30 \text{ days}\}$

(c) $\{x: x \text{ is a multiple of } 2 \text{ between } 2 \text{ and } 10\}$

(d) $\{x: x \text{ is an integer and } -3 < x < 0\}$

3. Insert the correct symbol $\in, \notin, \subset, \not\subset, =$ between the following pairs of sets:

(a) $\{-2\}$ $\{$integers$\}$ 　　　　　　　(b) $\{3\}$ $\{$even numbers$\}$

(c) 5 $\{$real numbers$\}$ 　　　　　　　(d) -7 $\{$natural numbers$\}$

(e) $\{$even prime numbers$\}$ $\{2\}$ 　　(f) 3 $\{$irrational numbers$\}$

4. If $A = \{2, 3, 4, 5, 6, 7, 8\}$, find

(a) $P = \{x: x \in A \text{ and } x \text{ is an even number}\}$

(b) $Q = \{x: x \in A \text{ and } x \text{ is an odd number}\}$

(c) $R = \{x: x \in A \text{ and } x \text{ is a prime number}\}$

5. Find the number of elements in each of the following sets:

(a) $\{a, b, c, d\}$ 　　　　(b) $\{-1, 0, 1\}$ 　　　　(c) $\{-3, -2, -1, 0, 1, 2, 3\}$

6. List all the subsets of $\{a, b, c\}$

7. How many subsets has the set $\{1, 2, 3, 4\}$

8. How many elements has a set that has 128 subsets?

9. Determine whether each of the following sets are finite or infinite.

(a) $\{x: x \text{ is a multiple of } 2\}$ 　　　　(b) $\{x: 3x + 2 = 8\}$

(c) $\{x: x \in R, 5 < x < 6\}$ 　　　　　　(d) $\{x: x \text{ is a factor of } 20\}$

10. If $\xi = \{1, 2, 3, 4, 5\}$ and $A = \{1, 3\}$, find

(a) A' 　　　　(b) ξ' 　　　　(c) $(A')'$ 　　　　(d) $\emptyset'$

11. If $\xi = \{1, 2, 3, 4, 5, 6\}$, $A = \{2, 4, 5\}$ and $B = \{1, 3, 6\}$ find

(a) A' 　　　　(b) B' 　　　　(c) $(A')'$ 　　　　(d) $(B')'$

12. State whether the following sets are empty sets or not.

(a) {even prime numbers} (b) {integers between 2 and 3}

(c) {real numbers between 2 and 3} (d) $\{x: x \in N, 3x + 2 = 9\}$

3.2 Union and Intersection of Sets

Union of Sets

The union of two sets A and B, denoted by $A \cup B$, is the set containing all the elements of A or B or both A and B. For example, if A = {1, 2, 4, 6} and B = {2, 3, 5}, then

$$A \cup B = \{1, 2, 3, 4, 5, 6\}$$

Intersection of Sets

The intersection of two sets A and B, denoted by $A \cap B$, is the set containing all the elements that are common to both A and B. For example, if A = {1, 3, 5, 7} and B = {2, 3, 4, 5, 6, 7, 8} then $A \cap B = \{3, 5, 7\}$.

Graphing Unions and Intersections

The two examples below illustrate how the union and intersection of two sets may be obtained by using number lines.

Example 1

(a) If $A = \{x: -2 \leq x \leq 3\}$ and $B = \{x: 1 < x < 6\}$ where x is a real number find $A \cup B$.

Solution Using graphs the solution set of $A \cup B$ is obtained as illustrated as follows:

$A = \{x: -2 \leq x \leq 3\}$

$B = \{x: 1 < x < 6\}$

$A \cup B = \{x: -2 \leq x < 6\}$

(b) If $A = \{x: -4 \leq x < 2\}$ and $B = \{x: -3 \leq x < 5\}$ where x is a real number find $A \cap B$.

Solution

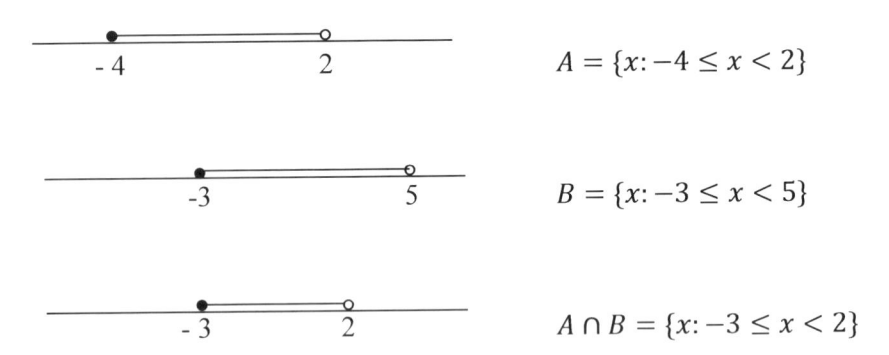

$A = \{x: -4 \leq x < 2\}$

$B = \{x: -3 \leq x < 5\}$

$A \cap B = \{x: -3 \leq x < 2\}$

Disjoint Sets

Two sets A and B are said to be disjoint if they have no element in common i.e. $A \cap B = \emptyset$. For example if $A = \{4, 6\}$ and $B = \{1, 3, 5\}$, then A and B are disjoint sets since $A \cap B = \emptyset$.

Properties of Sets

If A, B and C are subsets of the universal set ξ, then

1. $A \cup \emptyset = A$

   $A \cap \emptyset = \emptyset$

2. If $A \subset B$, then $A \cup B = B$ and $A \cap B = A$

3. $A \cup A = A$

   $A \cap A = A$

4. $A \cup B = B \cup A$ Commutative property

   $A \cap B = B \cap A$

5. $(A \cup B) \cup C = A \cup (B \cup C)$ Associative property

   $(A \cap B) \cap C = A \cap (B \cap C)$

6. $A \cup (B \cap C) = (A \cup B) \cap (A \cup C)$ Distributive property

   $A \cap (B \cup C) = (A \cap B) \cup (A \cap C)$

7. $(A')' = A$

8. $\phi' = \xi$ and $\xi' = \phi$

9. $A \cap A' = \phi$ and $A \cup A' = \xi$

10. $(A \cup B)' = A' \cap B'$ $(A \cap B)' = A' \cup B'$ De Morgan's laws

Exercise 3.2

1. If A = {-2, -1, 3, 5}, B = {-3, 0, -1, 4} and C = {-1, 2, 6}, find

(a) $A \cup B$ (b) $B \cap C$ (c) $(A \cup B) \cup C$ (d) $A \cap (B \cap C)$

2. If A = {1, 2, 3, 4, 5, 6}, B = {2, 3} and C = {3, 4, 5}, find

(a) $B \cap A$ (b) $A \cap B$ (c) $A \cup B$ (d) $B \cup A$ (e) $A \cap (B \cup C)$

(f) $(A \cap B) \cup (A \cap C)$ (g) $A \cup (B \cap C)$ (h) $(A \cup B) \cap (A \cup C)$

3. If U = {1, 2, 3, 4, 5, 7} and A = {2, 4, 7}, find:

(a) $A \cap A'$ (b) $A \cup A'$

4. If $U = \{a, b, c, d, e\}$, $A = \{c, e\}$ and $B = \{a, b, c\}$, find:

(a) $(A \cup B)'$ (b) $A' \cap B'$ (c) $A' \cup B'$ (d) $(A \cap B)'$

5. If $A = \{x: -3 \leq x \leq 4\}$ and $B = \{x: 2 < x < 7\}$ where x is a real number find:

(a) $A \cup B$ (b) $A \cap B$

6. If $A = \{x: -1 < x < 6\}$ and $B = \{x: x > 4\}$ where x is a real number find:

(a) $A \cup B$ (b) $A \cap B$

7. If $A = \{x: -4 \leq x < 1\}$ and $B = \{x: -1 < x \leq 3\}$ where x is a real number find:

(a) $A \cup B$ (b) $A \cap B$

8. If $A = \{x: -3 < x < 5\}$ and $B = \{x: x < 4\}$ where x is a real number find:

(a) $A \cup B$ (b) $A \cap B$

9. If $A = \{x: x \text{ is a factor of } 20\}$, $B = \{x: x > 3\}$ and $C = \{x: 15 > x\}$, where x is an integer. Find $A \cap B \cap C$

10. If $U = \{x: 1 \leq x \leq 10\}$, where x is an integer, $A = \{\text{multiples of 3}\}$, $B = \{\text{multiples of 2}\}$ and $C = \{\text{factors of 6}\}$, find:

(a) $A \cap (B \cup C)$ (b) $(A \cap B) \cup (A \cap C)$

11. If $U = \{x: 1 \leq x \leq 15\}$, where x is an integer, $A = \{\text{factors of 15}\}$ and $B = \{\text{factors of 12}\}$, find:

(a) $A \cap B$ (b) $(A \cap B)'$ (c) A' (d) $A \cup B$ (e) B' (f) $A' \cup B'$

12. Let $D_{(n)}$ denote the set of all factors of the natural number n. For example $D_6 = \{1, 2, 3, 6\}$. Find:

(a) $D_{(12)}$ (b) $D_{(20)}$ (c) $D_{(12)} \cup D_{(20)}$ (d) $D_{(12)} \cap D_{(20)}$

3.3 Venn Diagram

Relationship between sets can be express visually by drawing diagrams called Venn diagrams. A Venn diagram consists of a rectangle representing the universal set, and circles drawn inside the rectangle representing subsets of the universal sets.

Set Problems

A, B and C are three finite sets. To find the total number of elements in the three overlapping sets you must note that the number of elements in exactly two sets will be counted twice and the number of elements common to all three sets will be counted three times. We can illustrate this by the use of a Venn diagram.

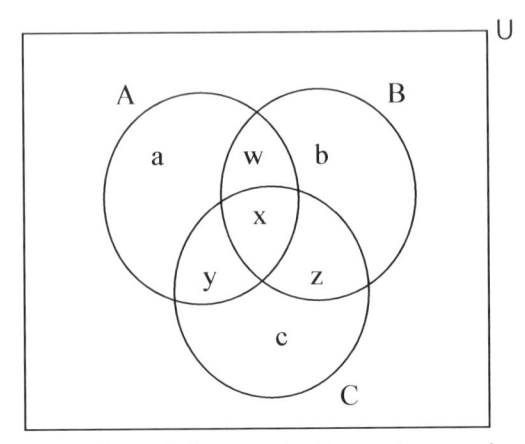

Let a, b, c, w, x, y and z be the number of elements in the various regions as indicated in the Venn diagram and $N = w + y + z$.

Adding the elements in each set we have

$n(A) = a + w + x + y$

$n(B) = b + w + x + z$

and $n(C) = c + y + x + z$

So $n(A) + n(B) + n(C) = a + b + c + 2w + 2y + 2z + 3x$

$= (a + b + c + w + x + y + z) + (w + y + z) + 2x$

$= n(A \cup B \cup C) + N + 2n(A \cap B \cap C)$

Hence $n(A \cup B \cup C) = n(A) + n(B) + n(C) - N - 2n(A \cap B \cap C)$, where N is the sum of the number of elements in exactly two sets.

Example 2

(a) In a survey of 76 teachers 20 indicated they were going to buy a new car, 26 said they were going to buy a new refrigerator, and 24 said they were going to buy a new stove. Of these 9 were going to buy a car and a refrigerator, 10 were going to buy a car and a stove, and 12 were going to buy a stove and refrigerator. Three teachers indicated they were going to buy all three items. How many will buy none of these items.

Solution Begin by constructing a Venn diagram with three overlapping circles labelled C, R and S to represent set of teachers buying cars, refrigerators and stoves respectively. We work from the intersection of all three sets outwards. Since 3 teachers were going to buy all three items, we place 3 in region $C \cap B \cap S$. Next we determine the number to be placed in region $C \cap R \cap S'$. Notice that that region $C \cap R$ consist of regions $C \cap B \cap S$ and $C \cap R \cap S'$. Since 3 have already been placed in region $C \cap B \cap S$ we must placed $9 - 3 = 6$ in region $C \cap R \cap S'$. Likewise, we place 9 in region $C' \cap R \cap S$ and 7 in region $C \cap R' \cap S$.

To determine the number to be placed in region $C \cap R' \cap S'$ subtract from 20 the sum of the numbers already placed in set C. That is, $20 - 16 = 4$ must be placed in region $C \cap R' \cap S'$. Similarly, we must place 8 and 5 in regions $C' \cap R \cap S'$ and $C' \cap R' \cap S$ respectively.

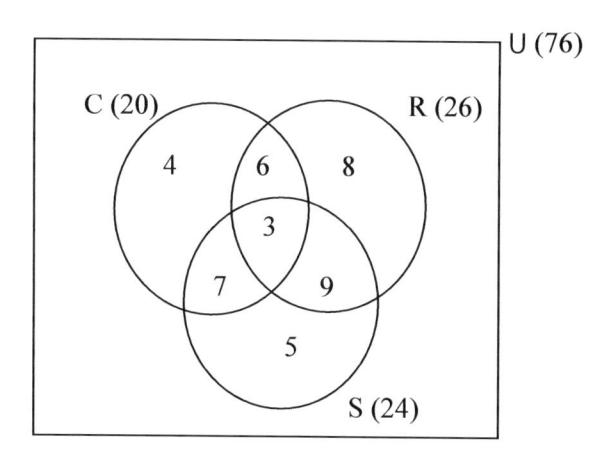

The number of teachers who were going to buy one item or more are $4 + 6 + 3 + 7 + 9 + 8 + 5 = 42$, so $76 - 42 = 34$ were going none of the items.

(b) Out of 46 students surveyed 20 are studying Mathematics, 25 are studying Physics. 18 are studying Biology, 9 are studying only Physics, 3 are studying only Biology and 5 are studying all three subjects. 10 are studying none of the three subjects. Find the number of students who are studying:

(a) exactly two of the subjects   (b) only Mathematics

Solution

(a) We put the given information in a Venn diagram. Let M, P and B denote Mathematics, Physics and Biology respectively.

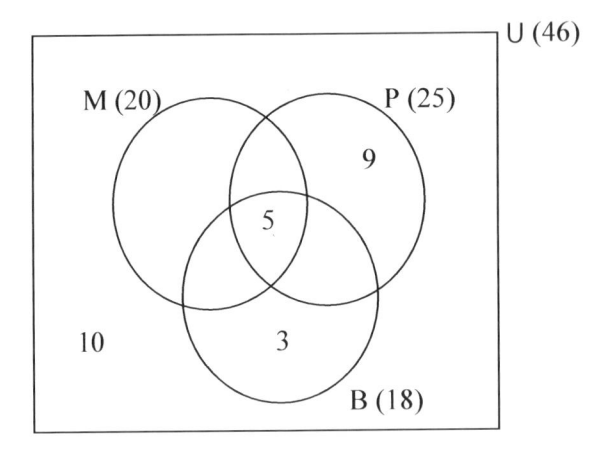

If N represents the number of students studying exactly two of the subjects then

$$n(M \cup P \cup B) = n(M) + n(P) + n(B) - N - 2n(M \cap P \cap B)$$

$$36 = 20 + 25 + 18 - N - 10$$

$$N = 17$$

Hence 17 students are studying exactly two subjects.

(b) Those studying only Mathematics are $36 - (17 + 5 + 3 + 9) = 2$

That is 2 students are studying only Mathematics

Exercise 3.3

1. Of a group of students surveyed 15 are studying Mathematics, 24 are studying Physics and 20 are studying Biology, 8 are studying Mathematics and Physics, 10 are studying

Physics and Biology, 6 are studying Mathematics and Biology but not Physics and 3 are studying all three subjects. How many students were surveyed?

2. Out of 1500 student at a certain senior high school, 350 are studying Geography, 300 are studying History, and 270 are taking both Geography and History. How many students are taking Geography or History?

3.

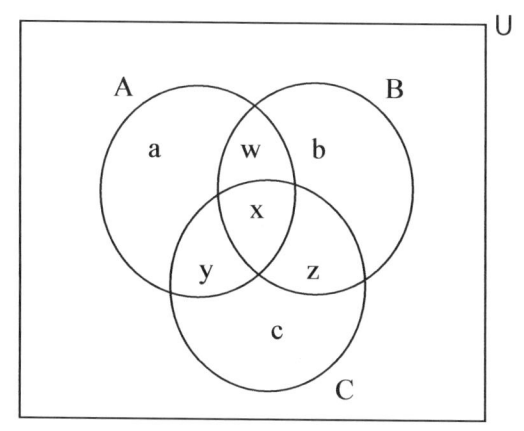

A, B and C are three intersecting sets. The regions of the Venn diagram are labelled as shown in the diagram above. Using the Venn diagram list the element(s) in the region that corresponds to:

(a) $A \cap (B \cup C)'$ (b) $(A \cap B)'$ (c) $A \cap (B \cup C)$

(d) $(A \cup B) \cap C'$ (e) $(A \cup C) \cap B'$ (f) $(A \cup B)' \cap C$

4. A survey taken at a school with an enrolment of 500 students revealed the following information.

173 are studying Elective Mathematics

246 are studying Accounting

290 are studying Economics

81 are studying both Elective Mathematics and Accounting

136 are studying both Economics and Accounting

120 are studying both Elective Mathematics and Economics

56 students are studying all three subjects

(a) How many students are not studying any of the three subjects?

(b) How many students are studying Elective Mathematics only?

(c) How many students are studying Economics only?

(d) How many students are studying Accounting only?

(e) How many students are studying only Accounting or only Elective Mathematics or only Economics?

(f) How many students are studying Accounting and Economics but not Elective Mathematics?

(g) How many students are studying Elective Mathematics and Economics but not Accounting?

5. Of the 150 cars imported into a country during the month of September, 80 had air conditioning, 96 had automatic transmission and 64 had power steering. Seven cars had all three of these extras. Twenty three cars had only automatics transmission, twelve cars had only power steering and four had none of these extras.

(a) How many cars had exactly two of these extras?

(b) How many cars had only air conditioning?

6. In a survey of 75 adults, it was found that of the three private news papers, Guide, Observer and Insight

32 read Guide

30 read Observer

27 read Insight

14 read Guide alone

10 read Observer alone

8 read Insight alone

12 read Guide and Observer

5 read all three news papers

(a) How many read Guide and Insight alone?

(b) How many read Insight and Observer alone?

(c) How many read none of these three news papers?

7. Of 100 children at a party: 25 like Pepsi, 35 like Coca Cola, 38 like Fanta, 9 like Pepsi and Coca Cola but not Fanta, 7 like Coca Cola and Fanta but not Pepsi, 8 like Pepsi and Fanta but not Coca Cola, 36 did not like any of these soft drinks.

(a) How many children like all three soft drinks?

(b) How many children like only Pepsi or only Coca Cola or Fanta?

Review Exercise 3

1. Let A = {1, 2, 3, 4, 5}, B = {2, 3, 5} and C = {1, 2, 4}. Insert the symbol, $\in, \notin, \subset, \supset, \not\subset$ to make each of the following a true statement.

(a) 4   A   (b) 6   B   (c) B   C

(d) A   B   (e) C   A   (f) 3   B

2. Given that U = {1, 2, 3, 4, 5, 6, 7, 8, 9}, A = {3, 4, 6, 8} and B = {3, 5, 7, 8} find:

(a) $A \cap B$ (b) $A \cup B$ (c) A' (d) B'

3. Given U = {1, 2, 3, 4, 5, 6, 7, 8, 9, 10}, A = {2, 3, 4, 5, 6} and B = {5, 6, 7, 8, 10} find:

(a) $A \cap B$ (b) $A \cup B$ (c) A' (d) B'

4. Given U = {1, 2, 3, 4, 5, 6, 7}, A = {1, 2, 3, 4, 5} and B = {5, 6, 7} find:

(a) $A \cap B$ (b) $A \cup B$ (c) B'

(d) A' (e) $A' \cup B$ (f) $A \cap B'$

5. If A = {1, 3, 5, 6, 8}, B = {2, 3, 6, 7} and C = {6, 8, 9} find:

(a) $(A \cap B) \cup C$ (b) $(A \cap B) \cap C$ (c) $(A \cup B) \cap B$

6. If U = {1, 2, 3, 4, 5, 6, 7}, A = {1, 3, 5, 6}, B = {2, 3, 6} and C = {4, 6, 7} find:

(a) $(A \cap B)'$ (b) $(B \cap C)' \cap A$ (c) $A' \cup B'$

7. If U = {1, 2, 3, 4, 5, 6, 7, 8}, A = {1, 3, 4, 7} and B = {2, 4, 5, 8}. List the elements of:

(a) $A' \cap B$ (b) $A \cup B'$ (c) $A' \cap B'$ (d) $(A \cap B)'$

Which of the sets are equal?

8. If $A = \{x: -2 < x < 4\}$ and $B = \{x: x > -1\}$ where x is a real number find:

(a) $A \cup B$ (b) $A \cap B$

9. If $A = \{x: -3 < x < 5\}$ and $B = \{x: -2 \leq x < 7\}$, where x is a real number find:

(a) $A \cup B$ (b) $A \cap B$

10. Of 50 families surveyed, 26 have television set, 27 have CD player, 24 have DVD player, 13 have both television set and DVD player, 15 have both television set and CD player, 12 have both CD player and DVD player, 5 have all three and 8 have none of these items. How many families have exactly one of these items?

11. Out of a group of 68 students, 30 are taking Mathematics, 35 are taking Physics, 40 are taking Biology, 9 are taking only Mathematics, 12 are taking only Physics and 6 are taking all three subjects.

(a) How many are taking exactly two of these subjects?

(b) How many are taking only Biology?

4 Surds

The positive root of a number a, denoted by $\sqrt{a}$ is a positive number whose square is a. For example, 3 is the square root of 9, since $3^2 = 9$. The symbol $\sqrt{}$ is called the radical sign.

Certain square roots are rational numbers. For example $\sqrt{4}$, $\sqrt{\frac{16}{25}}$ and $\sqrt{144}$ represent the rational numbers 2, $\frac{4}{5}$ and 12 respectively. However not all square roots are rational numbers. For example $\sqrt{2}$, $\sqrt{3}$ and $\sqrt{5}$ are said to be irrational numbers, generally called surds. The decimal representations of surds are always non – repeating and none terminating. You can approximate surds to a given number of decimal places.

4.1 Simplifying Surds

A surd is said to be in simplest form if the whole number under the radical sign has no perfect square factor. For example, $\sqrt{15}$, is in the simplest form since 15 has no perfect square factor, but $\sqrt{12}$ is not in the simplest form since one factor of 12, i.e. 4 is a perfect square. To simplify surds, we remove the perfect square factors by using the following properties.

1. $\sqrt{a^2} = a$

2. $\sqrt{a} \times \sqrt{a} = (\sqrt{a})^2 = a$

3. $\sqrt{ab} = \sqrt{a} \times \sqrt{b}$

4. $\sqrt{\frac{a}{b}} = \frac{\sqrt{a}}{\sqrt{b}}$

Example 1 Simplify:

(a) $\sqrt{18}$

Solution To simplify this surd, we remove the largest perfect square factor of 18.

$\qquad \sqrt{18} = \sqrt{9 \times 2}$ Remove the perfect square factor

$\qquad \quad\; = \sqrt{9} \times \sqrt{2}$ Use property 3

$\qquad \quad\; = 3\sqrt{2}$ Simplify

Note that it would not help to write $\sqrt{18} = \sqrt{6 \times 3}$ since neither factor is a perfect square.

An alternative method is to write the prime factors of 18, i.e. $3 \cdot 3 \cdot 2$ and then pair each group of like factors. Notice that each pair of like factors is a perfect square.

$$\sqrt{18} = \sqrt{3^2 \times 2}$$
$$= \sqrt{3^2} \times \sqrt{2} \qquad \text{Use property 3}$$
$$= 3 \times \sqrt{2} \qquad \text{Use property 2}$$
$$= 3\sqrt{2} \qquad \text{Simplify.}$$

(b) $\sqrt{108}$

Solution $\sqrt{108} = \sqrt{36 \times 3}$ Remove the perfect square factor.

$\qquad\qquad\qquad = 6\sqrt{3}$ Simplify.

The following table shows the most common roots. It would be helpful to memorize these roots.

Square Roots

$\sqrt{1} = 1$	$\sqrt{36} = 6$	$\sqrt{121} = 11$
$\sqrt{4} = 2$	$\sqrt{49} = 7$	$\sqrt{144} = 12$
$\sqrt{9} = 3$	$\sqrt{64} = 8$	$\sqrt{169} = 13$
$\sqrt{16} = 4$	$\sqrt{81} = 9$	$\sqrt{196} = 14$
$\sqrt{25} = 5$	$\sqrt{100} = 10$	$\sqrt{225} = 15$

Exercise 4.1

Simplify each of these surds:

1. $\sqrt{20}$
2. $\sqrt{28}$
3. $\sqrt{27}$
4. $\sqrt{40}$
5. $\sqrt{45}$
6. $\sqrt{48}$
7. $\sqrt{50}$
8. $\sqrt{54}$
9. $\sqrt{60}$
10. $\sqrt{72}$
11. $\sqrt{75}$
12. $\sqrt{80}$
13. $\sqrt{96}$
14. $\sqrt{98}$
15. $\sqrt{147}$
16. $\sqrt{150}$

17. $\sqrt{162}$ 18. $\sqrt{180}$ 19. $\sqrt{192}$ 20. $\sqrt{245}$

21. $\sqrt{288}$ 22. $\sqrt{320}$ 23. $\sqrt{405}$ 24. $\sqrt{588}$

25. $\sqrt{605}$ 26. $\sqrt{675}$ 27. $\sqrt{726}$ 28. $\sqrt{540}$

29. $\sqrt{847}$ 30. $\sqrt{720}$ 31. $\sqrt{338}$ 32. $\sqrt{567}$

4.2 Adding and Subtracting Surds

To add (or subtract) surds, we add (or subtract) their coefficients.

Example 2 Simplify:

(a) $2\sqrt{3} + 4\sqrt{3}$

Solution $2\sqrt{3} + 4\sqrt{3} = (2 + 4)\sqrt{3}$

$= 6\sqrt{3}$

In practice it is not necessary to show the intermediate step.

(b) $5\sqrt{2} - 4\sqrt{2}$

Solution $5\sqrt{2} - 4\sqrt{2} = \sqrt{2}$

Certain unlike surds can be added or subtracted when we rewrite them in simplest form.

Example 3 Simplify:

(a) $\sqrt{24} + \sqrt{96}$

Solution $\sqrt{24} + \sqrt{96} = 2\sqrt{6} + 4\sqrt{6}$

$= 6\sqrt{6}$

(b) $2\sqrt{45} - 3\sqrt{20}$

Solution $2\sqrt{45} - 3\sqrt{20} = 6\sqrt{5} - 6\sqrt{5}$

$= 0$

Exercise 4.2

Work out the following:

1. $5\sqrt{3} + 2\sqrt{3}$ 2. $3\sqrt{2} + 5\sqrt{2}$

76   Further Mathematics

3. $10\sqrt{5} - 3\sqrt{5}$

4. $6\sqrt{7} - 4\sqrt{7}$

5. $2\sqrt{3} - 5\sqrt{3}$

6. $2\sqrt{5} - 4\sqrt{5}$

7. $2\sqrt{2} + \sqrt{2} + 3\sqrt{2}$

8. $3\sqrt{3} + 2\sqrt{3} + \sqrt{3}$

9. $5\sqrt{5} - 2\sqrt{5} + \sqrt{5}$

10. $2\sqrt{7} + 3\sqrt{7} - 8\sqrt{7}$

11. $2\sqrt{3} + \sqrt{12}$

12. $5\sqrt{2} + \sqrt{18}$

13. $3\sqrt{12} - \sqrt{48}$

14. $2\sqrt{45} - 2\sqrt{20}$

15. $5\sqrt{8} + 2\sqrt{18}$

16. $\sqrt{50} + \sqrt{32} - \sqrt{8}$

17. $\sqrt{20} + 2\sqrt{5} - \sqrt{45}$

18. $5\sqrt{8} + 3\sqrt{18} - 4\sqrt{32}$

19. $9\sqrt{2} - 20\sqrt{2} + 11\sqrt{2}$

20. $\sqrt{12} + 2\sqrt{3} - \sqrt{75}$

21. $\sqrt{63} - 2\sqrt{28} + 5\sqrt{7}$

22. $2\sqrt{50} + 3\sqrt{18} - \sqrt{32}$

23. $\sqrt{96} - \sqrt{150} + \sqrt{6}$

24. $3\sqrt{8} + 5\sqrt{50} - 4\sqrt{32}$

25. $\frac{3}{5}\sqrt{125} - \sqrt{20}$

26. $\sqrt{243} - \frac{2}{3}\sqrt{27}$

27. $\frac{2}{3}\sqrt{18} - \frac{1}{2}\sqrt{72}$

28. $\frac{3}{4}\sqrt{128} - \frac{3}{5}\sqrt{50}$

29. $3\sqrt{125} - 5\sqrt{20} - 3\sqrt{45}$

30. $\sqrt{54} - 2\sqrt{24} + \sqrt{216}$

31. $3\sqrt{12} - 5\sqrt{147} + 4\sqrt{75}$

32. $\frac{2}{3}\sqrt{18} - \frac{3}{4}\sqrt{32} - \sqrt{50}$

4.3 Multiplying Surds

To multiply surds, we make use of the property $\sqrt{ab} = \sqrt{a} \cdot \sqrt{b}$ in the reverse i.e. $\sqrt{a} \cdot \sqrt{b} = \sqrt{ab}$.

Example 4 Simplify

(a) $\sqrt{2} \times \sqrt{10}$

Solution $\sqrt{2} \times \sqrt{10} = \sqrt{2 \times 10}$

$$= \sqrt{20}$$

$$= 2\sqrt{5}$$

(b) $2\sqrt{3} \times 3\sqrt{6}$

Solution $2\sqrt{3} \times 3\sqrt{6} = 2 \cdot \sqrt{3} \times 3 \cdot \sqrt{6}$

$2\sqrt{3} \times 3\sqrt{6} = 2 \cdot 3 \cdot \sqrt{3} \cdot \sqrt{6}$ Using Commutative Property

$= 6\sqrt{18}$

$= 18\sqrt{2}$ Simplify

In practice it is not necessary to show the first step.

Often the work is made easier when you simplify the surds before you multiply.

Example 5   Simplify $\sqrt{32} \times \sqrt{45}$

Solution $\sqrt{32} \times \sqrt{45} = 4\sqrt{2} \times 3\sqrt{5}$

$= 12\sqrt{10}$

Exercise 4.3

Work out:

1. $\sqrt{2} \times \sqrt{6}$
2. $\sqrt{5} \times \sqrt{15}$
3. $3\sqrt{2} \times \sqrt{3}$
4. $5\sqrt{3} \times 2\sqrt{5}$
5. $3\sqrt{2} \times 2\sqrt{5}$
6. $5\sqrt{3} \times 3\sqrt{2}$
7. $2\sqrt{5} \times 3\sqrt{3}$
8. $\sqrt{3} \times \sqrt{12}$
9. $\sqrt{8} \times \sqrt{10}$
10. $\sqrt{5} \times \sqrt{32}$
11. $\sqrt{15} \times \sqrt{27}$
12. $\sqrt{12} \times \sqrt{18}$
13. $\sqrt{50} \times \sqrt{30}$
14. $\sqrt{98} \times \sqrt{75}$
15. $\sqrt{60} \times \sqrt{8}$
16. $\sqrt{3} \times \sqrt{2} \times \sqrt{10}$
17. $\sqrt{6} \times 2\sqrt{2} \times \sqrt{15}$
18. $\sqrt{20} \times \sqrt{15} \times \sqrt{6}$
19. $\sqrt{30} \times \sqrt{2} \times \sqrt{27}$
20. $\sqrt{72} \times \sqrt{75} \times \sqrt{18}$

4.4 Rationalising the Denominator

To simplify a fraction whose denominator contains surds we use a technique called rationalising the denominator illustrated in the example below.

Consider the fraction $\frac{1}{\sqrt{3}}$

Multiplying the numerator and denominator by $\sqrt{3}$, we have

$$\frac{1}{\sqrt{3}} = \frac{1}{\sqrt{3}} \times \frac{\sqrt{3}}{\sqrt{3}} = \frac{\sqrt{3}}{3}$$

Note that we can do this because we are multiplying the given fraction by $\frac{\sqrt{3}}{\sqrt{3}}$ or 1, which does not change its value.

This technique transfers the surd expression from the denominator to the numerator and changes the denominator to a rational number.

Example 6 Write each expression in simplest form:

(a) $\frac{8}{\sqrt{2}}$

Solution $\frac{8}{\sqrt{2}} = \frac{8}{\sqrt{2}} \times \frac{\sqrt{2}}{\sqrt{2}} = \frac{8\sqrt{2}}{2} = 4\sqrt{2}$

(b) $\frac{15}{\sqrt{50}}$

Solution One approach is to simplify $\sqrt{50}$ first.

$$\frac{15}{\sqrt{50}} = \frac{15}{5\sqrt{2}} \times \frac{\sqrt{2}}{\sqrt{2}} = \frac{3\sqrt{2}}{2}$$

Exercise 4.4

Rationalise the denominator of:

1. $\frac{2}{\sqrt{3}}$
2. $\frac{10}{\sqrt{5}}$
3. $\frac{4\sqrt{3}}{\sqrt{6}}$
4. $\frac{9}{\sqrt{3}}$
5. $\frac{3\sqrt{5}}{\sqrt{6}}$
6. $\frac{12\sqrt{2}}{\sqrt{6}}$
7. $\frac{15}{\sqrt{10}}$
8. $\frac{18}{\sqrt{12}}$
9. $\frac{14}{\sqrt{7}}$
10. $\frac{3}{4\sqrt{2}}$
11. $\frac{8}{\sqrt{32}}$
12. $\frac{5\sqrt{6}}{\sqrt{15}}$
13. $\frac{3\sqrt{6}}{\sqrt{12}}$
14. $\frac{\sqrt{5}}{\sqrt{3}}$
15. $\sqrt{\frac{3}{2}}$

16. $\dfrac{\sqrt{12}}{\sqrt{27}}$ 17. $\dfrac{\sqrt{18}}{\sqrt{12}}$ 18. $\sqrt{\dfrac{12}{5}}$ 19. $\dfrac{\sqrt{15}}{2\sqrt{3}}$ 20. $\dfrac{6}{\sqrt{27}}$

4.5 Multiplication of brackets containing surds

You can multiply brackets containing surds by using the distributive property;

$$a(b + c) = ab + ac$$

Example 7 Work out:

(a) $3\sqrt{2}(\sqrt{3} + 2\sqrt{2})$

Solution $3\sqrt{2}(\sqrt{3} + 2\sqrt{2}) = 3\sqrt{2} \cdot \sqrt{3} + 3\sqrt{2} \cdot 2\sqrt{2}$

$$= 3\sqrt{6} + 12$$

(b) $(2 + \sqrt{3})(4 - \sqrt{3})$

Solution $(2 + \sqrt{3})(4 - \sqrt{3}) = 8 - 2\sqrt{3} + 4\sqrt{3} - 3$

$$= 5 + 2\sqrt{3}$$

Example 8 Work out:

(a) $(3\sqrt{3} - 5)(3\sqrt{3} + 5)$

Solution One approach is to use the identity $(a - b)(a + b) = a^2 - b^2$

$$(3\sqrt{3} - 5)(3\sqrt{3} + 5) = (3\sqrt{3})^2 - 5^2$$

$$= 27 - 25$$

$$= 2$$

(b) $(2\sqrt{5} - 3)^2$

Solution You can use the identity $(a - b)^2 = a^2 - 2ab + b^2$

$$(2\sqrt{5} - 3)^2 = (2\sqrt{5})^2 - 2(2\sqrt{5})(3) + 3^2$$

$$= 20 - 12\sqrt{5} + 9$$

$$= 29 - 12\sqrt{5}$$

Exercise 4.5

Work out:

1. $\sqrt{3}(\sqrt{5} - \sqrt{3})$
2. $\sqrt{3}(2\sqrt{5} - 3\sqrt{3})$
3. $\sqrt{3}(\sqrt{3} + \sqrt{5})$
4. $\sqrt{7}(2\sqrt{3} + 3\sqrt{7})$
5. $\sqrt{6}(\sqrt{6} + \sqrt{5})$
6. $\sqrt{2}(\sqrt{6} - \sqrt{2})$
7. $5\sqrt{2}(3\sqrt{2} - 2\sqrt{3})$
8. $2\sqrt{12}(\sqrt{3} + \sqrt{2})$
9. $(\sqrt{3} + 5)(\sqrt{3} + 2)$
10. $(\sqrt{5} - 2)(\sqrt{5} - 1)$
11. $(3\sqrt{2} - \sqrt{6})(2\sqrt{3} + \sqrt{2})$
12. $(2\sqrt{2} + \sqrt{3})(\sqrt{12} - \sqrt{8})$
13. $(1 - \sqrt{3})(2 + 3\sqrt{2})$
14. $(\sqrt{6} - \sqrt{5})(\sqrt{8} + \sqrt{15})$
15. $(\sqrt{5} + \sqrt{15})(2\sqrt{3} - 1)$
16. $(2\sqrt{2} + \sqrt{5})(\sqrt{10} - 2\sqrt{2})$
17. $(\sqrt{5} - 2)(\sqrt{5} + 2)$
18. $(\sqrt{7} + 5)(\sqrt{7} - 5)$
19. $(\sqrt{10} + 5)(\sqrt{10} - 5)$
20. $(\sqrt{8} - 2)(\sqrt{8} + 2)$
21. $(\sqrt{12} - 3)(\sqrt{12} + 3)$
22. $(2\sqrt{3} + 4\sqrt{2})(2\sqrt{3} - 4\sqrt{2})$
23. $(\sqrt{3} + 2)^2$
24. $(\sqrt{5} - 3)^2$
25. $(\sqrt{3} + \sqrt{2})^2$
26. $(\sqrt{3} - 2\sqrt{2})^2$
27. $(2 + 3\sqrt{2})^2$
28. $(3\sqrt{2} - 2\sqrt{3})^2$
29. $(\sqrt{6} - 2\sqrt{3})^2$
30. $(\sqrt{3} + 2\sqrt{2})^2$

4.6 Fractions whose denominators are sums or difference of surds

If a and b are both irrational numbers or either a is an irrational number and b is a rational number or a is a rational number and b is an irrational number, the product $(a + b)(a - b)$ will always be a rational number.

For example, $(\sqrt{7} + \sqrt{3})(\sqrt{7} - \sqrt{3}) = 7 - 3 = 4$

and $(\sqrt{6} + 3)(\sqrt{6} - 3) = 6 - 9 = -3$.

Surds 81

Notice that the expressions in the brackets differ only in the sign between the terms. These expressions are conjugates of each other. As the examples show the product of two conjugates is a difference of two squares.

To rationalise a fraction whose denominator is the sum or difference of surds, we multiply both the numerator and denominator by the conjugate of the denominator.

Example 9 Work out:

(a) $\dfrac{2\sqrt{3}}{2+\sqrt{3}}$

Solution
$$\dfrac{2\sqrt{3}}{2+\sqrt{3}} = \dfrac{2\sqrt{3}}{2+\sqrt{3}} \times \dfrac{2-\sqrt{3}}{2-\sqrt{3}}$$

$$= \dfrac{2\sqrt{3}(2-\sqrt{3})}{2^2-(\sqrt{3})^2}$$

$$= \dfrac{4\sqrt{3}-6}{4-3}$$

$$= 4\sqrt{3} - 6$$

(b) $\dfrac{4\sqrt{5}+2\sqrt{3}}{\sqrt{5}-\sqrt{3}}$

Solution
$$\dfrac{4\sqrt{5}+2\sqrt{3}}{\sqrt{5}-\sqrt{3}} = \dfrac{4\sqrt{5}+2\sqrt{3}}{\sqrt{5}-\sqrt{3}} \times \dfrac{\sqrt{5}+\sqrt{3}}{\sqrt{5}+\sqrt{3}}$$

$$= \dfrac{(4\sqrt{5}+2\sqrt{3})(\sqrt{5}+\sqrt{3})}{(\sqrt{5})^2-(\sqrt{3})^2}$$

$$= \dfrac{26+6\sqrt{15}}{2}$$

$$= 13 + 3\sqrt{15}$$

Notice that each term in the numerator is divided by 2.

Exercise 4.6

Rationalise the denominator of:

1. $\dfrac{1}{\sqrt{2}+1}$
2. $\dfrac{2}{\sqrt{3}-1}$
3. $\dfrac{1}{2-\sqrt{3}}$
4. $\dfrac{4}{3+\sqrt{7}}$
5. $\dfrac{\sqrt{5}}{\sqrt{15}-\sqrt{10}}$
6. $\dfrac{5}{4+\sqrt{6}}$
7. $\dfrac{3\sqrt{7}}{5-\sqrt{7}}$
8. $\dfrac{2\sqrt{6}}{2\sqrt{3}-1}$
9. $\dfrac{2-\sqrt{3}}{2+\sqrt{3}}$

10. $\dfrac{3+\sqrt{2}}{3-\sqrt{2}}$

11. $\dfrac{\sqrt{6}+\sqrt{2}}{\sqrt{6}-\sqrt{2}}$

12. $\dfrac{2\sqrt{3}-\sqrt{5}}{2\sqrt{3}+\sqrt{5}}$

13. $\dfrac{3-2\sqrt{2}}{3+2\sqrt{2}}$

14. $\dfrac{\sqrt{3}+\sqrt{5}}{\sqrt{3}-\sqrt{5}}$

15. $\dfrac{\sqrt{7}-3\sqrt{2}}{\sqrt{7}-\sqrt{2}}$

16. $\dfrac{\sqrt{6}+2\sqrt{2}}{\sqrt{6}-2\sqrt{2}}$

17. $\dfrac{\sqrt{5}+3}{4-\sqrt{20}}$

18. $\dfrac{\sqrt{3}+\sqrt{6}}{3\sqrt{2}-3}$

19. $\dfrac{2\sqrt{5}+\sqrt{3}}{5\sqrt{3}-3\sqrt{5}}$

20. $\dfrac{\sqrt{7}+3\sqrt{3}}{2\sqrt{7}+\sqrt{3}}$

21. $\dfrac{8-2\sqrt{6}}{2\sqrt{3}+3\sqrt{2}}$

4.7 Irrational Equations

An equation which contains a radical expression is called an irrational equation. Examples include $\sqrt{3x-2} = 4$ and $\sqrt{4x+3} = \sqrt{2x-3} + 5x$.

Solving Irrational Equations

To solve an irrational equation with one radical expression, we first isolate the radical expression and then square each side of the equation. If the equation contains more than one radical expression, you may have to repeat the process more than once.

Example 10 Solve:

(a) $\sqrt{2x-3} + x = 3$

Solution $\sqrt{2x-3} + x = 3$ Original equation

$\sqrt{2x-3} = 3 - x$ Isolate the radical expression

$2x - 3 = 9 - 6x + x^2$ Square each side

$x^2 - 8x + 12 = 0$ Simplify

$(x-2)(x-6) = 0$ Factorise the quadratic expression

$x - 2 = 0$ or $x - 6 = 0$

$x = 2$ $\quad x = 6$

Squaring each side of an equation can introduce extraneous solutions, so it is necessary to check all solutions in the original equation.

Check:

If $x = 2$, $\sqrt{4-3} + 2 = 1 + 2 = 3$

and if $x = 6$, $\sqrt{12-3} + 6 = 3 + 6 = 9$

Only $x = 2$ satisfies the original equation.

The solution is $x = 2$.

(b) $\sqrt{x+5} + \sqrt{5-x} = 4$

Solution $\sqrt{x+5} + \sqrt{5-x} = 4$ Original equation

Rewrite the equation with a radical term on each side.

$$\sqrt{x+5} = 4 - \sqrt{5-x}$$

$$x + 5 = 16 - 8\sqrt{5-x} + 5 - x \quad \text{Square each side}$$

$$x - 8 = -4\sqrt{5-x} \quad \text{Simplify}$$

$$x^2 - 16x + 64 = 16(5 - x) \quad \text{Square each side}$$

$$x^2 = 16 \quad \text{Simplify}$$

$$x = \pm 4$$

Check

If $x = 4$, $\sqrt{4+5} + \sqrt{5-4} = 3 + 1 = 4$

and $x = -4$, $\sqrt{-4+5} + \sqrt{5+4} = 1 + 3 = 4$.

Both $x = -4$ and $x = 4$ satisfy the original equation.

The solutions are $x = -4$ and $x = 4$.

Exercise 4.7

Solve each of these irrational equations:

1. $\sqrt{4x-3} = 3$ 2. $\sqrt{5x+1} = 4$

3. $\sqrt{5-4x} = 5$ 4. $x = \sqrt{3x^2 - 18}$

5. $\sqrt{4x^2 + 3} = 2x + 1$ 6. $\sqrt{3x-5} = x - 3$

7. $\sqrt{10-3x} + 2 = x$ 8. $\sqrt{7+x^2} - x = 1$

9. $\sqrt{20-x^2} - x = 6$ 10. $x - \sqrt{5x-9} = 1$

11. $\sqrt{2x+5} + 5 = x$

12. $\sqrt{x+5} - \sqrt{x} = 1$

13. $\sqrt{x-5} + 1 = \sqrt{x}$

14. $\sqrt{x+1} = 2 - \sqrt{x}$

15. $1 + 2\sqrt{x-3} = \sqrt{4x-3}$

16. $\sqrt{x-6} + 3 = \sqrt{x+9}$

17. $\sqrt{3x-2} + 1 = \sqrt{2x+5}$

18. $\sqrt{2x-4} = \sqrt{3x-1} - \sqrt{x-1}$

19. $\sqrt{2x+2} - \sqrt{x-6} = \sqrt{x+2}$

20. $\sqrt{2x+1} - \sqrt{4-2x} = \sqrt{4x-5}$

Review Exercise 4

1. Simplify:

(a) $3\sqrt{6} - 8\sqrt{6}$

(b) $2\sqrt{2} + 3\sqrt{2} - \sqrt{2}$

(c) $\sqrt{5} - 3\sqrt{5} - 4\sqrt{5}$

(d) $2\sqrt{2} + 3\sqrt{2} - 7\sqrt{2}$

(e) $\sqrt{18} - \sqrt{72} + \sqrt{32}$

(f) $2\sqrt{8} - 3\sqrt{50} + 5\sqrt{18}$

2. Calculate:

(a) $\sqrt{18} \times \sqrt{6}$

(b) $\sqrt{8} \times \sqrt{10}$

(c) $\sqrt{14} \times \sqrt{8}$

(d) $\sqrt{20} \times \sqrt{27}$

(e) $\sqrt{75} \times \sqrt{50}$

(f) $\sqrt{12} \times \sqrt{45}$

(g) $\sqrt{72} \times \sqrt{27}$

(h) $\sqrt{12} \times \sqrt{3} \times \sqrt{6}$

(i) $2\sqrt{5} \times \sqrt{20} \times \sqrt{45}$

3. Rationalise the denominator:

(a) $\frac{1}{5\sqrt{2}}$

(b) $\frac{8}{3\sqrt{2}}$

(c) $\frac{8}{3\sqrt{18}}$

(d) $\frac{9}{4\sqrt{6}}$

(e) $\frac{\sqrt{7}}{\sqrt{21}}$

(f) $\frac{2+\sqrt{3}}{\sqrt{3}}$

4. Find each of the following products:

(a) $(\sqrt{3} + 2\sqrt{2})(\sqrt{3} + 5\sqrt{2})$

(b) $(3\sqrt{2} - \sqrt{7})(3\sqrt{2} + 2\sqrt{7})$

(c) $(\sqrt{3} + \sqrt{2})^2$

(d) $(5\sqrt{3} - 2)^2$

(e) $(2\sqrt{3} + \sqrt{2})(\sqrt{3} - 2\sqrt{2})$

(f) $(2\sqrt{5} + 3\sqrt{2})(2\sqrt{5} - 3\sqrt{2})$

5. Rationalise the denominator:

(a) $\frac{1}{3\sqrt{2}-\sqrt{5}}$

(b) $\frac{3\sqrt{2}}{2\sqrt{2}-3}$

(c) $\frac{\sqrt{3}+2}{2\sqrt{3}-1}$

(d) $\frac{\sqrt{3}+\sqrt{2}}{\sqrt{3}-\sqrt{2}}$

(e) $\frac{2-\sqrt{3}}{\sqrt{6}+\sqrt{2}}$

(f) $\frac{3\sqrt{2}-2\sqrt{3}}{3\sqrt{3}-2\sqrt{2}}$

6. Solve each of the following irrational equation:

(a) $\sqrt{x+1} + 3 = 5$

(b) $x + 2 = \sqrt{8x+1}$

(c) $7 - \sqrt{x-1} = 2$

(d) $\sqrt{3x+4} - \sqrt{x+1} = 3$

5 Binary Operations

5.1 Binary Operations

In arithmetic we learned to do calculations with counting numbers by using basic operation of addition. The operation of addition assigns a definite number called the sum to each pair of numbers. For example, 7 added to 3 produce the number $7 + 3$, i.e., 10. Also, the operation of multiplication assigns a number called the product to a pair of numbers. Generally when any two counting numbers are added (or multiplied) we obtain a single number called the sum (or the product).

An operation performed on two elements of a set such that for any two elements of the set the result is a single element is called a binary operation. Addition and multiplication are both binary operations.

A binary operation is generally given by a symbol and a rule that defines the symbol. For example a binary operation $*$ may by defined by the rule $a * b = a + b - ab$. Using this rule we can evaluate $-3 * 2$ as follows: replacing a and b by -3 and 2 respectively we have

$$-3 * 2 = -3 + 2 - (-3)(2) = 5$$

Exercise 5.1

1. Evaluate the following using the given operations:

(a) $2 * 3$, $\ a * b = ab - b$

(b) $-3 * -2$, $\ x * y = 2xy$

(c) $-4 * 3$, $\ x * y = x^y$

(d) $5 * -4$, $\ x * y = xy - y$

(e) $-2 * 3$, $\ a * b = a^2 + b^2 - ab$

(f) $2 * \sqrt{3}$, $\ \frac{2b - 3a}{ab}$

(g) $-8 * -5$, $\ x * y = (x - y)^2$

(h) $(2 * 3) * 4$, $\ x * y = x + y - 2xy$

(i) $2 * (3 * -5)$, $\ p * q = p^2 + 2q - 5$

(j) $(3 * 2) * (4 * 3)$, $\ m * n = m^2 + n^2 - mn$

2. Given that $x * y = x^2 + y^2 + xy$ find x when $x * 2 = 12$

3. If $x * y = \sqrt{x^2 - 2xy}$ find x such that $x * 3 = 4$

4. Given that $a * b = a^2 + 2ab + b^2$ find a when $a * 3 = 16$

5. Given that $x * y = \sqrt{2x - y}$ find x when $x * 1 = x$

6. If $a * b = \frac{3a+2b}{4a-1}$ find a such that $a * -1 = \frac{1}{2}$

7. Given that $a * b = a + ab - b$ find the truth set of $3 * a = a * 2$

8. If $a * b = 4b^2 + 9a^2 - 12ab$ express b in terms of a given that $a * b = 0$

9. Two operations $*$ and $\circ$ are defined on the set R of real numbers by

$m * n = m + n - mn$ and $m \circ n = \frac{1}{2} mn$.

Evaluate:

(a) $2 \circ (4 * 3)$

(b) $(2 \circ 4) * (2 \circ 3)$

10. Two operations $*$ and $\circ$ are defined on the set R of real numbers by

$m * n = 3m + 2n - mn$ and $m \circ n = 2n - m$. Find a if $2 \circ (a * 4) = 8$

5.2 Properties of Binary Operations

There are many other systems which consist of a set of elements together with a rule for combining two elements in the set to obtain a third element of the set that possess the same properties of operations of addition and multiplication. A binary operation is a generalisation of addition and multiplication. We consider five properties of binary operation common to some system of numbers.

Closure

When we add any two integers, say 3 and 4, the sum $3 + 4 = 7$ is an integer. The operation of addition is said to be closed in the set of integers. Generally if a binary operation is performed on any two elements of a set and the result is an element of the set, then that set is said to be closed under the given binary operation. For example, the product of any two integers is an integer. Therefore the set of integers is closed under the operation of multiplication.

Commutative Property

We may add 5 and 9 as follows:

$5 + 9 = 14$

or $9 + 5 = 14$

You may have noticed that changing the order gives the same result. This property of integers is called the commutative property of addition. Similarly we can demonstrate that multiplication is commutative. Generally given that a and b belong to a set then the operation $*$ is commutative if $a * b = b * a$.

Associative Property

We may add 4, 5 and 6 as follows:

$(4 + 5) + 6 = 9 + 6 = 15$

or $4 + (5 + 6) = 4 + 11 = 15$

Although the two procedures are different the result of each addition is 15. When three numbers are added it makes no difference which two are added first. This property of addition is called associative property. Similarly we can demonstrate that multiplication is associative. Generally given that a, b and c belong to a set then the operation $*$ is associative if $(a * b) * c = a * (b * c)$.

Identity Elements

For any counting number a, $a + 0 = a$. For example, $2 + 0 = 2$, 0 is called the additive identity. Also for any number b, $b \times 1 = b$. For example, $-3 \times 1 = -3$, 1 is called the multiplicative identity. In general, given an element a and the identity element e, then $a * e = e * a = a$.

Inverse Elements

For any counting number a there exists a number $-a$ such that $a + (-a) = 0$. $-a$ is called the additive inverse of a. Notice that 0 is the additive identity. For example, -2 is the additive inverse of 2 since $2 + (-2) = 0$. 2 is the additive inverse of -2. Also for any counting number a where $a \neq 0$ there exists a number $\frac{1}{a}$ such that $a \times \frac{1}{a} = 1$. $\frac{1}{a}$ is called the multiplicative inverse or the reciprocal of a. For example, $\frac{1}{5}$ is the multiplicative inverse of 5 since $5 \times \frac{1}{5} = 1$. Generally, when a binary operation is performed on two elements in a set and the result is the identity element for the binary operation then each element is said to be the inverse of the other. Given that a' is the inverse of a then $a * a' = a' * a = e$.

Example 1

(a) A binary operation $\circ$ is defined on the set, R, of real numbers by $a \circ b = a^2 - ab + b^2$. Show that the operation $\circ$ is commutative.

Solution $a \circ b = a^2 - ab + b^2$

$\qquad\qquad b \circ a = b^2 - ba + a^2$ \hfill Changing the order of operation

$\qquad\qquad\qquad = a^2 - ab + b^2$ \hfill Using the commutative property

Since $a \circ b = b \circ a$ the operation $\circ$ is commutative.

(b) Determine whether or not the operation $*$ defined on the set, R, of real numbers by $a * b = ab + 2$ is associative.

Solution   Let $a, b, c \in R$ then

$$(a * b) * c = (ab + 2) * c$$
$$= (ab + 2)c + 2$$
$$= abc + 2c + 2$$

Also $a * (b * c) = a * (bc + 2)$

$$= a(bc + 2) + 2$$
$$= abc + 2a + 2$$

Since $(a * b) * c \neq a * (b + c)$ the operation is not associative.

(c) A binary operation $*$ is defined on the set, R, of real numbers by $a * b = a + b - 2ab$, find the identity element $e \in R$.

Solution $a * e = a$ \hfill By definition

$\qquad\qquad a * e = a + e - 2ae$ \hfill By the rule

Hence $\qquad a + e - 2ae = a$

$\qquad\qquad e(1 - 2a) = 0$

$\qquad\qquad e = 0$

The identity element is 0.

(d) A binary operation $*$ is defined on the set, R, of real numbers by $a * b = a + b - ab$, find the inverse a' of an element $a \in R$.

Solution   We first look for the identity element e.

$$a * e = a$$

$$a * e = a + e - ae$$

$$a + e - ae = a$$

$$e(1 - a) = 0$$

$$e = 0$$

Now $a * a' = e$ By definition

$$a * a' = a + a' - aa'$$ By the rule

Hence $a + a' - aa' = 0$ Since $e = 0$

$$a'(1 - a) = -a$$

$$a' = \frac{a}{a-1}, a \neq 1$$

Exercise 5.2

1. The operation $*$ is defined on the set of real numbers, R, by the relation $a * b = a - 2ab + b$. Show whether the operation $*$ is commutative or not.

2. Determine whether or not the operation $\circ$ defined on the set of real numbers, R, by the relation $a \circ b = a^2 - b^2$ is commutative or not.

3. Show that the operation $\otimes$ defined on the set of real numbers, R, by the relation $a \otimes b = a^b$ is not commutative.

4. Show that the operation defined on the set P of ordered pairs of real numbers by the relation $m * n = (m + n, mn)$ is associative.

5. The operation Δ is defined on the set of real numbers by $m \Delta n = \frac{m-n}{n}, n \neq 0$. Determine whether or not the operation is associative.

6. The operation * is defined on the set P = {1, 2, 3, 4} by the table below:

*	1	2	3	4
1	0	3	4	1
2	3	0	-1	-2
3	4	-1	-3	1
4	1	-2	2	3

Find whether * is closed under set P.

7. The operation $\oplus$ defined on the set of real numbers, R, by the relation $x \oplus y = 4xy$. Find the identity element $e \in R$.

8. Find the identity element e of the operation * defined on the set of real numbers by the relation $a * b = 2a - ab + 3b$.

9. Find the inverse of the element x under the binary operation $\circ$ defined by the relation $x \circ y = \frac{1}{2}xy$.

10. A binary operation * is defined on the set, R, of real numbers by the relation $a * b = a + b + 2ab$. Find the inverse element of a.

11. A binary operation * is defined on the set of real numbers, R. If x is the identity element of R and $x * 6 = y(y + 5)$ find y.

12. The identity element of a binary operation * defined on the set of real numbers R is 8. Find x if y is the inverse of the element x and $x * y = x^2 + 2x$.

Review Exercise 5

1. A binary operation * is defined on the set S = {0, 1, 2, 3} by $a * b = a + b - ab$ where $a, b \in S$.

(a) Form a table for the operation * on the set S

(b) Is the operation * closed on the set S?

(c) State the identity element for *

(d) Find the inverse of the element 2

2. The operation * is defined on the set P = {2, 3, 4, 5} by the table below

*	2	3	4	5
2	4	2	4	3
3	2	3	4	5
4	3	4	4	4
5	3	5	4	2

(a) State the identity element for *

(b) Which of the element has no inverse?

(c) Evaluate (3 * 2) * (5 * 4)

3. A binary operation ⊕ defined on the set R of real numbers by $a \oplus b = \dfrac{ab}{a+b}$, where $a + b \neq 0$. Show that the operation is

(a) Commutative

(b) Associative

4. A binary operation ∘ is defined on the set R of real numbers by $x \circ y = x + y - 2xy$. Find:

(a) the identity element e

(b) the inverse of an element $x \in R$

State the value of x for which no inverse exists.

5. A binary operation * is defined on the set R of real numbers by $a * b = 3a - 2b + ab$

(a) Determine whether or not * is commutative

(b) Find c such that $a * c = a$

(c) State the value of a for which c is not defined

6. A binary operation * is defined on the set of real numbers by $x * y = xy - y$. Find:

(a) the identity element under the operation

(b) the inverse of 3 under *

7. A binary operation $\oplus$ is defined on the set R of real numbers by $a \oplus b = \dfrac{4+a}{3-b}$, $b \neq 3$.

Evaluate $\sqrt{5} \oplus \sqrt{20}$ leaving your answer in the form $p + q\sqrt{5}$, where p and q are rational numbers.

8. A binary operation $*$ is defined on the set R of real numbers by $a * b = \dfrac{ab+3}{ab}$, $ab \neq 0$

(a) Find the truth set of the operation $a * 2 = \dfrac{a+4}{2}$

(b) Evaluate $\sqrt{3} * 2$ leaving your answer in the form $p + q\sqrt{3}$, where p and q are rational numbers.

9. The binary operations $*$ and ∇ are defined on the set R of real numbers by $a * b = a + b + 2$ and $a \nabla b = 2ab$ respectively where $a, b \in R$.

(a) Evaluate:

   (i) $2\nabla(4 * 3)$

   (ii) $(2\nabla 4) * (2\nabla 3)$

(b) Solve $x\nabla(x * 3) = 12$ for x

10. The operation ∇ is defined on the set of real numbers R by $m\nabla n = m^2 + n^2 - mn$.

(a) Evaluate $(1 + \sqrt{2})\nabla(1 - \sqrt{2})$

(b) Solve $(\sqrt{2x+3})\nabla 2 = 3$ for x

6 Functions

6.1 Functions

A function from a set A to a Set B, written $f: A \to B$, is a relation that associates each element in Set A with exactly one element in set B. Set A is called the domain of f and set B is called the co-domain (or range).

The function from set A to set B can be illustrated by a diagram, called the arrow diagram, as shown below:

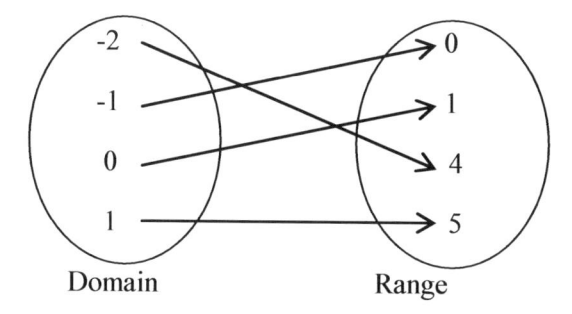

Figure 6.1

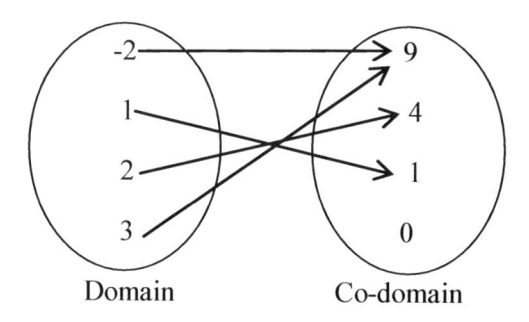

Figure 6.2

Notice from Figure 6.1 and Figure 6.2 that each element in the domain is associated with only one element in the co-domain (or range). For each element in the domain the corresponding element in the co-domain is called the image. The set of all images is called the range of the function.

There may be elements in the co-domain which are not images of elements in the domain (see Figure 6.2). The range of a function is often a proper subset of the co-domain.

For any element x, in the domain, the corresponding image is denoted by $f(x)$, read "f of x" or "the value of f at x."

Consider the function represented by the diagram below.

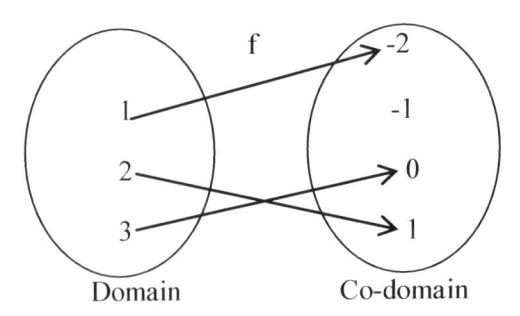

Figure 6.3

You can see from the diagram shown in Figure 6.3 that -2 is the image of the element 1, written briefly as $f(1) = -2$. Similarly, $f(2) = 1$ and $f(3) = 0$. The range of the function f is $\{-2, 0, 1\}$.

Our discussion suggests the following characteristics of functions.

1. Each element in the domain relates to only one element in the co-domain (or range).

2. Some elements in the co-domain may not have corresponding elements in the domain.

3. Two or more elements in the domain may have the same image in the co-domain.

A function can be expressed in one of the following forms.

1. a table or a list

2. a rule or equation

3. a graph

Function Notation

Functions are often named by letters such as f, g and h. For instance the function represented by the equation $y = 3x + 2$ can be named f and written as $f(x) = 3x + 2$ or $f: x \to 3x + 2$.

Recall that an element x in the domain of a function has only one image y in the range. This means that the graph of a function cannot have two or more different points with the same x-coordinate. This observation suggests the following vertical line test.

Vertical Line Test

A graph represents a function if and only if no vertical line intersects the graph at more than one point.

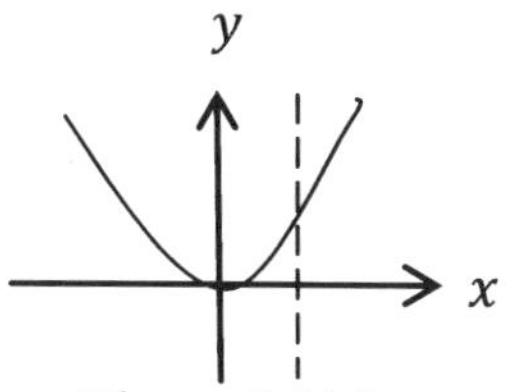

Figure 6.4(a)

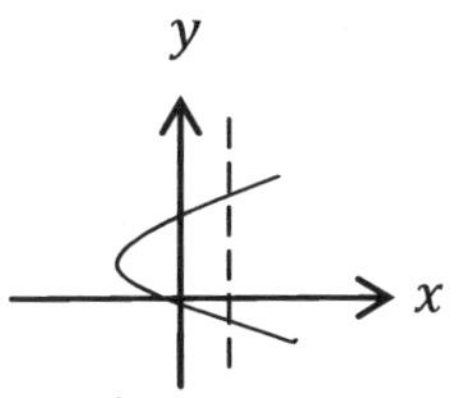

Figure 6.4(b)

A vertical line drawn anywhere in the graph shown in Figure 6.4(a) intersects the graph in at most one point, so this is a graph of a function. On the other hand, a vertical line drawn in the graph shown in Figure 6.4(b) intersects the graph twice, so this is not a graph of a function.

Evaluating a Function

The process of finding the value of a function $y = f(x)$ for a given value of x is called evaluating the function. To evaluate a function we replace x with the given value of x.

Example 1

(a) Evaluate the function $f(x) = 3x^2 - 5$ at $x = -2$

Solution $f(-2) = 3(-2)^2 - 5$ Replace x with -2

$= 12 - 5$

$= 7$

(b) Given that $f(x) = 2x - 3$, find $f(a+1)$

Solution $f(a+1) = 2(a+1) - 3$

$= 2a - 1$

(c) Given that $h(x) = \dfrac{5x+8}{x+7}, x \neq -7$ find the positive value of x, when $h(x) = x$

Solution $\dfrac{5x+8}{x+7} = x$

$x^2 + 2x - 8 = 0$

$(x-2)(x+4) = 0$

$x - 2 = 0$ or $x + 4 = 0$

$x = 2 \qquad x = -4$

Therefore $x = 2$

Domain of Functions

The set of all possible values of the independent variable, x of a function is called the domain of the function. Often the domain of a function defined by an equation is the largest set of real numbers that have corresponding images in the co-domain. For example, any real number can be used to replace x in the function $f(x) = 2x - 3$, so the domain of this function is the set of all real numbers.

In some cases, x cannot take on all real numbers. For example, the domain of the function $f(x) = \frac{2x}{x-3}$ cannot include 3, which makes the denominator equal to 0. Remember that division by zero is undefined. The domain of this function is the set of all real numbers except 3, which is written in set notation as $\{x: x \in R, x \neq 3\}$.

The function $f(x) = \sqrt{x}$ is defined on the set of real numbers only when $x \geq 0$, making the domain of this function $\{x: x \in R, x \geq 0\}$.

Example 2

(a) Find the largest possible domain for the function $f(x) = \dfrac{5x+4}{x^2 - x - 6}$

Solution Set the denominator equal to zero and solve for x.

$$x^2 - x - 6 = 0$$

$$(x - 3)(x + 2) = 0$$

$$x = -2 \text{ or } x = 3$$

The domain is $\{x: x \in R, x \neq -2, x \neq 3\}$

(b) Find the largest possible domain for the function $g(x) = \sqrt{5 - x}$

Solution Set the expression under the radical sign to greater than or equal to zero

$$5 - x \geq 0$$

$$x \leq 5$$

The domain is the set of all real numbers less than or equal to 5, written as $\{x: x \in R, x \leq 5\}$.

Functions 97

Exercise 6.1

1. Each of the following arrow diagrams describes a relation. Which of the relations are functions?

(a)

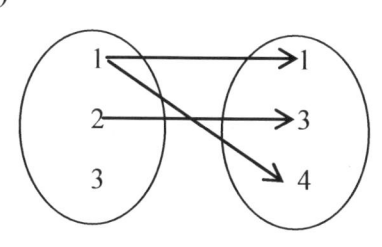

(b)

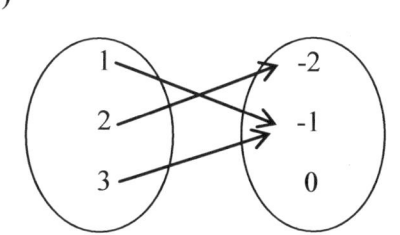

(c)

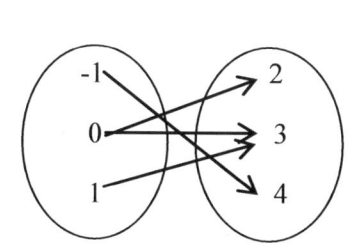

(d)

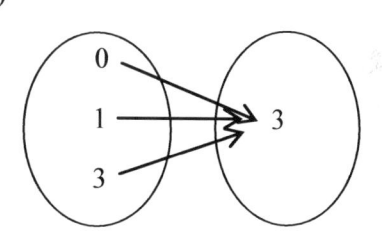

2. State the domain and range of each of the functions described by the arrow diagrams.

(a)

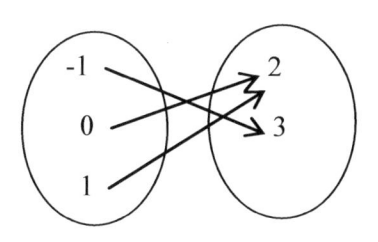

(b)

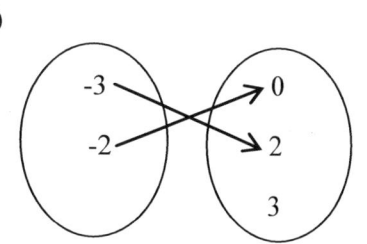

(c)

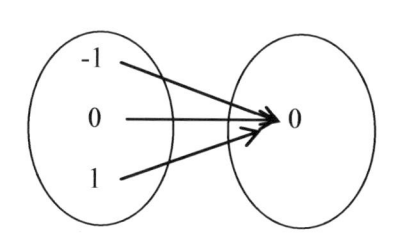

(d)

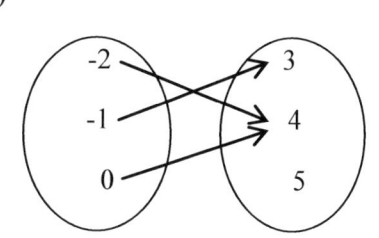

3. Determine whether each of the following is the graph of a function.

(a)

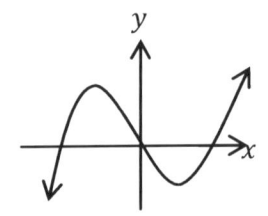

(b)

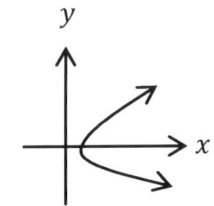

(c)

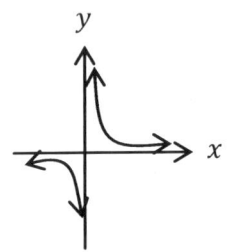

(d)

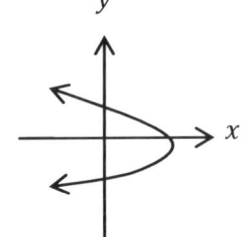

4. Find the domain of the functions represented by the following graphs.

(a)

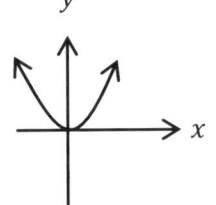

(b)

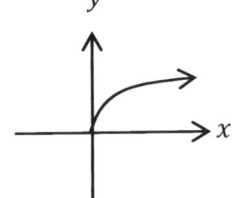

(c)

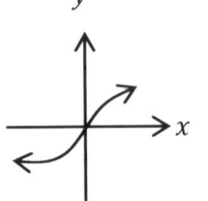

(d)

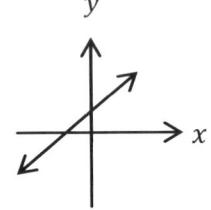

(d)

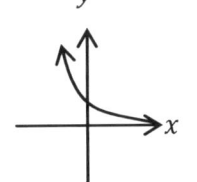

(e)

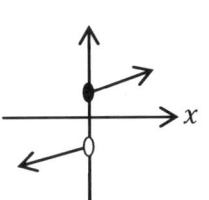

5. If $f(x) = 3x^2 - 2x + 5$ evaluate each of the following

(a) $f(-1)$　　　(b) $f(0)$　　　(c) $f(2)$　　　(d) $f(3)$

6. If $g(x) = 3 + 2x - x^2$ evaluate each of the following

(a) $g(-2)$　　　(b) $g(-1)$　　　(c) $g(1)$　　　(d) $g(2)$

7. If $f(x) = x^2$ find an expression for each of the following

(a) $f(a+1)$　　　(b) $f(2\sqrt{a})$　　　(c) $f(-2)$　　　(d) $f(x+y)$

8. If $g(x) = 2 - 3x$ find x when $g(x) = 8$

9. If $f(x) = x^2$ find the values of a when $f(a+1) = 16$

10. Given that $g(2x - 1) = 3x^2 - 5x + 2$ what is the value of $g(3)$?

11. Find the largest domain of each of the following functions defined on the set of real numbers

(a) $f(x) = 5x - 6$　　　(b) $g(x) = 2x + 3$

(c) $h(x) = \dfrac{3x+2}{x-5}$　　　(d) $f(x) = \dfrac{3x+2}{x^2+x-6}$

12. Find the largest domain of each of the following functions defined on the set of real numbers

(a) $f(x) = \sqrt{7-x}$　　　(b) $g(x) = \sqrt{x-6}$　　　(c) $h(x) = \sqrt{25-x^2}$

6.2 Special Functions

One-to-One Functions

A function f from set A to set B is said to be one-to-one if each element in set A corresponds to exactly one element in set B. In other words f is one-to-one if and only if $f(a) = f(b)$ implies $a = b$ for all a and b in A.

Horizontal Line Test

A graph represents a one-to-one function if and only if no horizontal line intersects the graph at more than one point.

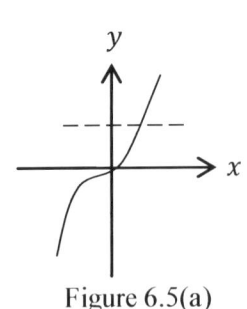

Figure 6.5(a)

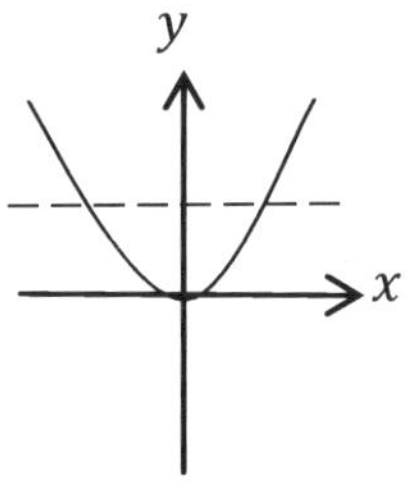

Figure 6.5 (b)

A horizontal line drawn anywhere in the graph shown in Figure 6.5 (a) intersects the graph only once. This graph represents a one- to-one function. However, a horizontal line intersects the graph shown in Figure 6.5 (b) twice, so the graph does not represent a one-to-one function.

Example 3

The function f, is defined on the set of real numbers by $f(x) = 5x + 7$. Determine whether or not f is one-to-one.

Solution Let $a, b \in R$

$f(x) = 5x + 7$ Write the original function.

$f(a) = 5a + 7$ Substitute a for x.

$f(b) = 5b + 7$ Substitute b for x.

Equating the two values, we get

$5a + 7 = 5b + 7$

$a = b$

Because $a = b$ if $f(a) = f(b)$ the function f is one-to-one.

Exercise 6.2

1. Determine whether each graph is that of one-to-one function

(a)

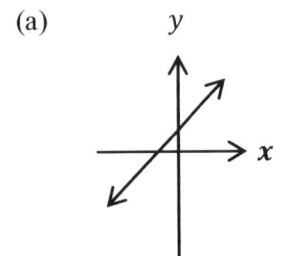

(b)

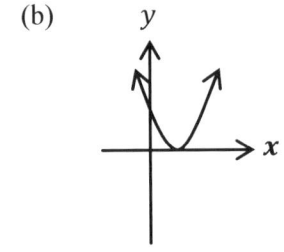

(c)

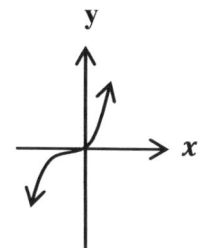

(d)

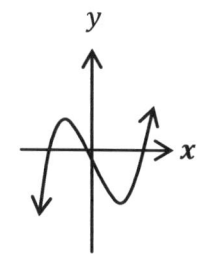

2. Show that the following functions are one-to-one

(a) $f(x) = 7x + 5$

(b) $g(x) = \frac{1}{4}x - 3$

(c) $h(x) = \frac{3x}{2x-1}, x \neq \frac{1}{2}$

(d) $g(x) = \frac{3x+2}{x+4}, x \neq -4$

3. Determine whether or not the following functions are one-to-one

(a) $f(x) = 3x^2 - 2$

(b) $g(x) = 5x + 3$

(c) $h(x) = 5 - x^2$

(d) $f(x) = \frac{2x+3}{x-5}, x \neq 5$

6.3 Inverse Functions

Here are two illustrations of functions:

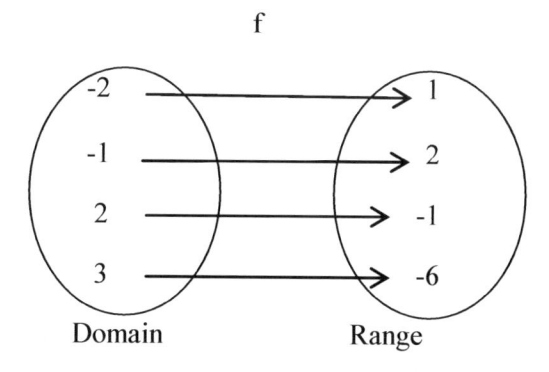

Figure 6.6

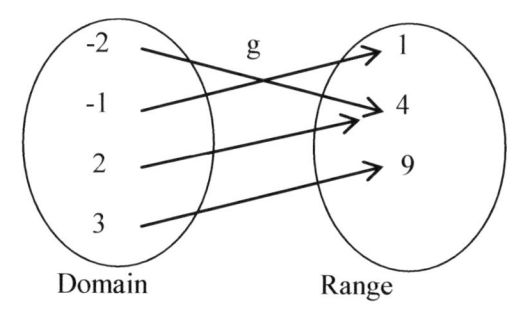

Figure 6.7

The function f (see Figure 6.6) maps each element in the domain to only one element in the range. The reverse mapping also forms another function called the inverse function of f, denoted by f^{-1}. However, the reverse mapping of the function g (see Figure 6.7) does not represent a function because 4 is matched with -2 and 2. The function g has no inverse function. Notice that a function has an inverse function if the function is one-to-one.

Observe that the domain of the function, f is the same as the range of the inverse function f^{-1}, and the range of f is the same as the domain of the inverse function.

You can use the horizontal line test to determine whether a function has an inverse or not. A function has an inverse function if and only if no horizontal line intersects the graph of the function at more than one point.

Sometimes it may be possible to obtain an inverse of a function by restricting the domain of the function. For example if we restrict the domain of the function $f(x) = x^2$ to only nonnegative values of x the function will have the inverse function $f^{-1}(x) = \sqrt{x}$.

Finding an Inverse Function Algebraically

To find the inverse function we interchange the roles of x and y and then solve for y.

Example 4

(a) Find the inverse function of the function $f(x) = 4x + 3$ defined on the set of real numbers.

Solution $f(x) = 4x + 3$ Write the original function.

$y = 4x + 3$ Replace $f(x)$ by y.

$x = 4y + 3$ Interchange x and y.

$x - 3 = 4y$ Isolate the y term.

$\frac{x-3}{4} = y$ Divide each side by 4.

$f^{-1}(x) = \frac{x-3}{4}$ Replace y by $f^{-1}(x)$

(b) Find the inverse of the function $g(x) = \frac{2x-3}{x+2}$, $x \neq -2$ defined on the set of real numbers.

Solution $g(x) = \frac{2x-3}{x+2}$ Write the original function.

$$y = \frac{2x-3}{x+2}$$ Replace g(x) by y.

$$x = \frac{2y-3}{y+2}$$ Interchange x and y.

$$y = \frac{2x+3}{2-x}$$ Solve for y.

$g^{-1}(x) = \frac{2x+3}{2-x}, x \neq 2$ Replace y by $g^{-1}(x)$.

Graphs of Inverse Functions

The preceding examples shows that to obtain an inverse function we interchange x and y coordinate in the ordered pair(x, y). So if a point (x, y) lies on the graph of a function then the point (y, x) must lie on the graph of its inverse function. This means that the graphs of a function f and its inverse function f^{-1} are reflections of each other in the line $y = x$. Recall from your study of transformation that the point (x, y) maps onto (y, x) under a reflection in the line $y = x$.

Range of Functions

The range is the y - values we get after substitution of all possible x - values. To find the range of a function we find the inverse function and then state its domain.

Example 5

(a) Find the range of the function $f(x) = 3 - x^2$

Solution First write the function as an equation and then isolate x

$$y = 3 - x^2$$

$$x = \sqrt{3-y}$$

The range of f is $\{y : y \in R, y \leq 3\}$

(b) Find the range of the function $g(x) = 5 - 2x$ in the domain $\{-2 \leq x \leq 4\}$.

Solution To find the range substitute the smallest and largest numbers in the domain. In this case the smallest number is -2 and the largest number is 4.

$g(-2) = 5 - 2(-2) = 9$

104   Further Mathematics

$g(4) = 5 - 2(4) = -3$

The range of g is $\{y: y \in R, -3 \leq y \leq 9\}$

Exercise 6.3

1. Each of the following graphs represents a function. Which of the function has an inverse?

(a)

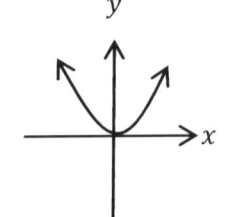

(b)

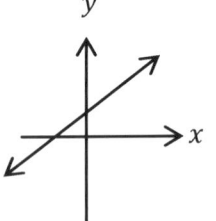

(c)

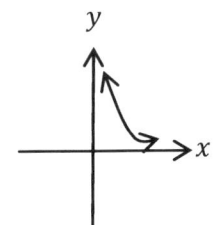

(d)

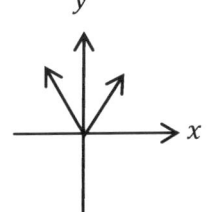

(e)

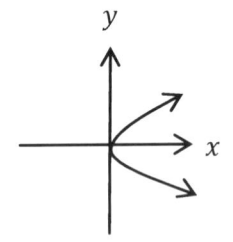

(f)

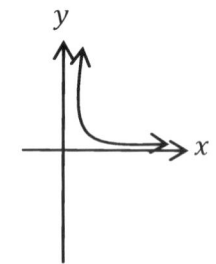

2. Each of the following graphs represents a function. Find the range of the function.

(a)

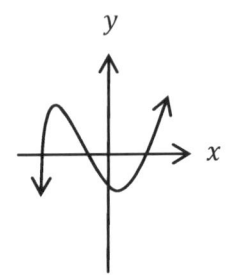

(b)

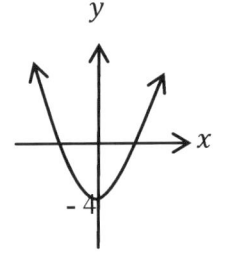

(c)

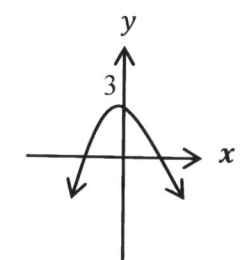

(d)

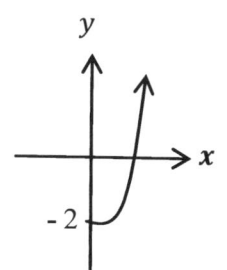

(e)

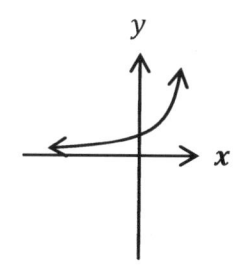

(f)

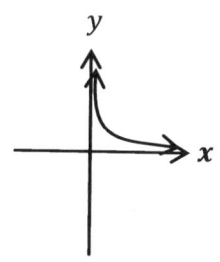

3. Sketch the graph of each function and its inverse function on the same axes

(a) $f(x) = x + 3$

(b) $f(x) = x^2 + 2,\ x \geq 0$

(c) $f(x) = (x - 3)^2,\ x \geq 3$

(d) $f(x) = x^3 - 1$

(e) $f(x) = |x - 5|, x \geq 5$

(f) $f(x) = \sqrt[3]{x} + 2$

4. Sketch the graph of each of the following functions in the given domain. Find the domain of the inverse function from the graph.

(a) $f(x) = (x - 3)^2,\ x \geq 3$

(b) $f(x) = |x| + 2$

(c) $f(x) = 4 - x^2,\ x \geq 0$

(d) $f(x) = |x - 5|,\ x \geq 5$

5. Find the inverse function of each of the following functions defined on the set of real numbers

(a) $f(x) = 3x - 8$

(b) $g(x) = 4 - x^2, x \geq 0$

(c) $h(x) = \frac{5}{2x+1}, x \neq -\frac{1}{2}$

(d) $f(x) = \frac{2x+3}{x-4}, x \neq 4$

6. Given that $g(x) = 3x - 12$ find $g^{-1}(3)$

7. Given that $h(x) = \frac{x+4}{2x-3}, x \neq \frac{3}{2}$ find $h^{-1}(-2)$

8. Given that $f(x) = 3x - 5$ find x if $f^{-1}(x) = 2$

9. Given that $h(x) = \frac{2x+5}{x-3}, x \neq 3$ find x if $h^{-1}(x) = \frac{1}{4}$

10. Find the range of the following functions

(a) $f(x) = 3x + 4$

(b) $f(x) = x^2 + 2$

(c) $g(x) = 8 - 2x - x^2$

(d) $h(x) = |4 - 3x|$

6.4 Operations on Functions

You can form a new function by adding, subtracting, multiplying or dividing two functions. These operations can be performed using the following rules. If f and g are two functions then

1. $(f + g)(x) = f(x) + g(x)$ Addition

2. $(f - g)(x) = f(x) - g(x)$ Subtraction

3. $(f \cdot g)(x) = f(x) \cdot g(x)$ Multiplication

4. $\left(\dfrac{f}{g}\right)(x) = \dfrac{f(x)}{g(x)}, g(x) \neq 0$ Division

Example 6   Given that $f(x) = 3x - 2$ and $g(x) = 2x + 3$, find

(a) $2f + g$

Solution $(2f + g)(x) = 2f(x) + g(x)$

$$= 2(3x - 2) + (2x + 3)$$

$$= 6x - 4 + 2x + 3$$

$$= 8x - 1$$

(b) $3f - 2g$

Solution $(3f - 2g)(x) = 3f(x) - 2g(x)$

$$= 3(3x - 2) - 2(2x + 3)$$

$$= 9x - 6 - 4x - 6$$

$$= 5x - 12$$

Example 6   Given that $f(x) = x^2 + 2x$ and $g(x) = x^2 - 4$, find

(a) $f \cdot g$

Solution $(f \cdot g)(x) = f(x) \cdot g(x)$

$$= (x^2 + 2x)(x^2 - 4)$$

$$= x^4 + 2x^3 - 4x^2 - 8x$$

(b) $\dfrac{f}{g}$

Solution $\left(\dfrac{f}{g}\right)(x) = \dfrac{f(x)}{g(x)}$

$$= \dfrac{x^2 + 2x}{x^2 - 4}$$

$$= \dfrac{x(x+2)}{(x-2)(x+2)}$$

$$= \dfrac{x}{x-2}, x \neq 2$$

Exercise 6.4

In Questions 1 – 4, use the given function f and g to find:

(a) $f + g$ (b) $f - g$ (c) $f \cdot g$ (d) $\dfrac{f}{g}$

1. $f(x) = 3x - 4, \ g(x) = 2x + 3$

2. $f(x) = 2x^2 + 3x, \ g(x) = 2x + 3$

3. $f(x) = x + 3, \ g(x) = x^2 - 9$

4. $f(x) = \dfrac{2x-3}{3x+2}, \ g(x) = \dfrac{4}{3x+2}$

5. Given $f(x) = 5x + 3$ and $g(x) = x^2 - 2x + 1$ find:

(a) $2f + 3g$ (b) $f - 2g$

6. Given $f(x) = x^2 - 3x - 10$ and $g(x) = 2x^3 - 5x^2$ find:

(a) $f \cdot g$ (b) $\dfrac{f}{g}$

6.5 Composite Functions

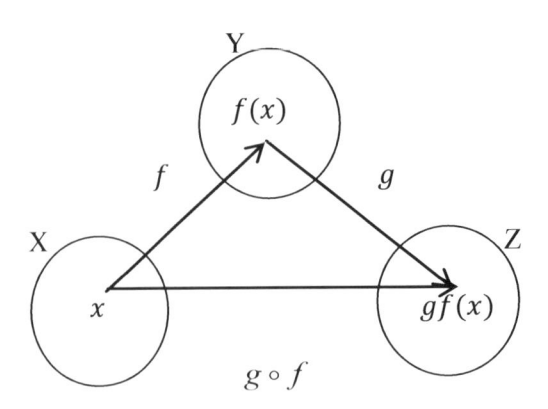

Figure 6.8

The function f maps x in the domain of f to an image $f(x)$. Then the function g maps $f(x)$ to an image $g(f(x))$, as shown in Figure 6.8. Notice that $f(x)$ is in the domain of the function g. The two functions f and g can be combined to form a single function denoted by $g \circ f$, which maps x to $g(f(x))$, without going through the two processes, f followed by g. The function $g \circ f$, read "f composite g" is called the composition of f and g or the composite function of "f followed by g."

In general, given two functions f and g, the composite function denoted by $f \circ g$ is defined by $(f \circ g)(x) = f(g(x))$

You can compose a function with itself. The composition of the function f with itself, i.e. $f \circ f$, can be written briefly as f^2.

Example 7 If $f(x) = 2x - 3$ and $g(x) = 3x^2 + 2$ are defined on the set of real numbers, find

(a) $f \circ g$

Solution $(f \circ g)(x) = f(g(x))$ Definition of $f \circ g$.

                   $= f(3x^2 + 2)$ Replace $g(x)$ by $3x^2 + 2$.

                   $= 2(3x^2 + 2) - 3$ Substitute $3x^2 + 2$ for x.

                   $= 6x^2 + 1$ Simplify.

(b) $g \circ f$

Solution $(g \circ f)(x) = g(f(x))$ Definition of $g \circ f$

$$= g(2x - 3) \qquad \text{Replace } f(x) \text{ by } 2x - 3.$$

$$= 3(2x - 3)^2 + 2 \qquad \text{Substitute } 2x - 3 \text{ for } x.$$

$$= 12x^2 - 36x + 29 \qquad \text{Simplify.}$$

Notice that $(f \circ g)(x) \neq (g \circ f)(x)$. In general the composition of a function is not commutative. However, it is possible to find two functions f and g such that $f[g(x)] = g[f(x)]$.

Example 8 If $f(x) = 2x + 3$ and $f^{-1}(x) = \frac{x-3}{2}$ are defined on the set of real numbers find:

(a) f^2

Solution
$$f^2(x) = f(f(x)) \qquad \text{Definition of } f^2.$$
$$= f(2x + 3) \qquad \text{Replace } f(x) \text{ by } 2x + 3.$$
$$= 2(2x + 3) + 3 \qquad \text{Substitute } 2x + 3 \text{ for } x.$$
$$= 4x + 9 \qquad \text{Simplify.}$$

(b) $f \circ f^{-1}$

Solution
$$f \circ f^{-1} = f(f^{-1}(x)) \qquad \text{Definition of } f \circ f^{-1}.$$
$$= f\left(\frac{x-3}{2}\right) \qquad \text{Replace } f^{-1}(x) \text{ by } \frac{x-3}{2}.$$
$$= 2\left(\frac{x-3}{2}\right) + 3 \qquad \text{Substitute } \frac{x-3}{2} \text{ for } x.$$
$$= x \qquad \text{Simplify.}$$

You can show that $(f^{-1} \circ f)(x) = x$. This is left as an exercise. These results suggest that two functions f and g are inverses if and only if $f(g(x)) = g(f(x)) = x$. The functions $f^{-1} \circ f$ and $f \circ f^{-1}$ are called identity functions.

Exercise 6.5

1. If $f(x) = 5x + 3$ and $g(x) = 3x - 2$ find:

(a) $f^2(x)$ (b) $g^2(x)$ (c) $f(g(x))$ (d) $g(f(x))$

2. If $f(x) = 2x + 3$ and $g(x) = x^2 - 1$ find:

(a) $f(g(x))$ (b) $g(f(x))$ (c) $f^2(x)$ (d) $g^2(x)$

3. If $f(x) = x^2$, $g(x) = 3x + 2$ and $h(x) = x - 2$ find:

(a) $(f \circ g) \circ h$ (b) $(f \circ h) \circ g$ (c) $(h \circ f) \circ g$ (d) $(h \circ g) \circ f$

4. If $f(x) = 2x - 3$ and $g(x) = \frac{3x}{x+4}, x \neq -4$ find:

(a) $f \circ g$ (b) $g \circ f$

5. If $f(x) = \frac{1}{2}x + 3$ and $g(x) = 3x - 2$ find:

(a) $f^{-1} \circ f$ (b) $f \circ f^{-1}$ (c) $g \circ g^{-1}$

(d) $g^{-1} \circ g$ (e) $g^{-1} \circ f^{-1}$ (f) $f^{-1} \circ g^{-1}$

6. Determine whether the functions f and g are inverse functions of each other.

(a) $f(x) = 3x + 4, g(x) = \frac{1}{3}(x - 4)$

(b) $f(x) = \frac{1}{2}(x + 5), g(x) = 2x - 5$

(c) $f(x) = \frac{2x-7}{x}, x \neq 0, g(x) = \frac{7}{2-x}\ x \neq 2$

(d) $f(x) = \frac{3x+5}{x-4}, x \neq 4, g(x) = \frac{4x+5}{x-3}, x \neq 3$

7. Given that $f(x) = 3x^2 - 2$ and $g(x) = 2x + 3$ find:

(a) $f(g(x))$ (b) $f(g(-1))$ (c) $g(f(x))$ (d) $g(f(2))$

8. If $f(x) = 3x - 1$ and $g(x) = x^2$ find:

(a) $f(g(3))$ (b) $g(f(x))$ (c) $g(f(5))$ (d) f(g(-2))

9. If $f(x) = \frac{2x-1}{5}$ and $g(x) = \frac{3x+2}{x-1}, x \neq 1$ find:

(a) $g(f(2))$ (b) $f(g(2))$ (c) $g^{-1}(f(-3))$ (d) $g(f^{-1}(-3))$

10. If $f(x) = 2x + 3$ and $g(x) = x^2 - 1$ find:

(a) $f(g(-2))$ (b) $g(f(2))$ (c) $f^2(-1)$ (d) $g^2(3)$

11. The functions f and g are defined on the set of real numbers. If $f(g(x)) = x^2 + 6x + 9$ and $f(x) = x^2$ find $g(x)$

12. If $f(x) = 2x + 1$ and $g(x) = 3x - 2$ find x when

(a) $f(g(x)) = 9$ (b) $g(f(x)) = 19$

13. If $f(x) = x^2$ and $g(x) = 2x + 1$ find x when

(a) $f(g(x)) = 9$ (b) $g(f(x)) = 51$

14. If $f(x) = 2x - 1$ and $g(x) = \frac{3x+2}{x+2}, x \neq -2$ find x when $f(g(x)) = x$

Review Exercise 6

1. The functions f and g are defined by

$$f: x \to x^2 - 25$$

$$g: x \to \frac{2x}{3x-2}$$

where x is a real number

(a) State the largest possible domain of each of the two functions.

(b) Find the range of f.

(c) Determine whether or not g is a one-to-one function.

2. A function g is defined on the set R of real numbers by $g: \to \frac{3x-4}{x+5}, x \neq -5$.

Find: (a) g^{-1} (b) the range of g^{-1}

3(a) State the largest possible domain of each of the functions

$$f: x \to \frac{3x+2}{x-4}$$

$$g: x \to \frac{4x+3}{2x+1}$$

where x is a real number

(b) Find:

   (i) the inverse of f^{-1}.

   (ii) the composite function $g \circ f^{-1}$.

4. The functions f and g are such that $f: x \to \frac{4x}{x-3}, x \neq -3$ and $g: x \to 2x + 1$. Find:

(a) $f \circ g$ (b) $g \circ f$

5. A function h defined on the set R of real numbers, is given by $h: x \to \frac{x+1}{x-2}, x \neq 2$.

(a) Show that h is a one-to-one function

(b) Find the inverse h^{-1} of h

6. Given that $f: x \to x^2 + 1$ and $g: x \to \frac{2x+1}{x+1}, x \neq -1$, find:

(a) $g \circ f$ (b) $g \circ f(-1)$

7. The functions f and h on the set R of real numbers are defined by

   $f: x \to 2x^2 + 1$

   $h: x \to x - 2$

If $f \circ h(a) = h \circ f(a)$, find the value of a.

8. The functions g and h are defined on the set of real numbers by:

   $g: x \to x^2 - 1$

   $h: \to \frac{x+2}{x}, x \neq 0$

Find: (a) h^{-1} (b) $g \circ h^{-1}$, when $x = -\frac{1}{2}$

9. The functions g and h are defined on the set of real numbers, by $g: x \to x^2 + p$ and $h: x \to px + q$ respectively, where p and q are positive real numbers. If $g \circ h(x) = 9x^2 + 12x + 7$, find the values of p and q.

10. Two functions f and g are defined by $f: x \to -2x + 3$ and $g: x \to px + q$, where p and q are constants. Given that $g^{-1}(4) = 2$ and $g \circ f(3) = -7$, find the values of p and q.

7 Linear Functions

7.1 The Distance between Two Points

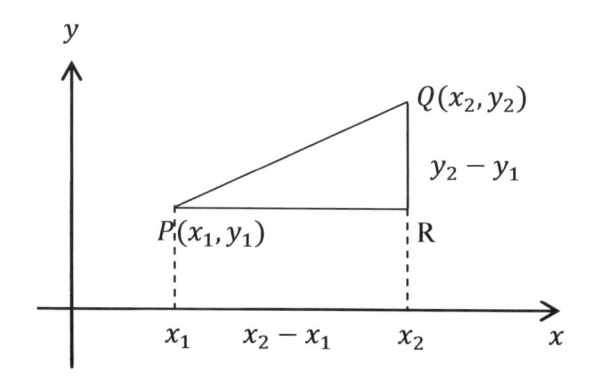

Figure 7.1

Figure 7.1 shows a straight line joining the points $P(x_1, y_1)$ and $Q(x_2, y_2)$. PR and QR are drawn parallel to the x-axis and y-axis respectively to form triangle PQR. Using the Pythagoras' theorem we have

$$PQ^2 = PR^2 + RQ^2$$
$$= (x_2 - x_1)^2 + (y_2 - y_1)^2$$

Taking the positive square root of both sides we get

$$PQ = \sqrt{(x_2 - x_1)^2 + (y_2 - y_1)^2}$$

In general, the distance between the two points (x_1, y_1) and (x_2, y_2) is given by $\sqrt{(x_2 - x_1)^2 + (y_2 - y_1)^2}$.

Example 1

Find the distance between the points A (2, 3) and B (5, 7).

Solution Let $(x_1, y_1) = (2, 3)$ and $(x_2, y_2) = (5, 7)$. Then

$$AB = \sqrt{(5 - 2)^2 + (7 - 3)^2}$$
$$= \sqrt{9 + 16}$$
$$= \sqrt{25}$$
$$= 5 \text{ units}$$

Exercise 7.1

1. Find the distance between the following pairs of points:

(a) (3, 2), (8, 14)   (b) (1, 3), (4, 7)   (c) (-1, -1), (-1, 7)

(d) (2, 4), (5, 2)   (e) (4, -1), (6, 2)   (f) (-4, -1), (-2, -3)

2. A triangle has vertices A (1, 1), B (4, 4) and C (9, -1). Calculate the lengths of the sides of the triangle and prove that the triangle is right-angled.

3. Show that the triangle whose vertices are (3, -2), (-1, 1) and (2, 5) is isosceles.

4. The point $(5, t)$ is equidistant from (2, -1) and (1, 6). Find the value of t.

5. The point $(t, -1)$ lies on a circle with a radius of 5 units and its centre at (-1, 3). Find the possible values of t

6. Three points A, B and C have coordinates (9, 6), (3, -2) and $(-5, h)$. Given that h is positive and $AB = BC$, find the value of h.

7.2 The Mid-point of a Straight Line

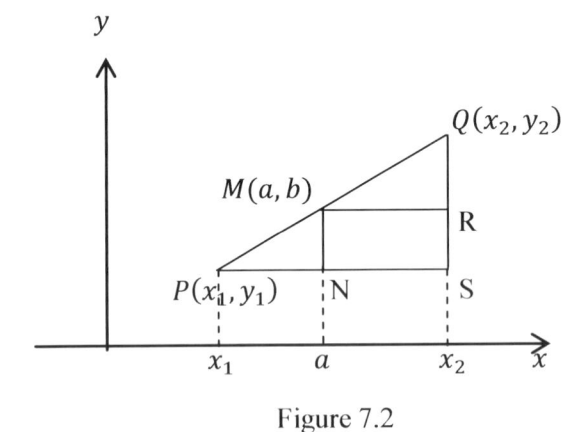

Figure 7.2

The mid-point of a line segment that joins two points is the point that divides the line into two equal parts.

Figure 7.2 shows a line joining the points $P(x_1, y_1)$ and $Q(x_2, y_2)$. The mid-point M of PQ has coordinates (a, b). MN and MR are drawn parallel to QS and PS respectively. Because M is the mid-point of PQ and MN is parallel to QS, N is the mid-point of PS.

Hence $a - x_1 = x_2 - a$

$$a = \frac{x_1 + x_2}{2}$$

Similarly, $b - y_1 = y_2 - b$

$$b = \frac{y_1 + y_2}{2}$$

Therefore, the mid-point of PQ has coordinates. $\left(\frac{x_1+x_2}{2}, \frac{y_1+y_2}{2}\right)$

Example 2

Find the mid-point of the line joining the points (3, 2) and (5, -8)

Solution The coordinates of the mid-point $= \left(\frac{3+5}{2}, \frac{2+(-8)}{2}\right) = (4, -3)$

Exercise 7.2

1. Find the coordinates of the mid-point of the lines joining the following pairs of points

(a) (1, 2), (5, 4)   (b) (4, 2), (6, 10)   (c) (-2, 3), (8, -4)

(d) (6, -5), (2, 3)   (e) (-1, -6), (-4, 3)   (f) (7, -4), (-3, 8)

2(a) Find the coordinates of the mid-points of the sides of the triangle whose vertices are A(3, 2), B(7, 4) and C(-6, -5)

(b) Show that the length of the line segment joining the mid-point of two sides of triangle ABC in (a) is one-half of the length of the third side.

3. The mid-point of a line segment AB is (3, 2). Given that the coordinates of A are (2, -4), find the coordinates of B.

4. If (6, 2) is the mid-point of the line segment connecting (3, -1) to $P(x, y)$, find the values of x and y.

7.3 The Gradient of a Line

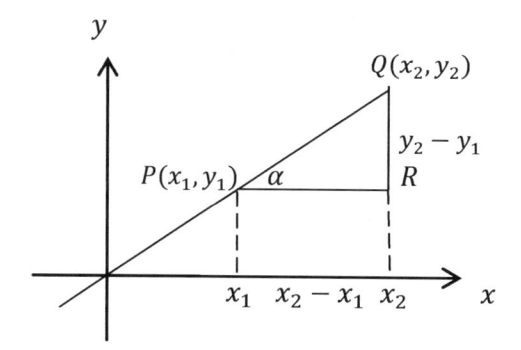

Figure 7.3

Figure 7.3 shows a straight line PQ through the origin. The line PQ makes angle α with the x-axis. Because PR is parallel to the x-axis, $\angle QPR = \alpha°$. The gradient of the line is defined as $\tan \alpha$. From triangle PQR

$$\tan \alpha = \frac{QR}{PR} = \frac{y_2 - y_1}{x_2 - x_1}$$

In general, the gradient of a line passing through two points (x_1, y_1) and (x_2, y_2) is given by

$$\text{Gradient} = \frac{y_2 - y_1}{x_2 - x_1}$$

Notice that the gradient is the ratio of the change in y-values to the change in x-values.

The gradient of a line that slants upward from left to right is positive, and the gradient of a line that slants from left to right is negative. The gradient of a horizontal line is 0, and the gradient of a vertical line is undefined.

Example 3

Find the gradient of the line joining the point A (3, 2) and B (5, 6).

Solution Let $(x_1, y_1) = (3, 2)$ and $(x_2, y_2) = (5, 6)$. Using the gradient formula, we have

$$\text{Gradient} = \frac{6 - 2}{5 - 3} = 2$$

The order of subtraction is important. You must form the numerator and denominator using the same order of subtraction.

Exercise 7.3

1. Find the gradients of the lines joining the following pairs of points

(a) (5, 4), (9, 11)   (b) (-4, 9), (2, -3)   (c) (4, -3), (-7, -4)

(d) (-1, -2), (5, 7)   (e) (6, 7), (11, 3)   (f) (-3, 5), (9, -5)

2. Prove, using gradients, that the points (-6, 6), (-2, 4) and (6, 0) are collinear (that is, the points lie on the same straight line)

3. Determine whether each of the following sets of points are collinear.

(a) (-1, -2), (4, 3), (5, 4)   (b) (-4, 2), (-6, 5), (5, 10)

(c) (3, 5), (-2, -5), (-3, 9) (d) (0, 6), (3, -3), (-1, 9)

4. If the points (-4, 3), (-1, 4) and $(x, 9)$ are collinear then find x

5. The points (-1, -1), (3, 11) and $(1, t)$ lie on the same line. What is the value of t?

6. The points $A(3, x)$ and $B(7, 5)$ lie on a straight line. If the gradient of the line is $\frac{3}{4}$, find the value of x

7. The points $(3, 2)$ and $(x, 5)$ lie on the same line. If the gradient of the line is $-\frac{3}{2}$, find the value of x

8. Three points $A(a, b)$, $B\left(\frac{4}{3}, a\right)$ and $C\left(b, \frac{5}{2}\right)$ lie on a straight line. If the gradient of the line is $-\frac{3}{2}$, find the values of a and b.

7.4 The Equation of a Line

There are three forms of the equation of a straight line.

The Gradient-Intercept Form

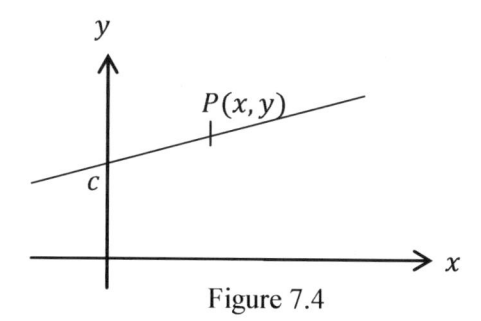

Figure 7.4

Let m represents the gradient of a straight line that cuts the y-axis at c, and passes through the point $P(x, y)$ as shown in Figure 7.4. Using the points $(0, c)$ and (x, y) in the gradient formula we have

$$\frac{y - c}{x} = m$$

Hence, $y = mx + c$

The equation $y = mx + c$, where m is the gradient and $(0, c)$ the intercept on the y-axis is called the gradient-intercept form.

If $c = 0$, then the line passes through the origin. Hence the equation of a line through the origin is $y = mx$.

Example 4

(a) Find the gradient and y-intercept of the line $2x + 3y - 6 = 0$

Solution Rewrite the equation in the gradient-intercept form.

$2x + 3y - 6 = 0$ Original equation.

$3y = -2x + 6$ Isolate the y term.

$y = -\frac{2}{3}x + 2$ Divide each side by 3.

The gradient is $-\frac{2}{3}$ and the y-intercept is $(0, 2)$

(b) A straight line passes through the point $(4, -2)$. If the gradient of the line is $\frac{3}{4}$, find the y-intercept of the line.

Solution Substituting $x = 4$, $y = -2$ and $m = \frac{3}{4}$ in the formula $y = mx + c$, we have

$$-2 = 4\left(\frac{3}{4}\right) + c$$

$$c = -5$$

Hence, the y-intercepts is $(0, -5)$

(c) The gradient of a line passing through the point $(0, 2)$ is -3. Find the equation of the line.

Solution Let (x, y) represents a point on the line.

Hence $\dfrac{y - 2}{x} = -3$

$y = -3x + 2$

An alternative solution is to substitute -3 for m and 2 for c in $y = mx + c$ to get $y = -3x + 2$

Exercise 7.4(a)

1. Find the gradient and the y-intercept of each of the following equations.

(a) $3x - 2y = 4$ (b) $4y - 3x = 10$ (c) $6x + 2y - 5 = 0$

(d) $4x - 5y = 13$ (e) $4x + 3y = 2$ (f) $3x + 6y - 9 = 0$

2. Find the equation for the line with the indicated gradient and y-intercept.

(a) 3, (0, 4) (b) −2, (0, 5) (c) 4, (0, 0)

(d) $-\frac{3}{4}$, (0, −4) (e) $\frac{2}{3}$, (0, 1) (f) $-\frac{3}{2}$, (0, −3)

3. Find the y-intercept of the line with the indicated gradient and the named point on it.

(a) 3, (1, 2) (b) −4, (3, −8) (c) 2, (−3, −1)

(d) $\frac{1}{2}$, (−4, 2) (e) $-\frac{3}{4}$, (−4, −2) (f) $\frac{3}{2}$, (6, 5)

Point-Gradient Form

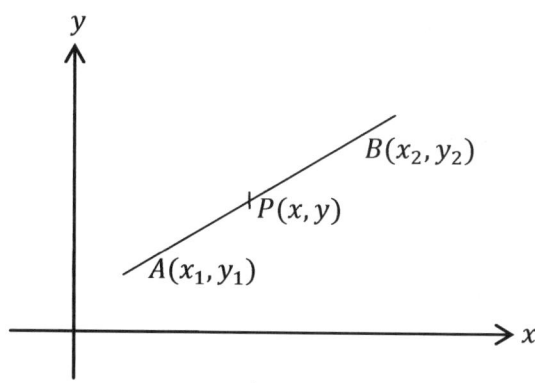

Figure 7.5

Figure 7.5 shows a straight line through the points $A(x_1, y_1)$ and $B(x_2, y_2)$. Let m represents the gradient of the line. Using the point $P(x, y)$ on the line and the point $A(x_1, y_1)$, we have

$$\frac{y - y_1}{x - x_1} = m$$

$$y - y_1 = m(x - x_1)$$

Similarly, $y - y_2 = m(x - x_2)$

This form of the equation of a line is called the point-gradient form.

Example 5

(a) The line through the point (2, −1) has gradient 3. Find the equation of this line.

Solution Here $x_1 = 2$, $y_1 = -1$ and $m = 3$. Using the point-gradient form we have

$$y + 1 = 3(x - 2)$$

$$y = 3x - 7$$

The equation of the line is $y = 3x - 7$.

(b) Find the equation of the line which passes through the points A (-1, 3) and B (-3, 7).

Solution Begin by finding the gradient of the line.

$$\text{Gradient} = \frac{7-3}{-3+1} = -2$$

Using the point A (-1, 3) and the equation

$$y - y_1 = m(x - x_1),$$

we have

$$y - 3 = -2(x + 1)$$

$$y = -2x + 1$$

The equation of the line is $y = -2x + 1$

The following is an alternative solution. Using the points A (-1, 3), B (-3, 7) and the equation $y = mx + c$, we can form two simultaneous equations and solve for m and c.

$$3 = -m + c \qquad (1)$$

$$7 = -3m + c \qquad (2)$$

Subtracting equation (2) from (1), gives

$$-4 = 2m$$

So $\quad m = -2$

Substituting $m = -2$ into (1), we have

$$3 = 2 + c$$

So $\quad c = 1$

Therefore the equation of the line is $y = -2x + 1$.

Exercise 7.4(b)

1. Find the equation of the line through each of the given points with the indicated gradient.

(a) (2, 4), 3　　　　　　　　(b) (-1, -2), -3　　　　　　　(c) (2, -3), -4

(d) (4, -1), 2　　　　　　　　(e) (-2, 2), -7　　　　　　　　(f) (2, 3), 4

2. Find the equations of the lines passing through the given pair of points

(a) (3, 4), (6, 7)　　　　　　(b) (4, -1), (7, 5)　　　　　　(c) (0, 4), (5, -1)

(d) (-2, 3), (5, 3)　　　　　　(e) (4, -5), (6, 1)　　　　　　(f) (0, 3), (3, -3)

(g) (-2, 4), (-4, -2)　　　　　(h) (4, 5), (5, 2)　　　　　　(i) (-1, 6), (-2, 4)

General Equation of a Line

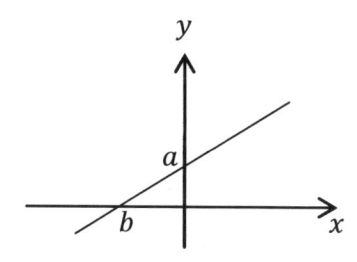

Figure 7.6

Figure 7.6 shows a line which cuts the x-axis at the point $(b, 0)$ and the y-axis at the point $(0, a)$. Let $P(x, y)$ be a point on the line.

The gradient of the line is $\dfrac{0-a}{b-0} = \dfrac{-a}{b}$

Hence the equation of the line is

$\dfrac{y-a}{x} = \dfrac{-a}{b}$

$by - ab = -ax$

$ax + by - ab = 0$

Replacing $-ab$ with c, we have

$ax + by + c = 0$

This equation is called the general equation of a line

From $ax + by - ab = 0$ we obtain

$\dfrac{x}{b} + \dfrac{y}{a} = 1$, where $(b, 0)$ and $(0, a)$ are the intercepts on the x and y axes respectively.

Example 6

Find the equation of the line which passes through (-1, 2) and (3, 5).

Solution Begin by finding the gradient m.

$$m = \frac{5-2}{3+1} = \frac{3}{4}$$

Using the point (-1, 2) and equation $y - y_1 = m(x - x_1)$ we have

$$y - 2 = \frac{3}{4}(x + 1)$$

$$4y - 8 = 3x + 3$$

$$3x - 4y + 11 = 0$$

The equation of the line is $3x - 4y + 11 = 0$.

The equation of a line can be written in any of the three forms. The gradient-intercept form provides the most information about a line. You can determined from this equation the line's gradient and its y- intercept.

Exercise 7.4(c)

1. Find the equation of the line which passes through each of the stated points and has the indicated gradient.

(a) (6, 3), $-\frac{3}{2}$ (b) (-3, -2), $\frac{5}{3}$ (c) (4, -2), $\frac{2}{5}$

(d) (-3, -5), $-\frac{3}{2}$ (e) (2, 3), $\frac{4}{3}$ (f) (-4, 5), $-\frac{3}{5}$

2. Find the equation of the line which passes through each of the following pairs of points.

(a) (-2, 2), (-4, 5)   (b) (-3, 6), (3, -2)   (c) (2, -7), (6, -10)

(d) (4, 3), (8, -2)   (e) (4, 3), (8, 6)   (f) (-1, -6), (2, -4)

3. Find the equations of the lines which make the following intercepts on the x- and y- axes respectively.

(a) 3, 4   (b) -3, 2   (c) -2, -3   (d) $\frac{1}{3}, \frac{1}{2}$ (e) $-\frac{1}{3}, \frac{1}{4}$

Horizontal and Vertical Lines

Any two points on a horizontal line parallel to the x-axis have the same y-coordinates, so its gradient is 0. Therefore the equation of a horizontal line through (o, b) is $y = b$. The graph of $y = b$ is shown in Figure 7.7(a).

A vertical line through $(a, 0)$ has an equation of the form $x = a$. Because the gradient is undefined, the equation of this line cannot be written in the form $y = mx + c$. The graph of $x = a$ is shown in Figure 7.7(b).

Figure 7.7(a)

Figure 7.7(b)

7.5 Parallel and Perpendicular Lines

Parallel Lines

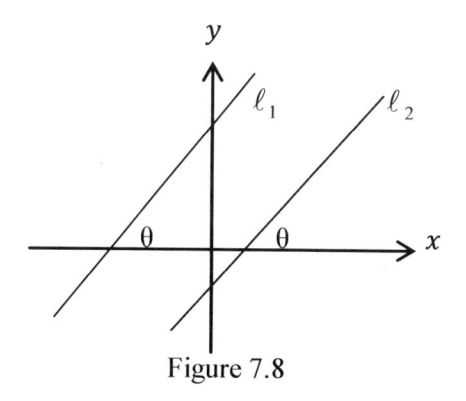

Figure 7.8

Figure 7.8 shows two parallel lines ℓ_1 and ℓ_2. Each line makes an angle θ with the x-axis and has the same gradient $\tan \theta$, In general, two parallel lines have the same gradient.

Example 7

(a) Determine whether the lines whose equations are $2x - 3y - 12 = 0$ and $6x - 9y + 45 = 0$ are parallel

Solution Rewrite each equation in the gradient-intercept form.

$$2x - 3y - 12 = 0$$

$$y = \frac{2}{3}x - 4$$

Also, $6x - 9y + 45 = 0$

$$y = \frac{2}{3}x + 5$$

Since the lines have the same gradient the lines are parallel.

(b) Find the equation of the line through (3, -2) parallel to the line $4x + 3y - 12 = 0$.

Solution First, rewrite $4x + 3y - 12 = 0$ in the gradient-intercept form

$$y = -\frac{4}{3}x + 4$$

Since the lines are parallel the line through (3, -2) has gradient $-\frac{4}{3}$.

Using the point-gradient formula, we have

$$y + 2 = -\frac{4}{3}(x - 3)$$

$$3y + 6 = -4x + 12$$

$$4x + 3y - 6 = 0$$

The equation of the line is $4x + 3y - 6 = 0$.

Exercise 7.5(a)

1. Determine whether the line through the first pair of points and the line through the second pair of points are parallel.

(a) (-3, 2), (1, 4) and (4, 3), (6, 4)

(b) (8, -5), (11, -3) and (1, 1), (-3, 7)

(c) (-2, -3), (4, 3) and (3, 5), (5, 7)

(d) (-2, -3), (3, -1) and (-3, 1), (7, 5)

2. Determine whether each pair of equations represents parallel lines.

(a) $x + 2y = 4$, $2x + 4y = 5$

(b) $3x - 4y + 1 = 0$, $3x + 4y - 10 = 0$

(c) $2x + 3y + 2 = 0$, $2x + 3y - 3 = 0$

(d) $2x + 3y + 8 = 0$, $4x + 6y - 21 = 0$

3. Find the equation of the line through each given point parallel to the indicated line.

(a) (4, 3), $4x - 3y + 12 = 0$

(b) (-1, 4), $5x - 2y - 1 = 0$

(c) (-3, 2), $2x + 3y - 5 = 0$ (d) (-2, -1), $3x - 2y + 7 = 0$

(e) (-2, 5), x-axis   (f) (5, 3), y-axis

4. A line passing through (-2, 1) and (3, y) is parallel to a line with gradient 2. Find the value of y.

5. A line passing through (5, 5) and (x, 8) is parallel to the line $3x + 2y + 8 = 0$, find the value of x.

6. Find the equation of the line which passes through the mid-point of the line joining the points A (2, 1) and B (5, 3) and parallel to the line $3x + 2y - 8 = 0$.

Perpendicular Lines

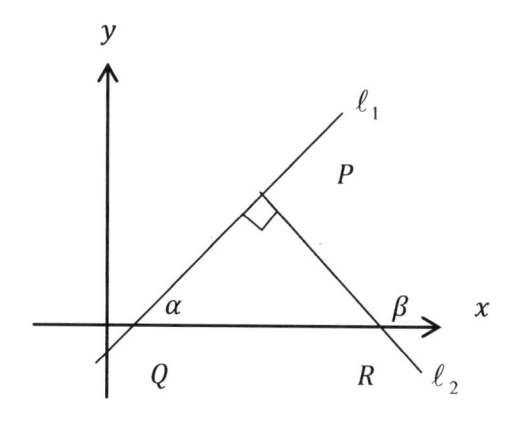

Figure 7.9

Figure 7.9 shows two perpendicular lines ℓ_1 and ℓ_2 which make angles α and β respectively with the x-axis. Let the gradients of the lines ℓ_1 and ℓ_2 be m_1 and m_2 respectively. From triangle QPR,

$$\tan \alpha = \frac{PR}{PQ}$$

$$\frac{PQ}{PR} \tan \alpha = 1$$

But $\dfrac{PQ}{PR} = \tan(180 - \beta)$ since $\angle QRP = 180 - \beta$

$\qquad\qquad = -\tan \beta$ since $\tan(180 - \beta) = -\tan \beta$

Hence, $\tan \alpha \tan \beta = -1$

Replacing $\tan \alpha$ and $\tan \beta$ with m_1 and m_2 respectively we have $m_1 \times m_2 = -1$. So $m_2 = -\frac{1}{m_1}$.

Notice that the gradients of two perpendicular lines are negative reciprocal of each other.

Example 8

(a) Determine whether the following lines $4x - 3y + 16 = 0$ and $3x + 4y - 29 = 0$ are perpendicular

Solution First, rewrite each equation in the gradient-intercept form

$$4x - 3y + 16 = 0$$

$$y = \frac{4}{3}x + \frac{16}{3}$$

Also $3x + 4y - 29 = 0$

$$y = -\frac{3}{4}x + \frac{29}{4}$$

Since the gradients of the lines are negative reciprocal of each other the lines are perpendicular.

(b) Find the equation of the line through (2, 3) perpendicular to the line $3x - 2y - 9 = 0$.

Solution First, rewrite $3x - 2y - 9 = 0$ in the gradient-intercept form

$$y = \frac{3}{2}x - \frac{9}{2}$$

Since the lines are perpendicular the line through (2, 3) has gradient $-\frac{2}{3}$.

Using the point-gradient formula, we have

$$y - 3 = -\frac{2}{3}(x - 2)$$

$$3y - 9 = -2x + 4$$

$$2x + 3y - 13 = 0$$

The equation of the line is $2x + 3y - 13 = 0$.

Exercise 7.5(b)

1. State the gradient of the line perpendicular to the line with gradient:

   (a) 3   (b) −2   (c) $\frac{1}{2}$ (d) $-\frac{1}{3}$ (e) $\frac{3}{4}$ (f) $-\frac{3}{2}$

2. Determine whether the lines through the first pair of points and the lines through the second pair of points are perpendicular.

   (a) (4, 3), (8, 4) and (7, 1), (6, 5)   (b) (7, -6), (10, -4), (2, 2), (-2, 8)

   (c) (8, -5), (11, -3) and (1, 3), (-3, 7)   (d) (-1, -1), (-5, 2) and (-2, 4), (1, 8)

3. Determine whether each pair of equations represents perpendicular lines.

   (a) $x - 3y = 6$, $3x + y = 3$ (b) $4x + 3y - 6 = 0$, $3x - 4y + 4 = 0$

   (c) $2x - 3y + 12 = 0$, $4x + 3y - 6 = 0$ (d) $3x - 4y + 1 = 0$, $3x + 4y - 10 = 0$

4. Find the equation of the line through each given point perpendicular to the indicated line.

   (a) (4, 3), $5x + 3y - 10 = 0$ (b) (-2, 5), $x + 3y + 2 = 0$

   (c) (-3, 2), $3x + 2y - 7 = 0$ (d) (3, -4), $2x - 3y + 8 = 0$

   (e) (1, 2), $2y + 3x - 1 = 0$ (f) (2, 3), $2x - 3y + 5 = 0$

5. Find the equation of the perpendicular bisector of the line joining the points A (2, 1) and B (6, 5).

6. A line passing through (3, 2) and (y, 5) is perpendicular to a line with gradient $\frac{4}{3}$. Find the value of y.

7. A line passing through $(x, -2)$ and (3, -1) is perpendicular to the line $3x + 4y - 8 = 0$. Find the value of x.

8. Find the equation of the perpendicular bisector of the line joining the points A (- 4, - 1) and B (8, 7).

9. Three points A (1, 1), B (4, 3) and C (- 1, 4) are the vertices of a triangle. Show that triangle ABC is a right-angled triangle. What is its area?

7.6 Point of Intersection

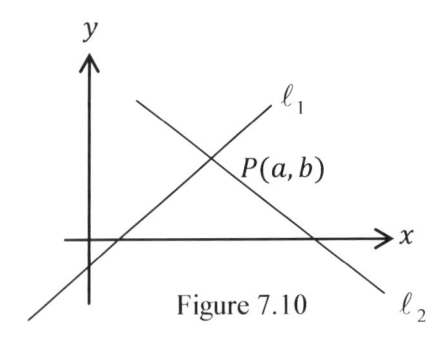

Figure 7.10

Two lines are said to intersect if they have exactly one point in common. The common point is called the point of intersection.

Figure 7.10 shows two lines ℓ_1 and ℓ_2 which intersect at the point $P(a, b)$. Since the point P lies on both lines, the coordinates of P satisfies the equations of the two lines. Hence to find the point of intersection of two lines we solve two simultaneous equations as illustrated in Example 9.

Example 9

Find the point of intersection of the lines $3x + 2y = 4$ and $2x - 3y = 7$.

Solution We solve two simultaneous equations as follows:

$$3x + 2y = 4 \quad (1)$$
$$2x - 3y = 7 \quad (2)$$

(1) × 3 $\quad 9x + 6y = 12 \quad (3)$

(2) × 2 $\quad 4x - 6y = 14 \quad (4)$

(3) + (4) $\quad 13x = 26$

$$x = 2$$

Substitute $x = 2$ into (1)

$$3(2) + 2y = 4$$
$$y = -1$$

The point of intersection is (2, -1)

An alternative method is to rewrite each equation in the gradient-intercept form and then set up an equation as shown below.

$$3x + 2y = 4$$

$$y = -\frac{3}{2}x + 2$$

Also $\quad 2x - 3y = 7$

$$y = \frac{2}{3}x - \frac{7}{3}$$

So, $\quad -\frac{3}{2}x + 2 = \frac{2}{3}x - \frac{7}{3}$

$$-9x + 12 = 4x - 14$$

$$-13x = -26$$

$$x = 2$$

Substituting $x = 2$ into $y = -\frac{3}{2}x + 2$ gives

$$y = -\frac{3}{2}(2) + 2 = -1$$

The point of intersection is (2, -1)

Exercise 7.6

1. Find the points of intersection of each of the following pairs of lines.

(a) $2x + y = 5$, $x - y = 1$ \hspace{2em} (b) $3x + y = 11$, $2x + y = 8$

(c) $2x - 5y + 9 = 0$, $3x + 2y - 15 = 0$ \hspace{1em} (d) $3x + 2y - 4 = 0$, $5x + 4y - 3 = 0$

(e) $5x + 2y - 2 = 0$, $2x + 3y + 8 = 0$ \hspace{1em} (f) $3x + 2y - 10 = 0$, $4x - y - 6 = 0$

(g) $3x - 2y + 5 = 0$, $3x + y + 2 = 0$ \hspace{1em} (h) $3x + 2y - 3 = 0$, $4x + 5y + 3 = 0$

2. The perpendicular bisector of the line joining the points A (-3, 7) and B (5, 3) meets the line $2y = x + 15$ at the point N. Find the coordinates of N.

3. The line from the point (1, 2) parallel to the line $3x + 4y - 8 = 0$ meets the line $3y = 4x + 27$ at P. Find the coordinates of P.

4. The line from the point (2, 1) perpendicular to the line $2x + 3y - 6 = 0$ meets the x-axis at P and the y-axis at Q. Find the coordinates of P and Q, and the area enclosed by the axes and this line.

5. The points P, Q and R have coordinates (-2, 1), (4, 5) and (-1, 6) respectively. The line from R perpendicular to PQ meets PQ at N. Find the coordinates of N.

6. A parallelogram has vertices P (2, 3), Q (5, 7), R (x, y) and S (-2, 6). Find the coordinates of R.

7. Two points A and B have coordinates (1, 5) and (3, 3) respectively. The line through the mid-point of the line AB parallel to the line $4x - 3y - 12 = 0$ meets the x-axis at the point P and the y-axis at the point Q. Find coordinates of P and Q, and the area enclosed by the axes and this line.

8. The points A and B have coordinates $\left(2.4\frac{1}{3}\right)$ and $\left(6, 1\frac{2}{3}\right)$ respectively. The line through the mid-point of the line AB perpendicular to $2x + 3y - 12 = 0$ meets the x-axis at P and the y-axis at Q. Find the coordinates of P and Q, and the area enclosed by the axes and this line.

7.7 Length of a perpendicular from a point to a line

We can show that the length of the perpendicular from the point $P(x_1, y_1)$ to the line $Ax + By + C = 0$ is given by

$$\frac{|Ax_1 + By_1 + C|}{\sqrt{A^2 + B^2}}$$

Example 10 Find the length of the perpendicular from the point $P(3, -2)$ to the line $12x + 5y - 13 = 0$.

Solution Here $A = 12$, $B = 5$, $C = -13$, $x_1 = 3$ and $y_1 = -2$. Using the formula we have

Length of line $= \dfrac{|12(3) + 5(-2) - 13|}{\sqrt{12^2 + 5^2}} = 1$

Exercise 7.7

1. In each of the following, find the length of the perpendicular from the given point to the given line.

(a) (0, 0) $3x + 4y + 10 = 0$

(b) (0, 0) $2x + 3y - 7 = 0$

(c) (8,7) $4x + 3y + 12 = 0$

(d) (3,5) $4x - 3y - 12 = 0$

(e) (- 2,- 3) $3x - 4y = 1$

(f) (1,3) $8x - 15y + 13 = 0$

(g) (-2,0), $2x + y = 6$

(h) (3,2) $4y = 3x + 9$

2. Given the triangle ABC with vertices A (3, 5), B (-2, -3) and C (2, 1),

(a) calculate the length of the altitude from A to BC.

(b) calculate the area of triangle ABC.

3. The points A (- 7, 1), B (4, - 2) and C are the vertices of a triangle, where angle BCA is 90°. The line $4x + 6y + 9 = 0$ is parallel to the line BC.

(a) Find the coordinates of C.   (b) Find the area of triangle ABC.

(c) Find length of the perpendicular from the point C to the line AB.

4. Find the distance between the parallel lines $4x - 3y - 8 = 0$ and $4x - 3y = 12$.

[Hint: Find a point P in one line, and then find the distance from the point P to the other line]

5. Calculate the length of the radius from the centre (3, 2) to the tangent $12x - 5y + 13 = 0$.

7.8 The Angle between Two Straight Lines

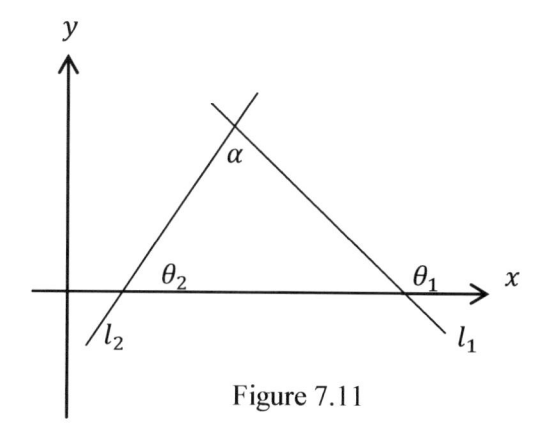

Figure 7.11

Consider two lines l_1 and l_2 with gradients $m_1 = tan\theta_1$ and $m_2 = tan\theta_2$ respectively (see Figure 7.11). The angle α between the lines is given by $\alpha = (\theta_1 - \theta_2)$. Hence

$tan\alpha = tan(\theta_1 - \theta_2)$

$= \dfrac{tan\theta_1 - tan\theta_2}{1 + tan\theta_1 tan\theta_2}$

$= \dfrac{m_1 - m_2}{1 + m_1 m_2}$

Therefore $\alpha = tan^{-1}\left(\dfrac{m_1 - m_2}{1 + m_1 m_2}\right)$

Example 11

Find the angle between straight lines l_1 and l_2 having equations $3x - 2y = 5$ and $4x + 5y = 1$ respectively.

Solution The gradient of $l_1 = m_1 = \frac{3}{2}$ and the gradient of $l_2 = m_2 = -\frac{4}{5}$. Thus if α is the angle between l_1 and l_2 then

$$tan\alpha = \frac{\frac{3}{2} - \left(-\frac{4}{5}\right)}{1 + \left(\frac{3}{2}\right)\left(-\frac{4}{5}\right)}$$

$$= -\frac{23}{2}$$

Note: If as in this example $tan\alpha$ is negative α is the obtuse angle between the lines. The acute angle between the lines is $tan^{-1}\left(\frac{23}{2}\right)$.

Exercise 7.8

1. Find the tangents of the acute angles between the following pairs of lines:

(a) $2x + 3y = 7, x - 6y + 5$

(b) $x + 4y - 1 = 0, 3x + 7y = 2$

(c) $3x + 4y = 2, 2x - 3y = 4$

(d) $2y = 4x + 3, 2y = 6x + 1$

2. Find the acute angles between the following pairs of lines:

(a) $x - y + 7 = 0, 2x - y - 4 = 0$

(b) $y = 3 - x, y = 3x - 2$

(c) $4x + 3y = 7, 3x + 4y = 1$

(d) $3x + 5y = 0, 3y = 2x + 1$

7.9 Sketching Graphs of Linear Functions

1. Using the intercepts

The graph of a linear function is straight line with gradient m. There are two possible graphs of a linear function as shown in Figure 7.12(a) and Figure 7.12(b).

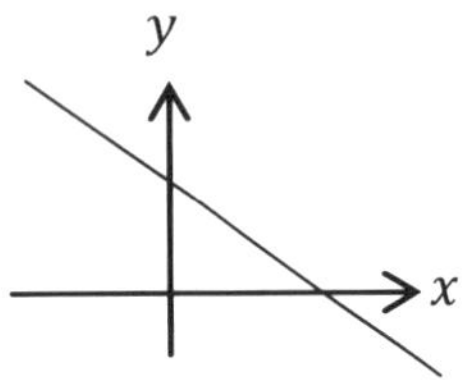

Figure 7.12(a) 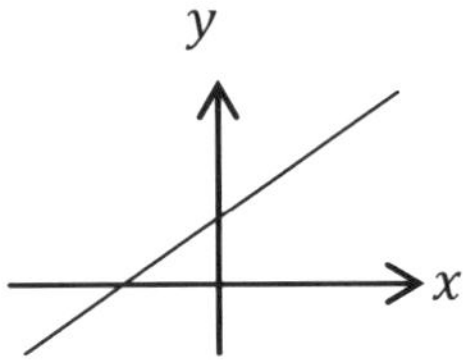 Figure 7.12(b)

In Figure 7.12(a) the function values get smaller from left to right, so we say that the function is decreasing. This line has a negative gradient. In Figure 7.12(b) the function values get larger from left to right, we say that the function is increasing. This line has a positive gradient.

Example 12

Sketch the graph of $3x + 2y = 6$.

Solution First find the x-intercept and the y-intercept.

To find the x-intercept let $y = 0$ and solve the equation for x.

$$3x + 2(0) = 6$$

$$3x = 6$$

$$x = 2$$

So the x-intercept is (2, 0).

To find the y-intercept let $x = 0$ and solve the equation for y.

$$3(0) + 2y = 6$$

$$2y = 6$$

$$y = 3$$

So the y-intercept is (0, 3).

Next obtain the graph by drawing a line through these points as shown in Figure 7.13.

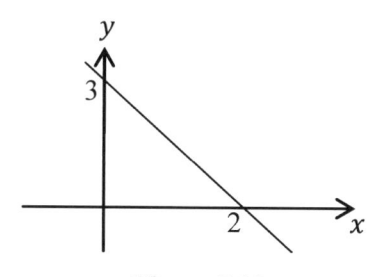

Figure 7.13

2. Using the y- intercept and the gradient

Example 13

Sketch the graph of $3x - 2y = -4$.

Solution First write the equation in the gradient-intercept form.

$$3x - 2y = -4$$
$$-2y = -3x - 4$$
$$y = \frac{3}{2}x + 2$$

You can see that the gradient of the line is $\frac{3}{2}$ and the y-intercept is (0, 2). Next plot the point (0, 2). Finally use the gradient to locate a second point. From (0, 2) move 2 units to the right and then 3 units up as illustrated in Figure 7.14.

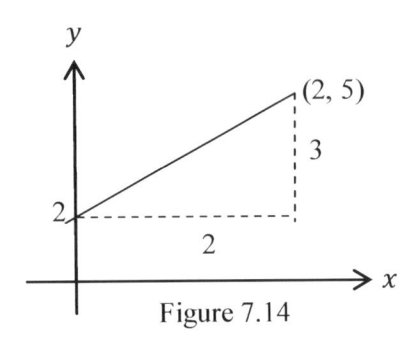

Figure 7.14

Exercise 7.9

Sketch the graph of each of the following linear functions:

1. $x + 3y = 6$
2. $2x - y = 6$
3. $2x - 3y = 6$
4. $x - 4y + 8 = 0$
5. $3x + 2y = 12$
6. $4x + 5y = 20$
7. $8x - 12y = 24$
8. $3x - 4y - 48 = 0$
9. $3x - 4y - 12 = 0$
10. $5x - 6y = 0$
11. $2x + 7y = 0$
12. $8x = 12y$

Review Exercise 7

1. Find the point on the y-axis equidistant from A (-3, 5) and B (6, 4).

2. Find the point on the x-axis equidistant from the points (-2, 4) and (6, 8).

3. Prove that (-2, 4), (10, 9), (15, -3) and (3, -8) are the vertices of a square.

4. Prove that the polygon with vertices (3, 2), (0, 5), (-3, 2) and (0, -1) is a rhombus.

5. For the triangle with vertices A(2, 3), B(4, 7), C(-6, -5), verify that the line joining the midpoints of two sides is parallel to the third side.

6. The vertices of a quadrilateral ABCD are A(1, 4), B(2, -7), C(-3, -2) and D(-4, 3).

(a) Find the midpoint of the sides.

(b) Prove that the midpoints of the sides are vertices of a parallelogram.

7. Given the quadrilateral A (-3, -3), B (9, 3), C (12, 8), D (0,2),

(a) prove that ABCD is a parallelogram.

(b) prove that the diagonals of ABCD bisect each other.

8. The midpoints of the sides of a triangle are (3, 1), (5, -2) and (7, 0). Find the coordinates of the vertices of the triangle.

9. Given the triangle with vertices A (5, 1), B (-1, -1), C (-2, 2),

(a) prove that the triangle is right angled.

(b) prove that the midpoint of AC is equidistant from the three vertices.

10. Find an equation of the line parallel to $y = 3x + 2$ having the same x-intercept as the line $4x + 3y = 8$.

11. Find an equation of the line perpendicular to $y = -3x + 4$ having the same y-intercept as the line $5x - 3y = 6$.

12. In triangle ABC the vertices are A(1, 2), B(3, 4) and C(7,6).

(a) Find the equations of the sides of triangle ABC.

(b) Find the equations of the altitudes of triangle ABC.

13. Find an equation of the line containing the point (-1, 3) and the point of interception of the lines $x - 4y = 5$ and $x - 2y = 4$.

14. The base of a triangle is the line segment joining (0,0) to (8,0). If the gradient of the other sides are $\frac{3}{2}$ and $\frac{3}{4}$, find the coordinates of the third vertex.

15. Find the equations of the lines which pass through the point of intersection of the lines $x - 3y = 4$ and $3x + y = 2$ and are respectively parallel and perpendicular to the line $3x + 4y = 0$.

16. The points P (-5, 4), Q (3, 5) and T are the vertices of a triangle, where angle PTQ is 90°. The line through R (0,3) on the line QT parallel to the line PQ meet the line PT at S.

(a) Find the coordinates of S and T.

(b) Find the area of the quadrilateral PQRS.

17. Find the length of the perpendicular from the indicated points to the lines:

(a) (2, 3) $3x - 4y - 4 = 0$
(b) (2, -4), $3y = 2x - 3$
(c) (7, -1), $4x = 3y + 1$
(d) (2, 5) $x + 2y + 8 = 0$

18. Find the acute angle between the following pairs of lines:

(a) $2x + y + 3 = 0, x + y = 0$
(b) $y = 5x + 2, y = 3x - 1$
(c) $3x + 2y - 1 = 0, 4x + 5y + 3 = 0$
(d) $3x + 4y = 12, 5x - 12y = 10$

19. Sketch the graphs of each of the following functions:

(a) $2x + 3y = 6$
(b) $5x - 4y = 20$
(c) $2y - 3x = 0$

(d) $3x + 2y = 0$
(e) $x - 3y = 6$
(f) $\frac{x}{3} - \frac{y}{2} = 4$

8 Linear Inequalities in Two Variables

If we replace the equal sign in the equation $ax + by = c$ with any one of the following inequality symbols, $<$, $>$, $\leq$ and $\geq$, we obtain a linear inequality in two variables. For example the expressions $3x + 2y \geq 4$, $4x + 5y \leq 3$, $2x - 3y < 0$ and $3x + 5y > -8$ are each inequalities in x and y. The first and the second inequalities are called non-strict inequalities and the third and fourth inequalities are called strict inequalities.

8.1 Graphs of Linear Inequalities

Consider the linear inequality

$$3x + 2y < 6$$

A point (x_1, y_1) is a solution of the linear inequality if the inequality is true when x_1 and y_1 are substituted for x and y respectively. For instance the point (2, -1) is a solution of the inequality because $3(2) + 2(-1) = 4 < 6$ is a true statement.

The graph of the linear inequality is the set of all points that satisfy the inequality. This set consists of all points in an entire region of the plane either above or below the graph of the corresponding linear equation.

To sketch $3x + 2y < 6$ we begin by sketching the graph of the corresponding linear equation $3x + 2y = 6$. The graph of the equation separates the plane into regions called half-planes. In each half-plane either all points are solutions of the inequality or no point is a solution of the inequality.

We determine which half-plane satisfy the inequality by simply testing one point called the test point in each half-plane. If a test point in a half-plane satisfies the inequality all points in that half-plane will do so, and therefore all points in that half-plane will satisfy the inequality. Then we shade that region. However, if a test point in a half-plane does not satisfy the inequality then we shade the other half-plane. For easy computation we use (0, 0) as a test point if the graph of the corresponding equation does not pass through the origin.

Example 1 Sketch the graph of each linear inequality

(a) $3x + 4y < 12$.

Solution First we draw the graph of the corresponding equation.

$$3x + 4y = 12$$

To decide on the appropriate half-plane take a test point not on the line. For easy computation we take (0, 0).

$$3 \cdot 0 + 4 \cdot 0 < 12$$

$$0 < 12 \qquad \text{True.}$$

The point (0, 0) is a solution and it appears in the region below the boundary line. Thus the graph consists of the half-plane lying below the line as shown in Figure 8.1.

Note that when the inequality is non-strict the points on the graph of the corresponding equation are solutions of the inequality. However when the inequality is strict the points on the graph of the corresponding equation are not solutions of the inequality. In the latter case a dotted line may be drawn to indicate that the points on the line are not solutions of the inequality.

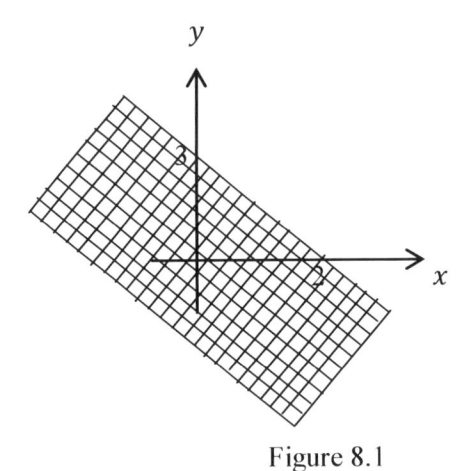

Figure 8.1

(b) $2y - 3x > 6$.

Solution First draw the graph of $2y - 3x = 6$.

Using (0, 0) as a test point we have

$$2 \cdot 0 - 3 \cdot 0 > 6$$

$$0 > 6 \qquad \text{False}$$

Because the origin (0, 0) does not satisfy the inequality the graph consists of the half-plane above the line as shown in Figure 8.2.

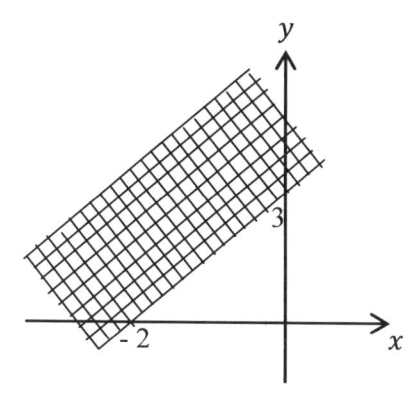

Figure 8.2

(c) $y - 2x \leq 0$.

Solution We proceed as before by drawing the graph of $y - 2x = 0$.

Because the graph passes through the origin we cannot use (0, 0) as a test point. We will choose any convenient point not on the line such as (1, 0). Then substituting 1 and 0 for x and y respectively we have

$$0 - 2 \cdot 1 \leq 0$$

$$-2 \leq 0$$

Because the point (1, 0) satisfies the inequality the graph consists of the half-plane lying below the line as shown in Figure 8.3.

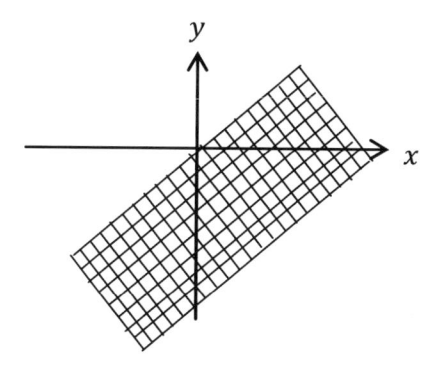

Figure 8.3

Exercise 8.1(a)

Graph the solution set of each of the following inequalities:

1. $3x + y > 6$
2. $x - 2y < 3$
3. $2x + 7 \leq 7$

4. $x + y < 6$
5. $3x - 4y \geq 12$
6. $2x + 3y < 6$
7. $2x + 5y \leq 10$
8. $x - 2y > 0$
9. $x + 4y \leq 0$
10. $3x - 6 \leq 0$
11. $2x - 3 > 0$
12. $2y + 6 < 0$

Using the Gradient-Form of Linear Inequalities

For a linear inequality in two variables we can simplify the graphing procedure by writing the inequality in the gradient-intercept form as shown in Example 2.

Example 2 Sketch the graph of each inequality.

(a) $2x - 3y \leq 12$.

Solution $2x - 3y \leq 12$

$$-3y \leq -2x + 12$$

$$y \geq \frac{2}{3}x - 4$$

In this form you can see that the solution points lie on or above the line $y = \frac{2}{3}x - 4$ as shown in Figure 8.4

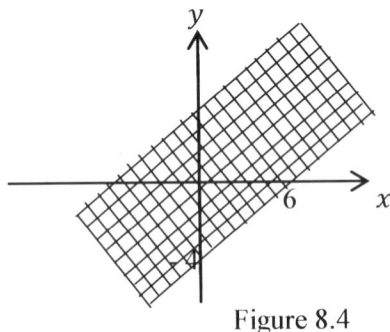

Figure 8.4

Note that for any inequality of the form $y > ax + b$ or $y \geq ax + b$ we shade the region above the boundary line.

(b) $5x + 3y < 15$.

Solution First sketch the graph of the corresponding equation. Rewrite the inequality in the gradient-intercept form.

$$5x + 3y < 15$$

$$3y < -5x + 15$$

$$y < -\frac{5}{3}x + 5$$

In this form you can see that the solution points lie below the line $y = -\frac{5}{3}x + 5$ as shown in Figure 8.5.

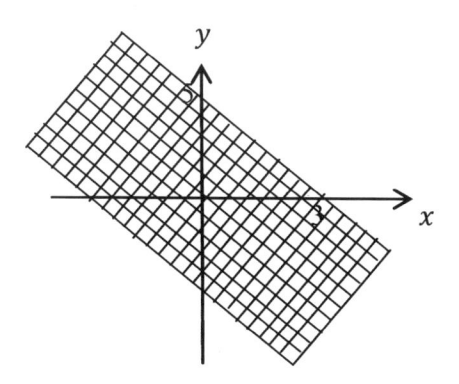

Figure 8.5

Note that for any inequality of the form $y < ax + b$ or $y \leq ax + b$ we shade the region below the boundary line.

Exercise 8.1(b)

Graph the solution set of each of the following linear inequalities:

1. $x \geq 6$
2. $x < -3$
3. $y < 5$
4. $y > -2$
5. $3x - 2y < 0$
6. $2x + 3y > 0$
7. $x - 3y \geq 0$
8. $x - 2y \geq 6$
9. $3x + y \leq 9$
10. $3x + 5y \leq 15$
11. $2x - 3y \geq 6$
12. $4x + 3y < 12$

8.2 Solution of Systems of Linear Inequalities

The solution set of a system of linear inequalities is all ordered pairs that satisfy each inequality. In Example 3, we show how to graph a solution set of a system of linear inequalities.

142   Further Mathematics

Example 3   Solve each system of the linear inequalities.

(a) $2x - y > 4$

   $2x + 3y > 6$

Solution   Begin by sketching the graph of the corresponding equation of each inequality. Then shade the half-plane that is the graph of each linear inequality. The graph of the system is the region that is common to every graph in the system. The graph of this system is the shaded region shown in Figure 8.6.

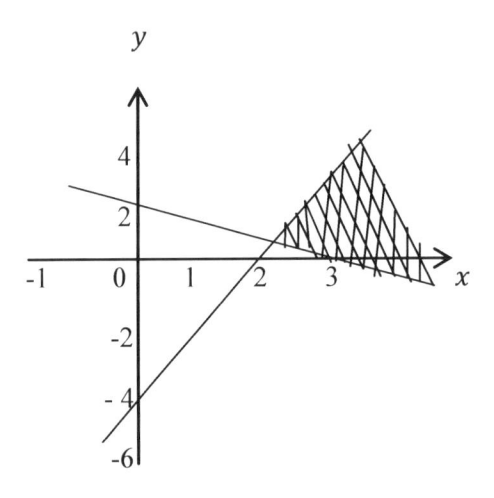

Figure 8.6

(b) $x + y > 5$

   $2x - y < 4$

   $y < 6$

   $x \geq 0$

Find the maximum and minimum value of $P = 3x + 2y$.

Solution   On the same axes draw the line of the corresponding equation of each inequality. Then choose the appropriate half-plane in each case and shade the region that represents the solution set as shown in Figure 8.7.

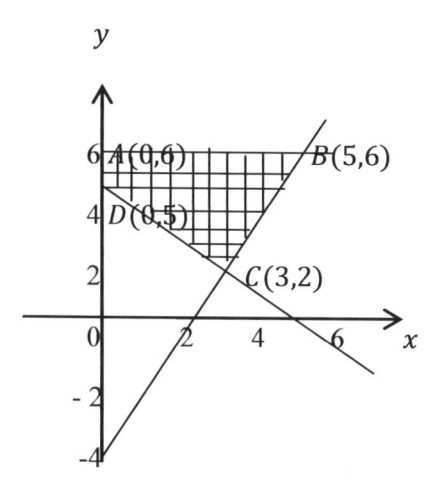

Figure 8.7

This region bounded on all sides by line segments contains the possible solution points, and is called the feasible region. The points where two or more boundaries intersect are called the vertices of the feasible region. You can see from the graph in Figure 8.7 that the coordinates of the vertices are A (0, 6), B (5, 6), C (3, 2) and D (0, 5).

The points on the boundary of the feasible region and the points inside the region are the solution set for the system of inequalities. However, the minimum or maximum value occurs at a vertex. To find the minimum value and the maximum value evaluate $P = 3x + 2y$ at each vertex point, as shown in Table 1, and choose the vertex point that produces the smallest and the largest numbers. The smallest number is the minimum value of P and the largest number is the maximum value.

Vertex	$P = 3x + 2y$
A (0, 6)	3(0) + 2(6) = 12
B (5, 6)	3(5) + 2(6) = 27
C (3, 2)	3(3) + 2(2) = 13
D (0, 5)	3(0) + 2(5) = 10

Table 1

Examining the values in Table 1 you can see that the maximum value of P is 27 and this occur when $x = 5$ and $y = 6$. The minimum value of P is 10 and this occurs when $x = 0$ and $y = 5$.

Exercise 8.2

1. Solve each of the following system of linear inequalities graphically:

(a) $2x - y < 0$

   $x + 2y < 0$

(b) $x + 2y \leq 4$

   $x - y \geq 1$

(c) $3x - y > 6$

   $x + y < 6$

(d) $3x + 4y + 8 < 0$

$2x + 3y + 6 > 0$

(f) $3x + 2y \leq 12$

$x \geq 2$

(h) $4x + 3y \leq 12$

$x + 4y \leq 8$

$x \geq 0$

(e) $x + y \leq 4$

$x + 3y \leq 6$

(g) $2x + y < 8$

$x > 1$

$y > 2$

(i) $x + 2y - 1 < 0$

$3y - 2x + 2 < 0$

$2 + y > 0$

$y \geq 0$

2. Solve the system of inequalities graphically:

$2x + 3y \geq 6$

$5y + 3x \leq 15$

$x > 0$

$y > 0$

Using your diagram find the values of x and y for which $2x + 3y$ is maximum.

3. Solve the system of inequalities graphically

$2x - y + 4 \geq 0$

$x + y \leq 10$

$y - x \geq 0$

$y \geq 2$

$x \geq 0$

Using your diagram find the maximum value of $10x + 5y$.

4. Solve the system of inequalities graphically:

$x + 2y - 8 > 0$

$$3x - y - 3 < 0$$
$$y - 6 < 0$$
$$x > 0$$

Use your diagram to find:

(a) the maximum value of x.

(b) the minimum value of $2x + 3y$.

8.3 Linear Programming

Linear programming uses method of graphing inequalities to find the optimum solution to problems in business, industries, military and other fields. For instance a manager may like to find how much to set his prices to make maximum profit or how to minimize cost due to some restrictions or constraints such as available raw material, labour and machine hour.

The constraints are represented as a system of linear inequalities. The quantity that can be minimized or maximized is called the objective function.

Solving a Linear Programming Problem

The guide lines for solving a linear programming problem are listed below:

1. Determine all necessary constraints.

2. Determine the objective function.

3. Graph the constraints and determine the feasible region.

4. Find the vertices of the feasible region.

5. Find the value of the objective function at each vertex.

Example 4

A manufacturer produces a standard and a deluxe model of an article. The standard model requires 2 hours of labour to produce while the deluxe model requires 3 hours. The labour available is limited to 60 hours per week. The total number of standard and deluxe models produced per week is 25. If the profit on a standard model is GH¢ 250.00 and the profit on a deluxe model is GH¢ 300.00 how many of each model should be produced to maximize the profit?

Solution Let x represent the number of standard model produced and y the number of deluxe model. Since the total labour can be less than or equal to 60 hours we have

$$2x + 3y \leq 60$$

Also the total production can be less than or equal to 25 so

$$x + y \leq 25$$

Since the number of standard model and deluxe model produced cannot be negative $x \geq 0$ and $y \geq 0$. If the total profit is P then $P = 250x + 300y$ since the profit on each standard model is GH¢ 250.00 and the profit on each deluxe model is GH¢ 300.00. Thus we are require to maximize

$$P = 250x + 300y$$

subject to the constraints

$$2x + 3y \leq 60$$

$$x + y \leq 25$$

$$x \geq 0$$

$$y \geq 0$$

The last two inequalities are included because x and y cannot be negative.

Now draw the graph of the system of inequalities as before (see Figure 8,8).

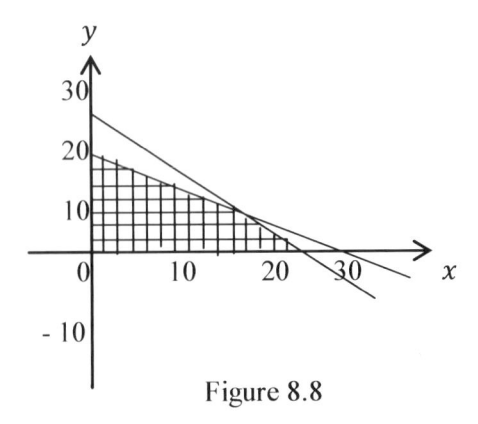

Figure 8.8

All the points in the shaded region meet the given conditions of the problem and represents possible production options. However the maximum profit will be attained at a vertex. Thus we evaluate the objective equation $P = 250x + 300y$ at each vertex as illustrated in Table 2.

Vertex	$P = 250x + 300y$
(0, 0)	$250(0) + 300(0) = 0$
(25, 0)	$250(25) + 300(0) = 6250$
(15, 10)	$250(15) + 300(10) = 6750$
(0, 20)	$250(0) + 300(20) = 6000$

Table 2

Thus the maximum profit is obtained if 15 of the standard model and 10 of the deluxe model are produced.

Exercise 8.3

1. A manufacturer produces both short sleeve and long sleeve shirts. The short sleeve shirt takes 6 hours of labour to produce and the long sleeve shirt 10 hours. The labour available is limited to 300 hours per week and the total production capacity is 40 shirts per week.

(a) Draw a graph of the feasible region given these conditions.

(b) If the manufacturer makes a profit of GH¢ 5.00 on a short sleeve shirt and GH¢ 6.50 on a long sleeve shirt how many of each should he produce to make the maximum profit.

2. A firm produces both AM and FM radios. The AM radios take 3 hours to produce and the FM radios take 4 hours. The number of production hours is limited to 60 hours per week. The plant's capacity is limited to a total of 18 radios per week and existing orders require that at least 4 AM radios and at least 3 FM radios be produced per week.

(a) Draw a graph of the feasible region given these conditions.

(b) If the firms makes a profit of GH¢ 2.50 on AM radio and GH¢ 3.00 on a FM radio how many AM and FM radios should be produced in order to make the maximum profit?

3. A factory manufactures two products. in the manufacturing process; each of the first product requires 1 hour of grinding and 1 hour of finishing and each of the second product needs 1 hour of grinding and 2 hours of finishing. The factory has two grinders and three finishers each of which work at most 20 hours per week. Each of the first product brings a profit of GH¢ 3.00 and each of the second product brings a profit of GH¢ 4.00. Assuming that every product made will be sold how many of each should be made to maximize profits?

4. A diet is to contain at least 400 units of vitamins, 500 units of minerals and 1,400 calories. Two foods are available: X which costs 50 Gp per unit, and Y which costs 30 Gp per unit. A unit of food X contains 2 units of vitamins, 1 unit of minerals, and 4 calories; a unit of food Y contains 1 unit of vitamins, 2 units of minerals, and 4 calories. Find minimum cost for a diet that consists of a mixture of these foods and also meets the minimal nutrition requirements.

5. A farmer can use two types of plant food, mix A and mix B. The amount (in kilograms) of nitrogen, phosphoric acid and potash in a cubic metre of each mix are given in the following table.

	Mix A	Mix B
Nitrogen	4	2
Potash	3	9
Phosphoric acid	6	4

Test performed on the soil in a large field indicate that the field needs at least 315 kg of potash and at least 140 kg of nitrogen. The tests also indicate that no more than 420 kg of phosphoric acid should be added to the field. A cubic metre of mix A cost GH¢ 70 and a cubic metre of mix B cost GH¢ 90. How many cubic metre of each mix should the farmer add to the field in order to supply the necessary nutrients at minimal cost?

6. A trucking firm wants to purchase a maximum of 15 new trucks that will provide at least 36 tons of additional shipping capacity. A model A truck holds 2 tons and costs GH¢ 18,000. A model B truck holds 3 tons and cost GH¢ 30,000. How many trucks of each model should the company purchase to provide the additional shipping capacity at minimal cost? What is the minimal cost?

Review Exercise 8

1. Draw the graph of the following inequalities:

$y \leq 8$

$2x + y \geq 10$

$x \geq 0$

$y \geq 0$

Shade the feasible region and list the vertices.

2. Using a graph sheet draw and shade the feasible region bounded by the inequalities:

$2y + x \leq 10$

$x - 4y \leq -2$

$4y + 3x \geq 12$

$x \geq 0$

$y \geq 0$

Find the minimum value of the function $P = 2xy - x^2$

3. Using a graph sheet draw and shade the feasible region which satisfies the following inequalities simultaneously:

$y - 2x \leq 4$

$x - y \leq 1$

$y + 4x \geq 4$

$y + 2x \leq 6$

$y \geq 0$

$x \geq 0$

Calculate the minimum and maximum values of $x^2 - 2y$.

4. Indicate on a diagram by shading the feasible region which satisfies the following inequalities:

$x + 2y \leq 4$

$x - y \leq 4$

$x \geq 1$

$y \geq -1$

Find subject to the four conditions the maximum and minimum values of $2y + x$.

5. A man has up to GH¢ 3000 to invest. He can invest in bonds or treasury bills: the bond yields 8 % per year, the treasury bills yields 12 % per year. After some consideration he decides to invest at most GH¢ 1,200 in bonds and at least GH¢ 600 in treasury bills. He also wants the amount invested in bonds to exceed or equal the amount invested in treasury bills. How much must he invest in bonds and treasury bills to maximize the return on his investment?

6. A factory manufactures two kinds of toy trucks: a standard model and a deluxe model, each requiring the use of three machines, A, B and C. Machine A can be used at most 70 hours, Machine B at most 40 hour and Machine C at most 90 hours. The standard model requires 2 hours on Machine A, 1 hour each on Machine B and 1 hour on Machine C; the deluxe model requires 1 hour each on Machine A and B, and 3 hours on machine C. If the profit made on each standard model is 40 Gp and each deluxe model is 60 Gp how many of each model should be made to maximize profits?

7. The minimum daily requirements from the liquid portion of a diet are 300 calories, 36 units of vitamin A and 90 units of vitamin C. A cup of dietary drink X cost 12 Gp and provides 60 calories, 12 units of vitamin A and 10 units vitamin C. A cup of dietary drink Y costs 15 Gp and provides 60 calories, 6 units of vitamin A and 30 units of vitamin C. How many cups of each drink should be consumed each day to minimize the cost and still meet the daily requirements?

8. A merchant plans to sell two models of computers at costs of GH¢1,500 and GH¢ 2,000. The computer model that cost GH¢ 1,500 yields a profit of GH¢ 250 per unit, and the model that cost GH¢ 2,000 yields a profit of GH¢ 400 per unit. The merchant estimates that the total monthly demand will not exceed 250 units. The merchant does not want to invest more than GH¢ 400,000 in the two products. Find the number of units of each model that should be stocked in order to maximize profit? What is the maximum profit?

9 Quadratic Functions

Quadratic Functions

The quadratic function is defined by the equation of the form $y = ax^2 + bx + c$ where a, b and c are real numbers and $a \neq 0$.

Zeros of Quadratic Functions

The zeros of a quadratic function f are the values of x for which $f(x) = 0$. They represent the points where the graph of the function crosses the x-axis. To find the zeros of a quadratic function we solve the equation $f(x) = 0$.

Example 1

Find the zeros of $f(x) = x^2 - x - 6$

Solution We solve the quadratic equation

$$x^2 - x - 6 = 0$$

Factorising the expression on the left gives

$$(x - 3)(x + 2) = 0$$

So $\qquad x - 3 = 0 \quad$ or $\quad x + 2 = 0$

$\qquad\qquad x = 3 \qquad\qquad x = -2$

Therefore the zeros of the function are -2 and 3.

9.1 Quadratic Equations

Any equation that can be written in the form $ax^2 + bx + c = 0$ where a, b and c are real numbers and $a \neq 0$ is called a quadratic equation.

Solution of a Quadratic Equation

A solution of a quadratic equation in one variable is any number that makes the equation true. The solution of a quadratic equation is also called the roots of the equation.

The following three methods can be used to solve quadratic equations.

1. Factoring

2. Completing the square and

3. Use of Quadratic Formula

Recall that in Chapter 1 we learned how to solve quadratic equations by factorisation. If a quadratic equation cannot be solved by factorisation, we use the method of completing the square or the Quadratic Formula. Remember that not all quadratic expression cannot be factorised.

Solving Quadratic Equations by Completing the Square

The method of completing the square is based on the relationship between the middle term and the constant term of the perfect square trinomial:

$$x^2 + 2ax + a^2 = (x + a)^2$$

$$x^2 - 2ax + a^2 = (x - a)^2$$

Notice that in each case the constant term is the square of one-half of the coefficient of x. This relationship is only true if the coefficient of x^2 is 1.

To solve a quadratic equation using the method of completing the square, add to each side of the equation the square of one-half of the coefficient of x. You must divide each side by the coefficient of x^2 before completing the square.

Example 2 Solve each quadratic equation.

(a) $x^2 + 6x - 40 = 0$

Solution First rewrite the equation as

$\qquad x^2 + 6x = 40$ \hfill Isolate 40

$\qquad x^2 + 6x + 9 = 40 + 9$ \hfill Add the square of 3 to each side.

$\qquad (x + 3)^2 = 49$ \hfill Factorise the quadratic expression.

$\qquad x + 3 = \pm 7$ \hfill Take square root of each side.

$\qquad x + 3 = -7 \text{ or } x + 3 = 7$ \hfill Solve for x.

$\qquad x = -10 \qquad x = 4$

The solutions are $x = -10$ and $x = 4$.

(b) $x^2 - 8x - 12 = 0$

Solution $x^2 - 8x - 12 = 0$

$\qquad x^2 - 8x = 12$ \hfill Isolate 12.

$$x^2 - 8x + 16 = 28 \qquad \text{Add the square of 4 to each side.}$$

$$(x - 4)^2 = 28 \qquad \text{Factorise the quadratic expression.}$$

$$x - 4 = \pm\sqrt{28} \qquad \text{Take the square root of each sides.}$$

$$x = 4 \pm \sqrt{28} \qquad \text{Solve for } x.$$

$$x = 4 \pm 5.29$$

$$x = 4 - 5.29 = -1.29 \quad \text{or} \quad x = 4 + 5.29 = 9.29$$

The solutions are $x = -1.29$ and $x = 9.29$

(c) $2x^2 - 3x - 9 = 0$

Solution $2x^2 - 3x - 9 = 0$

$$2x^2 - 3x = 9 \qquad \text{Isolate 9}$$

$$x^2 - \frac{3}{2}x = \frac{9}{2} \qquad \text{Divide each side by 2}$$

$$x^2 - \frac{3}{2}x + \frac{9}{16} = \frac{9}{2} + \frac{9}{16} \qquad \text{Add the square of } \frac{3}{4} \text{ to each side.}$$

$$\left(x - \frac{3}{4}\right)^2 = \frac{81}{16} \qquad \text{Factorise the quadratic expression.}$$

$$x - \frac{3}{4} = \pm\frac{9}{4} \qquad \text{Take square root of each side.}$$

$$x = \frac{3}{4} \pm \frac{9}{4} \qquad \text{Solve for } x.$$

So $x = \frac{3}{4} + \frac{9}{4} = 3$ or $x = \frac{3}{4} - \frac{9}{4} = -\frac{3}{2} = -1\frac{1}{2}$

The solutions are $x = -1\frac{1}{2}$ and $x = 3$

Exercise 9.1(a)

1. Find the constant that must be added to each binomial expression to make it a perfect square

(a) $x^2 - 12x$

(b) $y^2 - 14y$

(c) $x^2 + 8x$

(d) $x^2 - 16x$

(e) $y^2 + 3y$

(f) $x^2 + 5x$

(g) $x^2 - x$ (h) $x^2 - \frac{1}{2}x$ (i) $y^2 + \frac{1}{3}y$

2. Solve each equation by completing the square

(a) $x^2 + 12x - 2 = 0$ (b) $x^2 - 2x - 8 = 0$

(c) $x^2 + 10x + 13 = 0$ (d) $x^2 + 3x - 17 = 0$

(e) $x^2 + 3x - 27 = 0$ (f) $x^2 - 7x + 3 = 0$

(g) $2x^2 + x = 2$ (h) $3x^2 - x = 6$

(i) $3x^2 - 3x - 1 = 0$ (j) $4x^2 + 8x - 1 = 0$

(k) $3x^2 - 2x - 12 = 0$ (l) $7x^2 - 2x - 3 = 0$

Use of the Quadratic Formula

The Quadratic Formula is derived from the general quadratic equation $ax^2 + bx + c = 0$ by completing the square as shown below:

$ax^2 + bx + c = 0$

$ax^2 + bx = -c$ Isolate c.

$x^2 + \frac{b}{a}x = -\frac{c}{a}$ Divide each side by a.

$x^2 + \frac{b}{a}x + \frac{b^2}{4a^2} = \frac{b^2}{4a^2} - \frac{c}{a}$ Add the square of $\frac{b}{2a}$ to each side.

$\left(x + \frac{b}{2a}\right)^2 = \frac{b^2 - 4ac}{4a^2}$ Factorise the expression on the left

$x + \frac{b}{2a} = \pm\sqrt{\frac{b^2 - 4ac}{4a^2}}$ Take square root of each side.

$x = -\frac{b}{2a} \pm \sqrt{\frac{b^2 - 4ac}{4a^2}}$ Solve for x.

$x = \frac{-b \pm \sqrt{b^2 - 4ac}}{2a}$ Simplify.

The solution to the equation, $x = \frac{-b \pm \sqrt{b^2 - 4ac}}{2a}$, is called the quadratic formula.

To solve an equation using the quadratic formula substitute the values of a, b and c and then simplify the resulting expression as illustrated in Example 3.

Example 3 Solve the equation

(a) $x^2 + 4x - 45 = 0$.

Solution $x = -\frac{-b \pm \sqrt{b^2 - 4ac}}{2a}$

Here $a = 1, b = 4$ and $c = -45$. Substitute these values into the quadratic formula.

$$x = \frac{-4 \pm \sqrt{4^2 - 4(1)(-45)}}{2(1)}$$

$$x = \frac{-4 \pm \sqrt{196}}{2}$$

$$x = \frac{-4 \pm 14}{2}$$

This gives

$$x = \frac{-4 + 14}{2} = 5 \quad \text{and} \quad x = \frac{-4 - 14}{2} = -9$$

The solutions of the equation are $x = -9$ and $x = 5$.

(b) $3x^2 = 7x - 2$

Solution To determine a, b and c you must rewrite the equation in the general form.

$$3x^2 - 7x + 2 = 0$$

Here we have $a = 3$, $b = -7$ and $c = 2$.

$$x = \frac{-(-7) \pm \sqrt{(-7)^2 - 4(3)(2)}}{2(3)}$$

$$x = \frac{7 \pm \sqrt{25}}{6}$$

$$x = \frac{7 \pm 5}{6}$$

So $x = \frac{7+5}{6} = 2 \text{ and } x = \frac{7-5}{6} = \frac{1}{3}$

The solutions of the equation are $x = 2$ and $x = \frac{1}{3}$.

Exercise 9.1(b)

Solve each of the following quadratic equations:

1. $x^2 + 8x - 65 = 0$
2. $x^2 - 5x - 14 = 0$

3. $x^2 - 2x - 5 = 0$

4. $x^2 - 7x + 3 = 0$

5. $x^2 + 7x - 30 = 0$

6. $5x^2 + 4x - 1 = 0$

7. $16x^2 - 24x + 9 = 0$

8. $2x^2 - 6x + 1 = 0$

9. $3x^2 + 2x - 1 = 0$

10. $2x^2 - 3x - 1 = 0$

11. $2x^2 - \frac{1}{2}x - 5 = 0$

12. $3x^2 + \frac{1}{3}x - 3 = 0$

13. $3x^2 + 2x - \frac{3}{4} = 0$

14. $(x - 2)(x + 3) = 4$

15. $(x + 1)(2x - 4) = 7$

16. $3x - 5 = \frac{1}{x}$

17. $4x - \frac{1}{x} = 6$

18. $\frac{5}{x^2} + \frac{2}{x} = 1$

19. $2x - \frac{3}{x} - 3 = 0$

20. $\frac{6}{x^2} - \frac{2}{x} = 1$

21. $3 + \frac{8}{x} - \frac{5}{x^2} = 0$

22. $2x(x - 3) = x + 5$

9.2 Discriminant

The expression $b^2 - 4ac$ in the quadratic formula is called the discriminant and it gives us useful information about the nature of the roots. There are three possibilities:

1. If $b^2 - 4ac > 0$, the equation has two unequal real roots.

2. If $b^2 - 4ac = 0$, the equation has two equal real roots.

3. If $b^2 - 4ac < 0$, the equation has imaginary roots.

The link between the curve of $y = ax^2 + bx + c$ and the roots of the equation is illustrated in the following diagrams.

(a) $b^2 - 4ac > 0$

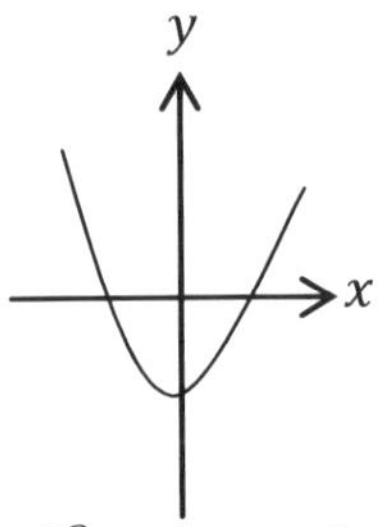

Figure 9.1(a)

(b) $b^2 - 4ac = 0$

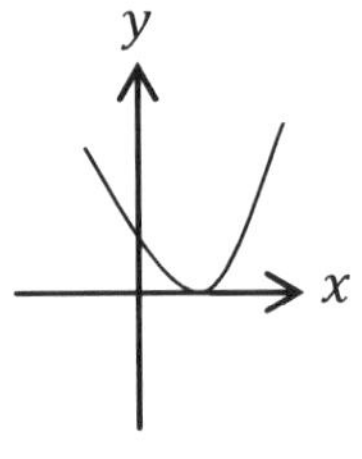

Figure 9.1(b)

(c) $b^2 - 4ac < 0$

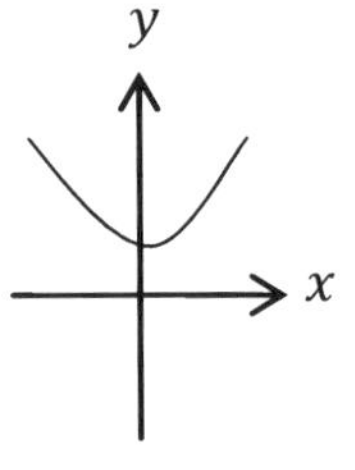

Figure 9.1(c)

The graph of a quadratic function that has two unequal roots intersects the x-axis at two points as shown in Figure 9.1(a).

The graph of a quadratic function that has two equal roots intersects the x-axis at one point, as shown in Figure 9.1(b). The x-axis is a tangent to the curve of the function.

The graph of a quadratic function with imaginary roots does not intersect the x-axis, as shown in Figure 9.1(c).

Example 4 State the nature of the roots of the following quadratic equations.

(a) $x^2 + 2x - 3 = 0$

Solution $x^2 + 2x - 3 = 0$ has two unequal real roots since $2^2 - 4(2)(-3) = 28$

(b) $9x^2 + 6x + 1 = 0$

Solution $9x^2 + 6x + 1 = 0$ has two equal roots since $6^2 - 4(9)(1) = 0$

(c) $2x^2 - 3x + 5 = 0$

Solution $2x^2 - 3x + 5 = 0$ has imaginary roots since $(-3)^2 - 4(2)(5) = -31$

Example 5

(a) Find the range of values of k for which $x^2 + 2x - k = 0$ has real roots.

Solution Since the equation has real roots $b^2 - 4ac \geq 0$.

$$2^2 - 4(1)(-k) \geq 0$$

$$4 + 4k \geq 0$$

$$k \geq -1$$

(b) Find the range of values of k for which the expression $2x^2 + 4x + k$ is always positive for all real values of x.

Solution The expression will always be positive when $b^2 - 4ac < 0$.

$$4^2 - 4(2)(k) < 0$$

$$16 - 8k < 0$$

$$k > 2$$

Exercise 9.2

1. Find the nature of the roots of each of the following equations:

(a) $x^2 - 4x + 3 = 0$

(b) $x^2 - 3x + 5 = 0$

(c) $x^2 - 4x + 4 = 0$

(d) $2x^2 - 3x + 4 = 0$

(e) $3x^2 + 7x - 6 = 0$

(f) $4x^2 + 4x + 1 = 0$

2. For each quadratic equation find the value of the discriminant and give the number of real solutions:

(a) $3x^2 + 8x = 0$

(b) $2x^2 - 5x = 0$

(c) $x^2 - 8x + 16 = 0$

(d) $4x^2 + 12x + 9 = 0$

(e) $3x^2 - 7x + 1 = 0$

(f) $2x^2 - x + 5 = 0$

(g) $3x^2 - x - 2 = 0$

(h) $2x^2 - 5x + 11 = 0$

3. Find the values of m for which the equation $x^2 + (m + 3)x + 4m = 0$ has equal roots

4. Find the value of m if the equation $(2m + 1)x^2 + 3mx + m = 0$ has equal roots.

5. If the equation $px^2 + (p + 1)x + p = 0$ has equal roots, find the value(s) of p.

6. Find the range of values of k for which $kx^2 + 4x + 2 = 0$ has real roots.

7. Find the range of values of k for which $x^2 + x - k = 0$ has two distinct roots.

8. Find the smallest possible integral value of k such that $2x^2 + 3x + k = 0$ will have imaginary roots.

9. Find the range of values of k for which the equation $x^2 + (3 - k)x + 1 = 0$ has real roots.

10. Find the range of values of k for which the expression $2x^2 + 4x + k$ is always positive for all real values of x.

9.3 Sum and Product of Roots of Quadratic Equations

Suppose the general quadratic equation $ax^2 + bx + c = 0$ has two roots α and β. Then $x - \alpha$ and $x - \beta$ are factors of $ax^2 + bx + c$. Hence

$(x - \alpha)(x - \beta) = 0$ which gives

$x^2 - (\alpha + \beta)x + \alpha\beta = 0$

Dividing the general equation by a we get

$$x^2 + \frac{b}{a}x + \frac{c}{a} = 0$$

Comparing like terms we have

$$\alpha + \beta = -\frac{b}{a}$$

and $\alpha\beta = \frac{c}{a}$

In general the sum of the roots of a quadratic equation is $-b/a$ and the product of the roots is c/a.

Example 6

Write down the sum and product of the roots of the quadratic equation $3x^2 + 2x - 7 = 0$.

Solution Here $a = 3, b = 2$ and $c = -7$

Hence the sum of the roots is $-\frac{2}{3}$ and the product is $-\frac{7}{3}$.

Example 7 If α and β are the roots of the quadratic equation $2x^2 - 3x + 5 = 0$ find the value of:

(a) $\alpha^2 + \beta^2$

Solution Since α and β are the roots of the equation $2x^2 - 3x + 5 = 0$ we have
$\alpha + \beta = \frac{3}{2}$ and $\alpha\beta = \frac{5}{2}$.

Now
$$\alpha^2 + \beta^2 = \alpha^2 + 2\alpha\beta + \beta^2 - 2\alpha\beta$$
$$= (\alpha + \beta)^2 - 2\alpha\beta$$
$$= \left(\frac{3}{2}\right)^2 - 2\left(\frac{5}{2}\right)$$
$$= \frac{9}{4} - 5$$
$$= -\frac{11}{4}$$

(b) $(\alpha - \beta)^2$

Solution
$$(\alpha - \beta)^2 = \alpha^2 - 2\alpha\beta + \beta^2$$
$$= \alpha^2 + 2\alpha\beta + \beta^2 - 4\alpha\beta$$

$$= (\alpha + \beta)^2 - 4\alpha\beta$$

$$= \left(\frac{3}{2}\right)^2 - 4\left(\frac{5}{2}\right)$$

$$= \frac{9}{4} - 10$$

$$= -\frac{31}{4}$$

Example 7 If α and β are the roots of the equation $3x^2 - 5x + 2 = 0$ find the equation whose roots are α^2 and β^2.

Solution $\alpha^2 + \beta^2 = (\alpha + \beta)^2 - 2\alpha\beta$

$$= \left(\frac{5}{3}\right)^2 - 2\left(\frac{2}{3}\right)$$

$$= \frac{25}{9} - \frac{4}{3}$$

$$= \frac{13}{9}$$

$$\alpha^2\beta^2 = (\alpha\beta)^2$$

$$= \left(\frac{2}{3}\right)^2$$

$$= \frac{4}{9}$$

Hence the equation is $x^2 - \frac{13}{9}x + \frac{4}{9} = 0$

i.e., $9x^2 - 13x + 9 = 0$

Exercise 9.3

1. Write down the sums and products of the roots of the following equations.

(a) $x^2 - 3x + 2 = 0$ (b) $x^2 + 7x + 10 = 0$ (c) $x^2 - 4x - 21 = 0$

(d) $2x^2 - x - 3 = 0$ (e) $5x^2 + 13x - 6 = 0$ (f) $x(x - 4) = 12$

(g) $x + \frac{1}{x} = 5$ (h) $\frac{3}{y^2} + \frac{5}{y} = 2$ (i) $\frac{3x}{x-2} = \frac{2}{x+1}$

2. Write down the equation, the sum and product of whose roots are:

(a) 4, 3   (b) −6, 5   (c) 2, -3   (d) $\frac{2}{3}, 0$ (e) $-\frac{1}{3}, \frac{2}{9}$ (f) 0, -5

3. Find in terms of $(\alpha + \beta)$ and $\alpha\beta$:

(a) $\dfrac{1}{\alpha^2} + \dfrac{1}{\beta^2}$
(b) $\dfrac{\beta}{\alpha} + \dfrac{\alpha}{\beta}$
(c) $\alpha^2\beta + \alpha\beta^2$

(d) $\alpha - \beta$
(e) $\alpha\beta^3 + \alpha^3\beta$
(f) $\alpha^3 + \beta^3$

4. The roots of the equation $2x^2 - 4x + 5 = 0$ are α and β. Find the value of:

(a) $\alpha^2 + \beta^2$
(b) $\dfrac{1}{\alpha} + \dfrac{1}{\beta}$
(c) $(\alpha + 1)(\beta + 1)$

(d) $\alpha^2\beta + \alpha\beta^2$
(e) $\dfrac{\alpha}{\beta} + \dfrac{\beta}{\alpha}$
(f) $(\alpha - \beta)^2$

5. The roots of $3x^2 - 4x + 2 = 0$ are α and β. Find the equation whose roots are:

(a) $\alpha + 2, \beta + 2$
(b) α^2, β^2
(c) $\alpha^2\beta, \alpha\beta^2$

(d) $\dfrac{\alpha^2}{\beta}, \dfrac{\beta^2}{\alpha}$
(e) $2\alpha + \beta, \alpha + 2\beta$
(f) $(\alpha - \beta), (\beta - \alpha)$

6. Without solving the given equation write down the equation whose roots are one more than those of $3x^2 + 2x - 1 = 0$.

7. Find the value of k if one root of $x^2 + kx + 27 = 0$ is the square of the other.

8. Find the value of k if the roots of $4x^2 - 12x + k = 0$ differ by two.

9. Prove that if one root of the equation $ax^2 + bx + c = 0$ is one-half the other then $2b^2 = 9ac$.

10. Given that the roots of the equation $ax^2 + bx + c = 0$ are a^2 and $a^2 - 2$ show that $b^2 = 4a(c + a)$.

9.4 Finding the Maximum or Minimum Value

You can use the method of completing the square to find the maximum or minimum value of a quadratic function.

Example 8 Find the minimum value of:

(a) $x^2 + 6x + 14$

Solution $\quad x^2 + 6x + 14 = x^2 + 6x + 9 + 5 \qquad$ Complete the square.

$$= (x + 3)^2 + 5 \qquad \text{Factorise.}$$

Because $(x + 3)^2$ is a square it is always positive or zero. The value of $x^2 + 6x + 14$ becomes larger when we substitute a value for x which is not -3. So the minimum value occur when $(x + 3)^2$ is zero. Hence the minimum value of $x^2 + 6x + 14$ is 5 and this occur when $x = -3$.

(b) $2x^2 - 5x + 7$

Solution $2x^2 - 5x + 7 = 2\left[x^2 - \frac{5}{2}x + \frac{7}{2}\right]$ Factor out 2.

$= 2\left[\left(x - \frac{5}{4}\right)^2 + \frac{31}{16}\right]$ Complete the square.

$= 2\left(x - \frac{5}{4}\right)^2 + \frac{31}{8}$ Remove the bracket.

The minimum value is $\frac{31}{8} = 3\frac{7}{8}$ and this occur when $x = \frac{5}{4} = 1\frac{1}{4}$

Example 9 Find the maximum value of:

(a) $5 - 3x - x^2$

Solution $5 - 3x - x^2 = -x^2 - 3x + 5$

$= -(x^2 + 3x - 5)$

$= -\left[\left(x + \frac{3}{2}\right)^2 - \frac{29}{4}\right]$

$= \frac{29}{4} - \left(x + \frac{3}{2}\right)^2$

The maximum value of the function is $\frac{29}{4}$, i.e. $7\frac{1}{4}$ and this occur when $x = -\frac{3}{2} = -1\frac{1}{2}$

(b) $5 + 2x - 3x^2$

Solution $5 + 2x - 3x^2 = -3x^2 + 2x + 5$

$= -3\left[x^2 - \frac{2}{3}x - \frac{5}{3}\right]$

$= -3\left[\left(x - \frac{1}{3}\right)^2 - \frac{16}{9}\right]$

$= \frac{16}{3} - 3\left(x - \frac{1}{3}\right)^2$

The maximum value is $\frac{16}{3}$, i.e. $5\frac{1}{3}$ and this occur when $x = \frac{1}{3}$.

Exercise 9.4

1. Find the maximum value of each of the following functions:

(a) $y = 2 + x - x^2$

(b) $y = 5 - 2x - x^2$

(c) $y = 6 + 5x - x^2$

(d) $y = 3 - 5x - 2x^2$

(e) $y = 2 + 4x - 3x^2$

(f) $y = 1 - 6x - 3x^2$

2. Find the minimum value of each of the following functions:

(a) $y = x^2 + 3x - 2$

(b) $y = x^2 - 8x + 6$

(c) $y = x^2 + 2x + 3$

(d) $y = 2x^2 + 3x - 4$

(e) $y = 3x^2 - 6x + 4$

(f) $y = 2x^2 + 6x + 5$

3. Find the minimum or maximum value of each of the following functions and state the value of x for which the function is minimum or maximum:

(a) $y = x^2 - 6x + 10$

(b) $y = 6 - x - x^2$

(c) $y = 2 - 2x - x^2$

(d) $y = 2x^2 + 8x + 9$

(e) $y = 3x^2 + 6x + 7$

(f) $y = -7 + 12x - 3x^2$

9.5 Sketching Quadratic Functions

The graph of the quadratic function $f(x) = ax^2 + bx + c$, $a \neq 0$ is a parabola which opens either upwards or downwards, as shown in Figure 9.2. When a is positive the parabola opens upwards with a minimum point and when a is negative the parabola opens downwards with a maximum point.

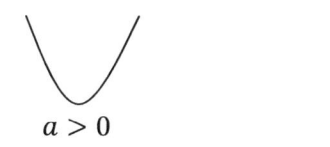

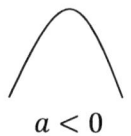

$a > 0$ $a < 0$

Figure 9.2

The maximum or minimum value of a quadratic function occurs at its vertex,. The vertex is also called the turning point and it is either the lowest or the highest point on the graph of the function (see Figure 9.3).

164 Further Mathematics

Every parabola has an axis of symmetry. The axis of symmetry is a vertical line midway between any pair of symmetric points on the parabola and intersects the parabola at its vertex.

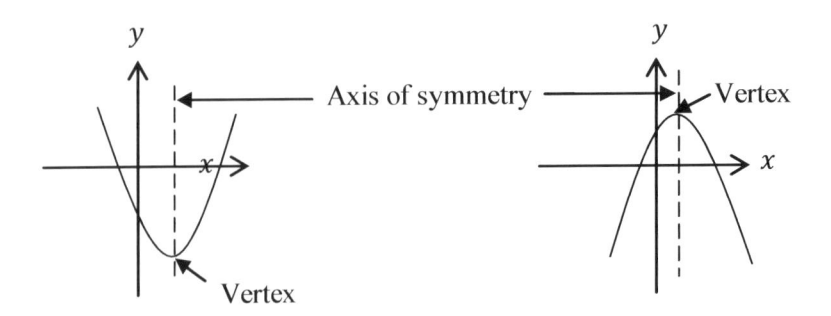

Figure 9.3

The graph of the quadratic function $y = ax^2 + bx + c$ intersects the y-axis at $(0, c)$. This point lies on the line $y = c$. Substituting c for y, we have

$c = ax^2 + bx + c$

$0 = ax^2 + bx$

$0 = x(ax + b)$

Setting each factor to 0 and solving for x, we have

$x = 0$ and $x = -\dfrac{b}{a}$

Hence $(0, c)$ and $\left(-\dfrac{b}{a}, c\right)$ are two symmetric points.

The x coordinate of the midpoint of the line joining these points is $\dfrac{0+(-b/a)}{2} = -\dfrac{b}{2a}$. Since the axis of symmetry passes through this point its equation is given by $x = -\dfrac{b}{2a}$.

The y coordinate of the vertex is obtained by substituting $-\dfrac{b}{2a}$ for x in the original equation.

It might help you to remember the following properties of graphs of quadratic functions.

1. If the quadratic expression factorises the two x-intercepts are symmetric points.

2. The x coordinates of the vertex of the curve of a function is $-\dfrac{b}{2a}$.

3. The y coordinate of the vertex of the graph of a quadratic function will be either a minimum value or a maximum value of the function.

Quadratic Functions

Example 10 illustrates the procedure for sketching graphs of quadratic functions.

Example 10 Sketch the graph of each function.

(a) $y = x^2 - 4x - 12$.

Solution First find the minimum or maximum value of y and then the minimum or maximum point of the curve.

The x coordinate of the vertex is given by $x = -\frac{b}{2a}$.

Here $a = 1$, $b = -4$ and $c = -12$.

Hence $x = -\frac{-4}{2 \times 1} = 2$.

The x-coordinate of the vertex is 2.

Substitute 2 for x in $y = x^2 - 4x - 12$

$y = 4 - 8 - 12$

$= -16$

Hence, the y coordinate of the vertex is -16.

Since a is positive, the graph opens upwards. So the graph has a minimum point at (2, -16).

Because $x^2 - 4x - 12$ can be factorised the two x-intercepts give two symmetric points. To find the x-intercept(s) we replace y with 0 and solve for x.

$x^2 - 4x - 12 = 0$

$(x - 6)(x + 2) = 0$

So $x - 6 = 0$ and $x + 2 = 0$

$x = 6$ $\qquad$ $x = -2$

The x-intercepts are (6, 0) and (-2, 0)

Finally, draw a smooth curve through the points found above. The sketch is shown in Figure 9.4.

166 Further Mathematics

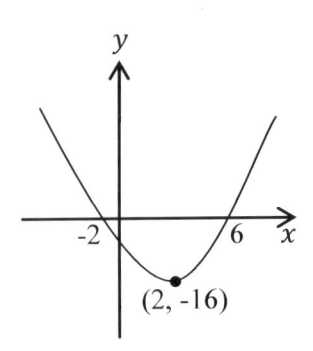

Figure 9.4

(b) $y = 5 + 6x - x^2$.

Solution Here $a = -1$, $b = 6$ and $c = 5$.

$$x = -\frac{b}{2a} = -\frac{6}{2 \times -1} = 3.$$

So $\qquad y = 5 + 6 \times 3 - 3^2$

$\qquad\qquad = 5 + 18 - 9$

$\qquad\qquad = 14$

Since a is negative, the graph opens downwards, so the graph has a maximum point at (3, 14). The quadratic expression does not factorise, so we can find a point symmetric to (0, 5).

Substitute 5 for y in $y = 5 + 6x - x^2$

$\qquad 5 + 6x - x^2 = 5$

$\qquad x(6 - x) = 0$

So $x = 0$ and $x = 6$

Hence (0, 5) and (6, 5) are symmetric points.

Draw a smooth curve through the points. The sketch is shown in Figure 9.5.

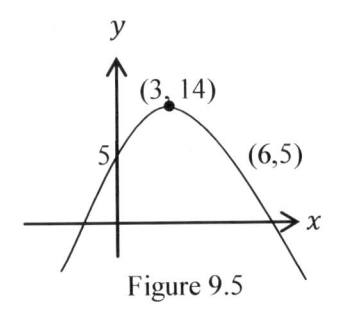

Figure 9.5

(c) $y = 2x^2 - 6x + 7$.

Solution Here $a = 2$, $b = -6$ and $c = 7$.

$$x = -\frac{b}{2a} = -\frac{-6}{2 \times 2} = \frac{3}{2}.$$

Substituting $x = \frac{3}{2}$ in $y = 2x^2 - 6x + 7$ gives

$$y = 2\left(\frac{3}{2}\right)^2 - 6\left(\frac{3}{2}\right) + 7 = \frac{5}{2}$$

Hence the graph has a minimum point at $\left(1\frac{1}{2}, 2\frac{1}{2}\right)$.

Now $2x^2 - 6x + 7 = 7$

$$2x(x - 3) = 0$$

So $x = 0$ and $x = 3$

The points (0, 7) and (3, 7) are symmetric points. Draw a smooth curve through the points. The sketch is shown in Figure 9.6.

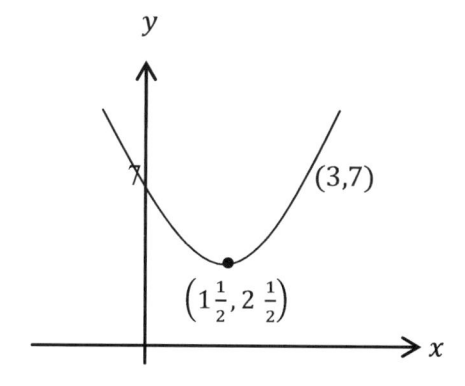

Figure 9.6

An alternative procedure for sketching graphs of quadratic functions is illustrated in Example 11.

Example 11 Sketch the graph of each function.

(a) $y = x^2 - 2x - 15$.

Solution Begin by finding the x- and y- intercepts. To find the y-intercept, replace x with 0 and solve for y. So $y = -15$.

The y-intercept is (0, -15).

To find the x-intercept, replace y with 0 and solve for x.

$$x^2 - 2x - 15 = 0$$

$$(x + 3)(x - 5) = 0$$

So $x + 3 = 0$ and $x - 5 = 0$

$x = -3$ $\qquad x = 5$

The x-intercepts are (-3, 0) and (5, 0).

Next, find the minimum or maximum value by completing the square.

$$x^2 - 2x - 15 = x^2 - 2x + 1 - 1 - 15$$

$$= (x - 1)^2 - 16$$

The minimum value is -16 and this occur at $x = 1$. So the minimum point is (1, -16).

Finally, complete the graph by using the fact that the curve opens upward since the coefficient of x^2 is positive. The sketch is shown in Figure 9.7.

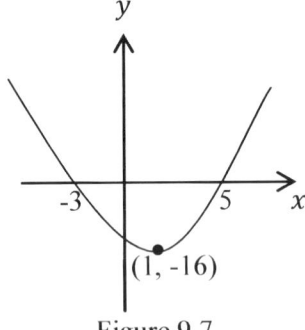

Figure 9.7

(b) $y = 12 - 5x - 2x^2$.

Solution Substituting $x = 0$ into $y = 12 - 5x - 2x^2$ gives $y = 12$.

Also, substituting 0 for y, we have

$$12 - 5x - 2x^2 = 0$$

$$(3 - 2x)(4 + x) = 0$$

So $\qquad x = \dfrac{3}{2}$ and $x = -4$.

Now $12 - 5x - 2x^2 = -2\left[x^2 + \frac{5}{2}x - 6\right]$

$= -2\left[\left(x + \frac{5}{4}\right)^2 - \frac{121}{16}\right]$

$= \frac{121}{8} - 2\left(x + \frac{5}{4}\right)^2$

The maximum point is $\left(-\frac{5}{4}, \frac{121}{8}\right)$.

The curve opens downward since the coefficient of x^2 is negative. The sketch is shown in Figure 9.8.

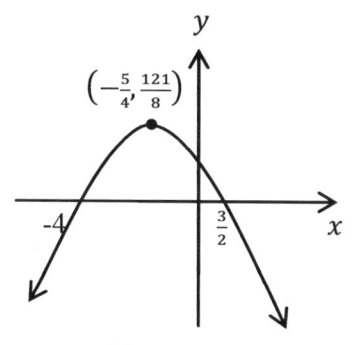

Figure 9.8

The following list summarises the procedure for sketching the curve of $y = ax^2 + bx + c$.

1. Find the maximum or minimum point by completing the square or by using the formula $x = -\frac{b}{2a}$.

2. Find the x- and y- intercepts or two symmetric points.

3. Plot the maximum point or minimum point.

4. Plot either the x- and y-intercepts or the symmetric points.

5. Complete the sketch, using the fact that the curve opens upwards when a is positive and opens downwards when a is negative.

Exercise 9.5

Sketch the graph of each of the following quadratic functions:

1. $y = x^2 + 6x + 5$
2. $y = 4 - 3x - x^2$
3. $y = x^2 + 2x - 10$
4. $y = x^2 - 2x - 15$

5. $y = -x^2 + 6x - 8$ 6. $y = x^2 + 6x + 8$

7. $y = 12 + x - x^2$ 8. $y = 6 + x - 2x^2$

9. $y = 3x^2 + 12x + 3$ 10. $y = -2x^2 - 4x - 1$

11. $y = 2x^2 + 4x - 1$ 12. $y = 1 + 6x - 3x^2$

13. $y = |x^2 + x - 6|$ 14. $y = |2x^2 - x - 6|$

15. $y = |4 - 3x - x^2|$ 16. $y = |3 - 5x - 2x^2|$

9.6 Quadratic Inequalities

An expression such as $ax^2 + bx + c < 0$, where $a \neq 0$ is called a quadratic inequality. Note that the inequality symbol $<$ can be replaced by the symbol $>$, $\leq$ or $\geq$.

Solving Quadratic Inequalities

The solution of a quadratic inequality can be obtained by graphical or algebraic method.

Graphical Solution of Quadratic Inequalities

To solve the quadratic inequality $ax^2 + bx + c < 0$ graphically, draw the graph of $y = ax^2 + bx + c$ and then determine from the graph the range of values of x that will give negative y values.

We know from the study of graphs that points on the y-axis above the x-axis are associated with positive y-values and the points below the x-axis are associated with negative y-values.

Example 12 Solve each of the following inequalities.

(a) $x^2 + 2x - 15 < 0$.

Solution First sketch the graph of $y = x^2 + 2x - 15$ as shown in Figure 9.9.

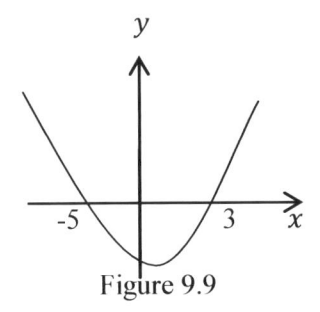

Figure 9.9

The graph cuts the x-axis at $x = -5$ and $x = 3$. You need to find the range of values of x such that $y < 0$. From the graph in Figure 9.9 you can see that this happens when x lies between -5 and 3. Hence the solution of $x^2 + 2x - 15 < 0$ is $-5 < x < 3$.

(b) $x^2 - 3x - 20 \geq 8$.

Solution First rewrite the inequality in the general form

$$x^2 - 3x - 28 \geq 0$$

Next sketch the graph of $y = x^2 - 3x - 28$. The graph is shown in Figure 9.10.

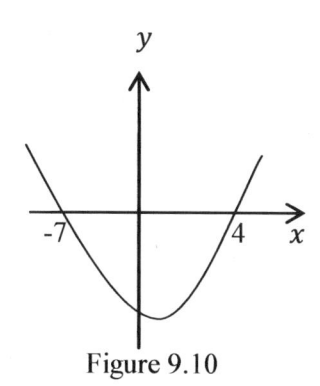

Figure 9.10

In this case look for values of x such that $y > 0$ or $y = 0$. From the graph you should see that the solution of the inequality $2x^2 - 3x - 20 \geq 8$ is $x \leq -7$ or $x \geq 4$.

Exercise 9.6(a)

Solve each of the following quadratic inequalities:

1. $x^2 + x - 12 < 0$
2. $x^2 + 3x - 10 > 0$
3. $x^2 + 4x - 12 > 0$
4. $x^2 + 2x - 15 < 0$
5. $x^2 - 5x + 6 \geq 0$
6. $x^2 + 7x + 10 \leq 0$
7. $2x^2 + x - 6 \leq 0$
8. $3x^2 - 10x - 8 \leq 0$
9. $4x^2 + x - 3 < 0$
10. $x(x + 5) \geq 24$

Algebraic Solution of Quadratic Inequalities

Example 13 Solve each of the following inequalities.

(a) $x^2 - 4x - 32 < 0$

Solution First find the solution of the corresponding quadratic equation

$$x^2 - 4x - 32 = 0$$

$$(x - 8)(x + 4) = 0$$

$$x = -4 \text{ or } x = 8$$

The quadratic equation has solutions – 4 and 8 called the critical numbers. The critical numbers divide a number line into three regions called the test regions as shown in Figure 9.11.

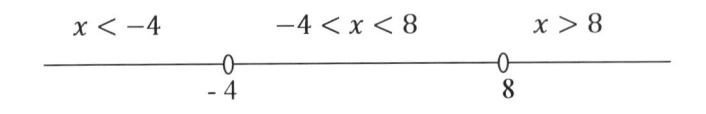

```
    x < -4         -4 < x < 8        x > 8
──────────○──────────────────○──────────────
          -4                 8
```

Figure 9.11

Next choose a number in each of these regions and substitute this number into the quadratic expression to determine whether your answer is positive or negative. If you get a negative answer, it means that the value of the expression is negative for all values of x in that region. Similarly if you get a positive answer it means that all values of the expression are positive for all values of x in that region.

Test interval	Chosen x value	$x^2 - 4x - 32$
$x < -4$	$x = -5$	$25 + 20 - 32 > 0$
$-4 < x < 8$	$x = 1$	$1 - 4 - 32 < 0$
$x > 8$	$x = 9$	$81 - 36 - 32 > 0$

Note that we do not need to find the actual value of the expression. All that we need is the sign of the value of the expression. We add the information obtained from the table to the number line shown in Figure 9.12.

```
──────────○──────────────────○──────────────
    +    -4         −        8       +
```

Figure 9.12

We are looking for values of x that will give negative values of the expression. This occur between – 4 and 8. Hence the solution of the inequality $x^2 - 4x - 32 < 0$ is $-4 < x < 8$.

Note that the critical numbers are found by factorising the quadratic expression. If the quadratic expression does not factorise, you can use the Quadratic Formula to find the critical numbers.

(b) $(x-4)(x+8) \geq 13$.

Solution First rewrite the inequality in the general form

$$(x-4)(x+8) \geq 13$$

$$x^2 + 4x - 45 \geq 0$$

Next find the critical numbers by solving the quadratic equation.

$$x^2 + 4x - 45 = 0$$

$$(x-5)(x+9) = 0$$

$$x = 5 \text{ or } x = -9$$

Test region	Chosen x value	$x^2 + 4x - 45$
$x < -9$	$x = -10$	$100 - 40 - 45 > 0$
$-9 < x < 5$	$x = 1$	$1 + 4 - 45 < 0$
$x > 5$	$x = 6$	$36 + 24 - 45 > 0$

The information is illustrated in Figure 9.13.

```
─────────○─────────○─────────
   +    -9    -    5    +
```

Figure 9.13

You can see from Figure 9.13 that the expression is positive when x is less than -9 or when x is greater than 5. Note that the expression is zero when x is -9 or 5. Hence the solution of the inequality is $x \leq -9$ or $x \geq 5$.

An alternative method can be used to find the solution of a quadratic inequality when the quadratic expression can be factorised as shown in Example 14.

Example 14 Solve each of the following inequalities.

(a) $x^2 + 2x - 15 < 0$.

Solution First factorise the quadratic expression.

$$(x-3)(x+5) < 0$$

Next find the critical numbers.

$$(x-3)(x+5) = 0$$

$$x = 3 \text{ or } x = -5$$

The value of $x - 3$ is positive when $x > 3$ and negative when $x < 3$. Also, the value of $x + 5$ is positive when $x > -5$ and negative when $x < -5$. The information is illustrated in Figure 9.14.

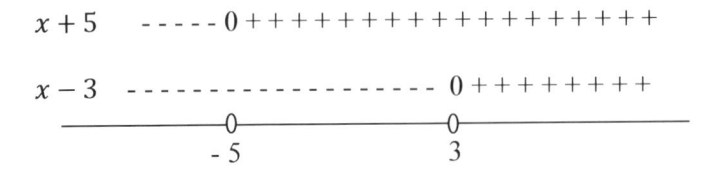

```
x + 5     -----0++++++++++++++++++
x - 3     ------------------0++++++++
         ————————0————————————0—————————
                -5                  3
```

Figure 9.14

The product $(x - 3)(x + 5)$ is negative in the region where one factor is negative and the other positive, and positive in the region where both factors are either negative or positive. Notice that the product between -5 and 3 is negative. Hence, the solution of the inequality $x^2 + 2x - 15 < 0$ is $-5 < x < 3$.

(b) $x^2 > 4x + 21$.

Solution First rewrite the inequality in standard form.

$$x^2 - 4x - 21 > 0$$

Next factorise the quadratic expression.

$$(x - 7)(x + 3) > 0$$

The sign graph is shown in Figure 9.15.

```
x + 3     -----0+++++++++++++++++++
x - 7     ------------------0++++++++
         ————————0————————————0—————————
                -3                  7
```

Figure 9.15

We are looking for a region where $(x - 7)(x + 3)$ is positive. So the solution of the inequality $x^2 > 4x + 21$ is $x < -3$ or $x > 7$.

Exercise 9.6(b)

Solve each of the following inequalities:

1. $x^2 - 3x - 4 > 0$
2. $x^2 - 2x - 8 < 0$
3. $x^2 - x - 12 \le 0$
4. $x^2 + 7x + 12 \ge 0$
5. $x^2 - 3x - 10 < 0$
6. $x^2 + 6x - 27 > 0$
7. $(x+3)(x-4) \le 0$
8. $x^2 - 3x > 18$
9. $2x^2 + 5x + 3 < 0$
10. $3x^2 + 8x + 5 \ge 0$
11. $1 - x - 2x^2 \le 0$
12. $6 + 7x \ge 3x^2$

Review Exercise 9

1. Solve each of the following quadratic equations by completing the square:

(a) $x^2 + 6x + 8 = 0$
(b) $2x^2 - x - 6 = 0$
(c) $x^2 - 5x + 3 = 0$
(d) $3x^2 + 4x - 1 = 0$

2. Solve each of the following quadratic equations by using the quadratic formula:

(a) $x^2 + 4x - 21 = 0$
(b) $3x^2 - 4x - 4 = 0$
(c) $x^2 - 3x - 5 = 0$
(d) $2x^2 + 3x - 5 = 0$

3. Solve each quadratic equation using any applicable method:

(a) $x^2 + 6x + 8 = 0$
(b) $x^2 - 8x - 20 = 0$
(c) $x^2 - 2x - 10 = 0$
(d) $2x^2 + 3x - 5 = 0$
(e) $3x^2 - 8x + 2 = 0$
(f) $2 + 6x - 5x^2 = 0$
(g) $x^2 - 10x + 23 = 0$
(h) $3x^2 - 2x = 9$
(i) $5x^2 + 7 = 15x$
(j) $4x^2 + 3x = 6$

4. Solve each of the following quadratic inequalities:

(a) $x^2 + 4x - 5 \ge 0$
(b) $x^2 - 6x - 7 < 0$
(c) $x^2 + 3x - 10 \le 0$
(d) $8x - x^2 > 12$
(e) $3x^2 + 2x - 8 \le 0$
(f) $2x^2 - 3x \ge 20$

5. Find the set of values of x that satisfy the inequality $x^2 + x - 6 \le 0$.

6. Find the set of values of x that satisfy the inequality $x^2 + 3x + 2 > 0$.

7. Find the range of values of k for which the equation $x^2 + (3 - k)x + 1 = 0$ has real roots.

8. Find the largest possible value of k such that $kx^2 + 4x + 2 = 0$ has real roots.

9. Find the value of k for which $x^2 - (k-2)x + 2k + 1 = 0$ has equal roots.

10. Find the smallest possible integral value of k such that $2x^2 - 3x + k = 0$ will have imaginary roots.

11. Find the range of values of k for which the equation $x^2 + (k+2)x + 3k - 2 = 0$ has an imaginary roots.

12. Find the range of values for k for which the expression $3x^2 + 2x + k$ is always positive for all real values of x.

13. Find the range of values of k for which the expression $3x^2 + 6x + k$ is always positive for all real values of x.

14. If α and β are roots of $2x^2 - 3x - 7 = 0$ find:

(a) the value of $\alpha^2 + \beta^2$ \qquad (b) the equation whose roots are $\frac{1}{\alpha^2}, \frac{1}{\beta^2}$

15. If α and β are the roots of the equation $3x^2 - 2x + 1 = 0$ find:

(a) the value of $\alpha\beta^2 + \alpha^2\beta$ \qquad (b) the equation whose roots are $\frac{\alpha}{\beta}, \frac{\beta}{\alpha}$

16. If α and β are the roots of the equation $3x^2 - 4x - 2 = 0$ find the equation whose roots are $\frac{1}{\alpha} + 1$ and $\frac{1}{\beta} + 1$.

17. Find the minimum or maximum value of each of the following functions and state the value of x for which function is minimum or maximum:

(a) $y = x^2 - x - 6$ \quad (b) $y = 20 + x - x^2$ \quad (c) $y = 2x^2 - 3x - 2$ \quad (d) $4 + 3x - 2x^2$

18. Express $2x^2 - 6x + 5$ in the form $a(x+b)^2 + c$. Hence state the minimum value of $2x^2 - 6x + 5$ and state the value of x for which it occurs.

19. Express $7 - 5x - 2x^2$ in the form $a - b(x+c)^2$. State the maximum value of $7 - 5x - 2x^2$ and the value of x for which it occurs.

20. Sketch the graph of each function:

(a) $y = x^2 - 3x + 5$ \qquad (b) $y = 1 + x - x^2$

(c) $y = 2x^2 - 5x + 3$ \qquad (d) $y = 5 - 2x - 4x^2$

10 Polynomials

You should be familiar with the polynomial function. A polynomial function of degree n is an expression of the form $a_n x^n + a_{n-1} x^{n-1} + \cdots + a_1 x + a_0$, $a_n \neq 0$ where $a_n \neq 0$ and $n \geq 0$ is an integer. The coefficients of the polynomial function, $a_n, a_{n-1}, \cdots, a_1$ and a_0, are real numbers. The degree of a polynomial is the largest power of x occurring in the expression. The following are examples of polynomials:

$f(x) = 3$ Constant function

$f(x) = 2x + 1$ Linear function

$f(x) = 3x^2 + 5x - 5$ Quadratic function

$f(x) = x^3 - 2x^2 + x + 3$ Cubic function

$f(x) = 2x^4 - x^3 + 3x^2 + 5x - 2$ Quartic function

The constant, linear, quadratic, cubic and quartic polynomials have degree 0, 1, 2, 3 and 4 respectively.

10.1 Division of Polynomials

You can divide a polynomial by another polynomial by a method similar to the process used for dividing positive integers in arithmetic.

Example 1

(a) Divide $x^2 + 2x - 8$ by $x - 2$.

Solution Begin by dividing the first term in the dividend x^2 by the first term in the divisor x. Then multiply the divisor by your answer, i.e. x to get $x^2 - 2x$. Next subtract $x^2 - 2x$ from $x^2 + 2x$. Continued the process until the degree of the remainder is less than that of the divisor. The process is illustrated below.

$$
\begin{array}{r}
x + 4 \\
x - 2 \overline{\smash{)} x^2 + 2x - 8} \\
\underline{x^2 - 2x} \\
4x - 8 \\
\underline{4x - 8} \\
0
\end{array}
$$

Multiply the divisor by x

Subtract and bring down -8

Multiply $x - 2$ by 4

Subtract

the quotient is $x + 4$ and the remainder is 0.

(b) Divide $3 + 5x^2 - 6x + 2x^3$ by $3 + x$.

Solution First write the terms of the divisor and dividend in descending powers of x before you divide.

$$x+3 \overline{) 2x^3 + 5x^2 - 6x + 3} \quad \text{quotient: } 2x^2 - x - 3$$

$\underline{2x^3 + 6x^2}$	Multiply $x + 3$ by $2x^2$
$-x^2 - 6x$	Subtract and bring down $-6x$
$\underline{-x^2 - 3x}$	Multiply $x + 3$ by $-x$
$-3x + 3$	Subtract and bring down 3
$\underline{-3x - 9}$	Multiply the divisor by -3
12	

The quotient is $2x^2 - x - 3$ and the remainder is 12.

(c) Divide $x^3 - 8$ by $x + 2$.

Solution Any missing term of the dividend can be written in using 0 for the coefficients.

$$x - 2 \overline{) x^3 + 0x^2 + 0x - 8} \quad \text{quotient: } x^2 + 2x + 4$$

$\underline{x^3 - 2x^2}$	Multiply $x - 2$ by x^2
$2x^2 + 0x$	Subtract and bring down $0x$
$\underline{2x^2 - 4x}$	Multiply $x - 2$ by $2x$
$4x - 8$	Subtract and bring down -8
$\underline{4x - 8}$	Multiply $x - 2$ by 4
0	

The quotient is $x^2 + 2x + 4$ and the remainder is 0.

Exercise 10.1

1. Work out each division and find the remainder.

(a) $(x^3 - 3x^2 - 5x + 2) \div (x - 3)$

(b) $(x^3 + 2x^2 - 3x - 4) \div (x + 2)$

(c) $(x^3 - 4x^2 + 5x - 2) \div (x - 2)$ (d) $(3x^3 + 4x^2 + x - 2) \div (x + 3)$

(e) $(2x^3 + 5x^2 - 11x + 4) \div (2x - 1)$ (f) $(x^3 + 15) \div (x + 2)$

(g) $(2x^3 + x^2 - 13x + 6) \div (x - 2)$ (h) $(2x^3 - 3x^2 + 4x + 4) \div (2x + 1)$

2. Find the quotients and the remainders in the following divisions:

(a) $(x^3 + 2x^2 + 3x + 1) \div (x - 1)$ (b) $(x^3 - 3x^2 + 4x - 2) \div (x - 2)$

(c) $(x^3 + 5x^2 + 3x + 1) \div (x + 3)$ (d) $(2x^3 + 5x^2 - 2x + 2) \div (x + 2)$

(e) $(3x^3 - 2x^2 + 5x - 2) \div (x - 1)$ (f) $(3x^3 + 6x - 2) \div (x + 4)$

(g) $(2x^3 - 7x^2 + 11x - 4) \div (2x - 1)$ (h) $(2x^3 + 5x^2 + 6x - 8) \div (2x + 1)$

10.2 Factorising Polynomials

Long division can be used to factorise polynomials. If the remainder is zero when we divide a polynomial, it shows that the divisor divides the polynomial exactly. The divisor and the quotient are then two factors of the polynomial. A polynomial can be written in an equivalent form as a product of two or more factors.

Remainder Theorem

Consider the division of $x^3 - 2x^2 - 5x + 8$ by $x - 3$.

The remainder can be found without performing the long division.

Suppose that when $x^3 - 2x^2 - 5x + 8$ is divided by $x - 3$ the quotient is $Q(x)$ and the remainder is R.

Then $x^3 - 2x^2 - 5x + 8 = (x - 3)Q(x) + R$

Substituting 3 for x gives

$3^3 - 2(3^2) - 5(3) + 8 = 0 \cdot Q + R$

$27 - 18 - 15 + 8 = R$

So $R = 2$

The remainder when $x^3 - 2x^2 - 5x + 8$ is divided by $x - 3$ is 2. You can verify this by performing the long division, and we leave this as an exercise.

Our discussion shows that the remainder can be obtained by substituting into the polynomial expression the value of x that will make the divisor 0.

So if a polynomial $P(x)$ is divided by $x - a$ the remainder will be $P(a)$.

This result is known as the Remainder Theorem.

A special case of the Remainder Theorem is the Factor Theorem.

Factor Theorem

If $P(\alpha) = 0$, then $(x - \alpha)$ is a factor of $P(x)$. α is called the zero of $P(x)$.

Example 2

(a) Find the remainder when $3x^3 + x^2 - 8x + 6$ is divided by $3x - 2$.

Solution Let $P(x) = 3x^3 + x^2 - 8x + 6$

$$\text{Then } P\left(\tfrac{2}{3}\right) = 3\left(\tfrac{2}{3}\right)^3 + \left(\tfrac{2}{3}\right)^2 - 8\left(\tfrac{2}{3}\right) + 6$$

$$= \tfrac{8}{9} + \tfrac{4}{9} - \tfrac{16}{3} + 6$$

$$= 2$$

The remainder is 2.

(b) Factorise $2x^3 + 3x^2 - 11x - 6$ and find the zeros.

Solution You can use the try and error method to find one factor of this expression and then find a second factor by long division.

$$\text{Let } P(x) = 2x^3 + 3x^2 - 11x - 6$$

Substitute values for x until you find a value that makes the expression zero.

We begin by trying 2.

$$P(2) = 2(2)^3 + 3(2)^2 - 11(2) - 6$$

$$= 16 + 12 - 22 - 6$$

$$= 0$$

This mean $x - 2$ is a factor of $P(x)$.

Now divide $2x^3 + 3x^2 - 11x - 6$ by $x - 2$ performing the long division.

$$\begin{array}{r}
2x^2+7x+3\\
x-2{\overline{\smash{\big)}\,2x^3+3x^2-11x-6}}\\
\underline{2x^3-4x^2}\\
7x^2-11x\\
\underline{7x^2-14x}\\
3x-6\\
\underline{3x-6}\\
0
\end{array}$$

So $P(x) = (x-2)(2x^2 + 7x + 3)$

Factorising the quadratic factor gives

$P(x) = (x-2)(2x+1)(x+3)$

Hence the zeros are -3, $-\frac{1}{2}$ and 2

(c) Find the value(s) of a if $x-2$ is a factor of $x^3 + ax^2 + x + 6$ and factorise the expression completely.

Solution Let $P(x) = x^3 + ax^2 + x + 6$

$$\begin{aligned}
P(2) &= 2^3 + a(2^2) + 2 + 6\\
&= 8 + 4a + 2 + 6\\
&= 16 + 4a
\end{aligned}$$

Since $x-2$ is a factor of $x^3 + ax^2 + x + 6$ we have

$$16 + 4a = 0$$
$$a = -4$$

So the polynomial is $x^3 - 4x^2 + x + 6$.

Dividing $x^3 - 4x^2 + x + 6$ by $x - 2$ gives $x^2 - 2x - 3$

So $P(x) = (x-2)(x^2 - 2x - 3)$

Factorising the quadratic factor gives

$P(x) = (x-2)(x-3)(x+1)$

(d) Find the values of a and b if $x^2 - 5x + 6$ is a factor of $2x^3 + ax^2 + bx + 6$ and state the third factor of the expression.

Solution Let $P(x) = 2x^3 + ax^2 + bx + 6$

Because $x^2 - 5x + 6$ is a factor of the expression then both $x - 2$ and $x - 3$ are factors of $P(x)$.

$$P(2) = 16 + 4a + 2b + 6$$
$$= 22 + 4a + b$$

So $\quad 22 + 4a + 2b = 0$ giving

$$2a + b = -11 \qquad (1)$$

and $\quad P(3) = 54 + 9a + 3b + 6$

$$= 60 + 9a + 3b$$

So $\quad 60 + 9a + 3b = 0$ giving

$$3a + b = -20 \qquad (2)$$

Subtracting Equation (2) from Equation (1) gives

$$a = -9$$

Substitute $a = -9$ into Equation (1) to get

$$-18 + b = -11$$
$$b = 7$$

So $P(x) = 2x^3 - 9x^2 + 7x + 6$

Dividing $2x^3 - 9x^2 + 7x + 6$ by $x^2 - 5x + 6$ gives $2x + 1$

The third factor is $2x + 1$

Exercise 10.2

1. Find the remainder when:

(a) $x^3 + 3x^2 - 4x + 2 \quad$ is divided by $x - 1$

(b) $x^3 + x^2 - 3x - 2 \quad$ is divided by $x + 2$

(c) $x^3 + 3x^2 - 8x + 1$ is divided by $x - 3$

(d) $4x^3 - 3x^2 + 2x - 7$ is divided by $x + 1$

(e) $x^3 - 5x^2 + 11x - 6$ is divided by $x - 2$

(f) $4x^3 - 6x^2 + 5$ is divided by $2x - 1$

2. Factorise completely:

(a) $x^3 - 3x^2 - x + 3$ given that $x - 1$ is a factor

(b) $x^3 - 5x^2 - x + 5$ given that $x + 1$ is a factor

(c) $x^3 + 2x^2 - 5x - 6$ given that $x - 2$ is a factor

(d) $2x^3 + 3x^2 - 5x - 6$ given that $x + 2$ is a factor

(e) $2x^3 - 3x^2 - 3x + 2$ given that $2x - 1$ is a factor

(f) $2x^3 - 17x - 9$ given that $x + 3$ is a factor

3. Factorise completely:

(a) $x^3 - 2x^2 + 3x - 6$

(b) $x^3 + 6x^2 + 11x + 6$

(c) $x^3 + 5x^2 - 2x - 24$

(d) $2x^3 + 3x^2 - 11x - 6$

(e) $5x^3 - 8x^2 - 11x + 14$

(f) $2x^3 + 9x^2 + 7x - 6$

4. If $x - 1$ is a factor of $3x^3 + kx^2 - 4x - 3$, find k.

5. If $3x^3 - kx + 7$ is divisible by $x + 1$ find k.

6. If $2x^3 + x^2 + px + 2p^2$ is divisible by $x + 1$ find p.

7. If $x^4 + px^3 + qx^2 - x + 2$ is divisible by $x^2 - 1$ find p and q.

8. If $x - 2$ and $x + 1$ are both factors of $ax^3 + 3x^2 - 9x + b$ find the values of a and b. State the third factor of the expression.

9. When $x^3 - x^2 + ax + b$ is divided by $x - 1$ the remainder is -8. If $x + 1$ is a factor of the expression, find a and b and factorise the expression completely.

184 Further Mathematics

10. If the polynomial $x^3 + ax^2 + bx - 2$ is divided by $x - 1$ the remainder is 3 and when divided by $x + 1$ the remainder is 4. Find the values of a and b.

11. When the expression $x^3 + x^2 + px + q$ is divided by $x^2 - 1$, the remainder is $2x + 3$. Find the values of p and q.

12. The expression $ax^4 + bx^3 + 3x^2 - 2x + 3$ has remainder $x + 1$ when divided by $x^2 - 3x + 2$. Find the values of a and b.

Review Exercise 10

1. Divide:

(a) $3x^3 + 4x^2 + 7x + 2$ by $3x + 1$

(b) $3x^3 + 2x^2 - 3x - 2$ by $3x + 2$

(c) $2x^3 + 7x^2 - x + 12$ by $x + 4$

(d) $2x^3 - x^2 - 2x + 1$ by $2x - 1$

2. Find the remainder when:

(a) $x^3 - 2x^2 + 5x + 8$ is divided by $x - 2$

(b) $12x^3 + 16x^2 - 5x - 3$ is divided by $2x - 1$

(c) $2x^3 + x^2 - 13x + 6$ is divided by $x - 2$

(d) $x^5 - 4x^3 + 2x + 3$ is divided by $x - 1$

3. Factorise fully:

(a) $x^3 - 5x^2 + 2x + 8$

(b) $3x^3 - 8x^2 - x + 10$

(c) $2x^3 - x^2 - 8x + 4$

(d) $x^3 - 7x + 6$

4. Show that $x - 3$ is a factor of $x^3 - 6x^2 + 11x - 6$. Find the other factors.

5. Show that $x + 1$ is a factor of $2x^3 - 3x^2 - 2x + 3$. Find the other factors.

6. Factorise the following polynomials. State the zeros of each of the polynomials.

(a) $x^3 - 2x^2 - 5x + 6$

(b) $x^3 - 4x^2 + x + 6$

(c) $x^3 - 6x^2 + 11x - 6$

(d) $2x^3 - x^2 - 8x + 4$

(e) $2x^3 - x^2 + 2x - 1$

(f) $x^4 - 8x^3 + 14x^2 + 8x - 15$

7. If $x^4 + 3x^3 + ax^2 + bx - 18$ is divisible by $x^2 - 9$ find the values of a and b.

8. The polynomial $f(x) = 2x^3 - x^2 + px + 1$ has $(2x - 1)$ as a factor. Find the value of value of p and the zeros of $f(x)$.

9. The polynomial $f(x) = x^3 + ax^2 + bx - 12$ has $(x - 2)$ as a factor. When it is divided by $(x - 1)$ the remainder is -12. Find:

(a) the values of a and b

(b) the other factors of $f(x)$

(c) the remainder when $f(x)$ is divided by $(x - 3)$

10. When the polynomial $f(x) = x^3 + ax^2 + bx + 6$ where a and b are constants is divided by $(x - 2)$, the remainder is -4. When divided by $(x + 1)$, the remainder is 8. Find:

(a) the values of a and b

(b) the zeros of the function $f(x)$

11. The polynomial $f(x) = x^3 - x^2 - 4kx + 3k^2$, where k is a constant has $(x - 2)$ as a factor.

(a) Find the possible values of k.

(b) For the integral value of k find the remainder when $f(x)$ is divided by $(x - 3)$.

12. If the polynomial $f(x) = ax^2 + bx + c$, where a, b and c are constant, is divided by $(x - 1)$ the remainder is -5 and if it is divided by $(x + 2)$ the remainder is 4. If $(x - 2)$ is a factor of $f(x)$, find the values of a, b and c. Hence find the zeros of $f(x)$.

13. The function $f(x) = ax^3 - 3x^2 + bx + 2$, where a and b are constants, is such that $2x - 1$ is a factor. Given that the sum of the remainder when $f(x)$ is divided by $x + 2$ and the remainder when $f(x)$ is divided by $x - 3$ is 0, find the values of a and b.

11 Rational Functions

If $P(x)$ and $Q(x)$ are polynomial functions of x then $f(x) = \frac{P(x)}{Q(x)}$ where $Q(x) \neq 0$ is called a rational function of x.

11.1 Zeros, Domain and Range of Rational Functions

Finding Zeros of Rational Functions

Finding the zeros of a rational function is the same as finding the solution of a rational equation.

Example 1

Find the zeros of $f(x) = \frac{x^2-2x-3}{(x-1)(x^2-4)}$

Solution Set the numerator equal to zero.

$$x^2 - 2x - 3 = 0$$

Factorising this equation we have

$$(x-3)(x+1) = 0$$

So $\qquad x = 3 \text{ or } x = -1$

Hence the zeros of $f(x)$ are 3 and -1

Exercise 11.1(a)

Find the zeros of the following rational functions:

1. $f(x) = \frac{2x+1}{x-1}$

2. $f(x) = \frac{x+4}{(x-2)(x+3)}$

3. $f(x) = \frac{x-1}{(x-4)(x+2)}$

4. $f(x) = \frac{(x-2)(x+2)}{x+1}$

5. $f(x) = \frac{2x^2+x-1}{3x^2+4x+3}$

6. $f(x) = \frac{x^2-9}{2x^2+3}$

Domain of Rational Functions

The denominator of a rational expression cannot be zero because division by zero is undefined. You learned in Section 6.1 that the domain of a rational function is the set of all real numbers for which the denominator is not equal to zero.

Example 2

(a) For what values of x is $\frac{3x+2}{x^2-x-6}$ is undefined?

Solution Set the denominator equal to zero and then solve for x.

$$x^2 - x - 6 = 0$$

$$(x-3)(x+2) = 0$$

So $\quad x = 3$ or $x = -2$

Hence $\frac{3x+2}{x^2-x-6}$ is undefined when $x = 3$ or $x = -2$

(b) Find the domain of the function $f(x) = \frac{5x-3}{x^2-x-2}$.

Solution Set the denominator equal to zero and then solve for x.

$$x^2 - x - 2 = 0$$

$$(x-2)(x+1) = 0$$

So $\quad x = -1$ or $x = 2$

The expression in the denominator is zero when $x = -1$ or $x = 2$. Hence the domain is the set of all real values of x except $x \neq -1$ and $x \neq 2$. In set notation we write $\{x: x \in R, x \neq -1, x \neq 2\}$.

Exercise 11.1(b)

State the value(s) of x for which the following rational functions are undefined:

1. $\frac{2}{2x-3}$
2. $\frac{x+5}{(x-2)(x-3)}$
3. $\frac{2x+3}{(x+4)(x-2)}$
4. $\frac{x^2+3x-2}{x+3}$
5. $\frac{3x^2+5x-2}{x^2-4}$
6. $\frac{x^2-2x-3}{2x^2-x-6}$

The Range of a Rational Function

Recall that the range of a function is the set of y-values we get after substituting all possible x-values. You can find the range of a rational function $f(x)$ by letting $f(x) = y$ and then solving for x.

Example 3

Find the range of the function $f(x) = \frac{x^2}{x-2}$.

Solution Let $\frac{x^2}{x-2} = y$

then $x^2 - xy + 2y = 0$

The values of x which produce corresponding values of y are the roots of this quadratic equation and these roots are real if and only if

$$y^2 - 4(2y) \geq 0$$

$$y(y-8) \geq 0$$

Hence $y \leq 0$ or $y \geq 8$

So $f(x)$ can take all values except those between 0 and 8 i.e. the range is $\{y: y \leq 0 \text{ or } y \geq 8\}$.

Exercise 11.1(c)

State the range of each of the following expressions:

1. $\frac{3}{x-1}$
2. $\frac{1}{x+1}$
3. $\frac{3x-5}{x}$
4. $\frac{3x+4}{x-1}$
5. $\frac{x^2}{x^2+4}$
6. $\frac{x^2}{x^2-4}$

Partial Fractions

In Chapter 1, you learned to combine two or more algebraic fractions into a single algebraic fraction. Often it is just as useful to reverse this process, that is to express an algebraic fraction as the sum of two or more simpler fractions. For example, $\frac{x+5}{(x-1)(x+2)}$ can be express in the form $\frac{2}{x-1} - \frac{1}{x+2}$. Splitting fractions into sum of two simpler fractions is called expressing the fraction in partial fractions.

In finding partial fractions of the rational expressions of the form $P(x)/Q(x)$, where $P(x)$ and $Q(x)$ are polynomials and $Q(x) \neq 0$, we assume that the degree of $P(x)$ is less than the degree of $Q(x)$. If the degree of $P(x)$ is greater than or equal to that of $Q(x)$, we divide to obtain a remainder $R(x)$ that has a degree less than that of $Q(x)$.

For example, if $Q(x)$ has a linear factor such as $ax + b$ the partial fraction of $P(x)/Q(x)$ contains a term of the form $\frac{A}{ax+b}$, where A is constant. If $Q(x)$ has a repeating factor such $(ax + b)^2$ then the partial fraction of $P(x)/Q(x)$ contains terms of the form $\frac{A}{ax+b} + \frac{B}{(ax+b)^2}$.

If $Q(x)$ has a non repeating quadratic factor such as $ax^2 + bx + c$ then the partial fraction of $P(x)/Q(x)$ contains a term of the form $\frac{Ax+B}{ax^2+bx+c}$, where A and B are constants.

Finding the Constants

Any one of the two methods illustrated below can be used to determine the constants.

Example 4

(a) Resolve $\frac{5x-8}{(x-1)(x-2)}$ in partial fractions.

Solution First write the fraction as

$$\frac{5x-8}{(x-1)(x-2)} = \frac{A}{x-1} + \frac{B}{x-2}$$

Adding the fractions on the right side of the equation and multiplying both sides by the common denominator gives

$$5x - 8 = A(x - 2) + B(x - 1)$$

Expanding the right hand side and grouping like terms we have

$$5x - 8 = (A + B)x - 2A - B$$

Equating the coefficients of x and the constant terms we obtain the following equations.

$$A + B = 5 \qquad (1)$$

$$2A + B = 8 \qquad (2)$$

Solving the simultaneous equations gives

$$A = 3 \text{ and } B = 2$$

Hence, $\frac{5x-8}{(x-1)(x-2)} = \frac{3}{x-1} + \frac{2}{x-2}$

An alternative method is shown on the next page.

$$\frac{5x-8}{(x-1)(x-2)} = \frac{A}{x-1} + \frac{B}{x-2}$$

Then $5x - 8 = A(x-2) + B(x-1)$

This equation is an identity and must hold for all values of x. Choose a value of x that will make one term on the right hand side zero. In this case if we substitute 1 for x we get

$$-3 = -A$$

So $A = 3$

Similarly, if we substitute 2 for x we get

$$2 = B$$

(b) Express $\frac{5x-3}{(x+1)(x^2+1)}$ in partial fractions.

Solution $\frac{5x-3}{(x+1)(x^2+1)} = \frac{A}{x+1} + \frac{Bx+C}{x^2+1}$

$$5x - 3 = A(x^2 + 1) + (Bx + C)(x + 1)$$

Substituting -1 for x we get $-4 = A$

We complete the solution by the method of equating coefficients. Equating the coefficients of x^2 gives

$$A + B = 0$$

Substituting -4 for A we get

$$-4 + B = 0$$

$$B = 4$$

Also equating the constant terms gives

$$A + C = -3$$

So $\quad -4 + C = -3$

$$C = 1$$

Hence, $\frac{5x-3}{(x+1)(x^2+1)} = -\frac{4}{x+1} + \frac{4x+1}{x^2+1}$

(c) Express $\dfrac{x^2+8x+9}{(x+1)(x+2)}$ in partial fractions

Solution Here the numerator has the same degree as the denominator. Begin by dividing the numerator by the denominator to obtain

$$\dfrac{x^2+8x+9}{(x+1)(x+2)} = 1 + \dfrac{5x+7}{(x+1)(x+2)}$$

We now express $\dfrac{5x+7}{(x+1)(x+2)}$ in partial fraction.

$$\dfrac{5x+7}{(x+1)(x+2)} = \dfrac{A}{x+1} + \dfrac{B}{x+2}$$

Then $5x + 7 = A(x + 2) + B(x + 1)$

If $x = -1$ we have $2 = A$

Also if $x = -2$ we have $3 = B$

Hence, $\dfrac{x^2+8x+9}{(x+1)(x+2)} = 1 + \dfrac{2}{x+1} + \dfrac{3}{x+2}$

Exercise 11.2

Resolve each of the following fractions in partial fractions:

1. $\dfrac{5x+7}{(x+1)(x+2)}$

2. $\dfrac{x+1}{(x-1)(x-2)}$

3. $\dfrac{x}{(x-4)(x-1)}$

4. $\dfrac{3}{(x+1)(x-1)}$

5. $\dfrac{x}{x^2-4}$

6. $\dfrac{3x+2}{x^2+x-2}$

7. $\dfrac{2x+3}{(x-2)(x-3)}$

8. $\dfrac{3-2x}{(x-2)(x+3)}$

9. $\dfrac{2}{(x-1)(x^2+1)}$

10. $\dfrac{2x+1}{(x^2+1)(x-2)}$

11. $\dfrac{x^2}{(x-1)(x+1)}$

12. $\dfrac{x^2-2}{(x+3)(x-1)}$

13. $\dfrac{x^3}{x^2-1}$

14. $\dfrac{x+3}{(x+1)^2}$

15. $\dfrac{x-1}{(x+2)^2}$

Review Exercise 11

1. For what value(s) of x is $\dfrac{x}{x^2-1}$ undefined?

2. Find the range of $\dfrac{2x-1}{3-x}$.

3. Find the values of x for which the expression $y = \dfrac{3x+2}{(x^2-4)(x+1)}$ is undefined.

4. Resolve each of the following expressions in partial fractions:

   (a) $\dfrac{x+7}{x^2-7x+10}$ (b) $\dfrac{5x+2}{3x^2+x-4}$

   (c) $\dfrac{5x-2}{x^2-3x-28}$ (d) $\dfrac{2x^2+6x-35}{x^2-x-12}$

5. Find A and B if $\dfrac{2x+1}{(x-1)^2} = \dfrac{A}{(x-1)^2} + \dfrac{B}{x-1}$.

6. Resolve each of the following in partial fractions:

   (a) $\dfrac{7x-21}{x^2+3x-10}$ (b) $\dfrac{3x+1}{x^2-4x+3}$

   (c) $\dfrac{x^2}{x^2-4}$ (d) $\dfrac{2x^2-4x-9}{x^2-x-2}$

   (e) $\dfrac{2x+5}{(1+x)^2}$ (f) $\dfrac{x^2}{(x-1)(x^2+1)}$

7(a) State the largest possible domain of the rational function

$$f: x \to \dfrac{x^2-x-6}{(x+1)(x^2-4)}$$

where x is a real number

(b) Find the zeros of f

(c) Find the values of the constants A, B and C such that

$$f(x) = \dfrac{A}{x-1} + \dfrac{Bx+C}{x^2-4}$$

8(a) State the largest possible domain of the rational function

$$f : x \to \frac{2x^2+x-1}{(x+2)(x^2-1)}$$

where x is a real number.

(b) Find the zeros of f in (a)

(c) Find the values of the constants A, B and C such that

$$f(x) = \frac{A}{x+2} + \frac{Bx+C}{x^2-1}$$

Test covering Chapter 1 – 11

These tests are provided to help you in the process of reviewing the previous chapters. If you missed any answer go back and review the appropriate chapter sections.

Take each test as you would take a test in class. Allow yourself 1 hour to take each test.

Test 1

1. Solve the equation:

(a) $7x - 6(x - 1) = 2(5 + x) + 11$ \hspace{2cm} (b) $(3x - 5)(x + 2) = 14$

2. Factorise each of the following completely:

(a) $9x^4 - 36y^6$ \hspace{2cm} (b) $4x^2 + 8xy - 5x - 10y$

3. Simplify each of the following rational expressions:

(a) $\dfrac{2x^2+13x+15}{6x^2+7x-3}$ \hspace{2cm} (b) $\dfrac{a^2-9}{a^2-a-12} - \dfrac{a^2-a-6}{a^2-2a-8}$

4. Simplify each of the following expressions:

(a) $\sqrt{18} + \sqrt{50} - 3\sqrt{32}$ \hspace{2cm} (b) $(3\sqrt{2} + 2)(2\sqrt{2} - 3)$

5. Solve the simultaneous equations

$2x + 3y = 6$

$5x + 3y = -24$

6. Find the identity element of the operation $x \oplus y = x + y - xy$ for all real numbers.

7. Find the inverse function g^{-1} of $g(x) = \dfrac{2x-1}{3-x}$.

8. If $f(x) = x^2 + 1$ and $g(x) = 2x$ find $f \circ g$.

9. State the largest domain of the function $f(x) = \dfrac{x}{x^2-1}$ defined on the set of real numbers.

10. Find the point of intersection of the lines $3x + 2y = 7$ and $2x - 3y = -4$.

11. Find the sum and product of the roots of the equation $3x^2 - 11x + 2 = 0$.

12. The roots of the equation $2x^2 + 7x - 4 = 0$ are α and β. Find the value of $\alpha^2 + \beta^2$.

13. Find the least value of $2x^2 + 8x + 9$.

14. Find the values of m for which the equation $x^2 + (m + 2)x + 3m - 2 = 0$ has equal roots.

15. Solve the following inequality $2x^2 + x - 3 \leq 0$.

16. Divide $2x^3 + 3x^2 - 32x + 15$ by $x - 3$.

17. If the remainder when $3x^2 + ax + b$ is divided by $x + 1$ and $x - 2$ are 4 and 16 respectively, find the values of a and b.

18. For what values of x is the expression $\frac{x^2 - 7x + 12}{x^2 - 9x + 18}$ undefined?

19. Resolve in partial fractions $\frac{2x - 3}{x^2 - x - 6}$.

20. Which of the following graphs represent one-to-one function?

(a)

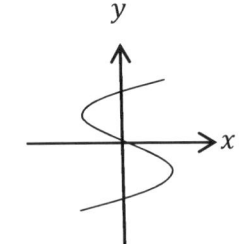

(b)

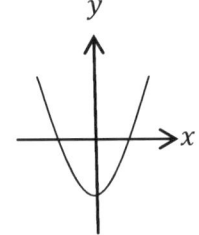

(c)

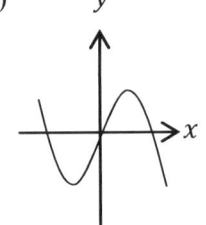

Test 2

1. Solve the equation $x + \sqrt{x + 1} = 5$.

2. Simplify:

(a) $\frac{a^2 - 4a}{a^2 - 6a + 8} \cdot \frac{a^2 - 4}{2a^2}$

(b) $\frac{x^2 - x - 6}{x^2 + 2x - 15} \div \frac{x - 2}{x + 5}$

3. Solve the simultaneous equations

$x + y = 3$

$x^2 + 3xy + y^2 = -1$

4. Simplify each of the following expressions:

(a) $\frac{5}{\sqrt{5} - \sqrt{2}}$

(b) $\frac{3 + \sqrt{2}}{3 - \sqrt{2}}$

Test covering Chapter 1 – 11   197

5. Find the inverse of an element x under the operation defined by $a \circ b = a - b + ab$.

6. Given that $f(4x - 5) = 5x^2 - 8x + 2$ what is the value of $f(3)$?

7. The functions f and g are defined on the set of real numbers. If $f \circ g(x) = x^2 + 2x + 1$ and $f(x) = x^2$ find $g(x)$.

8. Which of the following is the graph of a function f such that $f(x) = 0$ for exactly two values of x between -3 and 3?

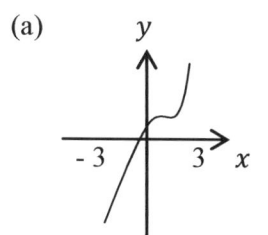

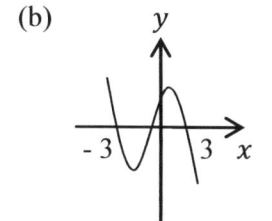

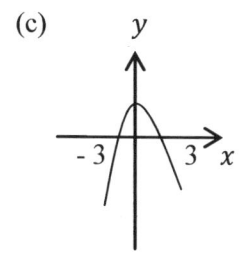

(a)

(b)

(c)

9. Find the equation of the line that has a y intercept of -6 and is parallel to the line $6x - 4y = 18$.

10. Find the equation of the perpendicular bisector of the line between the points P (1, 2) and Q (5, 6).

11. The roots of the equation $3x^2 + 5x - 4 = 0$ are α and β. Find the value of $\alpha\beta^2 + \alpha^2\beta$.

12. Find the greatest value of $15 - x - 2x^2$.

13. Find the range of the value of k for which the equation $x^2 + kx + (k + 3) = 0$ has real roots.

14. Solve the following inequality $15 - 2x - x^2 > 0$.

15. Select the false statement in the truth table below:

	p	q	~q	p ∨ ~q
A	T	T	F	T
B	T	F	T	T
C	F	T	F	T
D	F	F	T	T

16. When the expression $x^3 + x^2 + ax + b$ is divided by $x^2 - 1$ the remainder is $2x + 3$ find the values of a and b.

17. The expression $ax^4 + bx^3 + 3x^2 - 2x + 3$ has remainder $x + 1$ when divided by $x^2 - 3x + 2$. Find the values of a and b.

18. For what values of x is the expression $\dfrac{x^2+x-6}{x^2-x-12}$ defined?

19. Simplify $\dfrac{1-\frac{3}{x+3}}{\frac{1}{x^2-9}}$.

20. Resolve $\dfrac{x^2+3}{(x+1)(x-1)}$ in partial fractions.

12 Conic Sections

12.1 Circles

A circle is the set of all points in the plane equidistant from a fixed point called the centre of the circle. The distance between the centre of the circle and any point on the circle is called the radius of the circle. The perimeter of the circle is called the circumference.

Equation of a Circle

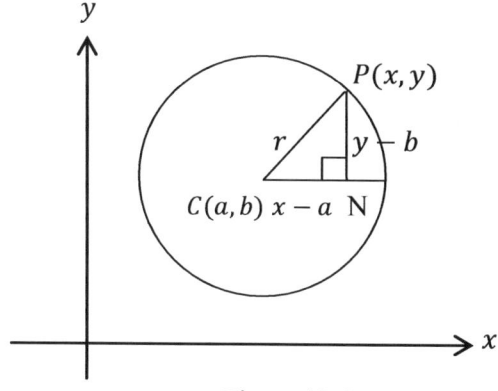

Figure 12.1

Consider a circle of radius r with centre at the point $C(a, b)$, as shown in Figure 12.1. Let $P(x, y)$ be any point on the circle. By the distance formula,

$$\sqrt{(x-a)^2 + (y-b)^2} = r$$

Squaring each side of this equation gives

$$(x-a)^2 + (y-b)^2 = r^2 \qquad (1)$$

This is called the equation of the circle.

If the centre is at the origin $a = b = 0$ and the equation (1) becomes

$$x^2 + y^2 = r^2$$

This is the equation of a circle with the centre at the origin.

Expanding the left hand side of equation (1), you obtain

$$x^2 + y^2 - 2ax - 2by + a^2 + b^2 - r^2 = 0$$

We write this equation in the form

$x^2 + y^2 + 2gx + 2fy + c = 0$ \hfill (2)

where $a = -g$, $b = -f$ and $c = a^2 + b^2 - r^2$

Notice that the centre of the circle is $(-g, -f)$ and the radius is $r = \sqrt{g^2 + f^2 - c}$

$x^2 + y^2 + 2gx + 2fy + c = 0$ is called the general equation of a circle.

Observe that the coefficients of x^2 and y^2 are equal and there is no term in xy.

Example 1

(a) Find the equation of a circle with centre (2, 3) and radius 2.

Solution $(x - a)^2 + (y - b)^2 = r^2$ \hfill Equation of circle.

$(x - 2)^2 + (y - 3)^2 = 2^2$ \hfill Substitute for a, b and r.

$x^2 - 4x + 4 + y^2 - 6y + 9 = 4$ \hfill Expand.

$x^2 + y^2 - 4x - 6y + 9 = 0$ \hfill Simplify.

Hence the equation of the circle is $x^2 + y^2 - 4x - 6y + 9 = 0$.

(b) Find the radius and coordinates of the centre of the circle $x^2 + y^2 + 6x - 4y - 3 = 0$

Solution $x^2 + y^2 + 6x - 4y - 3 = 0$ \hfill Original equation.

$x^2 + 6x + y^2 - 4y = 3$ \hfill Rearrange the equation.

$x^2 + 6x + 9 + y^2 - 4y + 4 = 3 + 9 + 4$ \hfill Complete squares.

$(x + 3)^2 + (y - 2)^2 = 16$ \hfill Simplify.

From this equation you can see that the circle has radius of 4 and centre at (- 3, 2).

If you compare the given equation with the general equation you can see that $2g = 6$, so $g = 3$, $2f = -4$, so $f = -2$, and $c = -3$. Since the centre of a circle is given by $(-g, -f)$, the centre of this circle is $(-3, 2)$. Substituting $g = 3$, $f = -2$ and $c = -3$ into $r = \sqrt{g^2 + f^2 - c}$ gives $r = \sqrt{(-3)^2 + (2)^2 - (-3)} = 4$.

(c) The point (1, 5) lies on a circle whose centre is at (4, 3). Find the equation of the circle.

Solution The radius of the circle is $\sqrt{(4 - 1)^2 + (3 - 5)^2} = \sqrt{13}$

Hence, the equation of the circle is $(x-4)^2 + (y-3)^2 = 13$

Simplifying we get $x^2 + y^2 - 8x - 6y + 12 = 0$

Exercise 12.1(a)

1. Which of the following equations represent circles?

(a) $x^2 - y^2 + 2x + 3y + 5 = 0$ \hspace{1em} (b) $x^2 + y^2 - 8x + 9y + 12 = 0$

(c) $3y^2 + 3y^2 + 7x - 8y - 20 = 0$ \hspace{1em} (d) $x^2 + y^2 + 3xy - 6 = 0$

2. Find the equation of the circle with:

(a) centre (3, 2), radius 4 \hspace{1em} (b) centre (-2, 3), radius 5

(c) centre (0, -5), radius 2 \hspace{1em} (d) centre (3, 0), radius 1

(e) centre (-3, 4), radius 5 \hspace{1em} (f) centre (5, 4), radius 3

3. Find the centre and radius of the following circles:

(a) $x^2 + y^2 - 6x + 2y - 6 = 0$ \hspace{1em} (b) $x^2 + y^2 - 6x + 8y - 11 = 0$

(c) $x^2 + y^2 - 2x - 6y - 6 = 0$ \hspace{1em} (d) $x^2 + y^2 - 4x + 2y + 1 = 0$

(e) $x^2 + y^2 + 2x = 15$ \hspace{1em} (f) $x^2 + y^2 + 8y - 9 = 0$

4. Find the coordinates of the centre and the radius of each of the following circles.

(a) $2x^2 + 2y^2 - 3x + 2y + 1 = 0$ \hspace{1em} (b) $3x^2 + 3y^2 + 6x - 3y - 2 = 0$

(c) $4x^2 + 4y^2 + 8x - 4y - 11 = 0$ \hspace{1em} (d) $9x^2 + 9y^2 - 12x - 18y - 23 = 0$

5. The point (3, 4) lies on a circle whose centre is at (1, 2). Find the equation of the circle.

6. A circle whose centre is at the point (-2, 3) passes through the mid-point of the line joining the points (-3, 1) and (5, 3). Find the equation of the circle.

7. Find the equation of the circle whose diameter is the line joining (-1, 2) to (7, 4).

8. Find the equation of the circle centre (1, -4) which passes through the point where the line $3x + 4y + 6 = 0$ cuts the x-axis.

9. A circle whose centre is in the second quadrant touches the y-axis at the point (0, 4) and also touches the x-axis at (-4, 0). Find the equation of the circle.

10. Find the equation of the circle whose centre lies on the line $x + y = 4$ and which passes through the points (-1, 0) and (3, 0).

11. Find the equation of the circle centre (-3, 2) which touches the y-axis.

12. Find the equation of the circle which passes through (9, 5), (8, -2) and (2, 6).

Equation of the tangent to a circle

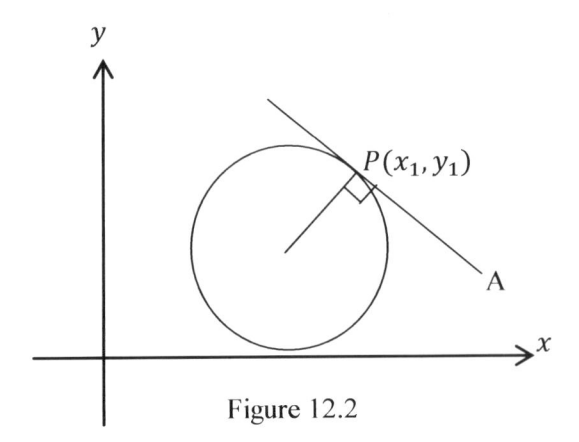

Figure 12.2

In Figure 12.2, AP is a tangent to the circle at the point $P(x_1, y_1)$. A line that joins the point of contact and the centre is perpendicular to the tangent. Recall that the product of the gradient of two perpendicular lines is -1. Therefore if the gradient of the line through the centre to the point of contact, called the normal, is m then the gradient of the tangent is $-\frac{1}{m}$. Hence the equation of the tangent at P is $y - y_1 = -\frac{1}{m}(x - x_1)$.

Example 2

Find the equation of the tangent to the circle $x^2 + y^2 + 6x + 4y + 4 = 0$ at the point (3, -4).

Solution Begin by finding the centre of the circle.

$$x^2 + y^2 + 6x + 4y + 4 = 0$$

$$(x + 3)^2 + (y + 2)^2 = 9$$

The centre of the circle is at (- 3, - 2).

The gradient of the normal is $\frac{-2+4}{-3-3} = -\frac{1}{3}$

Hence, the gradient of the tangent is 3

Then the equation of the tangent is $y + 4 = 3(x - 3)$

$y = 3x - 13$

The equation of the tangent is $y = 3x - 13$.

Exercise 12.1(b)

1. Find the equation of the tangent to the given circle at the indicated points.

(a) $x^2 + y^2 + 4x - 2y - 20 = 0$; (1, 5) (b) $x^2 + y^2 - 2x - 8y = 0$; (5, 3)

(c) $x^2 + y^2 - 8x + 3 = 0$; (7, -2) (d) $x^2 + y^2 + 4x + 2y - 12 = 0$; (-1, 3)

(e) $2x^2 + 2y^2 - 5x - 8y - 1 = 0$; (-1, 1)

2. Find the equation of the normal to the given circled at the indicated points.

(a) $x^2 + y^2 - 2x + 4y - 20 = 0$; (5, 1) (b) $x^2 + y^2 + 2x + 4y - 12 = 0$; (3, -1)

(c) $x^2 + y^2 - 8y + 3 = 0$; (-2, 7) (d) $3x^2 + 3y^2 - 6x - 2y - 5 = 0$; (2, -1)

3. Find the equation of the tangent of the circle $x^2 + y^2 - 4x - 2y - 5 = 0$ at the points where $x = 1$.

4. Find the equations of the tangents of the circle $x^2 + y^2 - x + 3y - 16 = 0$ at the points where $y = 2$.

5. The tangent to a circle centre (-3, -4) at a point where $x = 1$ is $4x + 3y - 1 = 0$. Find the equation of the circle.

6. The normal to a circle centre (4, 1) at the point where $x = 3$ is $4x + y = 17$. Find the equation of the circle.

7. Find the equations of the tangents to the circle $x^2 + y^2 - 2x + 4y - 15 = 0$ at the points where it meets the x-axis.

8. Find the equations of the tangents to the circle $x^2 + y^2 + 4x - 6y + 5 = 0$ at the points where it meets the y-axis.

9. If the line $y = x + 1$ is a tangent to the circle $x^2 + y^2 - 4x + 2y - 5 = 0$ find the coordinates of the point of contact.

10. Find the equations of the tangents of the circle $x^2 + y^2 = 4$ at the points (-2, 0) and (0, 2). If the tangents meet at A. Find the equation of the line joining A to the origin.

The length of a tangent from a point to a circle

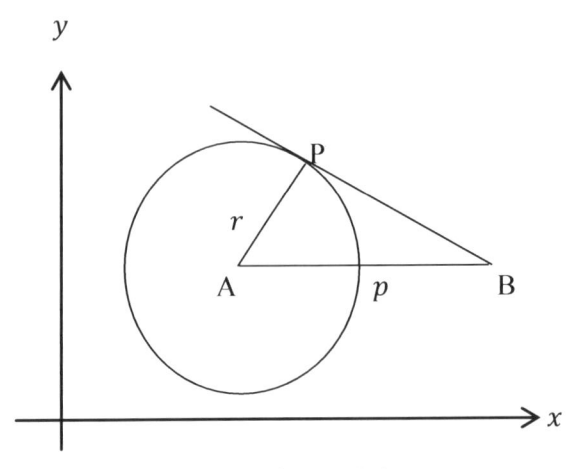

Figure 12.3

In Figure 12.3, PB is the tangent to the circle at P. Let r be the radius of the circle and $AB = p$. From $\triangle APB$ in Figure 12.3

$$AB^2 = AP^2 + PB^2$$

So $PB^2 = AB^2 - AP^2$

$$= p^2 - r^2$$

Example 3

Find the length of the tangent from the point A (7, 4) to the circle $x^2 + y^2 - 4x - 6y - 3 = 0$

Solution Begin by finding the centre and the radius of the circle.

$$x^2 + y^2 - 4x - 6y - 3 = 0$$

$$(x - 2)^2 + (y - 3)^2 = 16$$

The circle has centre (2, 3) and radius 4.

Let AT be the length of the tangent as shown in Figure 12.4.

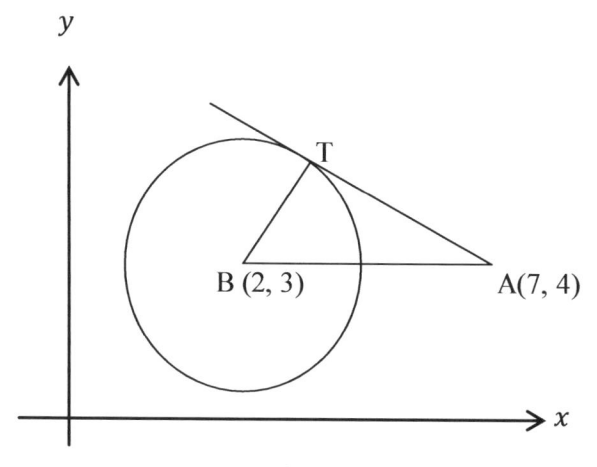

Figure 12.4

$AB^2 = (7-2)^2 + (4-3)^2$

$= 25 + 1$

$= 26$

Hence $AT^2 = 26 - 16$

$= 10$

$AT = \sqrt{10}$

Exercise 12.1(c)

1. Find the lengths of the tangents from the indicated points to the following circles.

(a) $x^2 + y^2 + 6x - 4y + 10 = 0$; (1, 1) (b) $x^2 + y^2 - 8x - 4y - 5 = 0$; (2, 8)

(c) $x^2 + y^2 + 8x - 10y + 5 = 0$; (4, 5) (d) $x^2 + y^2 + 10x + 6y - 2 = 0$; (3, -2)

2. The line $x + 2y + 6 = 0$ is a tangent to the circle $x^2 + y^2 - 6x + 4y + 8 = 0$. Find the length of the tangent to the circle from the point where it meets the y-axis.

3. The line $2x + y - 10 = 0$ is a tangent to the circle $x^2 + y^2 - 2x - 6y + 5 = 0$. Find the length of the tangent to the circle from the point where it meets the y-axis.

12.2 Loci

A locus is defined as the path of a moving point satisfying a set of given conditions. The plural of locus is loci.

Note that:

1. Every point on the locus satisfies the given condition.

2. Every point which satisfies the given condition lies on the locus.

The locus of points can be described by an equation as illustrated by the example below:

Example 4

Find the locus of a point which moves so that its distance from the point (3, -1) is 2 units.

Solution Let $P(x, y)$ be any point on the locus. Then by the distance formula

$$\sqrt{(x-3)^2 + (y+1)^2} = 2$$

$(x-3)^2 + (y+1)^2 = 4$ Square each side

$x^2 - 6x + 9 + y^2 + 2y + 1 = 4$ Expand.

$x^2 + y^2 - 6x + 2y + 6 = 0$ Simplify.

Therefore the locus of the point is $x^2 + y^2 - 6x + 2y + 6 = 0$.

Exercise 12.2

1. Find the locus of a point which moves so that its distance from the point (2, 1) is 3 units.

2. Find the locus of a point which is equidistant from the points (1, 2) and (-3, 1).

3. Find the locus of a point which is equidistant from the points (-1, 4) and (7, 3).

4. What is the locus of a point which moves so that its distance from the point (-3, 2) is equal to its distance from the point (1, 2).

5. Find the locus of a point which is equidistant from the origin and the line $x = -3$.

6. Find the locus of a point which is equidistant from the point (0, -1) and the line $y = 1$.

7. Find the locus of a point which moves so that its distance from the point A (-3, 0) is twice its distance from the origin.

8. A point P which moves so that its distance from A (1, 2) is twice its distance from B (5, -4). Find the locus of P.

9. Find the locus of a point which moves so that its distance from the point (0, 8) is twice its distance from the line $y = 2$.

10. Find the locus of a point which moves so that the sum of the squares of its distance from the points (-1, 1) and (3, 2) is 15 units.

11. A is at (3, -2) and B is at (5, 4). Find the locus of P which moves so that $PA^2 + PB^2 = 80$. Describe the locus of P.

12. A is (3, -2) and B is at (2, 5). P moves so that PA is perpendicular to PB. Describe the locus of P.

12.3 The Parabola

The graph of the equation $y = ax^2 + bx + c$ is a parabola. In this section we give the definition of a parabola and also the standard equation of a parabola.

Definition of a Parabola

A parabola is the set of all points that are equidistant from a fixed line called the directrix, and a fixed point called the focus.

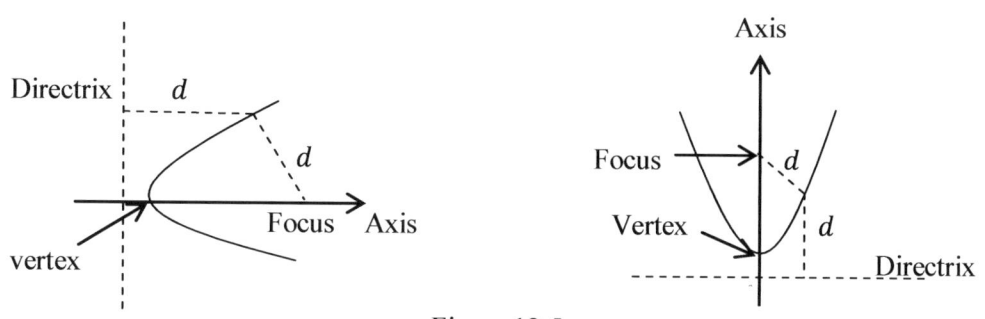

Figure 12.5

The line through the locus perpendicular to the directrix is called the axis of the parabola. A parabola is symmetric with respect to its axis. The midpoint between the focus and the directrix is called the vertex. Note that a parabola may open to the left or open downward.

Equation of a Parabola

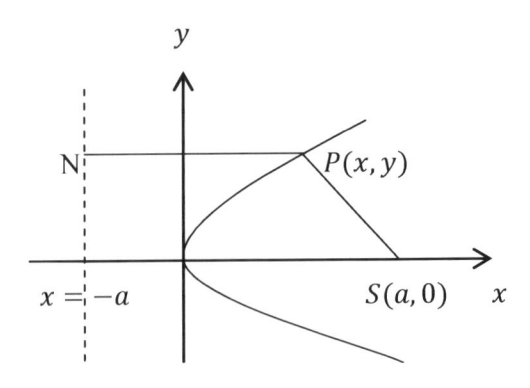

Figure 12.6

Figure 12.6 shows a parabola which opens to the right with its vertex at the origin and whose directrix is parallel to the y-axis. Let the focus of the parabola be $(a, 0)$ and the equation of the directrix be $x = -a$. By the definition of the parabola the distance PS and PN are equal. Thus

$$PS^2 = PN^2$$

So $(x - a)^2 + y^2 = (x + a)^2$

Simplifying this we have

$$y^2 = 4ax$$

The standard equation of a parabola with the vertex at the origin and focus at $(a, 0)$ is $y^2 = 4ax$. Similarly we can show that the equation of a parabola that opens up with vertex at the origin and focus at $(0, a)$ is $x^2 = 4ay$.

Using the definition of a parabola we can show that the standard equation of a parabola with a vertex at (p, q), focus $(p + a, q)$ and directrix $x = p - a$ is $(y - q)^2 = 4a(x - p)$. Also the equation of a parabola with vertex at (p, q), focus $(p, q + a)$ and directrix $y = q - a$ is $(x - p)^2 = 4a(y - q)$.

Note that if the focus of a parabola is above or to the right of the vertex, a is positive. If the focus is below or to the left of the vertex, a is negative.

Example 5

(a) Identify the vertex and focus of the parabola $y^2 = 20x$.

Solution Rewrite $y^2 = 20x$ as $y^2 = 4(5)x$

Comparing this with the standard equation you can see that $a = 5$. Hence the focus is $(5, 0)$ and the vertex is the origin.

(b) Identify the vertex and focus of the parabola $x^2 = -8y$.

Solution Rewrite $x^2 = -8y$ as $x^2 = 4(-2)y$.

You can see that $a = -2$. The focus is $(0, -2)$ and the vertex is at the origin.

(c) Identify the vertex, focus and the directrix of the parabola $y^2 - 6y - 4x + 17 = 0$.

Solution First complete the square

$$y^2 - 6y - 4x + 17 = 0$$

$$y^2 - 6y = 4x - 17$$

$$(y - 3)^2 = 4x - 8$$

$$(y - 3)^2 = 4(x - 2)$$

Note that $4a = 4$ so $a = 1$. The vertex is $(2, 3)$, the focus is $(3, 3)$ and the directrix is $x = 1$.

(d) Find the equation of the parabola with its vertex at the origin and focus $(2, 0)$.

Solution The sketch of the parabola is shown in Figure 12.7.

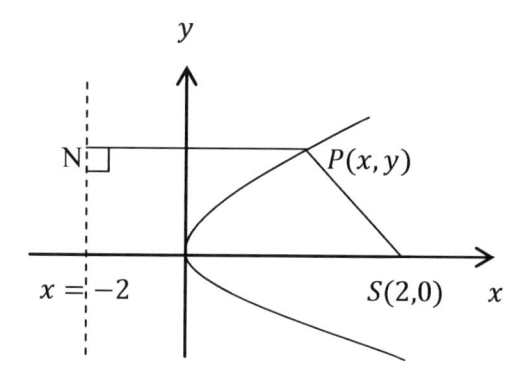

Figure 12.7

Using the definition of a parabola we get

$$(x - 2)^2 + y^2 = (x + 2)^2$$

$$x^2 - 4x + 4 + y^2 = x^2 + 4x + 4$$

$$y^2 = 8x$$

So the equation of the parabola is $y^2 = 8x$.

The same result is obtained if you use the standard equation $y^2 = 4ax$ and then substitute for a. Since the focus is (2, 0) we have $a = 2$.

So $y^2 = 4(2)x$

$y^2 = 8x$

(e) Find the equation of the parabola with its vertex at the origin and focus (0, 3).

Solution The sketch of the parabola is shown in Figure 12.8.

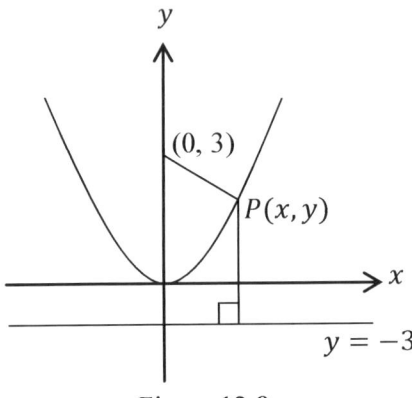

Figure 12.8

Using the definition of a parabola we get

$x^2 + (y - 3)^2 = (y + 3)^2$

$x^2 + y^2 - 6y + 9 = y^2 + 6y + 9$

$x^2 = 12y$

So the equation of the parabola is $x^2 = 12y$

An alternative method is to use the standard equation $x^2 = 4ay$ and then substitute for a. Since the focus is (0, 3) we have $a = 3$.

So $x^2 = 4(3)y$

$x^2 = 12y$

Tangent and Normal

Recall that a tangent to a curve at a point on a curve is a straight line that touches the curve at only one point, and the normal is the straight line through the point of contact of the tangent perpendicular to the tangent.

Let $y = mx + c$ be the tangent to the parabola $y^2 = 4ax$. Substituting $mx + c$ for y we get

$(mx + c)^2 = 4ax$

$m^2x^2 + 2(cm - 2a)x + c^2 = 0$

This equation gives the x- coordinates of the points of intersection of the line $y = mx + c$ and the parabola $y^2 = 4ax$. The equation has equal roots since the line $y = mx + c$ is a tangent to the parabola. Hence

$4(cm - 2a)^2 = 4m^2c^2$ which reduces to

$mc = a$ or $c = a/m$.

Thus $y = mx + c$ is a tangent to the parabola $y^2 = 4ax$ when $c = a/m$, i.e. the line $y = mx + a/m$ is a tangent to the parabola for all values of m. It can be shown that the normal to the parabola $y^2 = 4ax$ at the point (x_1, y_1) has equation $my = -x + c$ where $c = my_1 + x_1$.

Example 6

Find the equation of the tangent and normal to the parabola $y^2 = 18x$ at the point $(2, -6)$.

Solution The equation of the tangent to the parabola $y^2 = 4ax$ is $y = mx + a/m$ for all values of m.

Here $4a = 18$

So $a = \frac{9}{2}$

Therefore the line $y = mx + \frac{9}{2m}$ is a tangent to the parabola $y^2 = 18x$. The tangent touches the parabola at the point $(2, -6)$.

Hence $-6 = 2m + \frac{9}{2m}$

$4m^2 + 12m + 9 = 0$

Factorising this we get

$(2m + 3)^2 = 0$

So $m = -\frac{3}{2}$

Therefore $y = -\frac{3}{2}x - 3$

i.e. $3x + 2y + 6 = 0$

Hence the equation of the tangent at the point (2, - 6) is $3x + 2y + 6 = 0$.

The equation of the normal is

$y + 6 = \frac{2}{3}(x - 2)$

$3y + 18 = 2x - 4$

$2x - 3y - 22 = 0$

An alternative solution is given below.

Let m be the gradient of the tangent to the parabola $y^2 = 18x$ so $y + 6 = m(x - 2)$

Solving $y^2 = 18x$ for x yields

$x = \frac{1}{18}y^2$

Now substituting $x = \frac{1}{18}y^2$ we have

$y + 6 = m\left(\frac{1}{18}y^2 - 2\right)$

$my^2 - 18y - (36m + 108) = 0$

The equation has equal roots since the line is a tangent to the parabola.

So, $(-18)^2 = -4m(36m + 108)$

$4m^2 + 12m + 9 = 0$

$(2m + 3)^2 = 0$

So $m = -\frac{3}{2}$

Thus the tangent has equation

$y + 6 = -\frac{3}{2}(x - 2)$

$3x + 2y + 6 = 0$

Using differentiation to find tangents and normals

Differentiating the equation $y^2 = 4ax$ we have

$$2y\frac{dy}{dx} = 4a$$

So $\frac{dy}{dx} = \frac{2a}{y}$

Hence the tangent at any point on the parabola has gradient $2a/y$. Substituting the y-coordinate of a point gives the gradient of the tangent at that point. For example, the gradient of the tangent at the point (x_0, y_0) is $2a/y_0$.

Example 7

Find the equations of the tangent and the normal to the parabola $y^2 = 6x$ at the point $\left(\frac{3}{2}, 3\right)$.

Solution $y^2 = 6x$

$$2y\frac{dy}{dx} = 6$$

$$\frac{dy}{dx} = \frac{3}{y}$$

At the point $\left(\frac{3}{2}, 3\right)$ we have

$$\frac{dy}{dx} = \frac{3}{3} = 1$$

Thus the equation of the tangent at the point $\left(\frac{3}{2}, 3\right)$ is

$$y - 3 = x - \frac{3}{2}$$

$$2x - 2y + 3 = 0$$

The equation of the normal is

$$y - 3 = -\left(x - \frac{3}{2}\right)$$

$$2x + 2y - 9 = 0$$

Exercise 12.3

1. Find the focus of each of the following parabolas:

   (a) $y^2 = 4x$ (b) $y^2 = -6x$

   (c) $y^2 = 10x$ (d) $y^2 = -5x$

2. Find the focus of each of the following parabolas:

   (a) $x^2 = -4y$ (b) $x^2 = 6y$

   (c) $x^2 = -12y$ (d) $x^2 = 5y$

3. Write down the equation of the directrix of each of the following parabolas:

   (a) $y^2 = 8x$ (b) $y^2 = -4x$

   (c) $y^2 = 12x$ (d) $y^2 = -3x$

4. Write down the equation of the directrix of each of the following parabolas:

   (a) $x^2 = -8y$ (b) $x^2 = 4y$

   (c) $x^2 = -16y$ (d) $x^2 = 3y$

5. Write down the equation of the parabola with its vertex at the origin and the following focus:

   (a) (3, 0)   (b) (-2, 0)

   (c) (7, 0)   (d) (- 4, 0)

6. Write down the equation of the parabola with its vertex at the origin and the following focus:

   (a) (0, 1)   (b) (0, 2)

   (c) (0, -3)   (d) (0, -2)

7. Identify the vertex, focus and the directrix of the following parabolas:

   (a) $(y + 2)^2 + 8(x - 1) = 0$

   (b) $(y - 2) + (x + 3)^2 = 0$

   (c) $4y = x^2 - 2x + 5$

   (d) $x = \frac{1}{4}(y^2 + 2y + 33)$

(e) $x^2 + 6x + 8y + 25 = 0$

(f) $y^2 - 6y - 6x = 0$

8. Use the definition of a parabola to find the equation of a parabola with:

(a) directrix $y = 0$ and focus (-2, -2)

(b) directrix $y = 1$ and focus (3, 5)

(c) directrix $x = -1$ and focus (-3, 4)

(d) directrix $x = 2$ and focus (6, 3)

9. Write the standard form of the equation of the parabolas with the given vertex and focus:

(a) Vertex (4, 2); Focus (2, 2)

(b) Vertex (-1, 3); Focus (-1, 1)

(c) Vertex (0, 3); Focus (0, 5)

(d) Vertex (3, -1); Focus (4, -1)

10. Find the equations of the tangent and the normal to each of the following parabolas at the indicated points:

(a) $y^2 = -3x; (-3,3)$ (b) $y^2 = 8x; (2,-4)$

(c) $y^2 = 16x; (1,-4)$ (d) $x^2 = 8y; (4,2)$

(e) $x^2 = -9y; (6,-4)$ (f) $x^2 = 27y; (-9,3)$

11. Find the equation of the equation of the tangents from the given points to the parabolas:

(a) $\left(-2, \frac{3}{2}\right); y^2 = 2x$ (b) $(1, 1); y^2 = -3x$

(c) $(1,3); y^2 = 8x$ (d) $(2,1); y^2 = -12x$

Review Exercise 12

1. Find the equation of the circle with:

(a) centre (4, 3); radius 10   (b) centre (-2, 5); radius 6

(c) centre (5, -3); radius 9   (d) centre (-5, -2); radius $\frac{5}{2}$

2. Find the equation of the circle with:

(a) centre (-2, 1) which passes through the point (0, 1)

(b) centre (3, 2) which passes through the point (4, 6)

(c) centre (5, 3) which passes through the point (8, 7)

(d) centre (-3, -5) which passes through the point (-1, -3)

3. Find the centre and radius of each of the following circles:

(a) $x^2 + y^2 - 4x - 2y + 1 = 0$

(b) $x^2 + y^2 + 6x - 4y - 3 = 0$

(c) $x^2 + y^2 - 2x + 6y - 15 = 0$

(d) $x^2 + y^2 - 14x + 8y + 56 = 0$

4. Find the radius and the coordinates of the centre of the circle
$9x^2 + 9y^2 + 12x - 6y - 4 = 0$

5. Find the equation of the circle with centre (4, 5) which passes through the y-intercept of the line $3x - 2y + 6 = 0$.

6. The equation of a circle is of the form $x^2 + y^2 - 4x + 6y + c = 0$, where c is a constant. If the circle has radius 3 find the value of c.

7. The line $3y + 4x - 2 = 0$ is a tangent to a circle with centre (3, 5). Find:

(a) the coordinates of the point of contact

(b) the equation of the circle

8. Show that the line $2x + y = 7$ is a tangent to the circle $x^2 + y^2 + 4x - 2y - 15 = 0$ and find the coordinates of the point of contact.

9. The tangent to the circle $x^2 + y^2 - 6x - 2y + 15 = 0$ at the point (4, 3) meets the x-axis at P. Find the distance of P from the centre of the circle.

10. Find the equation of the tangent to the circle $x^2 + y^2 + x - 2y - 5 = 0$ at the point (-2, 3). If this tangent cuts the axes at P and Q find the area of the triangle OPQ.

11. Find the equation of the circle whose centre is at the point (4, 1) and which touches the line joining the points (-1, 4) and (7, 6).

12. The line $2x - y = 3$ meets the circle $x^2 + y^2 + 6x - 2y - 15 = 0$ at the points P and Q, where Q lies in the quadrant. Find:

(a) the coordinates of P and Q.

(b) the equation of the tangent at P

(c) the length of the tangent from T (5, 1).

13. The equation of the tangent to a circle at A (2, 6) is $4y + 3x = 30$. The line $y = 3x + 5$ passes through the centre C of the circle. Find:

(a) the coordinates of C

(b) the equation of the circle

(c) the points where the circle cuts the y-axis

14. P moves so that its distance from the origin is twice its distance from (2, -3). Prove that P moves in a circle and find its centre and radius.

15. A point P moves so that its distance from A (-1, 2) is three times its distance from B (0, -2).

16. A is at (-1, 3) and B is at (2, 1). P moves so that PA is perpendicular to PB. Prove that P moves in a circle and find its centre and radius.

17. A is the point (4, 3) and B is the point (-2, -3). Find the locus of a point P which moves so that $PA^2 + PB^2 = 50$.

18. Write the standard form of the equation of the parabola with its vertex at the origin:

(a) Focus $\left(0, -\frac{3}{4}\right)$

(b) Focus (3, 0)

(c) Focus (0, 1)

(d) Focus (-4, 0)

19. Identify the focus and the directrix of each of the following parabolas:

(a) $y = \frac{1}{2}x^2$

(b) $y^2 = 3x$

(c) $x^2 + 10y = 0$

(d) $x - y^2 = 0$

20. Find the equations of the tangent and normal to each of the following parabolas:

(a) $y^2 = -8x; (-2, 4)$

(b) $x^2 = -3y; \left(2, -1\frac{1}{3}\right)$

(c) $y^2 = 10x; \left(2\frac{1}{2}, 5\right)$

21. Show that the line $y - 3x - 1 = 0$ touches the parabola $y^2 = 12x$ and find the coordinates of the point of contact.

22. Find the equation of the normal to the parabola $y = \frac{1}{6}x^2$ at the point (6, 6). Find the coordinates of the point at which this normal meets the parabola again.

23. Find the equation of the tangents to the parabola $y^2 = 12x$ at (3, -6) and (3, 6). At what point will the two tangents meet?

24. Find the equations of the tangents from the point (-1, -2) to the parabola $y^2 = 5x$.

13 Binomial Expansions

13.1 Pascal's Triangle

In Chapter 1, you learned to expand the square of a binomial by multiplication. To expand $(a+b)^3$ by multiplication and using the fact that $(a+b)^2 = a^2 + 2ab + b^2$, work as follows:

$$(a+b)^3 = (a+b)^2(a+b)$$

$$= (a^2 + 2ab + b^2)(a+b)$$

$$= a^3 + 3a^2b + 3ab^2 + b^3$$

Similarly by working out $(a^3 + 3a^2b + 3ab^2 + b^3)(a+b)$ we get

$$(a+b)^4 = a^4 + 4a^3b + 6b^2b^2 + 4ab^3 + b^4$$

Expanding higher powers of $(a+b)$ by multiplication could be tedious and time consuming. As you will see soon higher powers of $(a+b)$ can be expanded without difficulty by using the Binomial Theorem.

Consider the following expansions

$$(a+b)^0 = 1$$

$$(a+b)^1 = a+b$$

$$(a+b)^2 = a^2 + 2ab + b^2$$

$$(a+b)^3 = a^3 + 3a^2b + 3ab^2 + b^3$$

$$(a+b)^4 = a^4 + 4a^3b + 6a^2b^2 + 4ab^3 + b^4$$

A triangular display of the numerical coefficient is shown below. This is called the Pascal's triangle.

```
                    1                        n = 0

                 1     1                     n = 1

              1     2     1                  n = 2

           1     3     3     1               n = 3

        1     4     6     4     1            n = 4
```

The first and last numbers in each row of Pascal's Triangle are 1. Also, every other number in each row is the sum of two numbers immediately above the number.

The succeeding row of the Pascal's Triangle is obtained as shown below.

1 $(1+4) = 5$ $(4+6) = 10$ $(6+4) = 10$ $(4+1) = 5$ 1

The preceding expansions suggest the following properties.

1. The expansion of $(a+b)^n$, where n is any positive integer, is a finite series with $(n+1)$ terms.

2. The sum of the powers of each term is n

3. The power of a decrease by 1 in successive terms, whereas the powers of b increase by 1.

4. The term in b^r is the $(r+1)$ th term

5. The first term is a^n and the last term is b^n.

Example 1 Expand

(a) $(2x - y)^4$

Solution Obtain the coefficients 1, 4, 6, 4, 1 from the fifth row of the Pascal's Triangle.

The expansion can be worked out as follows

$$(2x-y)^4 = (2x)^4 + 4(2x)^3(-y) + 6(2x)^2(-y)^2 + 4(2x)(-y)^3 + (-y)^4$$

$$= 16x^4 - 32x^3y + 24x^2y^2 - 8xy^3 + y^4$$

(b) $(3x + 2y)^3$

Solution The coefficients from the Pascal's Triangle are: 1, 3, 3, 1

$$(3x + 2y)^3 = (3x)^3 + 3(3x)^2(2y) + 3(3x)(2y)^2 + (2y)^3$$
$$= 27x^3 + 54x^2y + 36xy^2 + 8y^3$$

(c) Expand $(3 + x)^4$ and use the first three terms to find $(2.98)^4$

Solution $(3 + x)^4 = (3)^4 + 4(3)^3(x) + 6(3)^2(x)^2 + 4(3)(x)^3 + x^4$
$$= 81 + 108x + 54x^2 + 12x^3 + x^4$$

If x is made small successive terms may be so small that they do not affect the answer to the required degree of accuracy.

We write $(2.98)^4$ as $(3 - 0.02)^4 = [3 + (-0.02)]^4$

Replacing x with -0.02 in the expansion gives

$$(2.98)^4 = 81 + 108(-0.02) + 54(-0.02)^2$$
$$= 81 - 2.16 + 0.0216$$
$$= 78.8616$$

Note: When finding an approximation, take the first few terms of the expansion that will give you at least one place more than the required degree of accuracy.

Exercise 13.1

1. Expand:

(a) $(1 + 2x)^6$ (b) $(x - 2y)^4$ (c) $(1 - x)^5$

(d) $(2x + 3y)^4$ (e) $(4x - 1)^3$ (f) $(3x + y)^4$

(g) $(1 - 3x)^4$ (h) $(2a + x)^6$ (i) $(2a - 3x)^4$

2. Expand $(1 - x)^5$ and use the expansion to evaluate $(0.98)^5$ correct to the three significant figures.

3. Using the expansion of $(1 + x)^4$ find the value of $(1.01)^4$ correct to five significant figures.

4. Expand $(1 - 2x)^5$ and then use the expansion to evaluate $(0.98)^5$ correct to three significant figures.

13.2 The Binomial Theorem

The binomial theorem can be used to expand binomial expression of the form $(a + b)^n$, where n is a positive integer. The binomial theorem states that

$$(a+b)^n = a^n + {}^nC_1 a^{n-1}b + {}^nC_2 a^{n-2}b^2 + \cdots + {}^nC_r a^{n-r}b^r + \cdots + b^n$$

The coefficient of the expansion is given by the formula

$${}^nC_r = \frac{n!}{(n-r)!r!}$$

${}^nC_r$ is called the binomial coefficient.

The binomial coefficient is sometimes written as $\binom{n}{r}$.

The symbol $n!$, read as n factorial, is defined as

$$n! = n(n-1)(n-2)\cdots 3.2.1$$

For example

$3! = 3.2.1 = 6$

$6! = 6.5.4.3.2.1 = 720$

Note that $0! = 1$ and $1! = 1$

Example 2 Find the following binomial coefficients

(a) ${}^8C_3$

Solution ${}^8C_3 = \frac{8!}{5!3!} = \frac{8 \cdot 7 \cdot 6 \cdot 5!}{5! \cdot 3 \cdot 2 \cdot 1} = 8 \cdot 7 = 56$

(b) ${}^{10}C_4$

Solution ${}^{10}C_4 = \frac{10!!}{6!4!} = \frac{10 \cdot 9 \cdot 8 \cdots 7 \cdot 6!}{6! \cdot 4 \cdot 3 \cdot 2 \cdots 1} = 10 \cdot 3 \cdot 7 = 210$

Example 3

(a) Expand $(x - 2y)^4$

Solution $(x-2y)^4 = x^4 + {}^4C_1(x)^3(-2y) + {}^4C_2(x)^2(-2y)^2 + {}^4C_3(x)(-2y)^3 + (-2y)^4$

$$= x^4 - 8x^3y + 24x^2y^2 - 32xy^3 + 16y^4$$

(b) Write down the first five terms of the expansion $(1 + x)^7$

Solution $(1+x)^7 = 1 + {}^7C_1 x + {}^7C_2 x^2 + {}^7C_3 x^3 + {}^7C_4 x^4$

$= 1 + 7x + 21x^2 + 35x^3 + 35x^4$

(c) Write down the term in x^6 in the expansion of $(x - 2y)^9$

Solution The complete expansion is not required since we need only the term in x^6. The term in x^6 is the fourth term. We would use the fact that $(r + 1)th$ term is given by ${}^nC_r a^{n-r} b^r$.

Here $r + 1 = 4$, so $r = 3$. Also $a = x$, $b = -2y$ and $n = 9$

The term in x^6 is ${}^9C_3 (x)^6 (-2y)^3 = -672 x^6 y^3$

(d) Find the coefficient of the fifth term of the expansion of $(2x + 3y)^7$

Solution In this case $r + 1 = 5$, so $r = 4$.

The coefficient of the fifth term is ${}^7C_4 (2)^3 (3)^4 = 22680$

Exercise 13.2

1. Evaluate the following binomial coefficients:

(a) ${}^9C_6$　　(b) ${}^7C_0$　　(c) ${}^{10}C_1$　　(d) ${}^6C_4$

(e) ${}^7C_3$　　(f) ${}^{10}C_5$　　(g) ${}^{12}C_9$　　(h) ${}^{12}C_7$

2. Write down the term indicated in the binomial expansions of the following:

(a) $(1 + 4x)^6$, $3rd$ term　　(b) $(3x - 2)^7$, $4th$ term

(c) $(1 - \frac{x}{2})^{10}$, $6th$ term　　(d) $(2 + x)^9$, $5th$ term

(e) $(2x + 3y)^{12}$, $9th$ term　　(f) $(3 - 2x)^6$, $4th$ term

(g) $(2x - y)^{12}$, term containing x^4　　(h) $\left(2x - \frac{1}{x}\right)^9$, term containing x^3

(i) $(2x - 3y)^{10}$, term containing y^8　　(j) $\left(3x^2 - \frac{1}{2x}\right)^7$, term containing x^2

3. Write down the first four terms in the binomial expansion of:

(a) $(1-x)^{10}$ (b) $(1+2x)^9$ (c) $(2-x)^{10}$

(d) $\left(1+\frac{1}{2}x\right)^6$ (e) $\left(3-\frac{2}{3}x\right)^7$ (f) $\left(2+\frac{1}{2}x\right)^7$

4. Use the binomial theorem to find the values of:

(a) $(1.01)^{12}$, correct to three decimal places

(b) $(0.98)^{10}$, correct to three significant figures

(c) $(2.01)^9$, correct to six significant figures

(d) $(1.99)^{12}$, correct to three decimal places

5. Find the term independent of x in the expansion of $\left(x^2 - \frac{2}{x}\right)^6$

6. Find the term in x in the expansion of $\left(x - \frac{2}{x}\right)^5$

7. Find the middle term in the expansion of $\left(x - \frac{1}{y}\right)^8$

8. Write down the first four terms of the binomial expansion of $(1-2x)^{10}$ and use it to find the value of $(0.98)^{10}$, correct to four decimal places

9. Write down the first terms of the binomial expansion of $\left(1+\frac{x}{5}\right)^6$ and use it to find the value of $(1.01)^6$ correct to four significant figures.

10. Find each of the following, up to the term indicated:

(a) $(1+x-x^2)^7$, x^2 term   (b) $(1-2x+3x^2)^6$, x^2 term

(c) $(1-x)^7(1+x)$, x^3 term   (d) $(1+2x)^5(1-x)$, x^3 term

13.3 Binomial Theorem for any power

If n is not a positive integer the binomial theorem is given by

$$(1+x)^n = 1 + nx + \frac{n(n-1)}{2!}x^2 + \frac{n(n-1)(n-2)}{3!}x^3 + \cdots$$

In general this series continue indefinitely. Note that the expansion will have $(n+1)$ terms if n is a positive integer.

Binomial Expansions

We can adapt this expansion to expand the binomial of the form $(a + x)^n$.

Example 4

(a) Expand $(1 + 2x)^{-1}$ as far as the term in x^3

Solution $(1 + 2x)^{-1} = 1 + (-1)(2x) + \frac{(-1)(-1-1)}{2!}(2x)^2 + \frac{(-1)(-1-1)(-1-2)}{3!}(2x)^3$

$= 1 - 2x + 4x^2 - 8x^3$

(b) Expand $(1 - 3x)^{\frac{1}{3}}$ up to the term in x^3. Use your result to find $\sqrt[3]{0.97}$ to two decimal places.

Solution $(1-3x)^{\frac{1}{3}} = 1 + \frac{1}{3}(-3x) + \frac{\frac{1}{3}(\frac{1}{3}-1)}{2!}(-3x)^2 + \frac{\frac{1}{3}(\frac{1}{3}-1)(\frac{1}{3}-2)}{3!}(-3x)^3$

$= 1 - x - x^2 - \frac{5}{3}x^3$

Now $\sqrt[3]{0.97} = (1 - 0.03)^{\frac{1}{3}} = [1 - 3(0.01)]^{\frac{1}{3}}$

Substitute $x = 0.01$ in the expansion.

$\sqrt[3]{0.97} = 1 - 0.01 - (0.01)^2 - \frac{5}{3}(0.01)^3$

$= 1 - 0.01 - 0.0001 - 0.00000167$

$= 0.98989833$

$= 0.99$

(c) Expand $(2 + x)^5$

Solution Rewrite $(2 + x)^5$ in the form $\left[2\left(1 + \frac{x}{2}\right)\right]^5 = 32\left(1 + \frac{x}{2}\right)^5$

Using the $(1 + x)^n$ expansion with $n = 5$ and replacing x by $\frac{x}{2}$ we get

$(2+x)^5 = 32\left[1 + 5\left(\frac{x}{2}\right) + \frac{5 \cdot 4}{2!}\left(\frac{x}{2}\right)^2 + \frac{5 \cdot 4 \cdot 3}{3!}\left(\frac{x}{2}\right)^3 + \frac{5 \cdot 4 \cdot 3 \cdot 2}{4!}\left(\frac{x}{2}\right)^4 + \left(\frac{x}{2}\right)^5\right]$

$= 32 + 80x + 80x^2 + 40x^3 + 10x^4 + x^5$

Exercise 13.3

1. Expand each of the following expressions in ascending powers of x as far as the term in x^3:

(a) $(1-x)^{-2}$

(b) $(1+3x)^{-\frac{1}{3}}$

(c) $(1-x)^{\frac{1}{3}}$

(d) $(1+2x)^{\frac{1}{2}}$

(e) $\left(1-\frac{x}{2}\right)^{-2}$

(f) $(1+x)^{-\frac{2}{3}}$

(g) $(1+3x)^{-1}$

(h) $(1+x)^{-\frac{1}{2}}$

(i) $\left(1+\frac{x}{2}\right)^{-2}$

(j) $(1+2x)^{-\frac{1}{2}}$

(k) $(2-x)^{\frac{1}{2}}$

(l) $(3+x)^{-1}$

2. Use the binomial theorem to evaluate:

(a) $\sqrt{1.01}$ correct to five decimal places

(b) $(1.03)^{-2}$ correct to four decimal places

(c) $\sqrt{0.998}$ correct to three decimal places

(d) $\sqrt[3]{0.97}$ correct to four decimal places

3. Find the first four terms of the expansion of $(1+x)^{\frac{1}{3}}$ in ascending powers of x. Use your expansion to find $\sqrt[3]{64.8}$ correct to the four significant figures.

4. Find the expansion for $\sqrt{1+x}$ up to the term in x^3. Hence find an approximation for $\sqrt{17}$.

Review Exercise 13

1. Expand the following:

(a) $(x+2)^3$

(b) $(x+3)^5$

(c) $(x-y)^4$

(d) $(2x-1)^5$

(e) $(2x+y)^6$

(f) $(1-2x)^6$

(g) $(x+4)^3$

(h) $(2x-y)^5$

2. Find in ascending powers of the first 4 terms in the expansion of the following:

(a) $(2 + x)^9$

(b) $(3 - x)^7$

(c) $(1 + 2x)^{12}$

(d) $(1 - 3x)^8$

3. Find the specified terms in the expansion of the following:

(a) $(x - y)^{10}$, 4 th term

(b) $(3x + y)^{12}$, 10 th term

(c) $(3x + y)^7$ 3 rd term

(d) $(a - 4b)^9$, 6 th term

4. Find the coefficient of the terms indicated in the expansion of:

(a) $(x - 1)^{10}$, term containing x^7

(b) $(x + y)^{15}$, term containing y^{11}

(c) $(2x - y)^{12}$, term containing y^9

(d) $(x + y)^{10}$, term containing y^3

(e) $(x^2 + 3)^4$, term containing x^4

(f) $(x - 3)^{12}$, term containing x^9

5. Find the term independent of x in the expansion of the following:

(a) $\left(x^2 - \frac{2}{x}\right)^6$

(b) $\left(x^3 + \frac{1}{2x}\right)^8$

(c) $\left(2x - \frac{1}{x^4}\right)^{10}$

6. Write down the binomial expansion of $(x + y)^4$. Use your expansion to evaluate, correct to three decimal places $(1.99)^4$.

7(a) Using the Binomial Theorem write down the expansion of $(1 + x)^7$.

(b) Use your result in (a) to evaluate $(0.997)^7$ correct to five decimal places.

8(a) Write down and simplify the first five terms of the binomial expansion of $\left(1 - \frac{1}{3}x\right)^8$.

(b) Use your expansion to evaluate $(0.997)^8$, correct to five decimal places.

9. Expand $(1 - 2x)^6$ and then use the expansion to evaluate $(0.94)^6$ correct to three significant figures.

10. The coefficient of x^2 in the expansion of $(2 + px)^6$ is 15. Find the value of p.

11. The coefficient of x^2 in the expansion of $(1 + px)^5$ is 90. Find the value of p.

12. Find the coefficient of x^3 in each of the following expansions:

(a) $(3 - x)(2 + \frac{1}{4}x)^6$ (b) $(1 + 2x)(2 - x)^8$ (c) $(2 + x)^6(1 - x)^2$

13. Write down the expansion of $(1 + x)^{\frac{1}{3}}$ as far as the term in x^3. Use your expansion to approximate $\sqrt[3]{1010}$.

14. Expand $(1 + 2x)^{\frac{1}{2}}$ in ascending powers of x as far as the fourth term. Substituting 0.04 for x in your expansion find the value of $\sqrt{3}$.

15. Expand $\left(1 - \frac{1}{2}x\right)^{-\frac{1}{2}}$ in ascending powers of x as far as the fourth term. Substituting $\frac{1}{5}$ for x in your expansion find the value of $\sqrt{10}$.

14 Indices and Logarithms

14.1 Indices

In the expression of the form a^n, where $a \neq 0$, a is called the base and n is called the index (plural indices) or exponent. a^n is read " a raised to the n th power" or " a raised to the power n". For any non zero real number a and any positive integer n, a^n represents the product of n factors of a. That is

$$a^n = \underbrace{a \cdot a \cdot a \cdots a}_{n \; factors}$$

For example, $2^5 = 2 \cdot 2 \cdot 2 \cdot 2 \cdot 2$

Laws of Indices

Consider the product $a^2 \times a^3$

The product is obtained as follows:

$$a^2 \times a^3 = (a \cdot a)(a \cdot a \cdot a)$$
$$= a \cdot a \cdot a \cdot a \cdot a$$
$$= a^5$$

You can see that the index of the product is the same as the sum of the indices of the expressions on the left hand side. In general for any non zero real number a and positive integer m and n

$$a^m \times a^n = a^{m+n}$$

You can easily show that

$$a^m \div a^n = a^{m-n}$$

and $(a^m)^n = a^{mn}$

The definition of an index can be extended to include zero, negative and fractional index.

Zero Index

For any real number a where $a \neq 0$

$$a^0 \times a^n = a^n$$

$\qquad a^0 = 1 \qquad\qquad$ Divide each side by a^n

If a is any real number such that $a \neq 0$, then $a^0 = 1$

For example $125^0 = 1$ and $\left(\dfrac{1}{2}\right)^0 = 1$

Negative Index

For any real number a where $a \neq 0$

$a^{-n} \times a^n = a^0$

$\quad a^{-n} = \dfrac{a^0}{a^n} \qquad$ Divide each side by a^n

$\quad a^{-n} = \dfrac{1}{a^n} \qquad$ Since $a^0 = 1$

If a is any integer, then a^{-n} is defined as the reciprocal of a^n

Fractional Index

For any non zero real number a

$a^{\frac{1}{3}} \cdot a^{\frac{1}{3}} \cdot a^{\frac{1}{3}} = a$

$\quad \therefore a^{\frac{1}{3}} = \sqrt[3]{a}$

In general, if a is any real number and n is a positive integer then

$a^{\frac{1}{n}} = \sqrt[n]{a}$

For example, $3^{\frac{1}{2}} = \sqrt{3}$ and $2^{\frac{1}{5}} = \sqrt[5]{2}$

Also $a^{\frac{2}{3}} \cdot a^{\frac{2}{3}} \cdot a^{\frac{2}{3}} = a^2$

$\quad \therefore a^{\frac{2}{3}} = \sqrt[3]{a^2}$

Now $a^{\frac{2}{3}} = \left(a^{\frac{1}{3}}\right)^2 = (\sqrt[3]{a})^2$

Hence $a^{\frac{2}{3}} = \sqrt[3]{a^2} = \left(\sqrt[3]{a}\right)^2$

In general, if a is any real number and m and n $(n > 1)$ positive integers than

$a^{\frac{m}{n}} = \sqrt[n]{a^m} = \left(\sqrt[n]{a}\right)^m$

The rules of indices are summarized on the next page.

1. $a^m \times a^n = a^{m+n}$

2. $a^m \div a^n = a^{m-n}$

3. $(a^m)^n = a^{mn}$

4. $a^{\frac{1}{n}} = \sqrt[n]{a}$

5. $a^{\frac{m}{n}} = \sqrt[n]{a^m} = \left(\sqrt[n]{a}\right)^m$

6. $a^{-n} = \frac{1}{a^n}$

7. $a^0 = 1$

Example 1 Simplify

(a) $\left(\frac{64}{27}\right)^{-\frac{2}{3}}$

Solution

$\left(\frac{64}{27}\right)^{-\frac{2}{3}} = \left(\frac{27}{64}\right)^{\frac{2}{3}}$ Use rule 6

$= \left(\sqrt[3]{\frac{27}{64}}\right)^2$ Use rule 5

$= \left(\frac{3}{4}\right)^2$ Take the cube root

$= \frac{9}{16}$

The same result is obtain by using the following method.

$\left(\frac{64}{27}\right)^{-\frac{2}{3}} = \left(\frac{27}{64}\right)^{\frac{2}{3}}$ Use rule 6

$= \left(\frac{3^3}{4^3}\right)^{\frac{2}{3}}$ Write in index form

$= \left(\frac{3}{4}\right)^2$ Use rule 3

$= \frac{9}{16}$

232 Further Mathematics

(b) $\dfrac{\left(x^{-\frac{1}{4}} \cdot y^{-\frac{1}{3}}\right)^6}{x^{\frac{1}{2}} y^{-8}}$

Solution $\dfrac{\left(x^{-\frac{1}{4}} \cdot y^{-\frac{1}{3}}\right)^6}{x^{\frac{1}{2}} y^{-8}} = \dfrac{x^{-\frac{3}{2}} \cdot y^{-2}}{x^{\frac{1}{2}} \cdot y^{-8}}$ Use rule 3

$\qquad\qquad\qquad = x^{-2} \cdot y^6$ Use rule 2

$\qquad\qquad\qquad = \dfrac{y^6}{x^2}$ Use rule 6

(c) $\dfrac{x^2(x-2)^{-\frac{1}{2}} - x(x-2)^{\frac{1}{2}}}{x-2}$

Solution Multiply the numerator and denominator by $(x-2)^{\frac{1}{2}}$ to eliminate the negative index and simplify.

$\dfrac{x^2(x-2)^{-\frac{1}{2}} - x(x-2)^{\frac{1}{2}}}{x-2}$

$= \dfrac{x^2(x-2)^{-\frac{1}{2}} - x(x-2)^{\frac{1}{2}}}{x-2} \times \dfrac{(x-2)^{\frac{1}{2}}}{(x-2)^{\frac{1}{2}}}$

$= \dfrac{x^2 - x(x-2)}{(x-2)^{\frac{3}{2}}}$

$= \dfrac{2x}{(x-2)^{\frac{3}{2}}}$

Exercise 14.1(a)

1. Find the value of:

(a) $36^{\frac{1}{2}}$ 　　　　　　　　(b) $125^{\frac{1}{3}}$ 　　　　　　　　(c) $81^{\frac{1}{4}}$

(d) $32^{\frac{3}{5}}$ 　　　　　　　　(e) $64^{\frac{2}{3}}$ 　　　　　　　　(f) $8^{\frac{4}{3}}$

(g) $(-7)^0$ 　　　　　　　　(h) $25^{-\frac{1}{2}}$ 　　　　　　　　(i) $625^{-\frac{1}{4}}$

(j) $128^{-\frac{5}{7}}$ 　　　　　　　　(k) $16^{-\frac{3}{4}}$ 　　　　　　　　(l) $81^{\frac{3}{4}}$

(m) $\left(\dfrac{1}{32}\right)^{-\frac{3}{5}}$ 　　　　　　　　(n) $\left(\dfrac{1}{27}\right)^{-\frac{2}{3}}$ 　　　　　　　　(o) $\left(\dfrac{1}{16}\right)^{-\frac{3}{4}}$

2. Simplify:

(a) $\left(\dfrac{64}{27}\right)^{-\frac{2}{3}}$

(b) $\left(\dfrac{16}{81}\right)^{-\frac{3}{4}}$

(c) $\left(\dfrac{125}{216}\right)^{-\frac{2}{3}}$

(d) $\left(\dfrac{8}{27}\right)^{\frac{2}{3}}$

(e) $\left(1\dfrac{7}{9}\right)^{-\frac{3}{2}}$

(f) $\left(3\dfrac{3}{8}\right)^{\frac{2}{3}}$

(g) $\left(\dfrac{125}{27}\right)^{0}$

(h) $(0.0081)^{-\frac{1}{4}}$

3. Evaluate:

(a) $2^{-1} \cdot 3^2 \cdot 9^0$

(b) $20^{\frac{1}{2}} \cdot 5^{\frac{1}{2}}$

(c) $27^{\frac{1}{4}} \cdot 3^{\frac{1}{4}}$

(d) $243^{\frac{1}{2}} \cdot 3^{-\frac{1}{2}}$

(e) $\dfrac{27^{\frac{1}{3}} \cdot 125^{\frac{1}{2}}}{5^{\frac{1}{2}}}$

(f) $\dfrac{3^{\frac{1}{3}} \cdot 3^0 \cdot 9^{\frac{1}{3}}}{27^{\frac{1}{3}}}$

(g) $\dfrac{8^{\frac{1}{3}} \cdot 16^{\frac{1}{3}}}{32^{-\frac{1}{3}}}$

(h) $\dfrac{4^{\frac{1}{3}} \cdot 8^{-\frac{1}{2}}}{2^{-\frac{1}{2}} \cdot 4^{-\frac{2}{3}}}$

(i) $\dfrac{75^{\frac{1}{3}} \cdot 15^{\frac{1}{3}}}{9^{\frac{1}{3}}}$

4. Simplify::

(a) $\dfrac{\left(x^{\frac{1}{2}} \cdot y^{-1}\right)^3}{x^{-\frac{1}{2}} y}$

(b) $\dfrac{(x^{-1} y^3)^4}{(x^2 y^{-6})^{-2}}$

(c) $\left(\dfrac{8x^{-6} y^3}{z^{-9}}\right)^{\frac{1}{3}}$

(d) $\left(\dfrac{x^{\frac{3}{4}} y^{-\frac{1}{2}}}{z^{\frac{1}{4}}}\right)^4$

(e) $\dfrac{(x^2 y^6)^{\frac{1}{5}}}{\sqrt[5]{x^7 y}}$

(f) $\dfrac{\sqrt{x^3} \cdot x^5}{\sqrt{x^5}}$

(g) $\dfrac{x^{\frac{4}{3}} \cdot x^{-2}}{\sqrt[3]{x}}$

(h) $\left(\dfrac{x^{\frac{4}{3}} y^{\frac{1}{2}}}{z^{-4}}\right)^{\frac{3}{4}} \left(\dfrac{x^{-3} y^{\frac{15}{8}}}{z^6}\right)^{\frac{1}{3}}$

5. Simplify:

(a) $\dfrac{x^{\frac{1}{2}} + x^{\frac{3}{2}}}{x^{-\frac{1}{2}}}$

(b) $\dfrac{x^{\frac{5}{4}} - x^{\frac{1}{4}}}{x^{-\frac{3}{4}}}$

(c) $\dfrac{(2x+3)^{\frac{1}{2}} + (2x+3)^{-\frac{1}{2}}}{(2x+3)^{\frac{1}{2}}}$

(d) $\dfrac{2x(x+1)^{-\frac{1}{3}} + 3(x+1)^{\frac{2}{3}}}{(x+1)^{\frac{5}{3}}}$

(e) $\dfrac{x^2(1-x)^{-\frac{1}{2}} + x(1-x)^{\frac{1}{2}}}{x}$

(f) $\dfrac{3(1+x)^{\frac{1}{3}} - x(1+x)^{-\frac{2}{3}}}{(1+x)^{\frac{4}{3}}}$

Exponential Equations

An equation that contains an expression in index form is called an exponential equation. Some examples are $2^x = 8$ and $125^x = 5$.

Solving Exponential Equations

Example 2 Solve

(a) $3^{2x+1} = 243$

Solution Rewrite the equation in the form

$$3^{2x+1} = 3^5 \qquad \text{Write 243 as a power of 3}$$

$$2x + 1 = 5 \qquad \text{Equate the index}$$

$$2x = 4 \qquad \text{Subtract 1 from both sides}$$

$$x = 2 \qquad \text{Simplify}$$

(b) $4^x - 12 \cdot 2^x = -32$

Solution Rewrite the original equation in the form

$$2^{2x} - 12 \cdot 2^x + 32 = 0$$

Substituting y for 2^x, we have

$$y^2 - 12y + 32 = 0$$

Factorising the quadratic expression we have

$$(y - 8)(y - 4) = 0$$

So $y - 8 = 0$ or $y - 4 = 0$

$\qquad y = 8 \qquad\qquad y = 4$

But $y = 2^x$

So $2^x = 8 = 2^3 \qquad 2^x = 4 = 2^2$

$\qquad x = 3 \qquad\qquad x = 2$

The solutions of the equation are $x = 2$ and $x = 3$

Example 3 Solve the simultaneous equations

$$2^x \times 2^y = 8$$

$$3^x \div 3^y = 243$$

Solution Rewrite the two equations in the forms shown below.

$$2^{x+y} = 2^3$$

$$3^{x-y} = 3^5$$

Equating the indices we have

$$x + y = 3 \quad (1)$$

$$x - y = 5 \quad (2)$$

Adding equation (1) and (2) gives

$$2x = 8$$

$$x = 4$$

Substituting 4 for x in equation (1) we have

$$4 + y = 3$$

$$y = -1$$

The solutions of the equation are $x = 4$ and $y = -1$

Exercise 14.1(b)

1. Solve:

(a) $8^x = 64$

(b) $9^x = 27$

(c) $128^x = 32$

(d) $3^{2x-1} = 27$

(e) $2^{x-1} = \frac{1}{8}$

(f) $8^{x-1} = 16$

(g) $(x-1)^3 = 64$

(h) $x^{\frac{3}{4}} = 81$

(i) $\left(\frac{9}{25}\right)^{-x} = \left(\frac{5}{3}\right)^{3-x}$

(j) $9^{x-2} = 3^{3x-1}$

(k) $4^{2x} = \frac{1}{3}(8)^x$

(l) $3^{\sqrt{x}} = 27^x$

2. Solve:

(a) $\dfrac{25^x}{5^{5-x}} = \dfrac{5^{4x}}{125^{x-3}}$

(b) $\dfrac{3^{5x-2}}{9^x} = \dfrac{27^{x-1}}{9^{3-x}}$

(c) $\dfrac{36^{-2x}}{6^{3-5x}} = \dfrac{36^{3x-2}}{216^{2-x}}$

(d) $\dfrac{8^{3-x}}{4^{2x}} = \dfrac{2^{3x-5}}{4^{3x}}$

3. Solve the following equations:

(a) $4^x - 6 \cdot 2^x + 8 = 0$

(b) $9^x - 30 \cdot 3^x + 81 = 0$

(c) $5^{2x} - 6 \cdot 5^x + 5 = 0$

(d) $3^{2x} - 4 \cdot 3^{x+1} + 27 = 0$

(e) $2^{2x} - 5 \cdot 2^{x-1} = -1$

(f) $4^x - 6 \cdot 2^x = 16$

(g) $3^{2x+1} - 28 \cdot 3^x + 9 = 0$

(h) $2^x + 16 \cdot 2^{-x} - 10 = 0$

4. Solve the following simultaneous equations:

(a) $3^x \times 3^y = 27$

$3^x \div 3^y = \frac{1}{3}$

(b) $2^x \times 2^y = 128$

$3^{3x-y} = 27$

(c) $2^{x+y} = 32$

$2^x \div 2^y = 32$

(d) $2^{2x} \times 2^y = 16$

$3^{2x} \div 3^y = 1$

(e) $27^x \times 9^y = 3$

$2^x \times 4^y = 8$

(f) $5^{x+1} = 125^{y-1}$

$\left(\frac{1}{5}\right)^{2x} = 25^{1-y}$

14.2 Logarithms

The logarithm of any number to a given base is the power to which the base must be raised in order to give the number. Thus if $y = a^x$, x is called the logarithm of y to base a. This is abbreviated to $\log_a y = x$. For example, $9 = 3^2$ so $\log_3 9 = 2$.

Note that a logarithm is simply an index (or exponent).

Common Logarithms

The logarithm to base 10 is called the common logarithm. It is customary to omit the base in writing a common logarithm. When no other base is shown the base is assumed to be 10. That is $\log N$ means $\log_{10} N$.

Evaluating Logarithms

Certain logarithms can be calculated directly by writing $\log_a y = x$ in the form $y = a^x$.

Example 4 Evaluate each logarithm

(a) $\log_2 32$

Solution Let $x = \log_2 32$

$2^x = 32$ Write the logarithm in exponential form.

$2^x = 2^5$ Write 32 as a power of 2

$x = 5$ Equate the exponents

So $log_2 32 = 5$

(b) $log_3 \frac{1}{81}$

Solution Let $x = log_3 \frac{1}{81}$

$3^x = \frac{1}{81}$ Write the logarithm in exponential form

$3^x = 3^{-4}$ Write $\frac{1}{81}$ as a power of 3

$x = -4$ Equate the exponents

So $log_3 \frac{1}{81} = -4$

Exercise 14.2(a)

1. Express in logarithmic form:

(a) $5^3 = 125$ (b) $2^2 = 4$ (c) $32^{\frac{1}{5}} = 2$

(d) $e^0 = 1$ (e) $27^{-\frac{1}{3}} = \frac{1}{3}$ (f) $10^3 = 1000$

2. Express in exponent form:

(a) $log_2 8 = 3$ (b) $log_4 64 = 3$ (c) $log_9 9 = 1$

(d) $log_4 \frac{1}{16} = -2$ (e) $log_{10} 10000 = 4$ (f) $log_9 \frac{1}{27} = -\frac{3}{2}$

3. Evaluate:

(a) $log_2 32$ (b) $log_3 243$ (c) $log_4 64$ (d) $log_2 \frac{1}{128}$

(e) $log 1000$ (f) $log_6 \frac{1}{36}$ (g) $log_{25} 5$ (h) $log_{\sqrt{2}} 8$

(i) $log_{0.1} 100$ (j) $log_{\sqrt[3]{2}} \frac{1}{64}$ (k) $log_{\frac{1}{3}} 27$ (l) $log_{\frac{3}{4}} 1\frac{7}{9}$

Properties of Logarithms

Since a logarithm is by definition an index, we turn to the rules of indices discussed in the preceding section to derive the following properties of logarithms.

For any real number $a > 0$ and $a \neq 1$, $a^0 = 1$ and $a = a^1$. By the definition of logarithm, it follows that:

1. $\log_a 1 = 0$

2. $\log_a a = 1$

Further properties of logarithm can be derived using the rules of indices.

Product Rule

Let $\log_a x = p$ and $\log_a y = q$. Then $x = a^p$ and $y = a^q$

Hence $x \cdot y = a^p \cdot a^q = a^{p+q}$

Rewriting this as logarithmic equation we have

$\log_a xy = p + q$

Substituting $\log_a x$ and $\log_a y$ for p and q respectively, we have

$\log_a xy = \log_a x + \log_a y$

Notice that the logarithm of a product is the sum of the logarithm of the factors.

The following is a list of properties of logarithms. We leave the proof of properties 2 and 3 as exercises. You can easily proved them by arguments similar to those presented above.

Properties of Logarithms

1. $\log_a xy = \log_a x + \log_a y$

2. $\log_a \dfrac{x}{y} = \log_a x - \log_a y$

3. $\log_a x^n = n \log_a x$

4. $\log_a a = 1$

5. $\log_a 1 = 0$

Using the Properties of Logarithms

Writing a single logarithm as the sum of simpler logarithms

We can use the properties of logarithms to write a single logarithmic expression as sum (or difference) of two or more simpler logarithmic expressions.

Example 5 Expand the following expressions.

(a) $\log_a \dfrac{a^3 b^4}{c^2}$

Solution
$$\log_a \dfrac{a^3 b^4}{c^2} = \log_a a^3 b^4 - \log_a c^2$$
$$= \log_a a^3 + \log_a b^4 - \log_a c^2$$
$$= 3 + 4\log_a b - 2\log_a c$$

(b) $\log \sqrt[4]{\dfrac{x^2}{y^6 z^4}}$

Solution
$$\log \sqrt[4]{\dfrac{x^2}{y^6 z^4}} = \dfrac{1}{4}(\log x^2 - \log y^6 z^4)$$
$$= \dfrac{1}{4}\log x^2 - \dfrac{1}{4}\log y^6 - \dfrac{1}{4}\log z^4$$
$$= \dfrac{1}{2}\log x - \dfrac{3}{2}\log y - \log z$$

Exercise 14.2(b)

Expand the following:

1. $\log_a x^3 y^2$
2. $\log_a \dfrac{x^2}{y}$
3. $\log x^2 \cdot \sqrt[3]{y}$
4. $\log_2 \dfrac{1}{x^3}$
5. $\ln x^3 \sqrt{y}$
6. $\log_3 \dfrac{x^2 y^3}{z^2}$
7. $\log_5 \dfrac{x^2 y}{\sqrt[3]{z}}$
8. $\log_{10} \dfrac{x^4 \cdot \sqrt[3]{y}}{z^3}$
9. $\log \sqrt[4]{\dfrac{xy^2}{z^2}}$
10. $\log \sqrt[3]{\dfrac{xy^2}{z^4}}$
11. $\log_2 \dfrac{x^2 y^3}{\sqrt{z^5}}$
12. $\log_5 \sqrt[3]{\dfrac{1000 x^2}{y^3 z^4}}$

Writing the sums of simpler logarithms as a single logarithm

We reverse the process and use the properties of logarithms to write a single logarithm given a sum (or difference) of two or more logarithmic expressions.

Example 6 Write each expression as a single logarithm.

(a) $2\log_a x + 3\log_a y - 5\log_a z$

Solution
$$2\log_a x + 3\log_a y - 5\log_a z = \log_a x^2 + \log_a y^3 - \log_a z^5$$
$$= \log_a \dfrac{x^2 y^3}{z^5}$$

(b) $5\log a + 3 - \frac{1}{2}\log b$

Solution $5\log a + 3 - \frac{1}{2}\log b = \log a^5 + \log 1000 - \log \sqrt{b}$

$$= \log\left(\frac{1000 a^5}{\sqrt{b}}\right)$$

Exercise 14.2(c)

Express as single logarithms:

1. $\log_2 x^2 + \log_2 y^3$
2. $\log_3 x - 2\log_3 y$
3. $3\log x - \log y$
4. $\log_5 x + \frac{1}{3}\log_5 y^2$
5. $\frac{1}{3}\log x^2 - 2\log y$
6. $3\ln x + 2\ln y - \ln z$
7. $2\log_3 x - (3\log_3 y + \log_3 z)$
8. $\frac{1}{2}\log_5 x - \frac{2}{3}\log_5 y - 3\log_5 z$
9. $\frac{3}{2}(2\log x - 6\log y + 4\log z)$
10. $\frac{1}{4}(2\log_2 x + 3\log_2 y - 4\log_2 z)$
11. $3 + 2\log x - \frac{1}{4}\log y$
12. $\frac{2}{3}\log x - 3\log y + 1$

Simplifying Logarithmic Expressions

Example 7 Simplify:

(a) $\log_2 64$

Solution Writing 64 as a power of 2, we have

$\log_2 64 = \log_2 2^6$

$= 6\log_2 2$

$= 6$

(b) $\log_3 \sqrt[4]{27}$

Solution $\log_3 \sqrt[4]{27} = \log_3 27^{\frac{1}{4}}$

$= \log_3 3^{\frac{3}{4}}$

$= \frac{3}{4}$

(c) $\frac{log 32}{log 8}$

Solution $\frac{log\ 27}{log\ 9} = \frac{log 3^3}{log 3^2} = \frac{3 log 3}{2 log 3} = \frac{3}{2}$

Note that $\frac{log 27}{log 9} \neq log 27 - log 9$

Exercise 14.2(d)

Simplify:

1. $log 100{,}000$
2. $log_2 128$
3. $log_3 243$
4. $log_6 216$
5. $log_4 \frac{1}{16}$
6. $log_5 125$
7. $\frac{4}{5} log_2 32$
8. $log_{\frac{1}{3}} 27$
9. $log_8 \frac{1}{2}$
10. $log_2 \sqrt[3]{2}$
11. $log_{\frac{1}{4}} 64$
12. $-2 log_3 \frac{1}{3}$
13. $\frac{log_2 27}{log_2 9}$
14. $\frac{ln\ 8}{ln\ 4}$
15. $\frac{log 64}{log 8}$

Change of Base

Occasionally, we need to evaluate logarithms with bases which are not 10 or e. In such cases the change of base formula will be useful. The formula would help you use a table or calculator that has only the common logarithm and the natural logarithm to evaluate logarithms to any base.

Suppose that

$b^x = a$ where a, b and x are positive real numbers such that $a \neq 1$ and $b \neq 1$.

Taking logarithm to base c of both sides gives

$log_c b^x = log_c a$

$x log_c b = log_c a$

$x = \frac{log_c a}{log_c b}$

Since $x = log_b a$, we have

$log_b a = \frac{log_c a}{log_c b}$

Example 8 Evaluate:

(a) $log_2 25$

Solution $log_2 25 = \dfrac{log_{10} 25}{log_{10} 2}$

$= \dfrac{1.3979}{0.3010}$

$= 4.644$

(b) $log_5 16$

Solution $log_5 16 = \dfrac{log 16}{log 5}$

$= \dfrac{1.2041}{0.6990}$

$= 1.723$

Exercise 14.2(e)

Evaluate:

1. $log_2 26$ 2. $log_4 9$ 3. $log_5 10$ 4. $log_3 8$ 5. $log_7 12$

6. ln 15   7. $log_8 32$ 8. $log_9 8$ 9. $log_6 10$ 10. $log_5 16$

Exponential Equations

Recall that in Section 14.1 you learned to solve exponential equations by rewriting the expression on both side of the equation in exponent form with the same base.

Consider the exponential equation

$8^x = 64$

Rewrite each side of the equation as an index with the same base.

$2^{3x} = 2^6$

Equating the indices and solving for x we have

$3x = 6$

$x = 2$

So the solution of the equation is $x = 2$

However, for equations like $3^x = 25$ it would be impossible to write the expression on both sides in exponent form with the same base. We used logarithm to solve exponential equations like $3^x = 25$. We illustrate this in Example 9.

Example 9 Solve each equation for x

(a) $3^x = 7$

Solution Taking the logarithm to base 10 of each side we have

$$x \log 3 = \log 7$$

$$x = \frac{\log 7}{\log 3}$$

$$= \frac{0.8451}{0.4771}$$

$$= 1.7713$$

(b) $\left(\frac{1}{4}\right)^{x-2} = 5^{3x}$

Solution $\log\left(\frac{1}{4}\right)^{x-2} = 5^{3x}$

$$(-2x + 4)\log 2 = 3x \log 5$$

$$x = \frac{4\log 2}{2\log 2 + 3\log 5}$$

$$= \frac{1.2041}{2.6990}$$

$$= 0.4461$$

Exercise 14.2(f)

Solve for x:

1. $64^x = 32$
2. $4^{x+1} = 8$
3. $243^x = 9$
4. $2^x = \frac{1}{32}$
5. $3^{2x+1} = \frac{1}{81}$
6. $8^{x-1} = 16^{-x}$
7. $2^{\sqrt{x}} = 8^x$
8. $\left(\frac{1}{27}\right)^{-x} = 9^{2x+3}$
9. $5^x = 8$
10. $4^x = 20$
11. $3^{2x} = 4$
12. $7^{3x} = 50$
13. $2^{2x+1} = 15$
14. $5^{2x-3} = 15$
15. $4^x = 3^{x+1}$

14.3 Logarithmic Equations

A logarithmic equation is an equation that contains a logarithmic expression. To solve a logarithmic equation we rewrite the equation in exponential form and solve for the variable as illustrated in Example 10.

It is useful to check possible solutions to logarithmic equations in the original equation to avoid answers that would result in taking logarithm of zero or a negative number. Note that the logarithm of 0 or a negative number to any base is undefined

Example 10 Solve each logarithmic equation for x.

(a) $log_3 x + log_3 2 = 4$

Solution Write the expression on the left as a single logarithm

$$log_3 2x = 4$$

Rewrite the equation in exponential form

$$2x = 3^4$$

$$2x = 81$$

$$x = 40\frac{1}{2}$$

The solution is $x = 40\frac{1}{2}$. Check this in the original equation.

(b) $log_2 \frac{1}{2}x + log_2(x - 4) = 4$

Solution Write the expression on the left as a single logarithm

$$log_2 \frac{1}{2}x(x - 4) = 4$$

Rewrite the equation in the exponential form

$$\frac{1}{2}x(x - 4) = 2^4$$

Simplifying this equation we get

$$x^2 - 4x - 32 = 0$$

Factorise the quadratic expression

$$(x - 8)(x + 4) = 0$$

The possible solutions are $x = -4$ and $x = 8$.

You need to check each solution in the original equation.

Substituting -4 for x in the original equation gives

$$log_2(-2) + log_2(-8) = 4$$

Recall that the logarithm of a negative number is undefined, so we reject $x = -4$

The only solution is $x = 8$. Check this in the original equation.

(c) $log(2x + 6) - log(x + 3) = log x$

Solution Write the expression on the left as a single logarithm

$$log \frac{2x+6}{x+3} = log x$$

Hence $$\frac{2x+6}{x+3} = x$$

Simplifying this equation we get

$$x^2 + x - 6 = 0$$

Factorising the quadratic expression gives

$$(x + 3)(x - 2) = 0$$

The possible solutions are $x = -3$ and $x = 2$

The only solution for the equation is $x = 2$. Check this in the original equation.

Exercise 14.3

1. Solve each of the following equations:

(a). $log_3 x = 4$

(b). $log_x 25 = 2$

(c). $log_5 25 = x$

(d). $x = log_4 32$

(e). $log_x \frac{1}{8} = -3$

(f). $log_{32} x = \frac{3}{5}$

(g). $log_x \frac{9}{4} = -2$

(h). $log_5 x = 3$

(i). $log_2(x + 1) = 3$

(j). $log_5(2x - 1) = 2$

(k). $log_2 x + log_2 8 = 6$

(l). $log x + log 5 = 2$

(m). $\log_x 3 = 2 - \log_x 27$ (n). $\log_x 5 + 3 = \log_x 320$

(o). $\log_2 x + \log_x 2 = 2$ (p). $\log_2 x = \log_4(x+6)$

(q). $\log_3 x - 2\log_x 3 = 1$ (r). $\log_x 7 + 1 = \log_x 112 - 3$

2. Solve each of the following logarithmic equations:

(a). $\log(2x+1) - \log(x+2) = \log x$ (b). $\log_2 x + \log_2(x+2) = 3$

(c). $\log_3 x + \log_3(2x+3) = 2$ (d). $\log_2(x+2) + \log_2(x-5) = 3$

(e). $\log_3(x+1) - \log_3(x-2) = 2$ (f). $\log(x+2) - \log(2x-1) = 1$

(g). $\log(x+5) - \log(x-2) = \log 5$

(h). $\log_3(x+12) - \log_3(x-3) = \log_3 6$

14.4 Reducing non-linear Graphs to linear

Sometimes we reduce a non-linear equation to the linear form $y = mx + c$ to enable us determine some constants and the law relating the variables.

Consider the equation $y = ax^n$, where a and n are constants.

Taking logarithms to the base ten on both sides of the equation, we have

$\log y = \log a + n\log x$

Writing $\log y$ as y, $\log a$ as c and $\log x$ as x, we have

$y = c + nx$

In this form, if we plot $\log y$ against $\log x$, we obtain a straight line or a set of points lying nearly on a straight line. Then we draw the line of best fit. The gradient of the line determines the value of n, and $\log a$ is the intercept on the log y-axis.

Suppose y is a function of x of the form $y = ab^x$, where a and b are constants. Again taking logarithms we have

$\log y = \log a + x\log b$

This time if we plot $\log y$ against x, we again obtain a straight line graph with gradient $\log b$.

A function of the form $y = ax^2 + bx$ can be reduced to a linear form when we plot $\frac{y}{x}$, against x.

Example 11

(a) The table below gives the corresponding values of two related variables x and y.

x	1	2	3	4	5	6
y	5.75	22.9	51.3	91.2	144.5	208.9

The relation between x and y is of the form $y = ax^n$, where a and n are constants. Draw a suitable linear graph and use it to estimate the values of a and n.

Solution $y = ax^n$

Taking logarithms to the base ten on both sides of the equation we have

$$\log y = \log a + n\log x$$

A new table is constructed as shown below

$\log x$	0	0.30	0.47	0.60	0.70	0.78
$\log y$	0.76	1.36	1.71	1.96	2.16	2.32

Plotting $\log y$ against $\log x$ gives the graph below

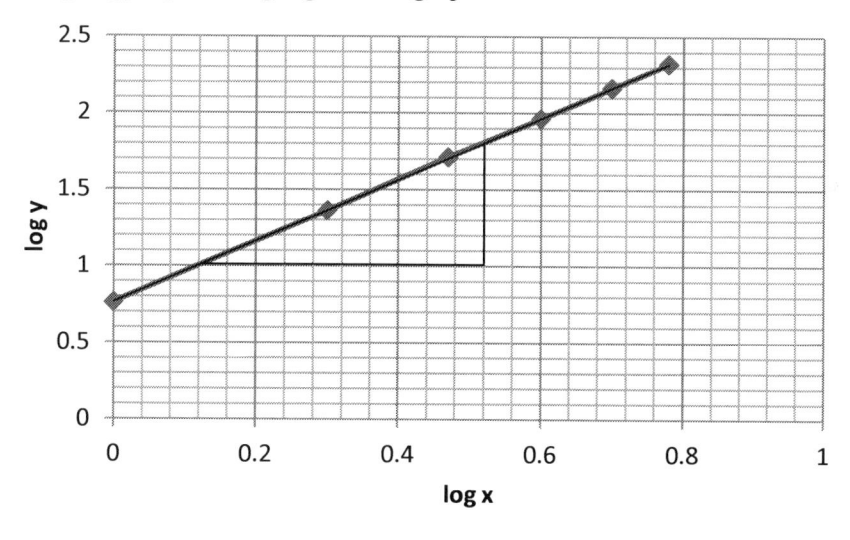

To find the gradient, we choose any two points on the line. We consider the points (0.12, 1) and (0.52, 1.8). The gradient, n, is given by

$$n = \frac{1.8-1}{0.52-0.12} = \frac{0.8}{0.4} = 2$$

The intercept on the vertical axis is 0.76,

Therefore $\log a = 0.76$

so, $a = 5.75$

(b) The relation between x and y is given as $y = ax^2 + bx$, where a and b are constants. The table below gives the corresponding values of x and y.

x	1	2	3	4	5
y	5	16	33	56	85

By drawing a suitable linear graph estimate the values of a and b.

$$y = ax^2 + bx$$

Dividing both sides by x, we get

$$\frac{y}{x} = ax + b$$

From the table above, we obtain the table shown below.

x	1	2	3	4	5
$\frac{y}{x}$	5	8	11	14	17

Plotting $\frac{y}{x}$ against x gives the graph below

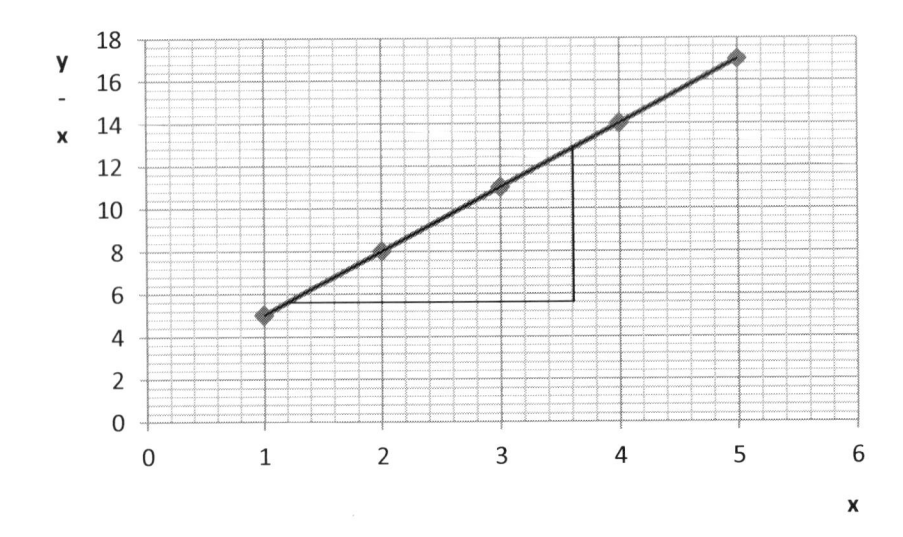

We choose the point $(1.2, 5.6)$ and $(3.6, 12.8)$ to calculate the gradient, a.

$$a = \frac{12.8 - 5.6}{3.6 - 1.2} = \frac{7.2}{2.4} = 3$$

The straight line does not intercept the vertical axis. In this case we use the equation $\frac{y}{x} = ax + b$ together with the value of a and the coordinates of any convenient point on the graph to find b.

Taking the point (3.6, 12.8), we have

$12.8 = 3(3.6) + b$

$b = 12.8 - 10.8 = 2$

Exercise 14.4

1. The data below fits an equation of the form $y = ax^n$

x	2	3	4	5
y	22.4	50.4	89.6	140.0

Draw a suitable graph and use it to find the values of a and n

2. The table below gives the corresponding values of two related variables x and y.

x	1	2	3	4	5
y	6	12	24	48	96

The relation between x and y is of the form $y = ab^x$, where a and b are constants. Draw a suitable linear graph and use it to estimate the values of a and b.

3. The relation between x and y is given as $y = ax^n$, where a and n are constants. The table below gives the corresponding values of x and y.

x	2	3	4	5	6
y	3.75	1.67	0.94	0.60	0.42

By drawing a suitable linear graph estimate the values of a and n. Find x when y is 2.

4. By using the given table of values to draw a graph, find the relationship between the variables x and y in the form $y = ax^n$.

x	2	3	4	5	6
y	12.9	26.0	40.0	63.2	73.5

5. The table shows experimental values of two variables x and y

x	0	1	2	3	4	5
y	8	9.5	14	21.5	32	45.5

It is known that x and y are related by the equation $y = ax^2 + b$, where a and b are constants. By drawing a suitable linear graph estimate the values of a and b. Find x when y is 35

6. The table shows experimental values of two variables x and y

x	2	3	4	5	6
y	9.2	10.2	8.8	5.0	-1.2

It is known that x and y are related by the equation $y = ax^2 + bx$, where a and b are constants. Using graph paper plot $\frac{y}{x}$ against x for the above data and use your graph to estimate the values of a and b.

7. The table below gives the values of two related variables x and y

x	0.2	0.4	0.6	0.8	1.2
y	15	9	7	6	5

The relation between x and y is of the form $y = a + \frac{b}{x}, x \neq 0$ and a and b are constants. Draw a suitable linear graph and use it to estimate the values of a and b.

8. The table shows experimental values of two variables x and y

x	1	2	3	4	5
y	1.65	1.50	1.75	2.10	2.49

It is known that x and y are related by the equation $y = ax + \frac{b}{x^2}, x \neq 0$ where a and b are constants. Draw a straight line graph and use it to estimate the values of a and b, correct to two significant figures.

9. The table shows experimental values of two variables x and y.

x	1	2	3	4
y	4.30	0.80	-0.97	-1.70

It is known that x and y are related by the equation $y = \dfrac{a}{x^2} + bx$, $x \neq 0$ where a and b are constants. Using graph paper, plot $x^2 y$ against x^3 for the above data and use your graph to estimate the values of a and b.

Review Exercise 14

1. Simplify:

(a) $27^{-2/3}$ (b) $4^{3/2}$ (c) $32^{-3/5}$ (d) $125^{2/3}$

(e) $\left(\dfrac{8}{27}\right)^{2/3}$ (f) $\left(\dfrac{16}{81}\right)^{-3/4}$ (g) $\left(\dfrac{1}{8}\right)^{-2/3}$ (h) $\left(\dfrac{125}{27}\right)^{-2/3}$

2. Simplify:

(a) $\left(\dfrac{x^3 \cdot y^{-6}}{z^{12}}\right)^{-2/3}$ (b) $\left(\dfrac{x^{1/4} \cdot y^{-1/2}}{x^{1/2} \cdot y^{3/4}}\right)^4$ (c) $\left(\dfrac{\sqrt[3]{x} \cdot y^{-1}}{\sqrt[3]{z^4}}\right)^3$

3. Simplify:

(a) $\dfrac{2x^{-2/3} + 3x^{1/3}}{\sqrt[3]{x^{-5}}}$ (b) $\dfrac{(1+x)^{2/3} + (1+x)^{-1/3}}{\sqrt[3]{(1+x)^5}}$

(c) $\dfrac{x^3(1-x^2)^{-1/2} + x^2(1-x^2)^{1/2}}{x^2}$ (d) $\dfrac{3x(2x+1)^{-1/3} + 4(2x+1)^{2/3}}{(2x+1)^{5/3}}$

4. Solve the following equations:

(a) $2^{x-4} = 8^2$ (b) $3^{3x} = 9$ (c) $4^{3x-1} = 64$

(d) $5^{2-x} = 25$ (e) $4^{x+3} = 32^x$ (f) $9^{x-2} = 234^{x+1}$

5. Use the properties of logarithms to expand:

(a) $\log\left(\dfrac{x^2 y}{z^3}\right)$ (b) $\log \dfrac{x^5 y^3}{z^2}$

(c) $\log_2 \dfrac{16 x^2 y^3}{z^4}$ (d) $\dfrac{1}{2} \ln\left(\dfrac{x^4 y^6}{z^8}\right)$

(e) $\log_5\sqrt{\dfrac{x^5}{y^8z^7}}$

(f) $\log\sqrt[3]{\dfrac{x^6y^2}{z^4}}$

6. Use the properties of logarithms to write each expression as a single logarithm:

(a) $3\log x + 5\log y$

(b) $2\log x - 3\log y$

(c) $5\log x + 3\log y - 2\log z$

(d) $\dfrac{1}{5}(3\log x - 4\log y)$

(e) $\dfrac{1}{3}(2\log x + 4\log y) - 3\log z)$

(f) $2\ln x + \dfrac{1}{4}\ln y$

(g) $\dfrac{1}{2}\ln x - 3\ln y + 2\ln z$

(h) $3\log_2 x - 8 - 2\log_2 y$

7. Evaluate the following logarithms:

(a) $\log 1000$

(b) $\log_2 64$

(c) $\log_4 \dfrac{1}{16}$

(d) $\log_3\sqrt{9}$

(e) $\log 0.001$

(f) $\log_8 2$

(g) $\log_a(\sqrt[3]{a})^4$

(h) $\log_5 125$

(i) $\log_9 \sqrt[3]{81}$

8. Simplify:

(a) $\dfrac{\log 128}{\log 32}$

(b) $\dfrac{\log_5 27}{\log_3 9}$

(c) $\dfrac{\ln 8}{\ln 16}$

(d) $\dfrac{\log_4 125}{\log_4 25}$

9. Solve the following exponential equations:

(a) $3^x = 91$

(b) $5^x = 8$

(c) $2^{x+1} = 6$

(d) $7^{3-x} = 15$

(e) $9^{3x} = 126$

(f) $8^{2x} = 35$

10. Solve the following logarithmic equations:

(a) $\log_6(x-5) + \log_6 x = 2$

(b) $2\log_4(x+5) = 3$

(c) $\log x + \log(x-3) = 1$

(d) $\log_3 x - \log_3 4 = 2$

(e) $\log_5(x+3) - \log_5 x = 1$

(f) $2\log_4 x - \log_4(x-1) = 1$

11. Without using tables, write down the values of:

(a) $\log_9 27\sqrt{3}$

(b) $\log_2 25 \times \log_5 8$

12. Without using tables, find the values of:

(a) $\log 75 + 3\log 2 - \log 6$

(b) $\dfrac{\log_4 64 + \log_3 9 - \log_2 16}{\log_9 81}$

(c) $\dfrac{2}{3}\log 64 - \log 27 - 3\log\left(\dfrac{2}{3}\right) + \log 50$

13. Given that $\log_3 5 = 1.465$, evaluate without using tables $\log_3 25 + \log_9 15$.

14. Find the values of x for which $\log_5 x - 2\log_x 5 = 1$.

15. Given that $\log 2 = 0.3010$ and $\log 3 = 0.4771$, find:

(a) $\log 15$ \qquad (b) $\log 1.8$

16. Find the truth set of the following simultaneous equations:

(a) $2\log_2 x - 3\log_2 y = 7$ \qquad (b) $\log_3 x + \log_3 y = 3$

$\log_2 x - 2\log_2 y = 4$ \qquad\qquad $\log_y x = 2$

17. Find the solution set of:

(a) $\log_2(x^2 - 1) - \log_2(x - 2) = 3$

(b) $2\log_3 x - \log_3 2 = \log_3(5x - 12)$

18. Given that a and b are both positive and unequal, find a in terms of b if $\log_a b^2 + \log_b a = 3$.

19. Given that $\log_a x = 3$, find:

(a) $\log_a x^2$ \qquad (b) $\log_a \dfrac{1}{\sqrt{x}}$

20. The table below gives the corresponding values of two variables x and y for the relation $y = ab^x$, where a and b are constants.

x	1	2	3	4	5	6
y	4.50	7.62	10.13	13.20	22.78	36.40

By means of a suitable graph estimate correct to one decimal place
(a) the values of a and b

(b) the value of y when $x = 4.5$

21. The table below gives the corresponding values of two related variables x and y

x	1.5	2.0	2.5	3.0	3.5
y	3.6	5.6	10.0	17.4	19.6

The relation between x and y is of the form $y = ax^n$, where a and n are constants. Draw a suitable linear graph and use it to estimate, correct to two decimal places, the values of a and n.

22. The table shows experimental values of two variables x and y

x	2	4	6	8
y	12.4	48.8	133.2	289.6

It is known that x and y are related by the equation $y = ax^3 + bx$, where a and b are constants.

(a) Draw a suitable linear graph and use it to estimate correct to one decimal place, the value of a and of b.

(b) Estimate the value of y when $x = 2.4$.

15 Sequences

A set of numbers is called a sequence if each number relates to other numbers in the set by a definite rule.

Consider the following sets of numbers

(a) 1, 3, 5, 7, 9, 11...

(b) 4, 8, 16, 32, 64, 128

For each set we will have little difficulty in finding how each number is obtained or in writing down as many subsequent numbers as we wish to find. For example, the next three numbers of (a) are 13, 15 and 17. You may have noticed that each number after the first exceeds the preceding one by 2. The numbers of the set are called terms of the sequence. It is customary to use the symbol u_n for the nth term of a sequence. For example, the first three terms of the sequence are $u_1 = 1$, $u_2 = 3$ and $u_3 = 5$.

If a sequence stops after a finite number of terms such as (b), the sequence is called a finite sequence. If the sequence does not stop but continues indefinitely such as (a), the sequence is called an indefinite sequence. Remember that the three dots indicate that the sequence continues and has an infinite number of terms.

Series

If the terms of a sequence are added together we get a series. In general, the sum of the terms of the sequence $u_1, u_2, u_3, \ldots, u_n$ gives the series $u_1 + u_2 + u_3 + \cdots + u_n$. This series can be written briefly using the Greek letter sigma as $\sum_{r=1}^{r=n} u_r$, where r is the index of summation, n is the upper limit of summation and 1 is the lower limit of summation. Note that the summation notation is an instruction to add the terms of a sequence.

15.1 Linear Sequence

A sequence in which each term after the first is found by adding the same number to the preceding term is called a linear sequence or arithmetical progression (AP). The difference of any two consecutive terms is a constant d,

$d = u_n - u_{n-1}.$

called the common difference.

For example, the common difference d, of the following sequence, 15, 12, 9, 6, ... is $12 - 15$ or $9 - 12$ or $6 - 9$, i.e., $d = -3$.

Exercise 15.1(a)

1. Find the common difference of each linear sequence.

(a) 4, 7, 10, 13 . . . (b) – 4, 0, 4, 8 . . . (c) 50, 44, 38, 32 . . .

(d) 10, -2, -14, -26 . . . (e) $4, \frac{9}{2}, 5, \frac{11}{2}, 6$ ... (f) $\frac{1}{2}, \frac{5}{4}, 2, \frac{11}{4}$ ...

2. Determine whether the sequence is a linear sequence. If so find the common difference.

(a) 3, 5, 7, 9 . . . (b) 1, 3, 4, 8 . . . (c) 8, 6, 4, 2, 0 . . .

(d) 45, 30, 26, 13 . . . (e) 5, 9, 13, 17 . . . (f) $2, \frac{7}{2}, 5, \frac{13}{2}$ ...

3. Write the next two terms of each linear sequence.

(a) 7, 10, 13 . . . (b) 1, 6, 11 . . . (c) 6, 4, 2 . . .

(d) $\frac{3}{2}, 4, \frac{13}{2}$ ... (e) $-\frac{5}{4}, -\frac{1}{2}, \frac{1}{4}$ ... (f) $4, \frac{15}{4}, \frac{7}{2}$ ...

The n th of a Linear Sequence

Recall that successive terms of a linear sequence differ by the same number. Hence if the first term of a linear sequence is a and the common difference is d then the second term is $a + d$. The third term is obtained by adding d to the second, the fourth by adding d to the third term and so on. So

$u_1 = a$

$u_2 = a + d$

$u_3 = a + 2d$

$u_4 = a + 3d$

The pattern suggests that the n th term u_n of a linear sequence is $u_n = a + (n-1)d$, where n represents the number of terms of the sequence.

Example 1

(a) Find the ninth term of the sequence 3, 5, 7 . . .

Solution Substituting $a = 3, d = 5 - 3 = 2$ and $n = 9$ into $u_n = a + (n-1)d$, we have

$u_9 = 3 + (9 - 1) \times 2 = 19$

(b) Find the number of terms of the sequence 50, 48 ... 2.

Solution Here $a = 50, d = -2$ and $u_n = 2$. Using $u_n = a + (n-1)d$ we have

$$2 = 50 + (n-1) \times -2$$

$$24 = n - 1$$

$$n = 25$$

(c) The third and sixth terms of a linear sequence are 17 and 35. Find the first term and the common difference.

Solution If the first term is a and the common difference is d, then the third and the sixth terms are $a + 2d$ and $a + 5d$ respectively. So

$$a + 2d = 17 \qquad (1)$$

$$a + 5d = 35 \qquad (2)$$

Subtracting (1) from (2) gives

$$3d = 18$$

$$d = 6$$

Substituting $d = 6$ into (1) we have

$$a + 12 = 17$$

$$a = 5$$

Thus the first term is 5 and the common difference is 6.

The same results are obtained as follows.

u_6 is the third term from u_3 so $u_6 = u_3 + 3d$

Substituting 17 for u_3 and 35 for u_6 we have

$$17 + 3d = 35$$

$$3d = 18$$

$$d = 6$$

Now $a + 2d = 17$

Substitute 6 for d and solve for a.

$$a + 12 = 17$$
$$a = 5$$

Exercise 15.1(b)

1. Write down the terms indicated in each of the following linear sequence.

(a) 2, 10, 18, 26 ...   15 th

(b) 9, 6, 3, 0 ...   24 th

(c) $1, 2\frac{1}{2}, 4, 5\frac{1}{2}$ ...   18 th

(d) 60, 58, 56, 54 ...   30 th

(e) $8, 7\frac{1}{2}, 7, 6\frac{1}{2}$ ...   21 st

2. Find the number of terms in each linear sequence.

(a) 3, 5, 7 ... 47

(b) 48, 45, 43 ... 12

(c) 3, -8, -19 ... -129

(d) $1, 3\frac{1}{2}, 6$ ... 101

(e) 10, 9, 8 ... -20

3. The second and fifth terms of a linear sequence are 7 and 19. Find the first term and the common difference.

4. The first term of a linear sequence is 26 and the fourth term is 11. Find the common difference.

5. The sixth term of a linear sequence is twice the third term and the first term is 5. Find the common difference and the tenth term.

6. The third term of a linear sequence is 8 and the sixth term is 17. Find the first three terms of the sequence.

7. The ninth term of a linear sequence is three times the second term and the fifth term is 13. Find the first three terms of the sequence.

8. A boy saved 20 Gp of his pocket money on the first day of June, 25 Gp on the second day, 30 Gp on the third day and so on. Find the amount he saved on the last day of June.

9. A job is advertised at a starting salary of GH¢ 25,000 with an annual increments of GH¢ 1,200. Find the salary of an employee during his eight year of work.

10. A man's starting pay was GH¢ 2,500 increasing to GH¢ 6,700 in 15 years. How much increase did he receive each year?

11. A man accepts a job in a factory that pays a starting salary of GH¢ 5, 428 with annual increments of GH¢ 450. If the man retired on a salary of GH¢ 10, 378 how many years did he work at the factory?

12. An auditorium has 20 seats in the first row. Each successive row has two more seats than the previous row. How many seats are in the twelfth row?

Sum of Linear Sequence

If a and u_n are the first term and the n th term respectively of a linear sequence $a, a + d, a + 2d \ldots u_n$ then the sum S_n of the first n term of the sequence is

$$S_n = a + (a + d) + (a + 2d) + \cdots + u_n \qquad (i)$$

Writing this backward we have

$$S_n = u_n + (u_n - d) + (u_n - 2d) + \cdots + a \qquad (ii)$$

Adding (i) and (ii) gives

$$2S_n = (a + u_n) + (a + u_n) + (a + u_n) + \cdots + (a + u_n)$$

Since there are n terms we have

$$2S_n = n(a + u_n)$$

Dividing each side by 2 we get

$$S_n = \frac{n}{2}(a + u_n) \qquad (1)$$

But $u_n = a + (n - 1)d$

Therefore $S_n = \frac{n}{2}[a + a + (n - 1)d]$

$$= \frac{n}{2}[2a + (n - 1)d] \qquad (2)$$

Example 2

(a) Find the sum of the linear sequence $3 + 7 + 11 + \cdots + 35$.

Solution Begin by finding the number of terms of the sequence.

Using $u_n = a + (n - 1)d$ we have

$$3 + (n - 1) \times 4 = 35$$

$$n - 1 = 8$$

$$n = 9$$

Now using $S_n = \frac{n}{2}(a + u_n)$ we have

$$S_9 = \frac{9}{2}(3 + 35)$$

$$= 171$$

(b) Find the sum of the first twelve terms of the linear sequence $5 + 8 + 11 + \cdots$

Solution Here $a = 5, d = 3$ and $n = 12$

Using $S_n = \frac{n}{2}[2a + (n-1)d]$ we get

$$S_{12} = \frac{12}{2}[2(5) + (12 - 1) \times 3]$$

$$= 6(10 + 33)$$

$$= 258$$

(c) In a linear sequence the sum of the first six terms is 135 and the eighth term is three times the second term. Find the first term and the sum of the first 20 terms.

Solution Using $S_n = \frac{n}{2}[2a + (n-1)d]$ we get

$$S_6 = \frac{6}{2}(2a + 5d) = 135$$

i.e. $\qquad 2a + 5d = 45 \qquad\qquad (1)$

Now using $u_n = a + (n-1)d$ we have $u_2 = a + d$ and $u_8 = a + 7d$.

Therefore $a + 7d = 3(a + d)$

$$-a + 2d = 0 \qquad\qquad (2)$$

Solving equation (1) and (2) gives $d = 5$ and $a = 10$.

So the first term is 10.

The sum of the first twenty terms is given by

$$S_{20} = \frac{20}{2}[2(10) + (20 - 1) \times 5]$$

$$= 10(20 + 95)$$

$$= 1150$$

Exercise 15.1(c)

1. Find the sum of each of the following linear sequences:

(a) $2 + 4 + 6 + \cdots + 100$

(b) $3 + 8 + 13 + \cdots + 78$

(c) $1 + 3\frac{1}{2} + 6 + \cdots + 101$

(d) $-9 - 6 - 3 + \cdots + 51$

(e) $76 + 72 + 68 + \cdots - 48$

2. Find the sum of each of the following linear sequences as far as the terms indicated:

(a) $5 + 11 + \cdots$ 12 th

(b) $17 + 15 + \cdots$ 20 th

(c) $7 + 8 + \cdots$ 10 th

(d) $19 + 12 + \cdots$ 16 th

(e) $1 + 2\frac{1}{2} + \cdots$ 12 th

3. Find the sum of the terms between 1 and 100 divisible by 3.

4. Find the sum of the numbers from 1 to 100 inclusive which are not divisible by 4.

5. The sum of a linear sequence is 144, the first term is 3 and the final term is 15. Find the common difference.

6. The sum of the first and last terms of a linear sequence is 42. The sum of all the terms is 420. The second term is 4. Find the common difference.

7. The fifth term of a linear sequence is 24 and the sum of the first five terms is 80. Find the first term, the common difference and the sum of the first fifteen terms of the sequence.

8. Find how many terms of the linear sequence $3 + 5 = 7 + 9 + \cdots$ have a sum of 624.

9. The sum of the first n terms of a linear sequence is 360 and the sum of the first $(n - 1)$ th terms is 308. If the first term of the sequence is 8, find the n th term and the number of terms of the sequence.

10. A man works in a company that pays GH¢ 8,750 the first year. If his salary is increased each year by GH¢ 450 how much money did he earn during his first six years of employment?

11. A boy who works part-time earn 25 Gp on the first day of the month, 50 Gp on the second day, 75 Gp on the third day and so on. Determine the total amount that he will earn during a 30-day mouth.

12. A free-falling object will fall 12 metre during the first second, 44 more metre during the second second, 76 more metre during the third second and so on. What is the total distance the object will fall in 10 seconds if this pattern continues?

13. A man starts work on an initial salary of GH¢ 2,500 per annum with the promise that his salary will increase yearly by GH¢ 250. In how many years will he received in total GH¢ 46,500?

14. Milk tins are stacked in a pile. The top row has 15 milk tins and the bottom row has 21 milk tins. If the pile has 7 rows how many milk tins are in the pile?

15. A boy stacks logs so that there are 20 logs in the bottom layer and each layer contains one log less than the layer below it. How many logs are in the pile?

16. A man accepted a job that offered him a starting salary of GH¢ 36,000 with an increase of GH¢ 2,000 at the end of each of the first 5 years. How much money will he have received during his first six years of work?

17. An auditorium has 18 seats in the first row. Each successive row has two more seats than the previous row. How many seats are in the first 15 rows?

18. Each swing of a pendulum is 25 centimetres shorter than the preceding swing. The first swing is 8 metres. Determine the distance travelled by the pendulum during the first 12 swings.

15.2 Exponential Sequence

A sequence in which each term after the first is obtained by multiplying the preceding term by the same number is called an exponential sequence or a geometric progression (GP). The ratio of any two consecutive term is a constant r,

$$r = \frac{u_n}{u_{n-1}}$$

called the common ratio.

For example, the common ratio r, of the following sequence, 4, 8, 16, 32, ... is 8/4 or 16/8 or 32/16, i.e., $r = 2$.

Exercise 15.2(a)

1. Find the common ratio of each sequence.

(a) 7, 14, 28, 56 ...

(b) 2, 6, 18, 54 ...

(c) -5, -0.5, -0.05, -0.005 ...

(d) $\frac{1}{3}, -\frac{1}{9}, \frac{1}{27}, -\frac{1}{81}$ ...

(e) $75, 15, 3, \frac{3}{5}$ ...

(f) $12, -4, \frac{4}{3}, -\frac{4}{9}$ ...

2. Determine whether each sequence is exponential sequence. If so find the common ratio.

(a) 64, 32, 16, 8 . . .   (b) 10, 15, 20, 25 . . .   (c) 5, 10, 20, 40 . . .

(c) 54, -18, 6, -2 . . .   (d) 1, 8, 27, 64, 125 . . .   (e) $1, -\frac{2}{3}, \frac{4}{9}, -\frac{8}{27}$ ...

3. Find the next terms of each exponential sequence.

(a) 4, 8, 16 . . .   (b) $6, 3, \frac{3}{2}$ ..   (c) 5, -10, 20 . . .

(d) $1, -\frac{1}{2}, \frac{1}{4}$ ...   (e) 3, 12, 48 . . .   (f) 1, 0.2, 0.04 . . .

The n th term of an Exponential Sequence

If a and r are the first term and common ratio of an exponential sequence then

$u_1 = a$

$u_2 = ar$

$u_3 = ar(r) = ar^2$

$u_4 = ar^2(r) = ar^3$

You can see that in each term the index of r is one less than the number of terms. From the above you can deduce that if a is the first term and r the common ratio of an exponential sequence then $u_n = ar^{n-1}$.

Example 3

(a) Write down the seventh term of the exponential sequence 6, 12, 24 . . .

Solution Substituting $a = 6, r = 2$ and $n = 7$ into $u_n = ar^{n-1}$, we get

$\qquad u_7 = 6 \times 2^6$

$\qquad\quad = 6 \times 64$

$\qquad\quad = 384$

(b) Find the number of terms of the exponential sequence $64, 32, 16, \cdots, \frac{1}{2}$.

Solution Substituting $a = 64, r = \frac{1}{2}$ and $u_n = \frac{1}{2}$ into $u_n = ar^{n-1}$ we get

$\qquad 64 \times \left(\frac{1}{2}\right)^{n-1} = \frac{1}{2}$

$$\left(\frac{1}{2}\right)^{n-1} = \frac{1}{128}$$

$$\left(\frac{1}{2}\right)^{n-1} = \left(\frac{1}{2}\right)^{7}$$

So $\qquad n - 1 = 7$

$\qquad\qquad n = 8$

(c) The fourth term of an exponential sequence is 3, the seventh term is 81. Find the first term and the common ratio.

Solution If the first term is a and the common ratio is r then $u_n = ar^{n-1}$.

Thus $\qquad ar^3 = 3 \qquad\qquad (1)$

and $\qquad ar^6 = 81 \qquad\qquad (2)$

Dividing Equation (2) by Equation (1) gives $r^3 = 27$

$$r = 3$$

Substituting $r = 3$ into Equation (1) gives

$$a \times 3^3 = 3$$

$$a = \frac{1}{9}$$

The first term is $\frac{1}{9}$ and the common ratio is 3.

The same results are obtained as follows.

u_7 is the third term from u_4 so $u_7 = u_4(r^3)$

Substituting 3 for u_4 and 81 for u_7 gives

$$3 \times r^3 = 81$$

$$r^3 = 27$$

$$r = 3$$

Now $u_4 = a \times 3^3$

Substituting 3 for u_4 we have

$$27a = 3$$

$a = \frac{1}{9}$

Exercise 15.2(b)

1. Find the indicated term of each exponential sequence.

(a) 2, 6, 18 . . .   6 th

(b) 8, 4, 2 . . .   7 th

(c) $6, 2, \frac{2}{3}$ ...   5 th

(d) 4, -8, 16 . . .   9 th

(e) $2, 1, \frac{1}{2}$ ...   5 th

2. Find the number of terms of each exponential sequence.

(a) 2, 4, 8... 256

(b) $9, 3, 1 ... \frac{1}{243}$

(c) $32, 16, 8 ... \frac{1}{8}$

(d) $\frac{1}{27}, \frac{1}{9}, \frac{1}{3}, ... 81$

(e) $\frac{1}{32}, \frac{1}{16}, \frac{1}{8} ... 16$

3. The sixth term of an exponential sequence is 192 and the third term is 24. Find the first term and the common ratio.

4. The third term of an exponential sequence is 1 and the fifth is 9. Find two possible values of the common ratio and hence find two possible values for the first term.

5. The second, fourth and eight terms of a linear sequence form consecutive terms of an exponential sequence and the first term is 8. Find the common ratio of the exponential sequence.

6. The second, third and fourth terms of an exponential sequence are $(n + 6), n$ and $(n - 4)$. Find n and hence find the first term of the sequence.

7. A man bought a machine for GH¢ 76,800. The machine depreciates at the rate of 25 % per year. Find the value of the machine at the end of 5 years.

8. A village of 5,000 people is growing at the rate of 2 % per year. Estimate the population of the village at the end of 29 years.

9. A deposit of GH¢ 500 is made in an account that earns 7 % interest compounded yearly. Find the balance in this account after 20 years.

10. At the end of each year the value of a machine with an initial cost of GH¢ 32,000 is three-fourths what it was at the beginning of the year. Find the value of the machine 3 years after it was purchased.

Sum of an Exponential Sequence

If a and r are the first term and the common ratio respectively of an exponential sequence then the sum S_n of the first n terms is

$$S_n = a + ar + ar^2 + ar^3 + \cdots + ar^{n-1} \qquad (1)$$

Multiplying each side of Equation (1) by r we get

$$rS_n = ar + ar^2 + ar^3 + ar^4 + \cdots + ar^n \qquad (2)$$

Subtracting Equation (2) from Equation (1) gives

$$S_n - rS_n = a - ar^n$$

Factorising the expression on both sides yields

$$(1-r)S_n = a(1-r^n)$$

Dividing each side by $(1-r)$ gives

$$S_n = \frac{a(1-r^n)}{1-r}, \quad r \neq 1$$

Multiplying both the numerator and the denominator by -1 we obtain

$$S_n = \frac{a(r^n - 1)}{r - 1}, \quad r \neq 1$$

This is more convenient if r is greater than 1.

Example 4

(a) Find the sum of the first six terms of the exponential sequence 16, 8, 4...

Solution Here $a = 16$, $r = \frac{1}{2}$ and $n = 6$. Using the formula $S_n = \frac{a(1-r^n)}{1-r}$ we get

$$S_6 = \frac{16\left[1 - \left(\frac{1}{2}\right)^6\right]}{1 - \frac{1}{2}}$$

$$= 32\left[1 - \frac{1}{64}\right]$$

$$= 31\frac{1}{2}$$

(b) Find the sum of the first seven terms of the exponential sequence 2, 6, 18...

Solution Here $a = 2$, $r = 3$ and $n = 7$. Using the formula $S_n = \frac{a(r^2 - 1)}{r - 1}$ we get

$$S_7 = \frac{2(3^7-1)}{3-1}$$

$$= 3^7 - 1$$

$$= 2186$$

(c) How many terms of the exponential sequence 8, 12, 18 ... must be taken for the sum to exceed 10,000?

Solution Let us find n for the sum to be exactly 10,000. Substituting $a = 8$, $r = \frac{3}{2}$ $S_n = 10,000$ in the formula $S_n = \frac{a(r^n-1)}{r-1}$ we have

$$\frac{8\left[\left(\frac{3}{2}\right)^n - 1\right]}{\frac{3}{2}-1} = 10,000$$

$$1.5^n - 1 = 625$$

$$1.5^n = 626$$

Taking logarithm of both sides we get

$$n \log 1.5 = \log 626$$

$$n = \frac{\log 626}{\log 1.5}$$

$$= \frac{2.798}{0.1761}$$

$$= 15.89$$

Since n must be an integer the number of terms required to make a total exceeding 10,000 is 16.

Exercise 15.2(c)

1. Find the sum of each exponential sequence:

(a) $2 + 4 + 8 + \cdots$ 8 terms   (b) $9 + 3 + 1 + \cdots$ 6 terms

(c) $5 + 1 + \frac{1}{5} + \cdots$ 5 terms   (d) $8 - 4 + 2 - \cdots$ 6 terms

(e) $2 + 1 + \frac{1}{2} + \cdots$ 10 terms   (f) $3 - 6 + 12 - \cdots$ 8 terms

2. The following are the n term of an exponential sequence. Find the common ratio and write down the first four terms. Also find the sum of the first n terms.

(a) $-5\left(\frac{1}{5}\right)^n$ (b) $\frac{3^{n-1}}{9}$ (c) $\frac{2^n}{4}$ (d) $\frac{3^{n-1}}{2^n}$

3. Find the sixth term and the sum of the first 8 terms of the sequence 3, 6, 12...

4. The sum of the first 3 terms of an exponential sequence is 21. If the first term is 3, find the possible values of the common ratio and hence find the possible values for the sum of the first 5 terms.

5. The fifth and second terms of an exponential sequence are 81 and 24. What is the first term and the sum of the first five terms?

6. How many terms of the sequence 1, 3, 9, 27 ... must be taken for the sum to exceed 1,000?

7. How many terms of the sequence $1, \frac{1}{2}, \frac{1}{4} \cdots$ must be taken for the sum to exceed 1.995?

8. At the beginning of each year an man invests GH¢ 100 at 5 % compound interest. What is the total value of his investment at the end of the tenth year?

9. A man accepts a job that pays a salary of GH¢ 3,000 the first year. During the next 19 years he received a 5 % raise each year. What would his total salary be over a 20-year period?

10. A man invests GH¢ 1,000 at 5 % interest compounded annually in a savings account. Determine the amount in his account and the amount of interest earned at the end of 10 years.

11. A substance losses half its mass each hour. If there are initially 300 grams of the substance, find the number of hours after which only 37.5 grams of the substance remain.

12. In February 2000 the population of a village was about 2,810 people. If the population grows at a rate of 2.2 % per year, find the population after 10 years.

Sum to Infinity

Consider the infinite exponential sequence $u_1, u_2, u_3, \ldots u_n, \ldots$

If $|r| < 1$ then the sum of the first n terms is $\frac{a(1-r^n)}{1-r}$. Now as n becomes very large r^n gets closer to 0, so the sum of an infinite series has a limiting value $\frac{a}{1-r}$, called the sum to infinity. Thus the sum to infinity of an infinite exponential sequence is $S_\infty = \frac{a}{1-r}$.

Notice that this term is not dependant on the n th term of the sequence.

Example 6

(a) Calculate the sum to infinity of the sequence $1 + \frac{2}{3} + \frac{4}{9} + \cdots$

Solution Here $a = 1$ and $r = \frac{2}{3}$. Using the formula $S_\infty = \frac{a}{1-r}$ we get

$$S_\infty = \frac{1}{1-\frac{2}{3}}$$

$$= 3$$

(b) Express the recurring decimal $0.\overset{..}{1}\overset{..}{5}$ as a fraction.

Solution The recurring decimal $0.\overset{..}{1}\overset{..}{5} = 0.151515\ldots$ can be written as

$$\frac{15}{100} + \frac{15}{10,000} + \frac{15}{100,000} + \cdots$$

Finding the sum to infinity we have

$$S_\infty = \frac{\frac{15}{100}}{1-\frac{1}{100}}$$

$$= \frac{15}{99}$$

$$= \frac{5}{33}$$

Hence $0.\overset{..}{1}\overset{..}{5} = \frac{5}{33}$

Exercise 15.2(d)

1. Calculate the sum to infinity of each exponential sequence:

(a) $9 + 6 + 4 + \cdots$

(b) $12 + 6 + 3 + \cdots$

(c) $20 - 10 + 5 - \cdots$

(d) $1 + \frac{1}{3} + \frac{1}{9} + \cdots$

(e) $1 - \frac{1}{2} + \frac{1}{4} - \cdots$

(f) $0.3 + 0.03 + 0.003 + \cdots$

(g) $1 + 0.9 + 0.81 + \cdots$

(h) $\frac{5}{10} + \frac{5}{100} + \frac{5}{1,000} + \cdots$

2. Express the following recurring decimals as fractions:

(a) $0.\dot{6}$

(b) $0.\dot{6}\dot{3}$

(c) $0.\dot{2}\dot{7}$

(d) $0.1\dot{8}\dot{1}$

(e) $5.\dot{2}$

(f) $3.\dot{7}\dot{2}$

3. If the sum to infinity of an exponential sequence is two times the first term, what is the common ratio?

4. The sum to infinity of an exponential sequence is 27 and the second term is 6. Find the two possible values of the common ratio and the corresponding first term.

5. On each swing a certain pendulum travels 90 % as far as on its previous swing. If the first swing is 10 metres determine the total distance travelled by the pendulum by the time it comes to rest.

15.3 Recurrence Sequence

A sequence can be defined by a rule and one or more terms defined in terms of previous terms, called the initial condition(s). Sequences defined this way are said to be defined recursively, and the rule is called a recursive formula.

Example 7

(a) Compute the first six terms of the sequence defined by the following recursive formula and initial condition;

$u_n = 2u_{n-1} + 1$, $u_0 = 1$ (for all integers $n \geq 1$).

Solution Here the initial condition is given as $u_0 = 1$. Therefore we need to compute u_1, u_2, u_3, u_4 and u_5. Replace n in turn with 1, 2, 3, 4 and 5.

Substitute $n = 1$ to get $u_1 = 2u_0 + 1 = 2(1) + 1 = 3$

Substitute $n = 2$ to get $u_2 = 2u_1 + 1 = 2(3) + 1 = 7$

Substitute $n = 3$ to get $u_3 = 2u_2 + 1 = 2(7) + 1 = 15$

Substitute $n = 4$ to get $u_4 = 2u_3 + 1 = 2(15) + 1 = 31$

Substitute $n = 5$ to get $u_5 = 2u_4 + 1 = 2(31) + 1 = 63$

The first six terms of this sequence are 1, 3, 7, 15, 31 and 63.

(b) Given that $u_1 = 1$ and $u_{n+1} = 2 + u_n$, find the first five terms of the sequence and an expression for u_n in terms of n. Use your expression to find u_{20}.

Solution Replace n in turn with 1, 2, 3 and 4.

If $n = 1$ $u_2 = 2 + u_1 = 2 + 1 = 3$

If $n = 2$ $u_3 = 2 + u_2 = 2 + 3 = 5$

If $n = 3$ $u_4 = 2 + u_3 = 2 + 5 = 7$

If $n = 4$ $u_5 = 2 + u_4 = 2 + 7 = 9$

The first five terms of this sequence are 1, 3, 5, 7 and 9.

You can see that this is a linear sequence so

$$u_n = 1 + (n - 1) \times 2$$
$$= 2n - 1$$

Hence $u_{20} = 2(20) - 1$
$$= 39$$

Exercise 15.3

1. Find the first five terms of the following sequence defined recursively for all integers $n \geq 1$.

(a) $u_n = u_{n-1} + n,\ u_0 = 1$

(b) $u_n = nu_{n-1} - 1,\ u_0 = 1$

(c) $u_n = u_{n-1} + 2^n,\ u_0 = -1$

(d) $u_n = u_{n-1} + n^2,\ u_0 = 1$

(e) $u_{n+1} = n - u_n,\ u_1 = 1$

(f) $u_{n+1} = n + 3u_n,\ u_1 = 1$

2. Find the first five terms of the recursive sequence and an expression for the n th term in terms of n.

(a) $u_{n+1} = u_n + 3,\ u_1 = 2$

(b) $u_{n+1} = u_n - 2,\ u_1 = 3$

(c) $u_n = u_{n-1} + 5,\ u_0 = 3$

(d) $u_n = 2u_{n-1},\ u_0 = 1$

3. Given that $u_n = 4n + 1$ for $n \geq 1$. Find u_1, u_2 and u_3. Find also, the sum of the first 20 terms of the sequence.

4. A sequence is defined by the recursive formula $u_n = au_{n-1} + b$. If $u_1 = 3$, $u_2 = 5$ and $u_3 = 7$, find

(a) a and b

(b) u_0 and u_4

Review Exercise 15

1. Determine whether each sequence is linear or exponential sequence. Find also the common difference or the common ratio.

(a) -3, 1, 5, 9, 13 ...

(b) 5, 2, -1, -4, -7 ...

(c) -4, 8, -16, 32, -64...

(d) 7, 5, 11, 13, 15...

(e) 48, 24, 12, 6, 3...

(f) $\frac{1}{3}$, 1, 3, 9, 27 ...

(g) -8, 2, 12, 22, 32...

(h) 18, 6, 3, $\frac{2}{3}$, $\frac{2}{9}$ ...

2. Write down the terms indicated in each sequence.

(a) 7, 9, 11...   20 th

(b) 3, 6, 12...   8 th

(c) 4, -2, 1...   6 th

(d) 5, 12, 19...   10 th

(e) 8, 6, 4...   15 th

(f) 12, 4, $\frac{4}{3}$ ...   5 th

(g) -5, 2, 9 ....   12 th

(h) 15, 3, $\frac{3}{5}$ ...   5 th

3. Find the number of terms of each sequence.

(a) 2, 6, 10 ... 38

(b) 6, 3, $\frac{3}{2}$ ... $\frac{3}{256}$

(c) 5, 8, 11... 50

(d) 3, 8, 13 ... 73

(e) 2, 6, 18 ... 486

(f) 10, 8, 6 ... -12

(g) 5, 20, 80 ... 1,280

(h) 4, 2, 1 ... $\frac{1}{128}$

4. Find the sum of each sequence as far as the term indicated.

(a) 1, -3, 9....   10 th

(b) 5, 12, 19 ...   12 th

(c) 9, 6, 4....   12 th

(d) 2, 8, 14, 20...   25 th

(e) 3, -6, 12...   12 th

(f) 4, 12, 36...   8 th

(g) -50, -38, -26... 50 th

(h) 3.2, 3.8, 4.4... 12 th

5. Calculate the sum to infinity of each exponential sequence.

(a) $9 + 3 + 1 + \cdots$

(b) $24 + 12 + 6 + \cdots$

(c) $0.5 + 0.05 + 0.005 + \cdots$

(d) $1 + \frac{2}{3} + \frac{4}{9} + \cdots$

6. Express each of the following recurring decimals as a fraction.

(a) $0.\dot{5}$ (b) $0.\dot{7}$ (c) $0.\dot{9}\dot{3}$

(d) $0.5\dot{1}\dot{6}$ (e) $0.\dot{8}\dot{1}$ (f) $0.4\dot{8}\dot{6}$

7. Find the first five terms of the following sequence defined recursively for all integers $n \geq 1$:

(a) $u_{n+1} = n + u_n, \quad u_1 = -2$

(b) $u_n = u_{n-1} + 2^n, \quad u_0 = -1$

(c) $u_{n+1} = n + 3u_n, \quad u_1 = -2$

(d) $u_n = u_{n-1} + n^2, \quad u_0 = 1$

8. The first term of an arithmetic sequence is 4, and the last term is 31. If the sum of the sequence is 175, find the number of terms in the sequence and the common difference.

9. Find the sum of the first n terms of the linear sequence $3 + 5 + 7 + \cdots$. Find the value of n for which the sum of the first $2n$ terms will exceed the sum of the first n terms by 161.

10. The sum of n terms of a series is $2n^2$ for $n = 1, 2, 3, \ldots$. Show that the series is an arithmetic progression and find its n th term.

11. The sum to n terms of a certain linear sequence is $3n(n + 2)$. What is:

(a) its n th term?

(b) its common difference?

12. Find how many terms of the sequence $5 + 9 + 13 + 17 + \cdots$ have a sum of 495.

13. An exponential sequence (GP) is given by 27, 9, 3.... Find:

(a) the n th term

(b) the sum of the first n terms

(c) the sum for large values of n of the sequence

14. The third and the fifth terms of an exponential sequence (GP) of positive terms are 2 and $\frac{1}{2}$ respectively. Find:

(a) the common ratio

(b) the first term

(c) the sum of the first n terms of the sequence

15. Find the least number of terms of the exponential sequence (GP) $3 + 5 + 8\frac{1}{3} + \cdots$ that will give a sum greater than 100.

16. The n th term u_n of a sequence is given by $u_n = 2 \times 3^{n-1}$.

(a) Write down the first four terms of the sequence.

(b) Calculate the least value of n for which $u_n > 900$.

17. The fifth, ninth and sixteenth terms of a linear sequence (AP) are consecutive terms of an exponential sequence (GP) and the sum of the first and fourth terms is 17. Find the common ratio.

18. The numbers $u_1, u_2, u_3 \ldots$ satisfy the recurrence relation

$$u_{n+1} = u_n + \left(\frac{3}{4}\right)^n$$

for all possible integers $n \geq 1$. Find:

(a) u_2, u_3 and u_4

(b) an expression for u_n in terms of n

(c) the value of u_n for large values of n

19(a) Given that $u_0 = 5$ and $u_n = u_{n-1} + 4$ for $n > 1$ find:

(i) u_1, u_2 and u_3

(ii) a formula for u_n in terms of n

(b) Use your formula for u_n to deduce a formula for S_n, the sum of the first n terms of the sequence $u_1, u_2, ..., u_n$. Hence find S_{32}.

20. The n th term u_n of a sequence is given by

$$u_n = 5 + 7 + 9 + 11 + \cdots + (2n + 3)$$

Express u_n in the form $u_n = pn(qn + r)$, where p, q and r are constants. Hence find the least value of n for which $u_n \geq 96$.

21. Each row in a small theatre has one more seats than the preceding row. The front row seats 20 people and there are 20 rows of seats. How much should you change per a ticket in order to obtain GH¢ 15,000 for the sale of all seats in the theatre?

22. A man is given a starting salary of GH¢ 12,750 and is told he will receive a GH¢ 900 raise each year. How much will the salary be during the 10 th year? How much will be the total sum earned after 10 years?

23. A boy stacks logs so that there are 20 logs in the bottom layer and each layer contains one log less than the layer below it. If he stopped stacking the logs after completing the layer containing eight logs, how many logs are in the pile?

24. Each swing of a pendulum is 25 cm shorter than preceding swing. The first swing is 8 m.

(a) Find the length of the twelfth swing.

(b) Determine the distance travelled by the pendulum during the first 12 swings.

25. The number of a certain type of bacteria doubles every hour. If there are initially 1,000 bacteria after how many hours will the number of bacteria reached 128,000?

26. A company buys a machine for GH¢ 75,000. During the next 4 years the machine depreciates at the rate of 15 % per year. Find the value of the machine at the end of the fourth year.

27. A job is advertised at a starting salary of GH¢ 12,000 with an annual increase of 5 %. What will the salary be during the 10 th year? What will be the total sum earned after 10 years?

28. A man increased GH¢ 10,000 in a savings account paying 6 % interest annually. Find the amount in his account at the end of 8 years.

29. A man invests GH¢ 500 each year at 9 % compound interest. How much will he have in the bank just before his tenth investment?

30. The temperature of water in an ice cube tray placed in a freezer is 70° F. The temperature of the water at any time is 20 % less than it was one hour earlier. Find the temperature of the water six hours after it is placed in the freezer.

16 Trigonometry

16.1 Trigonometric Ratios

A ratio of lengths of two sides of a right-angled triangle is called a trigonometric ratio. The trigonometric ratios are related to the acute angles. The trigonometric ratio for a given angle is constant; it does not depend on the size of the triangle.

The three basic trigonometric ratios are sine, cosine and tangent, written in brief as sin, cos and tan.

Consider the right-angled triangle shown in Figure 16.1.

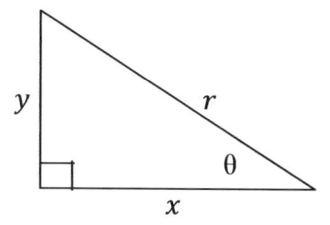

Figure 16.1

Let θ be an acute angle. Given the side opposite θ as y, the side adjacent as x and the hypotenuse r, we define sine, cosine and tangent of angle θ as:

$$sin\,\theta = \frac{y}{r}$$

$$cos\,\theta = \frac{x}{r}$$

and $tan\,\theta = \frac{y}{x}$

There are three other trigonometric ratios: cotangent (cot), secant (sec) and cosecant (cosec), defined by

$cot\,\theta = \frac{x}{y}$ or $cot\theta = \frac{1}{tan\theta}$

$sec\,\theta = \frac{r}{x}$ or $sec\theta = \frac{1}{cos\theta}$

$cosec\,\theta = \frac{r}{y}$ or $cosec\theta = \frac{1}{sin\theta}$

278　Further Mathematics

Figure 16.2(a)

Figure 16.2(b)

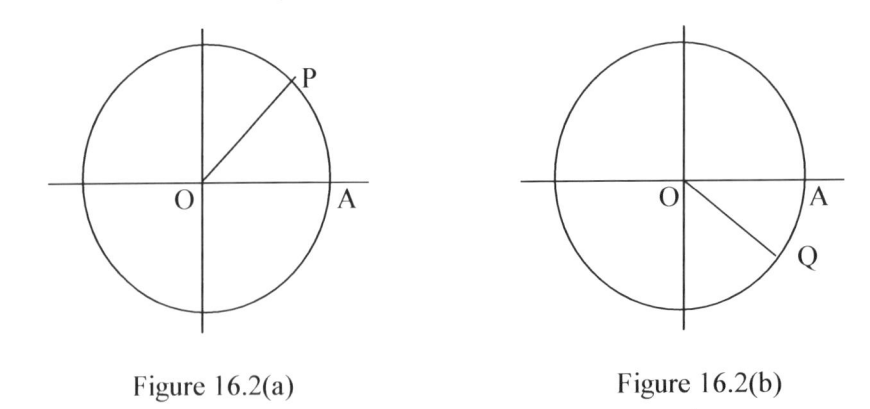

A straight line starting from OA can be rotated about the point O in an anticlockwise direction as shown in Figure 16.2(a) or in a clockwise direction as shown in Figure 16.2(b). The size of angle AOP is taken to be positive, and the size of angle AOQ is taken to be negative. By convention a positive angle is measured anticlockwise from the positive x-axis and a negative angle is measured clockwise from the positive x-axis.

Unit Circles

A unit circle is a circle with centre at the origin and radius 1 unit.

Consider the unit circle shown in Figure 16.3.

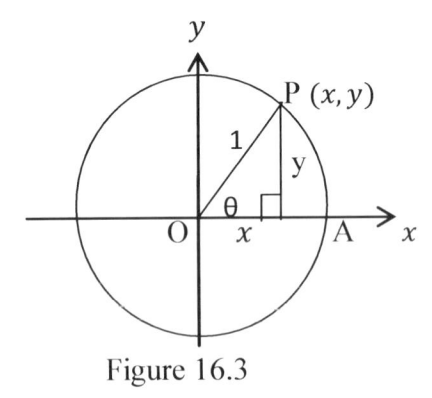

Figure 16.3

The point A with coordinates (1, 0) moves in an anticlockwise direction to the point $P(x, y)$. The size of the angle formed by the line OP and the x-axis is $\theta°$. Since the length of OP is 1 unit, you can see from the diagram that:

$\sin \theta = \frac{y}{1} = y$

$\cos \theta = \frac{x}{1} = x$

and $\tan \theta = \frac{y}{x} = \frac{\sin \theta}{\cos \theta}$

In general we define the sine of an angle to be the y-coordinate of a point on a unit circle and the cosine of the angle to be the x-coordinate of the point. The point P can be rotated several times around the circle. It follows that both $sin\theta$ and $cos\theta$ are defined for unlimited values of θ. Observe that both the x- and y- coordinates of any point on the unit circle vary from -1 to 1. Hence, for any angle θ we have $-1 \leq \sin\theta \leq 1$ and $-1 \leq \cos\theta \leq 1$.

Special Angles

Trigonometric functions of some angles have exact values and can be found without using a calculator. These angles are called special angles.

Trigonometric ratios of $0°, 90°, 180°, 270°$ and $360°$

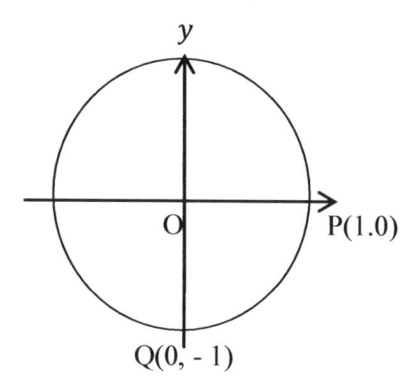

Figure 16.4

The point P starts at the point (1, 0) and moves in an anticlockwise direction to the point Q (0, -1). By the definition of the sine and cosine functions, we have $\sin 270° = -1$, $\cos 270° = 0$ and $\tan 270° = \frac{-1}{0}$, which is undefined. From the diagram shown in Figure 16.5 you can obtain the exact values of the trigonometric functions shown in the table on the next page.

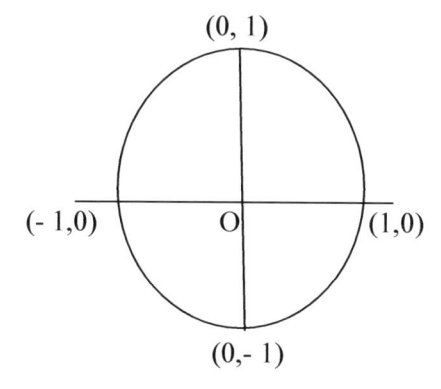

Figure 16.5

θ	sin	cos	tan
0	0	1	0
90	1	0	undefined
180	0	-1	0
270	-1	0	undefined
360	0	1	0

Trigonometric Ratios for 45°

The triangle shown in Figure 16.6 is a right-angled isosceles triangle. The equal sides are of length 1 unit.

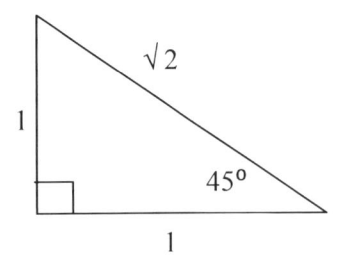

Figure 16.6

The Pythagoras's theorem gives the length of the hypotenuse as $\sqrt{1^2 + 1^2} = \sqrt{2}$.

You can obtain from the diagram the following trigonometric function values:

$\sin 45° = \dfrac{1}{\sqrt{2}} = \dfrac{\sqrt{2}}{2}$

$\cos 45° = \dfrac{1}{\sqrt{2}} = \dfrac{\sqrt{2}}{2}$

$\tan 45° = 1$

Trigonometric Ratios for 30° and 60°

The triangle shown in Figure 16.7 is an equilateral triangle of length 2 units. From plane geometry we know that any line drawn from a vertex perpendicular to a base bisects the base and the angle at the vertex.

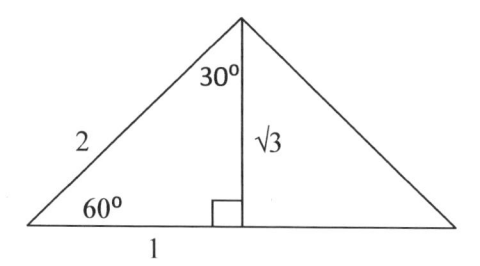

Figure 16.7

The Pythagoras' theorem gives the length of the altitude as $\sqrt{2^2 - 1^2} = \sqrt{3}$. You can obtain from the diagram the following trigonometric function values:

$\sin 60° = \frac{\sqrt{3}}{2}$ and $\sin 30° = \frac{1}{2}$

$\cos 60° = \frac{1}{2}$ $\cos 30° = \frac{\sqrt{3}}{2}$

$\tan 60° = \sqrt{3}$ $\tan 30° = \frac{1}{\sqrt{3}} = \frac{\sqrt{3}}{3}$

Trigonometric Functions of any Angle

Figure 16.8 shows a circle of radius r centre at the origin. A point P on the circle can be moved around the circle indefinitely. Hence the trigonometric functions are defined for unlimited values of angles.

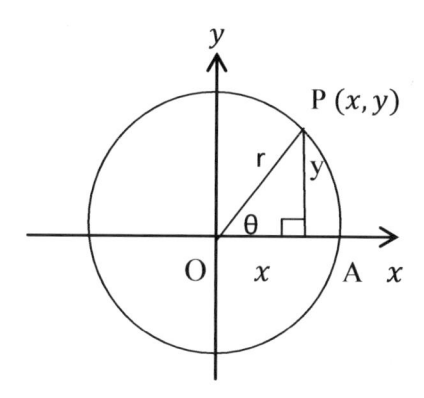

Figure 16.8

Suppose the point A moves to the point $P(x, y)$ in the first quadrant as shown in Figure 16.8. The line OP makes an angle θ with the positive x-axis. A line segment is drawn from P perpendicular to the x-axis to form a right-angled triangle with the vertex at O. From the triangle, we have

$\sin\theta = \frac{y}{r}$, $\cos\theta = \frac{x}{r}$ and $\tan\theta = \frac{y}{x}$

Since the signs of the x and y coordinates of P are both positive, $\sin\theta$, $\cos\theta$ and $\tan\theta$ are positive in the first quadrant. The signs of $\operatorname{cosec}\theta$, $\sec\theta$ and $\cot\theta$ are positive since these trigonometric ratios are the reciprocal of $\sin\theta$, $\cos\theta$ and $\tan\theta$ respectively.

The trigonometric functions of angles greater than 90° have the same values as their corresponding acute angles, called reference angles, but may differ in sign. When P moves into the second quadrant the signs of the coordinates will change; the x-coordinate is negative and the y-coordinate is positive. In this quadrant

$\sin\theta = \frac{y}{r} = +$, $\cos\theta = \frac{-x}{r} = -$ and $\tan\theta = \frac{y}{-x} = -$

The results in the following table were found in a similar way, using the definition of the trigonometric functions.

Quadrant	x	y	$\sin\theta = \frac{y}{r}$	$\cos\theta = \frac{x}{r}$	$\tan\theta = \frac{y}{x}$
1 st	+	+	+	+	+
2 nd	-	+	+	-	-
3 rd	-	-	-	-	+
4 th	+	-	-	+	-

You may observe that in the first quadrant all the trigonometric ratios of the angles are positive, in the second only sine is positive, in the third only tangent is positive and in the fourth only cosine is positive.

These results are summarised in the following diagram. The diagram shows the positive ratios in each of the quadrants

Sine	All
Tangent	Cosine

Starting from the fourth quadrant and moving in an anticlockwise direction the bold-face letters spell the word CAST.

Any angle can be placed in any one of the quadrants. Angles between 0° and 90° are in the first quadrant, angles between 90° and 180° are in the second quadrant, angles between 180° and 270° are in the third quadrant and finally angles between 270° and 360° are in the fourth quadrant.

The trigonometric functions of angles that are not acute angles have the same values as their reference angles but may differ in sign as illustrated in Example 1.

Example 1 Find the exact value of each of the trigonometric functions.

(a) sin 150°

Solution

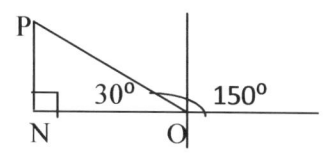

The angle 150° has its terminal side in the second quadrant. A line drawn from P perpendicular to the horizontal line forms a right-angled triangle. The reference angle of 150° is 180° − 150° = 30°. We know that sine is positive in the second quadrant. Thus sin 150° = sin 30° = 1/2.

(b) cos 225°

Solution

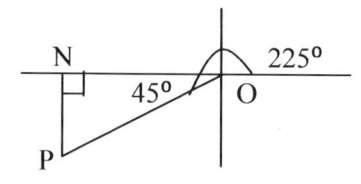

The angle 225° has its terminal side in the third quadrant as shown in the diagram above. The reference angle of 225° is 225° − 180° = 45°. We know that cosine is negative in the third quadrant. Thus cos 225° = − cos 45° = − √2/2.

(c) tan 330°

Solution

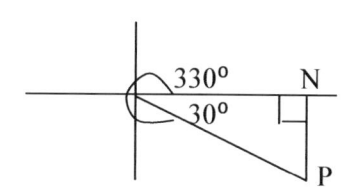

The angle 330° has its terminal side in the fourth quadrant, as shown in the diagram above. The reference angle of 330° is 360° − 330° = 30°. We know that tangent is negative in the fourth quadrant. Thus tan 330° = − tan 30° = − √3/2.

(d) cos 780°

Solution

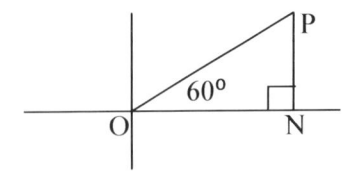

Because $780° - 2(360°) = 60°$, the angles $780°$ and $60°$ are coterminal. Since the terminal side of $780°$ is in the first quadrant then $cos 780°$ is positive, so $\cos 780° = \cos 60° = 1/2$. The results of Example 1 are summarised below:

1. For any acute angle θ in the second quadrant

$\sin(180 - \theta)° = \sin\theta$ $\qquad \cos(180 - \theta)° = -\cos\theta°$ $\qquad \tan(180 - \theta)° = -\tan\theta°$

2. For any acute angle θ in the third quadrant

$\sin(180 + \theta)° = -\sin\theta°$ $\qquad \cos(180 + \theta)° = -\cos\theta°$ $\qquad \tan(180 + \theta)° = \tan\theta°$

3. For any acute angle θ in the fourth quadrant

$\sin(360 - \theta)° = -\sin\theta°$ $\qquad \cos(360 - \theta)° = \cos\theta°$ $\qquad \tan(360 - \theta)° = -\tan\theta°$

Trigonometric Ratios of Negative Angles

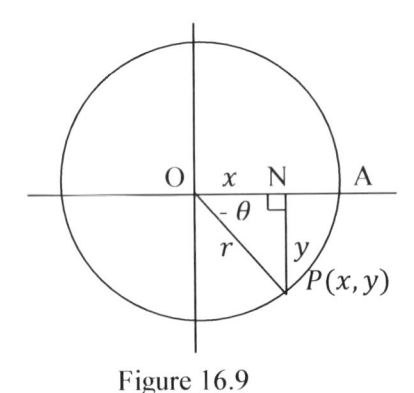

Figure 16.9

Let point A be moved clockwise to $P(x, y)$. PN is drawn perpendicular to form a right-angled triangle as shown in Figure 16.9. From triangle OPN

$\sin(-\theta) = \frac{-y}{r} = -\sin\theta$

$\cos(-\theta) = \frac{x}{r} = \cos\theta$

and $\tan(-\theta) = \frac{-y}{x} = -\tan\theta$

Example 2 Find the exact value of each trigonometric function.

(a) $\cos(-30°)$

Solution The angle $-30°$ is measured in a clockwise direction. This angle has its terminal side in the fourth quadrant as shown in the diagram below.

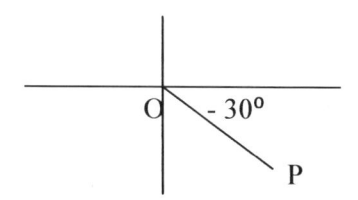

The angle $-30°$ is coterminal with angle $330°$. We know that cosine is positive in the fourth quadrant. Thus $\cos(-30°) = \cos 330° = \cos 30° = \sqrt{3}/2$

(b) $\sin(-210°)$

Solution The angle $-210°$ has its terminal side in the second quadrant as shown in the diagram below.

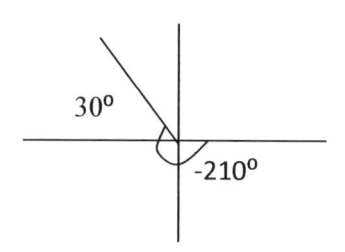

The angle -210 is coterminal with angle $150°$. The reference angle for both angle $-210°$ and angle $150°$ is $30°$. We know that sine is positive in the second quadrant. Thus $\sin(-210°) = \sin 150° = \sin 30° = 1/2$.

Exercise 16.1

1. Find the associated acute angle for:

(a) $110°$ (b) $200°$ (c) $260°$ (d) $320°$

(e) $216°$ (f) $480°$ (g) $-170°$ (h) $563°$

(i) $627°$ (j) $-401°$ (k) $-760°$ (l) $1270°$

2. State the quadrant in which the terminal side of each angle lies.

(a) $-135°$ (b) $265°$ (c) $315°$ (d) $-215°$

(e) $575°$ (f) $-330°$ (g) $780°$ (h) $-420°$

(i) $480°$ (j) $575°$ (k) $-315°$ (l) $-240°$

3. Find the values of the following, leaving surds in your answers:

(a) $\cos 150°$ (b) $\tan 120°$ (c) $\sin 210°$

(d) $\cot 225°$ (e) $\sin 135°$ (f) $\cos 270°$

(g) $\sec(-30°)$ (h) $\operatorname{cosec}(-120°)$ (i) $\cot(-135°)$

(j) $\tan(-60°)$ (k) $\cos 405°$ (l) $\sin(-270°)$

(m) $\sec 600°$ (n) $\tan 765°$ (o) $\sin(-540°)$

4. Find the maximum and minimum values of the following expressions, giving the smallest positive or zero value of θ for which they occur:

(a) $3\sin\frac{1}{2}\theta$ (b) $-2\cos\theta$ (c) $3 + 2\cos\theta$

(d) $1 - 2\sin 3\theta$ (e) $2\operatorname{cosec}\theta$ (f) $\frac{1}{2}\sec\theta - 1$

(g) $\frac{1}{3-2\sin\theta}$ (h) $\frac{3}{2\cos\theta+1}$ (i) $\frac{2}{4-3\operatorname{cosec}\theta}$

16.2 The Radian Measure

One unit of measure of angles is the radian. The radian is defined as the angle subtended at the centre of a circle by an arc of length equal to the radius of the circle. The abbreviation for radian is rad.

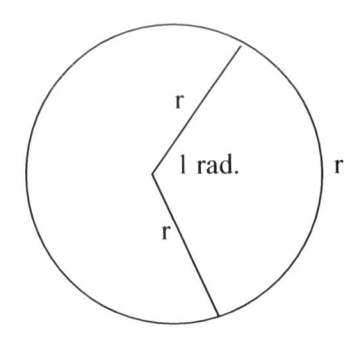

The length of the circumference of any circle is given by the formula $C = 2\pi r$. Then the angle representing one complete revolution of the circle is $2\pi \times 1\,rad = 2\pi\,rad$

Therefore $2\pi\,rad = 360°$

and $\pi\,rad = 180°$

The size of an angle can be measure in degree or radian. We can change angles in degrees to radians or angles in radians to degrees.

Changing degrees to radian

To change from degree to radian we multiply by $\frac{\pi}{180}$, since $1° = \frac{\pi}{180}$ radian.

Example 3 Convert the following angles to radian.

(a) $135°$

Solution $135° = 135 \times \frac{\pi}{180} = \frac{3}{4}\pi$ rad.

(b) $315°$

Solution $315° = 315 \times \frac{\pi}{180} = \frac{7}{4}\pi$ rad.

Often when we change degrees to radians we give the answer in multiples of π

Changing radians to degrees

To change radians to degrees we multiply by $\frac{180}{\pi}$, since 1 radian $= \frac{180°}{\pi}$.

Example 4 Convert the following angles to degrees

(a) $\frac{1}{4}\pi$ radian

Solution $\frac{1}{4}\pi$ rad. $= \frac{1}{4}\pi \times \frac{180}{\pi} = 45°$

(b) $\frac{3}{5}\pi$ radian

Solution $\frac{3}{5}\pi$ rad. $= \frac{3}{5}\pi \times \frac{180}{\pi} = 108°$

Exercise 16.2(a)

1. Convert each of the following angles to radians

(a) $30°$ (b) $60°$ (c) $90°$ (d) $120°$

(e) 150° (f) -210° (g) 225° (h) 270°

(i) 405° (j) -240° (k) -330° (l) 300°

2. Convert each of the following to degrees

(a) $\frac{1}{5}\pi$ rad. (b) $\frac{3}{4}\pi$ rad. (c) $\frac{2}{3}\pi$ rad. (d) 3π

(e) $-\frac{5}{4}\pi$ rad. (f) $\frac{4}{3}\pi$ rad. (g) $-\frac{7}{5}\pi$ rad. (h) $-\frac{5}{3}\pi$ rad.

(i) $\frac{4}{9}\pi$ rad. (j) $\frac{\pi}{10}$ rad. (k) -4π rad. (l) $\frac{5}{2}\pi$ rad.

3. Find the values of:

(a) $\cos\frac{\pi}{2}$ (b) $\sin\frac{\pi}{4}$ (c) $\tan\frac{\pi}{6}$ (d) $\sin\frac{\pi}{3}$

(e) $\cos\pi$ (f) $\sin\frac{3}{4}\pi$ (g) $\tan\frac{2}{3}\pi$ (h) $\sin\frac{3}{2}\pi$

(i) $\cos\frac{2}{3}\pi$ (j) $\sec(-2\pi)$ (k) $\cot(-\frac{4}{3}\pi)$ (l) $\csc(-\frac{5}{3}\pi)$

(m) $\cos(-\frac{16}{3}\pi)$ (n) $\sin(-\frac{9}{4}\pi)$ (o) $\tan(-\frac{5}{3}\pi)$

Length of arcs

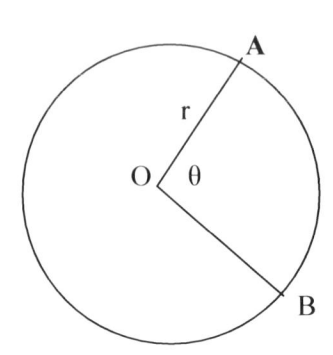

The arc AB subtends an angle θ radian at the centre of the circle. The length of the arc is a certain fraction of the length of the circumference. This fraction is equal to the ratio $\frac{\theta}{2\pi}$. Hence, the length of the arc $AB = \frac{\theta}{2\pi} \times 2\pi r = r\theta$.

Example 5

(a) Find the length of an arc which subtends an angle of 3 radians at the centre of a circle radius 4.5 cm.

Solution Length of arc = $r\theta$

$$= 4.5 \times 3$$

$$= 13.5 \text{ cm}$$

(b) An arc AB of a circle radius 10 cm subtends an angle 108° at the centre O. Find the perimeter of the sector containing the angle.

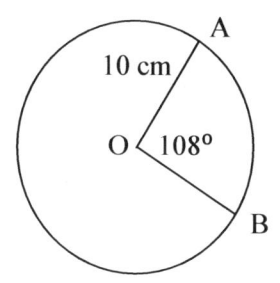

Solution Begin by converting 108° to radian

$$108° = 108 \times \frac{\pi}{180} = \frac{3}{5}\pi \text{ radian}$$

Length of arc AB = $10 \times \frac{3}{5}\pi = 6\pi = 18.85$ cm

The perimeter is the sum of the two radii and the length of the arc,.
i.e.   Perimeter = 10 + 10 + 18.85 = 38.85 cm

Area of Sectors

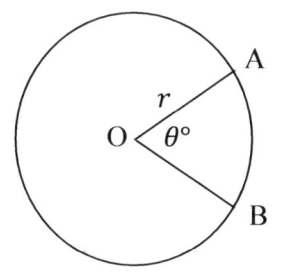

The area of a sector is a certain fraction of the total area of the circle. This fraction is $\frac{\theta}{2\pi}$. Hence, the area, A, of the sector is

$$A = \frac{\theta}{2\pi} \times \pi r^2 = \frac{1}{2}r^2\theta$$

Example 6

(a) Find the area of the sector containing an angle of 150° in a circle of radius 12 cm.

Solution Begin by converting 150° to radian

$$150° = 150 \times \frac{\pi}{180} = \frac{5}{6}\pi \text{ rad.}$$

Area of sector $= \frac{1}{2}r^2\theta$

$$= \frac{1}{2} \times 12^2 \times \frac{5}{6}\pi$$

$$= 60\pi \text{ cm}^2$$

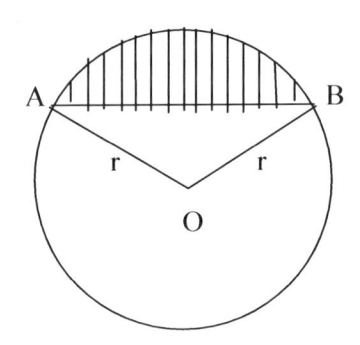

(b) The diagram shows two points A and B on the circumference of a circle of radius 6 cm, centre O. Given that angle AOB is $\frac{2}{3}\pi$ radian, find the area of the segment, shown shaded in the diagram.

Solution Area of sector AOB $= \frac{1}{2}r^2\theta$

$$= \frac{1}{2} \times 6^2 \times \frac{2}{3}\pi$$

$$= 37.7 \text{ cm}^2$$

Area of triangle AOB $= \frac{1}{2}r^2 \sin\theta$

$$= \frac{1}{2} \times 6^2 \times \sin\frac{2}{3}\pi$$

$$= 15.6 \text{ cm}^2$$

Area of segment = 37.7 − 15.6 = 22.1 cm^2

Exercise 16.2(b)

1. Find the length of an arc and area of circular sector whose radius and angle are:

(a) 15 cm, 0.72 rad.   (b) 8 m, 2.5 rad.   (c) 6 m, 1.2 rad.

(d) 12 cm, $\frac{2}{3}\pi$ rad.   (e) 20 cm, $\frac{3}{2}\pi$ rad.   (f) 15 m, 150°

2. Find in radian, the angle subtended at the centre of a circle with the indicated radius and arc length:

(a) 5 cm, 6cm   (b) 1.5 cm, 3.75 cm

(c) 21 cm, 44 cm   (d) 2.5 cm, 2 cm

3. The area of a sector of a circle radius 3 cm is 18 cm^2. Find the angle contained by the sector.

4. An arc of a circle subtends an angle of 1.5 radians at the centre. Find the radius of the circle if the length of the arc is 9 centimetres.

5. An arc of length 7.2 cm subtends an angle of 1.2 radians at the centre of a circle, what is the radius of the circle?

6. An arc subtends an angle of 2 radians at the centre of a circle and a sector of area 49 cm^2 is bounded by this arc and two radii. Find the radius of the circle.

7. The arc of a sector in a circle radius 3 cm is 4.5 cm long. Find the area of the sector.

8. An arc AB of a circle radius 4.2 centimetres subtends an angle 150° at the centre O. Find the perimeter of the sector containing the angle.

9. A chord AB 18 centimetres long subtend an angle of 90° at the centre O. Find the perimeter of the sector AOB containing the angle, correct to one decimal place.

10. A chord AB subtends an angle 120° at the centre of a circle radius 15 centimetres. Find the perimeter of the minor segment cut off by AB, correct to one decimal place.

11. A chord subtends 1.5 radians at the centre of a circle radius 12 centimetres. Find the areas of the minor and major segments.

12. A chord AB divides a circle of radius 4 centimetres into two segments. If AB subtends an angle of 135° at the centre of the circle, find the area of the major segment.

13. A circle of radius 1.5 centimetres has a sector with area 3.6 cm^2. Calculate the perimeter of the sector.

14.

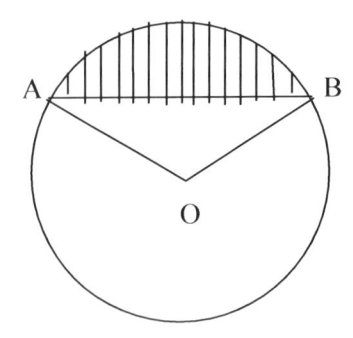

The diagram shows a circle with centre O and a chord AB. The radius of the circle is 15 centimetres and angle AOB is 1.2 radians.

(a) Find the perimeter of the region that is not shaded.

(b) Find the area of the region that is not shaded.

15.

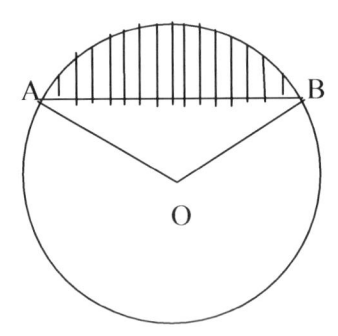

The diagram shows two points A and B on the circumference of a circle, centre O. Given that angle AOB is 1.5 radians, find the ratio of the area of the segment shown shaded in the diagram, to the area of the circle.

16. A chord AB subtends 120° at the centre O of a circle whose radius is 10 cm. Find:

(a) the length of the minor arc AB

(b) the area of triangle OAB

(c) the area of the minor segment cut off by AB

17. A chord AB, divides a circle of radius 6 centimetres into two segments. If AB subtends an angle of 60° at the centre of the circle, find the area of the minor segment.

18. Two circles, each of radius 10 centimetres have their centres 15 cm apart. Find the area which is common to both the circles.

16.3 Trigonometric Identities

If an angle θ corresponds to a point $P(x, y)$ on a unit circle $x^2 + y^2 = 1$, we obtain the following identity.

$$cos^2\theta + sin^2\theta = 1 \qquad (1)$$

Recall that we define $cos\theta$ and $sin\theta$ as the x- and y- coordinates respectively of a point on a unit circle.

Dividing (1) by $cos^2\theta$ gives

$$1 + tan^2\theta = sec^2\theta \qquad (2)$$

Also dividing (1) by $sin^2\theta$ gives

$$1 + cot^2\theta = cosec^2\theta \qquad (3)$$

Note that $cos^2\theta = (cos\theta)^2$ and $sin^2\theta = (sin\theta)^2$

Example 7 Simplify

(a) $\frac{1-cos^2\theta}{1-sin^2\theta}$

Solution From $cos^2\theta + sin^2\theta = 1$ we get $1 - cos^2\theta = sin^2\theta$ and $1 - sin^2\theta = cos^2\theta$

$$\frac{1-cos^2\theta}{1-sin^2\theta} = \frac{sin^2\theta}{cos^2\theta}$$

$$= tan^2\theta$$

(b) $\frac{sin^2\theta}{1-cos\theta}$

Solution $\frac{sin^2\theta}{1-cos\theta} = \frac{1-cos^2\theta}{1-cos\theta}$

$$= \frac{(1-cos\theta)(1+cos\theta)}{1-cos\theta}$$

$$= 1 + cos\theta$$

Exercise 16.3(a)

Simplify:

1. $sin\theta cot\theta$
2. $sec\theta cot\theta$
3. $(sec\theta - 1)(sec\theta + 1)$

4. $\dfrac{sec^2\theta - 1}{1 - cos^2\theta}$

5. $\dfrac{sin^2\theta}{1 - cos\theta}$

6. $\dfrac{cot^2\theta}{cosec\theta - 1}$

7. $\dfrac{1 - cos^2\theta}{1 - sin^2\theta}$

8. $\dfrac{cos\theta}{sin\theta\, cot^2\theta}$

9. $(1 - sin^2\theta)sec^2\theta$

10. $\dfrac{cos^2\theta}{1 + sin\theta} + \dfrac{cos^2\theta}{1 - sin\theta}$

11. $cot\theta + \dfrac{sin\theta}{1 + cos\theta}$

12. $(sin\theta + cos\theta)^2 - 1$

Proving an Identity

The proof of the equality of two expressions is called 'proving the identity'. The general method of procedure is to choose one side of the equation, and then express it in the same form as the expression on the other side. Sometimes it would be more convenient to reduce both sides to the same form.

Example 8

(a) Prove the following identity

$$\tan\theta + \cot\theta = \dfrac{1}{\sin\theta\,\cos\theta}$$

Solution We take the expression on the left hand side. It will be found convenient to express all the trigonometric ratios in terms of sine and cosine.

$$\tan\theta + \cot\theta = \dfrac{\sin\theta}{\cos\theta} + \dfrac{\cos\theta}{\sin\theta}$$

$$= \dfrac{\sin^2\theta + \cos^2\theta}{\sin\theta\,\cos\theta}$$

$$= \dfrac{1}{\sin\theta\,\cos\theta}$$

Hence $\tan\theta + \cot\theta = \dfrac{1}{\sin\theta\,\cos\theta}$

If we take the expression on the right hand side, we get the expression on left hand side.

$$\dfrac{1}{\sin\theta\,\cos\theta} = \dfrac{\sin^2\theta + \cos^2\theta}{\sin\theta\,\cos\theta}$$

$$= \dfrac{\sin^2\theta}{\sin\theta\,\cos\theta} + \dfrac{\cos^2\theta}{\sin\theta\,\cos\theta}$$

$$= \dfrac{\sin\theta}{\cos\theta} + \dfrac{\cos\theta}{\sin\theta}$$

$$= \tan\theta + \cot\theta$$

(b) Show that $\dfrac{\tan\theta}{\sec\theta-1} - \dfrac{\tan\theta}{\sec\theta+1} = 2\cot\theta$

Solution Taking the expression on the left hand side we have

$$\dfrac{\tan\theta}{\sec\theta-1} - \dfrac{\tan\theta}{\sec\theta+1} = \dfrac{\tan\theta(\sec\theta+1)-\tan\theta(\sec\theta-1)}{(\sec\theta-1)(\sec\theta+1)}$$

$$= \dfrac{2\tan\theta}{\sec^2\theta-1}$$

$$= \dfrac{2\tan\theta}{\tan^2\theta}$$

$$= \dfrac{2}{\tan\theta}$$

$$= 2\cot\theta$$

Hence $\dfrac{\tan\theta}{\sec\theta-1} - \dfrac{\tan\theta}{\sec\theta+1} = 2\cot\theta$

Exercise 16.3(b)

Prove the following identities:

1. $sin\theta cot\theta = cos\theta$

2. $cos\theta cosec\theta = cot\theta$

3. $(1-cos^2\theta)cosec^2\theta = 1$

4. $(1-cos^2\theta)sec^2\theta = tan^2\theta$

5. $(1+tan^2\theta)cos^2\theta = 1$

6. $sin^2\theta(1+cot^2\theta) = 1$

7. $\dfrac{1}{sec^2\theta} + \dfrac{1}{cosec^2\theta} = 1$

8. $sec\theta - tan\theta sin\theta = cos\theta$

9. $\dfrac{1}{1-sin\theta} + \dfrac{1}{1+sin\theta} = 2sec^2\theta$

10. $\dfrac{1-tan^2\theta}{1+tan^2\theta} = 1 - 2sin^2\theta$

11. $\dfrac{1}{1+sin^2\theta} + \dfrac{1}{1+cose^2\theta} = 1$

12. $\dfrac{2tan\theta}{1+tan^2\theta} = 2sin\theta cos\theta$

13. $(tan\theta + sec\theta)^2 = \dfrac{1+sin\theta}{1-sin\theta}$

14. $\dfrac{tan^2\theta}{1+tan^2\theta} \times \dfrac{1+cot^2\theta}{cot^2\theta} = sin^2\theta sec^2\theta$

15. $(cos\theta + sin\theta)^2 + (cos\theta - sin\theta)^2 = 2$

16. $(1 + tan\theta)^2 + (1 - tan\theta)^2 = 2sec^2\theta$

16.4 Solving Trigonometric Equations

An equation which contains at least one trigonometric function is called a trigonometric equation. You can use your knowledge in solving algebraic equations to solve trigonometric equations as shown in Examples 9.

Example 9

(a) Find the values of x between $0°$ and $360°$ which satisfy the equation $sin^2 x = \frac{3}{4}$.

Solution $sin^2 x = \frac{3}{4}$ Original equation

$sinx = \pm \frac{\sqrt{3}}{2}$ Take square of each side.

First we consider

$sinx = \frac{\sqrt{3}}{2}$

The angle whose sine is $\frac{\sqrt{3}}{2}$ is $60°$. The sine function is positive in the first and second quadrants. The angle in the second quadrant is $180° - 60° = 120°$. There are two solution between $0°$ and $360°$, namely $x = 60°$ and $120°$.

Next we consider

$sinx = -\frac{\sqrt{3}}{2}$

The sine function is negative in the third and fourth quadrants. In third quadrant the angle is $180° + 60° = 240°$. In the fourth quadrant the angle is $360° - 60° = 300°$. There are two solutions of this equation, namely, $240°$ and $300°$.

Hence the solutions of the equation between $0°$ and $360°$ are $x = 60°, 120°, 240°$ and $300°$.

(b) Solve, within the interval 0 and 2π, the equation $tanx + 2sinx = 0$

Solution $tanx + 2sinx = 0$ Original equation

$\frac{sinx}{cosx} + 2sinx = 0$ Replace $tanx$ by $\frac{sinx}{cosx}$

$sinx + 2sinxcosx = 0$ Multiply each side by $cosx$

$sinx(1 + 2cosx) = 0$ Factorise.

Therefore either $sinx = 0$ or $1 + 2cosx = 0$.

The solutions of $sinx = 0$ between 0 and 2π are $x = 0, \pi$ and 2π.

Now $1 + 2cosx = 0$

so $\qquad cosx = -\frac{1}{2}$

The acute angle whose cosine is $\frac{1}{2}$ is $\frac{\pi}{3}$. The cosine function is negative in the second and third quadrants. In the second quadrant the angle is $\pi - \frac{\pi}{3} = \frac{2\pi}{3}$. In the third quadrant the angle is $\pi + \frac{\pi}{3} = \frac{4\pi}{3}$.

Hence the solutions of the equation between 0 and 2π are $x = 0, \frac{2\pi}{3}, \pi, \frac{4\pi}{3}$ and 2π.

(c) Solve $2cos^2x + cosx - 1 = 0$ for values of x from -180° to +180° inclusive.

Solution $\quad 2cos^2x + cosx - 1 = 0 \qquad$ Original equation

This equation is a quadrant equation in $cosx$. We solve the equation by factorising.

$$(2cosx - 1)(cosx + 1) = 0$$

So $\qquad 2cosx - 1 = 0$ or $cosx + 1 = 0$

First we consider

$$2cosx - 1 = 0$$

$$cosx = \frac{1}{2}$$

The cosine function is positive in the first and fourth quadrants. In the first quadrant the angle is 60°. In the fourth quadrant the angle is - 60°.

Next consider

$$cosx + 1 = 0$$

$$cosx = -1$$

In this case the solutions are $x = 180°$ and - 180°.

Hence the solutions of the equation from - 180° to 180° are $x = \pm 60°, \pm 180°$.

Note: If the equation involves more than two functions, it will usually be best to express each function in terms of sine and cosine.

Example 16.4

1. Solve the following equations for values of x from $0°$ to $360°$ inclusive.

(a) $2\cos x = 1$

(b) $\tan x - 1 = 0$

(c) $2\sin^2 x = 1$

(d) $\cosec^2 x = 4\cot^2 x$

(e) $\sec^2 x + \tan^2 x = 7$

(f) $\cos^2 x + \sin x + 1 = 0$

(g) $2\sin^2 x + \cos x = 1$

(h) $\tan x = 4 - 3\cot x$

(i) $2\sin^2 x = 3\cos x$

(j) $2\cos^2 x + 4\sin^2 x = 3$

2. Solve in radians each of the following equations between 0 and 2π.

(a) $\sin x = \frac{1}{2}$

(b) $\cos x = \frac{1}{\sqrt{2}}$

(c) $\sin x = -1$

(d) $\tan x = -1$

(e) $\cos^2 x = \frac{1}{4}$

(f) $\frac{1}{3}\tan^2 x = 1$

(g) $\sec^2 x + \tan x = 1$

(h) $2\sin^2 x - 5\sin x + 2 = 0$

3. Solve each of the following equations for values of x from $-180°$ to $+180°$ inclusive.

(a) $\cos x = -\frac{1}{2}$

(b) $\tan x = 1$

(c) $\tan x = -\sqrt{3}$

(d) $\cos(x - 60)° = \frac{1}{2}$

(e) $\sin(x + 30)° = \frac{\sqrt{3}}{2}$

(f) $\tan x = 2\sin x$

(g) $\sec^2 x = 2\tan^2 x$

(h) $\sec x = 4\cos x$

(i) $4\sin x = 3\cosec x$

(j) $\sin^2 x + \sin x = 0$

(k) $\tan^2 x = \tan x$

(l) $2\cos^2 x - \cos x - 1 = 0$

(m) $2\sin^2 x + 3\sin x + 1 = 0$

(n) $\sin x = 4\cosec x + 3$

(o) $3\tan x + 2\cot x + 7 = 0$

(p) $\cos^2 x - \sin^2 x = \sin x$

16.5 Graphs of Trigonometric Functions

Graph of $\sin x$

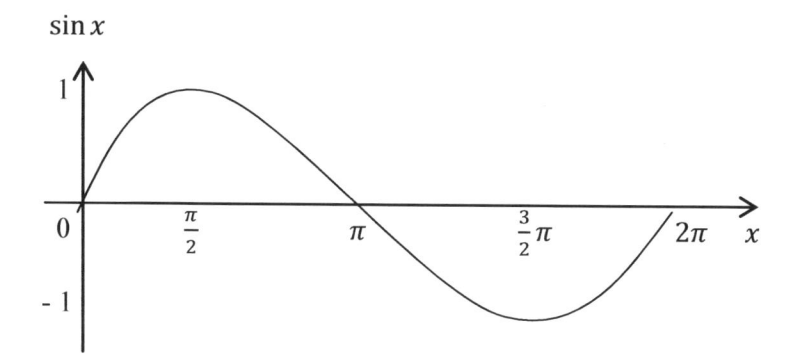

[Graph of sin x]

You can see that the sine curve has a maximum value of 1 when $x = \frac{\pi}{2}$ and a minimum value of −1 when $x = \frac{3}{2}\pi$. If further values are plotted outside the range $0 \leq x \leq 2\pi$, you will find that the curve repeats itself every 2π radians so the sine function is periodic with period 2π radians.

Graph of cos x

[Graph of cos x]

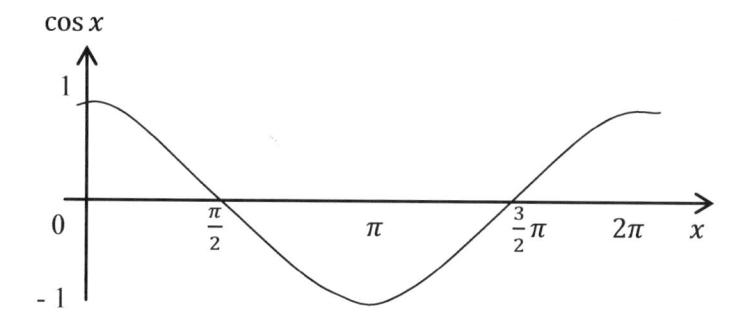

The cosine curve has a maximum value of 1 when $x = 0$ and a minimum value of −1 when $x = \pi$. The cosine curve is identical to the sine curve but is shifted $\frac{\pi}{2}$ radian to the left. The curve repeats itself every 2π radians so it is periodic with period 2π radians.

Graph of tan x

[Graph of tan x]

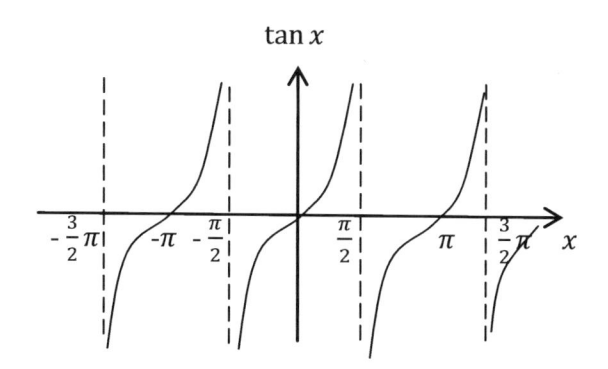

There is no maximum value of tan x, and no minimum value. Certain values of tan x are not defined. For example, the tangent of $\frac{\pi}{2}$ and $\frac{3}{2}\pi$ are not defined. The curve repeats itself every π radians.

Graphical Solution of Trigonometric Equations

By drawing up tables of values in a given interval, graphs of trigonometric functions can be plotted.

Example 10

Draw the graph of $y = 2\sin x + \cos x$ for $0° \leq x \leq 180°$ at intervals of $20°$. Find, from your graph the values of x to the nearest degree for which $2\sin x + \cos x = 1.5$

Solution Begin by making a table of values.

x	0	20	40	60	80	100	120	140	160	180
$\sin x$	0	.342	.643	.866	.985	.985	.866	.643	.342	0
$2\sin x$	0	.684	1.286	1.732	1.970	1.970	1.732	1.266	.684	0
$\cos x$	1	.940	.766	.500	.174	-.174	.500	-.766	-.940	-1
y	1	1.62	2.05	2.23	2.14	1.80	1.23	.52	-.26	-1

Next plot the ten points shown in the table. Finally connect the points with a smooth curve as shown in Figure 16.17.

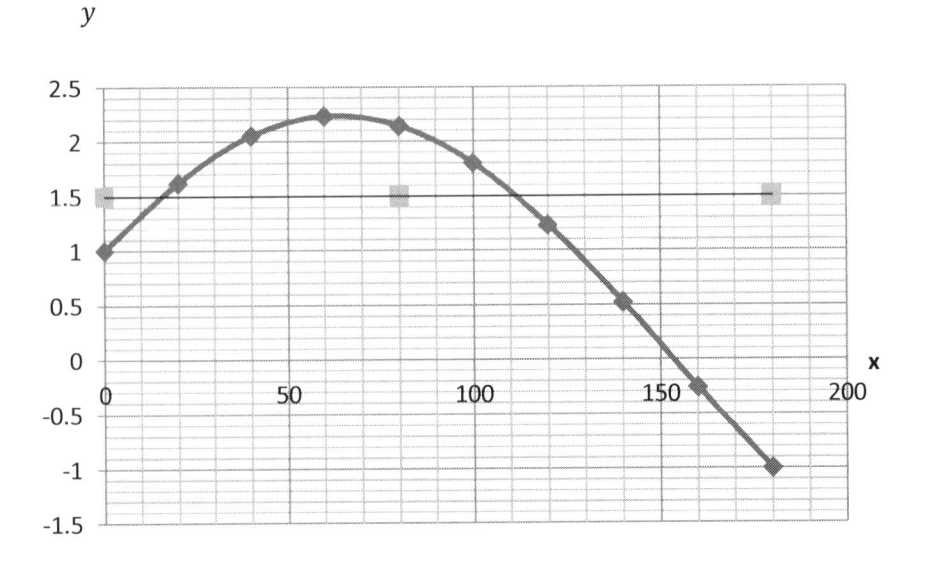

Figure 16.17

To solve $2\sin x + \cos x = 1.5$ we find the x- coordinates of points at which the graphs of $y = 2\sin x + \cos x$ and $y = 1.5$ intersect. Hence, from the graph $x = 16°$ and $x = 112°$.

Exercise 16.5

1. Draw the graph of the function $y = 1 - 2\cos x$ for values of x from $0°$ to $360°$, inclusive at intervals of $60°$. Use your graph to find:

(a) the maximum value of y

(b) the solution of the equation $4\cos x = -1$

2. Draw on the same axes the graphs of $y = \tan x$ and $y = 2 - \sin x$, for $0° \leq x \leq 60°$, at intervals of $10°$. Hence, solve the equation $\tan x + \sin x = 2$.

3. Draw on the same axes the graphs of $y = 3\tan x$ and $y = 1 + 2\cos x$ for $0° \leq x \leq 60°$ at intervals of $10°$. Use your graph to solve $2\cos x = 3\tan x - 1$

4. Draw the graph of the function $y = 3\cos x - \sin x$ for $0° \leq x \leq 80°$ at intervals of $10°$. Use your graph to find:

(a) the maximum value of y

(b) the truth set of the equation $3\cos x - \sin x = 2.75$

5. Draw the graph of the function $y = 2\sin x - \cos x$ for $90° \leq x \leq 160°$ at intervals of $10°$. Use your graph to find:

(a) the maximum value of y

(b) the solution of the equation $2\sin x - \cos x = 1.95$

6(a) Draw the graph of the function $y = 3\sin x + 2\cos x$ for values of x from $0°$ to $120°$ at intervals of $15°$.

(b) Use your graph to estimate, in the interval, the values of x for which $3\sin x + 2\cos x = 2.3$

7(a) Draw the graph of the function $y = 3\sin x + 4\cos x$ for $0° \leq x \leq 270°$ at intervals of $30°$.

(b) Find from your graph, the values of x, for which $3\sin x + 4\cos x = 4.5$

8. Draw the graph of $y = \sin x + 2\cos x$ for values of x from $-\pi$ to π at intervals of $\frac{\pi}{6}$. Find from your graph:

(a) the maximum and minimum values of y

(b) the values of x for which $y = 1$

9. Draw the graph of the function $y = 3sinx + 4cosx$ for $0° \leq x \leq 270°$ at intervals of $30°$. Use your graph to solve:

(a) $3sinx + 4cosx = 4.5$

(b) $cotx = -0.75$

10. Using a scale of 2 cm to $30°$ in the x-axis and 4 cm to 1 unit on the y-axis, draw on the same axes the graphs of $y = 2sinx$ and $y = \cos(x - 60°)$ for values of x from $0°$ to $360°$ at intervals of $30°$. Use your graph to solve $2sinx = \cos(x - 60°)$.

16.6 Compound Angles

When an angle is made up by the algebraic sum of two or more angles it is called a compound angle. Thus the trigonometric functions of angle $A + B$ and $A - B$ in terms of the function of A and B are stated below without prove.

$\sin(A + B) = sinAcosB + cosAsinB$

$\sin(A - B) = sinAcosB - cosAsinB$

$\cos(A + B) = cosAcosB - sinAsinB$

$\cos(A - B) = cosAcosB + sinAsinB$

Note that $\sin(A + B) \neq sinA + cosB$

A numerical example will illustrate this: if $A = 60°$, $B = 30°$ then
$\sin(A + B) = \sin(60 + 30)° = sin90° = 1$.

But $sinA + sinB = sin60° + sin30° = \frac{\sqrt{3}}{2} + \frac{1}{2} \neq 1$

Hence $\sin(A + B) \neq sinA + cosB$.

We can use the identities listed above to find $\tan(A + B)$.

$\tan(A + B) = \dfrac{\sin(A+B)}{\cos(A+B)}$

$= \dfrac{sinAcosB + cosAsinB}{cosAcosB - sinAsinB}$

Dividing both the numerator and denominator by $cosAcosB$ we get

$\tan(A + B) = \dfrac{\frac{sinA}{cosA} + \frac{sinB}{cosB}}{1 - \frac{sinAsinB}{cosAcosB}}$

$$= \frac{tanA+tanB}{1-tanAtanB}$$

Similarly, $\tan(A - B) = \frac{tanA-tanB}{1+tanAtanB}$.

The six identities above can be used to find the values of certain angles in terms of the standard angles.

Example 11 Find, without using a calculator, the values of:

(a) $cos75°$

Solution We write 75° as $45° + 30°$, so

$cos75° = \cos(45° + 30°)$

$\qquad = cos45°cos30° - sin45°sin30°$

$\qquad = \frac{\sqrt{2}}{2} \cdot \frac{\sqrt{3}}{2} - \frac{\sqrt{2}}{2} \cdot \frac{1}{2}$

$\qquad = \frac{\sqrt{2}}{4}(\sqrt{3} - 1)$

(b) $sin15°$

Solution We write 15° as $45° - 30°$, so

$sin15° = \sin(45° - 30°)$

$\qquad = sin45°cos30° - cos45°sin30°$

$\qquad = \frac{\sqrt{2}}{2} \cdot \frac{\sqrt{3}}{2} - \frac{\sqrt{2}}{2} \cdot \frac{1}{2}$

$\qquad = \frac{\sqrt{2}}{4}(\sqrt{3} - 1)$

(c) $\tan(-75°)$

Solution We write -75° as $45° - 120°$, so

$\tan(-75°) = \tan(45° - 120°)$

$\qquad = \frac{tan45°-tan120°}{1+tan45°tan120°}$

$\qquad = \frac{1+\sqrt{3}}{1-\sqrt{3}}$

$$= -\frac{1}{2}\left(1+\sqrt{3}\right)^2$$

(d) If $sinA = \frac{3}{5}$ and $cosB = \frac{5}{13}$, where A is obtuse and B is acute, find without using a calculator the value of $\cos(A - B)$.

Solution We draw two right triangles one with angle A and the second one with angle B as shown in Figure 16.18(a) and Figure 16.18(b).

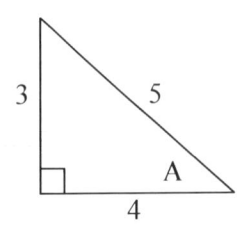

Figure 16.18(a)   Figure 16.18(b)

Using the appropriate right-angled triangle we have

$cosA = -\frac{4}{5}$ and $sinB = \frac{12}{13}$.

Now, $\cos(A - B) = cosAcosB + sinAsinB$

$$= -\frac{4}{5} \cdot \frac{5}{13} + \frac{3}{5} \cdot \frac{12}{13}$$

$$= -\frac{20}{65} + \frac{36}{65}$$

$$= \frac{16}{65}$$

Exercise 16.6

1. Find, leaving your answer in surds, the value of:

(a) $cos15°$
(b) $sin75°$
(c) $tan75°$

(d) $tan105°$
(e) $sin165°$
(f) $tan15°$

(g) $\sin(-15°)$
(h) $\cos(-75°)$
(i) $\tan(-15°)$

2. Simplify:

(a) $\sin(90° + x)$
(b) $\cos(360° + x)$
(c) $\tan(180° - x)$

3. Evaluate:

(a) $sin45°cos15° - cos45°sin15°$
(b) $cos80°cos20° + sin80°sin20°$

(c) $sin38°cos7° + cos38°sin7°$

(d) $cos72°cos63° - sin72°sin63°$

(e) $\dfrac{tan40°-tan10°}{1+tan40°tan10°}$

4. Simplify:

(a) $\dfrac{1}{2}cosx + \dfrac{\sqrt{3}}{2}sinx$

(b) $\dfrac{1}{\sqrt{2}}sinx - \dfrac{1}{\sqrt{2}}cosx$

(c) $\dfrac{\sqrt{3}+tanx}{1-\sqrt{3}tanx}$

(d) $\dfrac{1}{2}cos75° - \dfrac{\sqrt{3}}{2}sin75°$

5. If $sinA = \dfrac{4}{5}$ and $sinB = \dfrac{12}{13}$, where A is obtuse and B is acute, find without using a calculator the values of:

(a) $sin(A+B)$　　　(b) $cos(A+B)$　　　(c) $tan(A+B)$

6. If $sinA = \dfrac{3}{5}$ and $cosB = \dfrac{5}{13}$, where A is obtuse and B is acute, find without using a calculator, the values of:

(a) $sin(A-B)$　　　(b) $cos(A-B)$　　　(c) $tan(A+B)$

7. If $cosA = \dfrac{4}{5}$, where $180° < A < 270°$ and $tanB = \dfrac{5}{12}$, where $270° < B < 360°$, find without using a calculator the values of:

(a) $sin(A+B)$　　　(b) $tan(A-B)$　　　(c) $cos(A+B)$

8. Prove the following identities:

(a) $sin(A+B) + sin(A-B) = 2sinAcosB$

(b) $cos(A+B) - cos(A-B) = -2sinAsinB$

(c) $\dfrac{sin(A+B)}{cosAcosB} = tanA + tanB$

(d) $cos(A+B)cos(A-B) = cos^2A - sin^2B$

(e) $sin(A+B)sin(A-B) = cos^2B - cos^2A$

16.7 Double Angles

If we put A = B in the formula for the sums of angles we obtain results for sin 2A, cos 2A and tan 2A.

sin 2A = sin (A +A)

= sin A cos A + cos A sin A

$$= 2\sin A \cos A$$

$$\cos 2A = \cos(A + A)$$
$$= \cos A \cos A - \sin A \sin A$$
$$= \cos^2 A - \sin^2 A \qquad (1)$$

There are two other useful forms in which cos 2A may be expressed, one involving cos A only, the other sin A only.

From $\sin^2 A + \cos^2 A = 1$ we obtain $\cos^2 A = 1 - \sin^2 A$. Thus from (1) we get

$$\cos 2A = (1 - \sin^2 A) - \sin^2 A$$
$$= 1 - 2\sin^2 A \qquad (2)$$

Again from (1) we get

$$\cos 2A = \cos^2 A - (1 - \cos^2 A)$$
$$= 2\cos^2 A - 1 \qquad (3)$$

All three formulas for cos 2A are equivalent. Use the formula that works best with the information given.

$$\tan 2A = \tan(A + A)$$
$$= \frac{\tan A + \tan A}{1 - \tan A \tan A}$$
$$= \frac{2\tan A}{1 - \tan^2 A}$$

Expressing sin 2A and cos 2A in terms of tan A

$$\sin 2A = 2\sin A \cos A$$
$$= 2\sin A \cos A \times \frac{\cos A}{\cos A}$$
$$= 2\tan A \cos^2 A$$
$$= \frac{2\tan A}{1 + \tan^2 A}$$

$$\cos 2A = \cos^2 A - \sin^2 A$$

$$= \cos^2 A \left(1 - \frac{\sin^2 A}{\cos^2 A}\right)$$

$$= \frac{1 - \tan^2 A}{\sec^2 A}$$

$$= \frac{1 - \tan^2 A}{1 + \tan^2 A}$$

Example 12

(a) Use the formula for sin 2A to find sin 60°

Solution $\sin 60° = \sin(2 \times 30°)$

$$= 2\sin 30° \cos 30°$$

$$= 2 \times \frac{1}{2} \times \frac{\sqrt{3}}{2}$$

$$= \frac{\sqrt{3}}{2}$$

If $\cos\theta = \frac{3}{5}$, find $\cos 2\theta$.

Now $\cos 2\theta = 2\cos^2\theta - 1$

$$= 2\left(\frac{3}{5}\right)^2 - 1$$

$$= \frac{18}{25} - 1$$

$$= -\frac{7}{25}$$

(b) Prove the identity $\sin 3A = 3\sin A - 4\sin^3 A$

Solution Begin by writing 3A as 2A + A

$$sin3A = \sin(2A + A)$$

$$= sin2A cosA + cos2A sinA$$

$$= 2sinA cosA \cdot cosA + (1 - 2sin^2 A) \cdot sinA$$

$= 2sinAcos^2A + sinA - 2sin^3A$

$= 2sinA(1 - 2sin^2A) + sinA - 2sin^3A$

$= 2sinA - 2sin^3A + sinA - 2sin^3A$

$= 3sinA - 4sin^3A$

Similarly, writing cos 3A as cos(2A +A) and tan 3A as tan (2A + A) you can show that

Cos 3A = $4cos^3$ A – 3cos A and

$$\tan 3A = \frac{3\tan A - \tan^3 A}{1 - 3\tan^2 A}$$

We leave the proof of these identities as exercises.

Exercise 16.7

1. Evaluate without using a calculator or table:

(a) $2sin15°cos15°$

(b) $2cos^2 15° - 1$

(c) $1 - 2sn^2 22\frac{1}{2}°$

(d) $cos^2 67\frac{1}{2}° - sin^2 67\frac{1}{2}°$

(e) $1 - 2sin^2 75°$

(e) $\frac{2tan15°}{1-tan^2 15°}$

2. Find, without using a calculator, the values of $sin2\theta$ and $cos2\theta$ when:

(a) $cos\theta = \frac{12}{13}$

(b) $sin\theta = \frac{7}{25}$

(c) $tan\theta = \frac{12}{5}$

3. Solve the following equations for values of x from $0°$ to $360°$ inclusive.

(a) $cos2x + sinx + 2 = 0$

(b) $3sinx = 2sin2x$

(c) $cos2x = sinx$

(d) $3cos2x - cosx + 2 = 0$

(e) $3tan2x = 4$

(f) $tanxtan2x = 2$

4. Prove the following identities:

(a) $\frac{sin2A}{1+cos2A} = tanA$

(b) $\frac{sin2A}{1-cos2A} = cotA$

(c) $\frac{cotA-tanA}{cotA+tanA} = cos2A$

(d) $tanA + cotA = 2coses2A$

(e) $cos^4 A - sin^4 A = cos2A$

(f) $cotA - tanA = 2cot2A$

(g) $\dfrac{2-\sec^2 A}{\sec^2 A} = \cos 2A$ \qquad (h) $\dfrac{2\cot A}{1+\cot^2 A} = \sin 2A$

5. Prove the following identities:

(a) $\cos 3A = 4\cos^3 A - 3\cos A$ \qquad (b) $\tan 3A = \dfrac{3\tan A - \tan^3 A}{1 - 3\tan^2 A}$

(c) $\cot 3A = \dfrac{\cot^3 A - 3\cot A}{2\cot^2 A - 1}$ \qquad (d) $\dfrac{\sin 3A}{\sin A} - \dfrac{\cos 3A}{\cos A} = 2$

(e) $\dfrac{3\cos A + \cos 3A}{3\sin A - \sin 3A} = \cot^3 A$ \qquad (f) $\dfrac{\sin 3A + \sin^2 A}{\cos^3 A - \cos 3A} = \cot A$

16.8 An expression of the form $a\cos\theta + b\sin\theta$

We can rewrite the expression $a\cos\theta + b\sin\theta$ in the form $R\cos(\theta - \alpha)$

Suppose R and α exist such that

$R[\cos\theta\cos\alpha + \sin\theta\sin\alpha] = a\cos\theta + b\sin\theta$

Comparing the coefficients of $\cos\theta$ and $\sin\theta$ we have

$R\cos\alpha = a$ \qquad (1)

$R\sin\alpha = b$ \qquad (2)

Dividing (2) by (1) gives

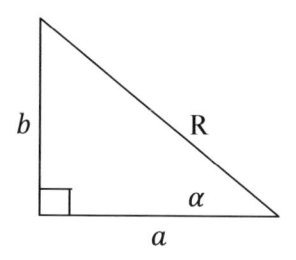

$\tan\alpha = \dfrac{b}{a}$

Figure 16.19

Figure 16.19 is a right-angled triangle, so by Pythagoras' theorem.

$R = \sqrt{a^2 + b^2}$

Thus $a\cos\theta + b\sin\theta = R\cos(\theta - \alpha)$

where $R = \sqrt{a^2 + b^2}$

and $\alpha = \tan^{-1}\left(\dfrac{b}{a}\right)$

Example 13

(a) Find the maximum and minimum values of $3\cos\theta + 4\sin\theta$, and the corresponding values of θ between $0°$ to $360°$.

Solution $R[\cos\theta\cos\alpha + \sin\theta\sin\alpha] = 3\cos\theta + 4\sin\theta$

So $\quad R\cos\alpha = 3$

$\quad\quad R\sin\alpha = 4$

Thus $R = 5$

And $\quad \tan\alpha = \dfrac{4}{3}$

$\quad\quad\quad\quad = 1.333$

$\quad\quad \alpha = 53.1°$

Thus $3\cos\theta + 4\sin\theta = 5\cos(\theta - 53.1°)$

The greatest value of $\cos\theta$ is 1, and this occurs when $\theta = 0°$. The least value of $\cos\theta$ is -1, this occur when $\theta = 180°$. Hence $5\cos(\theta - 53.1°)$ has a maximum value of 5 when $\theta - 53.1° = 0$, i.e. $\theta = 53.1°$. Also $5\cos(\theta - 53.1°)$ has a minimum value of -5 when $\theta - 53.1° = 180°$, i.e. $\theta = 233.1°$.

Hence the maximum and minimum values of $3\cos\theta + 4\sin\theta$ are 5 and -5; and are given by $\theta = 53.1°$ and $233.1°$ respectively.

(b) Solve the equation $\sqrt{3}\cos x - \sin x = 1$ for values of θ from $0°$ to $360°$, inclusive.

Solution $R[\cos x\cos\alpha - \sin x\sin\alpha] = \sqrt{3}\cos x - \sin x$

So $\quad R\cos\alpha = \sqrt{3}$

$\quad\quad R\sin\alpha = 1$

Thus $\quad R = 2$

and $\quad \tan\alpha = \dfrac{1}{\sqrt{3}}$

so $\quad \alpha = 30°$

Hence $\sqrt{3}\cos x - \sin x = 2\cos(x + 30°)$

Therefore $2\cos(x + 30°) = 1$

$$\cos(x + 30°) = \frac{1}{2}$$

We know that the cosine is positive in the first and fourth quadrants.

So $x + 30° = 60°$, i.e. $x = 30°$ or $x + 30° = 300°$, i.e. $x = 270°$

Hence $x = 30°, 270°$.

Exercise 16.8

1. Transform each of the following expressions into the form suggested:

(a) $\cos\theta - \sqrt{3}\sin\theta$ $R\cos(\theta + \alpha)$

(b) $3\cos\theta + \sin\theta$ $R\cos(\theta - \alpha)$

(c) $3\sin\theta - 4\cos\theta$ $R\sin(\theta - \alpha)$

(d) $2\sin\theta + 3\cos\theta$ $R\sin(\theta + \alpha)$

2. Find the maximum and minimum values of the following expressions, stating in each case the values from $0°$ to $360°$, of θ at which they occur.

(a) $\sin\theta + \cos\theta$ (b) $3\cos\theta + 2\sin\theta$ (c) $3\sin\theta - 4\cos\theta$ (d) $\sqrt{3}\cos\theta + \sin\theta$

3. Solve the following equations for values of x from $0°$ to $360°$, inclusive.

(a) $3\cos x - 4\sin x = 2$ (b) $3\cos x + 2\sin x = -1$ (c) $\sqrt{3}\cos x + \sin x = 1$

(d) $2\cos x + 7\sin x = 4$ (e) $\cos x + \sin x = \sqrt{2}$ (f) $2\cos x - \sin x = 2$

16.9 The Sine and Cosine Rules

In this section we would state, without prove, two important formulas that would enable us solve a triangle that does not contain a right angle.

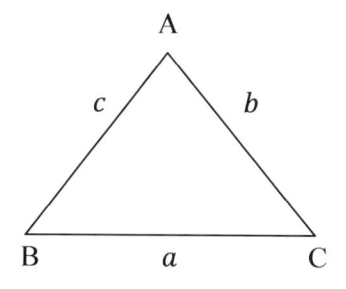

The sides of the triangle ABC are labelled as shown in the diagram above. Notice that a, b and c represent the sides opposite angles A, B and C respectively.

The Sine Rule

If ABC is a triangle with a, b and c representing the length of sides opposite angles A, B and C respectively, then

$$\frac{a}{\sin A} = \frac{b}{\sin B} = \frac{c}{\sin C}$$

The rule can be written as:

$$\frac{\sin A}{a} = \frac{\sin B}{b} = \frac{\sin C}{c}$$

The sine rule can be used to solve a triangle when given:

1. the sizes of two angles and the length of any side or

2. the lengths of two sides and the size of the angle opposite one of these sides of the triangle.

Example 14

Find the length of the side BC of triangle ABC given $B = 50°$, $C = 60°$ and $b = 8$ cm

Solution Draw and label the triangle

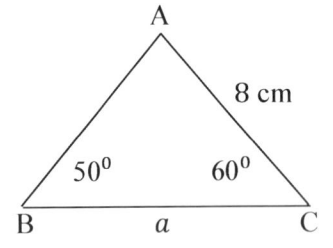

First find angle A.
$A + 50 + 60 = 180$

$$A = 70°$$

Next, use the sine rule

$$\frac{a}{\sin 70°} = \frac{8}{\sin 50°}$$

$$a = \frac{8 \sin 70°}{\sin 50°}$$

$$= 9.8 \text{ cm}$$

The length of BC is 9.8 cm

Cosine Rule

If a, b and c represents the sides opposite angles A, B and C respectively, then

$a^2 = b^2 + c^2 - 2bc\cos A$

Two other forms are:

$b^2 = a^2 + c^2 - 2ac\cos B$

$c^2 = a^2 + b^2 - 2ab\cos C$

The cosine rule can be used to solve a triangle when given:

1. the lengths of two sides and the size of the included triangle or

2. the lengths of the three sides

Example 15

Given triangle ABC with $A = 60°$, $B = 8$ cm and $c = 6$ cm, find the length of a.

Solution First, draw and label the triangle

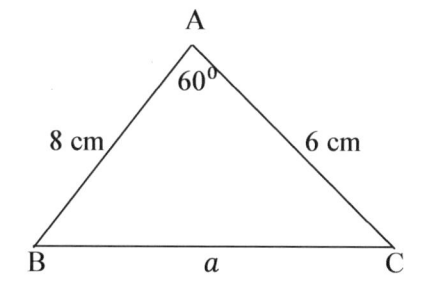

$a^2 = b^2 + c^2 - 2bc\cos A$

$= 6^2 + 8^2 - 2(6)(8)\cos 60°$

$= 52$

$a = 7.2$ cm

Solution of Triangles

A triangle has three sides and three angles. A triangle is said to be solved when all three sides and all the three angles are known.

Further Mathematics

The trigonometric ratios cannot be used to solve triangles which are not right-angled triangles. When at least two sides and an angle are given you can use the sine rule and the cosine rule to find three unknown quantities.

Solving Triangles

1. Given one side and two angles of triangle ABC

Example 16

Solve the triangle ABC given that $\angle A = 54°, \angle B = 61°, c = 12$.

Solution

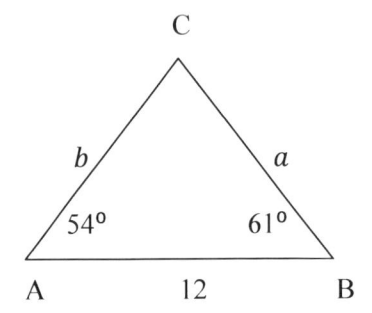

This is best solved as follows:

1. Find the third angle from $A + B + C = 180°$

2. Use the sine rule to find the other two sides

$$54° + 61° + C = 180°$$

$$C = 65°$$

By the Sine Rule

$$\frac{a}{\sin 54°} = \frac{12}{\sin 65°}$$

$$a = \frac{12 \sin 54°}{\sin 65°}$$

$$a = \frac{12 \times 0.8090}{0.9063}$$

$$= 10.7$$

Also $\dfrac{b}{\sin 61°} = \dfrac{12}{\sin 65°}$

$$b = \frac{12 sin 61°}{sin 65°}$$

$$= \frac{12 \times 0.8746}{0.9063}$$

$$= 11.6$$

Exercise 16.9(a)

Solve triangle ABC where

1. A = 60°, B = 40°, c = 7   2. A = 30°, C = 75°, b = 4.5   3. A = 45°, C = 70°, a = 8.2

4. A = 52°, B = 70°, a = 20   5. A = 87°, B = 63°, c = 3.2

2. Given two sides and an angle of triangle ABC

Example 17

Solve the triangle ABC when a = 8, b = 10 and $\angle A = 46°$.

Solution

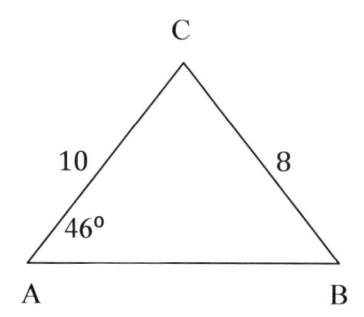

This is best solved as follows:

1. Use the sine rule to find one of the two remaining angles

2. Find the third angle from $A + B + C = 180°$

3. Use sine rule to find the third side

$$\frac{8}{sin 46°} = \frac{10}{sin B}$$

$$sin B = \frac{10 sin 46°}{8}$$

$$= \frac{10 \times 0.7193}{8}$$

$$= 0.8991$$

$$B = 64°$$

$$46° + 64° + C = 180°$$

$$C = 70°$$

$$\frac{c}{\sin 70°} = \frac{8}{\sin 46°}$$

$$c = \frac{8\sin 70°}{\sin 46°}$$

$$= 10.5$$

Exercise 16.9(b)

Solve triangle ABC where

1. A = 80°, a = 8, b = 5
2. A = 44°, a = 8, b = 5
3. B = 61.2°, b = 7.6, c = 5.7
4. A = 37°, a = 14, b = 12
5. C = 53.2°, b = 18.6, c = 16.1

3. Given two sides and the included angle of triangle ABC

Example 18

Solve the triangle ABC given that a = 8, b = 6 and $\angle C = 72°$

Solution

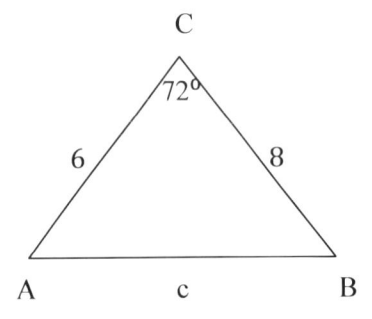

This is best solved as follows:

1. Use the cosine rule to find the third side

2. Use the sine rule to find the smaller of the two remaining angles

3. Find the third angle from $A + B + C = 180°$

Note that the smaller angle is the angle opposite the smaller side.

$$c^2 = a^2 + b^2 - 2ab\cos C$$

$$= 6^2 + 8^2 - 2(6)(8)\cos 72°$$

$$= 73.33$$

$$c = 8.6$$

$$\frac{\sin B}{6} = \frac{\sin 72°}{8.6}$$

$$\sin B = \frac{6\sin 72°}{8.6}$$

$$= 0.6636$$

$$B = 41.6°$$

$$A + 41.6° + 72° = 180°$$

$$A = 66.4°$$

Exercise 16.9(c)

Solve triangle ABC where

1. a = 32, b = 45, C = 87° 2. a = 50, b = 77, C = 78° 3. c = 13, a = 2, B = 30°

4. b = 5, c = 4, A = 60° 5. a = 13, b = 11, C = 120°

4. Given three sides of triangle ABC

Example 19

Solve the triangle ABC given that a = 7, b = 8 and c = 9

Solution

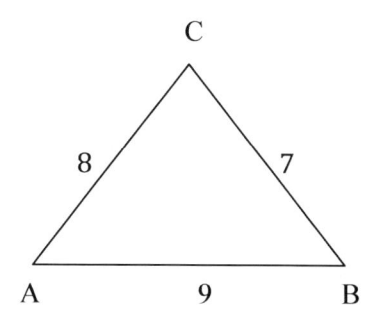

This is best solved as follows:

1. Use the cosine rule to find the smallest angle
2. Use the sine rule to find the smaller of the two remaining angles
3. Find the third angle from $A + B + C = 180°$

From $a^2 = b^2 + c^2 - 2bc\cos A$

$$\cos A = \frac{b^2 + c^2 - a^2}{2bc}$$

$$= \frac{8^2 + 9^2 - 7^2}{2(8)(9)}$$

$$= 0.6667$$

$$A = 48.2°$$

$$\frac{\sin B}{8} = \frac{\sin 48.2°}{7}$$

$$\sin B = \frac{8 \sin 48.2°}{7}$$

$$= \frac{8 \times 0.7455}{7}$$

$$= 0.8520$$

$$B = 58.4°$$

$$48.2° + 58.4° + C = 180°$$

$$C = 73.4°$$

Exercise 16.9(d)

Solve triangle ABC where

1. a = 6.0, b = 7.3, c = 4.8
2. a = 7, b = 9, c = 10
3. a = 7.8, b = 8.0, c = 12.2
4. a = 10, b = 5, c = 6
5. a = 10, b = 11, c = 13

Practical Application of Trigonometry

Some practical problem could be solved by solving triangles. This is illustrated in Example 20.

Example 20

Trigonometry

A man leaves a point walking at 5.6 km h^{-1} in a direction N 70° E. A cyclist leaves the same point at the same time in a direction S 60° E travelling a constant speed. Find the average speed of the cyclist if the walker and cyclist are 60 kilometres apart after 5 hours.

Solution

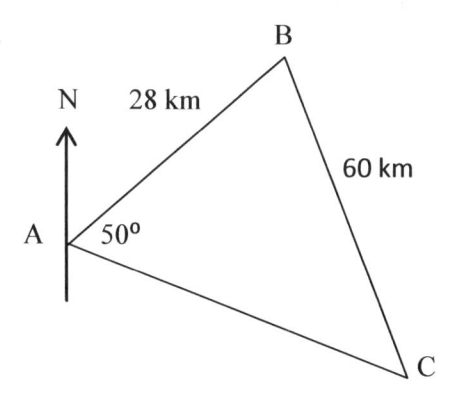

Figure 16.22

After 5 hours the walker has travelled $5 \times 5.6 = 28$ km (shown as AB in Figure 16.22). If AC is the distance the cyclist travels in 5 hours then BC = 60 km. Applying the sine rule gives

$$\frac{60}{\sin 50°} = \frac{28}{\sin C}$$

$$\sin C = \frac{28 \sin 50°}{60}$$

$$= 0.3575$$

$$C = 20.95°$$

Now $\quad 50° + 20.95° + B = 180°$

So $\quad\quad\quad\quad\quad B = 109.05°$

Applying the sine rule again

$$\frac{b}{\sin 109.05°} = \frac{60}{\sin 50°}$$

$$b = \frac{60 \sin 109.05°}{\sin 50°}$$

$$= 74.04 \text{ km}$$

Since the cyclist travels 74.04 km in 5 hours then the

$$Average\ speed = \frac{74.04}{5}$$

$$= 14.8$$

Hence the average speed of the cyclist is 14.8 km h^{-1}.

Exercise 16.10

1. A weight hangs by two strings AB, BC attached to two hooks A and C, fixed 6 metres apart in a horizontal beam. Given that AB = 4 m and BC = 5 m find:

(a) the angles which the strings make with the horizontal

(b) B's distance below the beam

2. A room 6.0 metre wide has a span roof which slopes at 33° on one side and 40° on the other. Find the length of the roof slopes correct to the nearest metre.

3. From a point Q in a plain, 210 metres away from a point P in the same plain, the bearing of P is N 40° E. The bearing from Q to a point R is S 65° E, and the points Q and R are 170 metres apart. Calculate the distance and bearing of Q from R.

4. The banks of a canal are straight, and parallel points A and B are marked on one bank 400 metres apart. C is a tree on the other bank. ∠BAC is 68° and ∠ABC is 73°. Find AC and the shortest distance between the banks.

5. A ship sails 15 kilometres on a bearing of 125°, and then for 12 kilometres on a bearing of 282°. How far is it away from its starting point?

6. Two sides of a triangular plot of land are 52 metres and 34 metres respectively. If the area of the plot is 620 m^2 find:

(a) the length of fencing required to enclose the plot and

(b) the angles of the triangular plot

7. A destroyer and a cruiser leave a harbour at 0800 hours, the destroyer at 24 knots on course 037° and the cruiser at 15 knots on course 139°. Find the bearing and the distance in nautical miles of the destroyer from the cruiser at 1300 hours (1 knot is a speed of 1 nautical mile per hour).

8. A ship P sails at a steady speed of 45 km h^{-1} in a direction of W 32° N (i.e. a bearing of 302°) from a port. At the same time another ship Q leaves the port at a steady speed of 35 km h^{-1} in the direction N 15° E (i.e. a bearing of 015°). Determine their distance apart after 4 hours.

9.

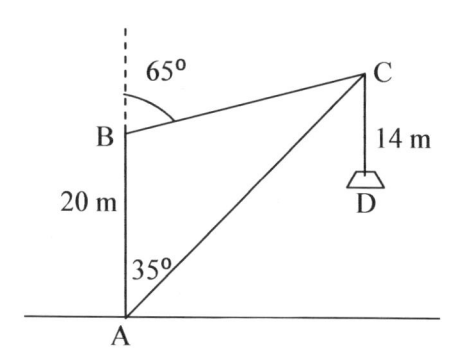

The figure above shows a crane with AB vertical and a head D. The tie BC makes 65° with the upward vertical. Find:

(a) the length of the jib, AC (b) the height of D above the level A.

10. A and B are two places on a straight level road, B being 7 kilometres due East of A. P is a church spine on the North side of the road such that AP = 3 km, BP = 5 km. Find the bearing of P from A.

Review Exercise 16

1. Give the associated acute angles of:

(a) 165° (b) 310° (c) -218° (d) -640°

2. State the quadrant in which the radius vector lies after describing the following angles:

(a) 183° (b) -208° (c) 697° (d) -385°

3. Find the values of the following leaving surds in your answers:

(a) $tan 135°$ (b) $sin 240°$ (c) $cos(-150°)$ (d) $sin(-570°)$

4. Find the maximum and minimum values of the following expressions, giving the smallest positive or zero value of θ for which they occur:

(a) $2cos\frac{1}{2}\theta°$ (b) $-2sin\theta°$ (c) $3 - 2sin\theta°$ (d) $1 - 2cos3\theta°$

5. Convert each of the following angle to radians:

(a) 30° (b) -45° (c) -150° (d) 330°

6. Convert each of the following to degrees:

(a) $\frac{\pi}{3}$ rad. (b) $\frac{5}{4}\pi$ rad. (c) $-\frac{7}{4}\pi$ rad. (d) 5π rad.

7. Find the length of an arc and area of a circular sector whose radius and angles are:

(a) 9 cm, 3 rad. (b) 15 m, 2.5 rad. (c) 8.5 cm, $\frac{3}{4}\pi$ rad.

8. Find, in radian, the angle subtended at the centre of a circle whose radius and arc length are:

(a) 15 cm, 9 cm (b) 12 m, 3 m (c) 14 cm, 0.33 m (d) 0.24 m, 4.8 cm

9. An arc of length 21.9 centimetres subtends an angle of 7.3 radians at the centre of a circle. Find the radius of the circle.

10. A pendulum 25 centimetres long swings through an arc of 47.5 centimetres. Find the number of radius through which the pendulum swings.

11. A wheel makes 12 revolutions a minute. Through how many radius does the wheel revolve in one minute?

12. The flywheel of an engine makes 35 revolutions in a second; how long will it take to turn through 10 radians? [Take $\pi = \frac{22}{7}$]

13. Simplify the following:

(a) $cos\theta tan\theta$

(b) $(cosec\theta - 1)(cosec\theta + 1)$

(c) $\frac{sin^2\theta}{1+cos\theta} + \frac{sin^2\theta}{1-cos\theta}$

(d) $\frac{1}{cos^2\theta} - \frac{1}{cot^2\theta}$

14. Prove the following identities:

(a) $(1 - sin^2\theta)sec^2\theta - 1$

(b) $(1 - sin^2\theta)cosec^2\theta = cot^2\theta$

(c) $\frac{1}{1-cos^2\theta} - \frac{1}{sec^2\theta-1} = 1$

(d) $\frac{tan\theta}{sec\theta} + \frac{tan\theta}{sec\theta+1} = 2cosec\theta$

15. Solve the following equations for values of x from $0°$ to $360°$ inclusive:

(a) $sec^2x + tanx = 1$ (b) $2cosecx = 5sinx$ (c) $6sin^2x - 5sinx - 1 = 0$

(d) $6cos^2x = 1 + cosx$ (e) $sec^2x + tan^2x = 7$ (f) $cotx + tanx = 2secx$

16. Solve the following equations for values of x from $-180°$ to $+180°$ inclusive:

(a) $2cos^2x + 3cosx + 1 = 0$ (b) $cosx = 4secx + 3$ (c) $sin^2x - cos^2x = cosx$

(d) $2sin^2x = 3cosx$ (e) $2cos^2x + 4sin^2x = 3$ (f) $sec^2x = 3tan^2x - 1$

17. Draw the graph of $y = \sin x + 2\cos x$ for $0° \leq x \leq 180°$ at intervals of $20°$. Use your graph to solve $\sin x + 2\cos x = -1$.

18. Draw the graph of $y = 2\sin x + \cos x$ for values of x from $0°$ to $180°$ at intervals of $20°$. Use your graph to find y when x is $106°$.

19. Find the values of the following in surd form:

(a) $\sin 75°$

(b) $\cos 105°$

(c) $\tan 75°$

(d) $\sin(-75°)$

(e) $\cos(-15°)$

(f) $\tan(-135°)$

20. Evaluate:

(a) $\sin 80° \cos 20° - \cos 80° \sin 20°$

(b) $\cos 45° \cos 15° + \sin 45° \sin 15°$

(c) $\dfrac{\tan 60° - \tan 15°}{1 + \tan 60° \tan 15°}$

21. If $\sin A = \dfrac{12}{13}$ and $\cos A = \dfrac{3}{5}$, where A and B are acute angles, find without using a calculator the values of:

(a) $\sin(A + B)$

(b) $\cos(A - B)$

(c) $\tan(A - B)$

22. If $\cos A = \dfrac{3}{5}$ and $\sin B = \dfrac{12}{13}$, where A and B are acute angles, find without using a calculator the values of:

(a) $\sin(A - B)$

(b) $\cos(A + B)$

(c) $\tan(A + B)$

23. Evaluate without using a calculator:

(a) $2\sin 30° \cos 30°$

(b) $2\cos^2 22\tfrac{1}{2}° - 1$

(c) $\dfrac{2\tan 60°}{1 - \tan^2 60°}$

(d) $1 - 2\sin^2 15°$

24. Solve the following equations for values of x from $0°$ to $360°$ inclusive:

(a) $\sin 2x + \cos x = 0$

(b) $2\cos 2x + 3\sin x - 3 = 0$

(c) $\cos 2x - 7\cos x = 3$

25. Prove the following identities:

(a) $\cot A - \cot 2A = \csc 2A$

(b) $\cot 2A + \tan A = \csc 2A$

(c) $1 + \tan 2A \tan A = \sec 2A$

(d) $\dfrac{\cot A - \tan A}{\cot A + \tan A} = \cos 2A$

26. Express the following functions in the form $A\sin(x + \alpha)$ and state their maximum and minimum values:

(a) $2\sin x + 3\cos x$ (b) $4\sin x - 3\cos x$

27. Solve the following equations for values of x between $0°$ and $360°$:

(a) $\sqrt{3}\sin x + \cos x = 1$ (b) $2\cos x - 3\sin x = 1$ (c) $4\sin x - 3\cos x = -2$

28. Solve each of the following triangles. Find angles correct to the nearest degree and sides to one decimal place:

(a) $\triangle ABC$, given $\angle A = 67°, \angle C = 42°, b = 25$

(b) $\triangle PQR$, given $\angle R = 80°, \angle Q = 49°, r = 8$

(c) $\triangle DEF$, given $\angle D = 52°, \angle E = 70°, d = 20$

29. Solve triangle ABC given:

(a) a = 32, b = 45, $\angle C = 87°$ (b) b = 50, c = 77, $\angle A = 78°$

(c) a = 6.3, c = 6.3, $\angle B = 54°$

30. Solve triangle ABC given:

(a) a = 5, b = 2.5, c = 3   (b) a = 3.9, b = 4.0, c = 6.1

(c) a = 9, b = 7.5, c = 5.5

31. The hour hand of a clock is 3 centimetres long and the minute hand is 4 centimetres. What is the distance between the tips of the two hands when the time shown is 4 o'clock?

32. Two ships sail from part in directions N 55° W and S 47° W at the rates of 7 and 8 kilometres per hour respectively. Find their distance apart at the end of an hour?

33. At noon a ship which is sailing a straight course due West at 10 kilometres an hour observes a lighthouse on a bearing of 327°. At 1:30 p.m. the lighthouse was on a bearing of 042°. Find the distance of the lighthouse from the first position of the ship.

34. From the roof of a house 30 metres high the angle of elevation of the top of a monument is 42°, and the angle of depression of its foot is 17°. Find its height.

35. A ship sailing due East at a constant speed was 12 kilometres from a lighthouse on a bearing of 020° at 1 p.m., and on a bearing of 315° at 2:10 p.m. Calculate her speed in kilometres per hour.

17 Matrices and Transformations

17.1 Matrix Algebra

A matrix (plural matrices) is a rectangular array of numbers arranged in rows and columns and commonly enclosed in square brackets. An example is $\begin{bmatrix} 3 & -1 & 5 \\ -2 & 0 & 1 \end{bmatrix}$.

Each number of the matrix is called an element or entry. The example above has two rows and three columns, written briefly as 2×3 (read 2 by 3 matrix). You can also say that the order of the matrix is 2×3. In general, if a matrix has m rows and n columns it is called $m \times n$ matrix, read "m by n" matrix. The expression $m \times n$ is called the order (or size) of the matrix. It is important to note that the number of rows is always given first.

Other examples are:

$$[5 \quad 6], \quad \begin{bmatrix} 4 \\ -3 \end{bmatrix}, \quad \begin{bmatrix} 2 & -3 \\ -1 & 0 \\ 3 & 4 \end{bmatrix} \quad \text{and} \quad \begin{bmatrix} 4 & 5 & 6 & 0 \\ -1 & 0 & 1 & -2 \\ 2 & -3 & 1 & 3 \end{bmatrix}$$

Square Matrix

A matrix that has the same number of rows as the number of columns is called a square matrix. For example, $\begin{bmatrix} -3 & 1 \\ 2 & 0 \end{bmatrix}$ is a 2×2 square matrix and $\begin{bmatrix} 0 & 4 & -1 \\ -2 & 0 & 1 \\ 3 & 2 & 1 \end{bmatrix}$ is a 3×3 square matrix. In general, a matrix with n rows and n columns is called a square matrix of order n.

Equal Matrices

Two matrices are said to be equal if they are of the same order and their corresponding elements are equal. Thus if $A = \begin{bmatrix} a & b \\ c & d \end{bmatrix}$ and $B = \begin{bmatrix} e & f \\ g & h \end{bmatrix}$ then $A = B$ if $a = e$, $b = f$, $c = g$ and $d = h$.

Example 1

Given that $\begin{bmatrix} 3x + y & 5 \\ -2 & -2x + y \end{bmatrix} = \begin{bmatrix} -1 & 5 \\ x & 3z \end{bmatrix}$ find the values of x, y and z

Solution Because the two matrices are equal

$$3x + y = -1 \qquad (1)$$

and $\quad x = -2 \quad$ (2)

Substituting $x = -2$ into (1) gives

$$3(-2) + y = -1$$

$$y = 5$$

Also $\quad -2x + y = 3z$

$$-2(-2) + 5 = 3z \qquad \text{Substitute } -2 \text{ for } x \text{ and } 5 \text{ for } y.$$

$$z = 3$$

Adding Matrices

If A and B are two matrices of the same order, their sum $A + B$ is obtained by adding the corresponding elements of A and B. For example, if $A = \begin{bmatrix} -1 & 2 \\ 9 & 4 \end{bmatrix}$ and $B = \begin{bmatrix} 2 & 3 \\ 4 & 8 \end{bmatrix}$, then

$$A + B = \begin{bmatrix} -1+2 & 2+3 \\ 9+4 & 4+8 \end{bmatrix} = \begin{bmatrix} 1 & 5 \\ 13 & 12 \end{bmatrix} \text{ and } B + A = \begin{bmatrix} 2+(-1) & 3+2 \\ 4+9 & 8+4 \end{bmatrix} = \begin{bmatrix} 1 & 5 \\ 13 & 12 \end{bmatrix}.$$

This example suggest that matrix addition is commutative

Negative Matrices

The negative of the matrix $A = \begin{bmatrix} a & b \\ c & d \end{bmatrix}$, denoted by $-A$ is $\begin{bmatrix} -a & -b \\ -c & -d \end{bmatrix}$.

Example 2

Find the negative of the matrix $A = \begin{bmatrix} -3 & 1 \\ 0 & -2 \end{bmatrix}$

Solution $\quad -A = \begin{bmatrix} 3 & -1 \\ 0 & 2 \end{bmatrix}$

Subtraction of Matrices

If A and B are two matrices of the same order their difference $A - B$ is obtained by subtracting each element in B from the corresponding elements in A. For example, if $A = \begin{bmatrix} -2 & 3 \\ -8 & 7 \end{bmatrix}$ and $B = \begin{bmatrix} -2 & 4 \\ -10 & 6 \end{bmatrix}$,

then $A - B = \begin{bmatrix} -2-(-2) & 3-4 \\ -8-(-10) & 7-6 \end{bmatrix} = \begin{bmatrix} 0 & -1 \\ 2 & 1 \end{bmatrix}$

and $B - A = \begin{bmatrix} -2-(-2) & 4-3 \\ -10-(-8) & 6-7 \end{bmatrix} = \begin{bmatrix} 0 & 1 \\ -2 & -1 \end{bmatrix}.$

Zero Matrix (Null Matrix)

The zero matrix O has all its elements being zero. For example $\begin{bmatrix} 0 & 0 \\ 0 & 0 \end{bmatrix}$ is a 2×2 zero matrix.

For any matrix A, and the zero matrix O of the same order

$$A + O = O + A = A \text{ and } A - A = O$$

Multiplying a Matrix by a Scalar

If $A = \begin{bmatrix} a & b \\ c & d \end{bmatrix}$, then kA, where k is a scalar is obtained by multiplying each element by k,

i.e. $kA = \begin{bmatrix} ka & kb \\ kc & kd \end{bmatrix}$

Example 3

If $A = \begin{bmatrix} -1 & 0 \\ 2 & -3 \end{bmatrix}$ find $-2A$

Solution $-2A = \begin{bmatrix} 2 & 0 \\ -4 & 6 \end{bmatrix}$

Exercise 17.1(a)

1. Find the values of x, y and z

(a) $\begin{bmatrix} 2 & x+y \\ 2y & 6 \end{bmatrix} = \begin{bmatrix} x-y & 4 \\ z & 6 \end{bmatrix}$

(b) $\begin{bmatrix} 3 & x-2y \\ 5 & -2 \end{bmatrix} = \begin{bmatrix} 3 & -4 \\ x+y & -2 \end{bmatrix}$

2. Work out:

(a) $\begin{bmatrix} 5 & 3 \\ 1 & 2 \end{bmatrix} + \begin{bmatrix} 6 & 4 \\ 1 & 3 \end{bmatrix}$

(b) $\begin{bmatrix} 0 & 2 \\ 4 & -1 \end{bmatrix} + \begin{bmatrix} 6 & -1 \\ 5 & 2 \end{bmatrix}$

(c) $\begin{bmatrix} 4 & 2 \\ 3 & -1 \end{bmatrix} + \begin{bmatrix} 1 & 3 \\ -2 & 2 \end{bmatrix}$

(d) $\begin{bmatrix} 5 & 3 \\ 2 & 4 \end{bmatrix} + \begin{bmatrix} -2 & 2 \\ -1 & 0 \end{bmatrix}$

(e) $\begin{bmatrix} 3 & -1 & 0 \\ -2 & 0 & 1 \end{bmatrix} + \begin{bmatrix} -5 & 2 & -1 \\ 4 & 3 & -4 \end{bmatrix}$

(f) $\begin{bmatrix} -4 & 2 & 3 \\ 5 & -3 & 1 \end{bmatrix} + \begin{bmatrix} 5 & -3 & -2 \\ -3 & 7 & -3 \end{bmatrix}$

(g) $\begin{bmatrix} 0 & 5 \\ 3 & -4 \\ -2 & 1 \end{bmatrix} + \begin{bmatrix} -1 & -6 \\ 2 & 6 \\ 3 & -2 \end{bmatrix}$

(h) $\begin{bmatrix} -7 & 8 & 6 \\ 5 & 4 & -3 \\ 2 & -5 & 1 \end{bmatrix} + \begin{bmatrix} 9 & -6 & -10 \\ -3 & 0 & 5 \\ -3 & 6 & -2 \end{bmatrix}$

3. Find the values of a, b, c and d so that

$$\begin{bmatrix} a & b \\ c & d \end{bmatrix} + \begin{bmatrix} -2 & 3 \\ 2 & 1 \end{bmatrix} = \begin{bmatrix} -1 & 2 \\ -3 & 4 \end{bmatrix}$$

4. Find the values of w, x, y and z so that

$$\begin{bmatrix} -1 & 5 \\ 1 & -2 \end{bmatrix} + \begin{bmatrix} w & x \\ y & z \end{bmatrix} = \begin{bmatrix} -2 & 3 \\ 6 & 1 \end{bmatrix}$$

5. Find the values of x and y so that

$$\begin{bmatrix} 2x & 5 \\ -1 & 3x \end{bmatrix} + \begin{bmatrix} 3y & -3 \\ -5 & -2y \end{bmatrix} = \begin{bmatrix} 7 & 2 \\ -6 & 4 \end{bmatrix}$$

6. Find the values of x and y so that

$$\begin{bmatrix} -3x & -3 \\ 2 & 4x \end{bmatrix} + \begin{bmatrix} 2y & 5 \\ -2 & -5y \end{bmatrix} = \begin{bmatrix} -1 & 2 \\ 0 & -1 \end{bmatrix}$$

7. Write down the negative matrix of each of the following matrices:

(a) $\begin{bmatrix} -2 & 3 \\ 1 & -4 \end{bmatrix}$
(b) $\begin{bmatrix} -3 & -2 \\ 5 & -1 \end{bmatrix}$
(c) $\begin{bmatrix} 6 & 0 \\ 0 & -6 \end{bmatrix}$

(d) $\begin{bmatrix} 3 & -2 & 0 \\ -1 & 5 & -2 \end{bmatrix}$
(e) $\begin{bmatrix} -4 & 0 \\ 3 & 1 \\ -2 & -1 \end{bmatrix}$
(f) $\begin{bmatrix} 2 & -5 & 3 \\ -3 & 0 & -2 \\ 4 & -3 & 2 \end{bmatrix}$

8. Work out:

(a) $\begin{bmatrix} -1 & 1 \\ 4 & 2 \end{bmatrix} - \begin{bmatrix} -2 & 0 \\ 4 & 1 \end{bmatrix}$
(b) $\begin{bmatrix} 4 & 2 \\ -1 & 3 \end{bmatrix} - \begin{bmatrix} 1 & 8 \\ 1 & 0 \end{bmatrix}$

(c) $\begin{bmatrix} 3 & -6 \\ -2 & 1 \end{bmatrix} - \begin{bmatrix} -4 & -2 \\ 1 & -3 \end{bmatrix}$
(d) $\begin{bmatrix} -2 & 4 \\ -8 & 5 \end{bmatrix} - \begin{bmatrix} -2 & 0 \\ -5 & 3 \end{bmatrix}$

(e) $\begin{bmatrix} 0 & 3 & -1 \\ 1 & -2 & 0 \end{bmatrix} - \begin{bmatrix} -1 & -5 & 2 \\ -4 & 4 & 3 \end{bmatrix}$
(f) $\begin{bmatrix} 5 & -3 & 1 \\ -4 & 2 & 1 \end{bmatrix} - \begin{bmatrix} -3 & -7 & -3 \\ -5 & 3 & 0 \end{bmatrix}$

(g) $\begin{bmatrix} 5 & 0 \\ -4 & 3 \\ 1 & -2 \end{bmatrix} - \begin{bmatrix} -6 & -1 \\ 6 & 2 \\ -2 & 3 \end{bmatrix}$
(h) $\begin{bmatrix} 2 & -5 & 1 \\ -7 & 8 & 6 \\ 2 & -5 & 1 \end{bmatrix} - \begin{bmatrix} -3 & -6 & 2 \\ -3 & 6 & 4 \\ -5 & 7 & -2 \end{bmatrix}$

9. Work out:

(a) $3 \begin{bmatrix} 1 & 2 \\ -2 & 1 \end{bmatrix}$
(b) $-2 \begin{bmatrix} -3 & 1 \\ 1 & -2 \end{bmatrix}$
(c) $4 \begin{bmatrix} -1 & 0 \\ 0 & 1 \end{bmatrix}$

(d) $-\frac{1}{2} \begin{bmatrix} -2 & 6 \\ 4 & 0 \end{bmatrix}$
(e) $\frac{2}{3} \begin{bmatrix} -6 & 0 \\ 12 & -9 \end{bmatrix}$
(f) $-\frac{3}{2} \begin{bmatrix} 4 & -2 \\ -6 & 2 \end{bmatrix}$

(g) $-2\begin{bmatrix} 2 & -1 & -3 \\ 0 & 3 & -2 \end{bmatrix}$ (h) $3\begin{bmatrix} 0 & 1 \\ -1 & 2 \\ -2 & 3 \end{bmatrix}$ (i) $-\frac{1}{3}\begin{bmatrix} 3 & 0 & -6 \\ -3 & 9 & -3 \\ 0 & -3 & 6 \end{bmatrix}$

10. Given that $A = \begin{bmatrix} -3 & 2 \\ 2 & 0 \end{bmatrix}$, $B = \begin{bmatrix} -2 & 1 \\ 1 & 5 \end{bmatrix}$ and $C = \begin{bmatrix} 0 & -3 \\ 1 & 2 \end{bmatrix}$, find

(a) $2A + B$ (b) $3C - 2B$ (c) $2C - 3A$ (d) $2(B - A) + C$

Multiplying Matrices

Two matrices can be multiplied if the number of columns of the first matrix is the same as the number of rows of the second matrix. The product will have the same number of rows as the first matrix and the same number of columns as the second matrix. For example, 3×2 matrix multiplied by a 2×3 gives a 3×3 matrix.

If A and B are two 2×2 matrices the product AB is obtained as follows:

Multiply 1st element in the 1st row in A by 1st element in the 1st column in B. Next multiply 2nd element in 1st row in A by 2nd element in 1st column in B. Then add the two results to give the 1st element in 1st row and 1st column of the matrix AB. The element in the 1st row and second column of AB is obtained in the same manner using the 1st row of matrix A and 2nd column of matrix B. Similarly, elements in 2nd row, 1st column and 2nd row, 2nd column are obtained by multiplying the 2nd row of matrix A by the 1st column and 2nd column of matrix B respectively.

For example, if $A = \begin{bmatrix} a & b \\ c & d \end{bmatrix}$ and $B = \begin{bmatrix} e & f \\ g & h \end{bmatrix}$ then

$$AB = \begin{bmatrix} a & b \\ c & d \end{bmatrix} \begin{bmatrix} e & f \\ g & h \end{bmatrix}$$

$$= \begin{bmatrix} ae + bg & af + bh \\ ce + dg & cf + dh \end{bmatrix}$$

Example 4

Given that $A = \begin{bmatrix} 2 & 0 \\ -3 & 1 \end{bmatrix}$ and $B = \begin{bmatrix} -1 & 3 \\ -2 & 4 \end{bmatrix}$ find AB and BA

Solution $AB = \begin{bmatrix} 2 & 0 \\ -3 & 1 \end{bmatrix} \begin{bmatrix} -1 & 3 \\ -2 & 4 \end{bmatrix}$

$= \begin{bmatrix} -2 & 6 \\ 1 & -5 \end{bmatrix}$

$BA = \begin{bmatrix} -1 & 3 \\ -2 & 4 \end{bmatrix} \begin{bmatrix} 2 & 0 \\ -3 & 1 \end{bmatrix}$

330 Further Mathematics

$$= \begin{bmatrix} -11 & 3 \\ -16 & 4 \end{bmatrix}$$

The order of multiplication of matrices is important. Notice that $AB \neq BA$. In general, matrix multiplication is not commutative.

If A is a square matrix then AA is defined and is denoted by A^2. Similarly, $AAA = A^3$.

It is possible to find two matrices, A and B, with non-zero elements such that $AB = 0$. For example,

$$\begin{bmatrix} -2 & 3 \\ -4 & 6 \end{bmatrix} \begin{bmatrix} 3 & -6 \\ 2 & -4 \end{bmatrix} = \begin{bmatrix} 0 & 0 \\ 0 & 0 \end{bmatrix}$$

Exercise 17.1(b)

1. Evaluate:

(a) $\begin{bmatrix} 1 & 2 \\ 0 & 3 \end{bmatrix} \begin{bmatrix} 3 & 2 \\ 4 & -1 \end{bmatrix}$

(b) $\begin{bmatrix} 2 & -3 \\ -5 & 1 \end{bmatrix} \begin{bmatrix} 1 & 2 \\ 3 & 2 \end{bmatrix}$

(c) $\begin{bmatrix} 1 & 3 \\ -2 & 5 \end{bmatrix} \begin{bmatrix} 0 & -2 \\ 2 & -4 \end{bmatrix}$

(d) $\begin{bmatrix} 3 & 4 \\ 0 & 1 \end{bmatrix} \begin{bmatrix} 6 & 2 \\ 5 & 0 \end{bmatrix}$

(e) $\begin{bmatrix} 1 & 2 \\ 2 & 4 \end{bmatrix} \begin{bmatrix} 2 \\ 3 \end{bmatrix}$

(f) $\begin{bmatrix} 1 & 3 \\ 2 & 5 \end{bmatrix} \begin{bmatrix} 2 \\ -1 \end{bmatrix}$

(g) $\begin{bmatrix} 1 & -3 \\ -2 & 1 \end{bmatrix} \begin{bmatrix} 1 \\ 5 \end{bmatrix}$

(h) $\begin{bmatrix} 4 & -3 \\ -2 & 1 \end{bmatrix} \begin{bmatrix} -4 \\ -5 \end{bmatrix}$

(i) $\begin{bmatrix} 3 & -2 & 1 \\ 1 & 0 & 2 \end{bmatrix} \begin{bmatrix} -1 & 0 \\ 2 & 1 \\ -3 & -2 \end{bmatrix}$

(j) $\begin{bmatrix} 4 & -2 \\ 2 & 1 \\ -1 & 0 \end{bmatrix} \begin{bmatrix} 3 & -1 \\ 2 & 0 \end{bmatrix}$

(k) $\begin{bmatrix} -2 & 1 & -1 \\ 3 & 4 & 0 \\ -1 & 2 & -2 \end{bmatrix} \begin{bmatrix} 0 & 2 \\ 1 & -2 \\ 0 & 1 \end{bmatrix}$

(l) $\begin{bmatrix} 4 & 5 & -3 \\ 0 & 2 & 1 \\ -1 & 0 & 2 \end{bmatrix} \begin{bmatrix} 0 & 1 & 2 \\ 4 & -1 & 0 \\ 3 & 2 & 1 \end{bmatrix}$

2. Evaluate:

(a) $\begin{bmatrix} 6 & -3 \\ -4 & 2 \end{bmatrix} \begin{bmatrix} 4 & 3 \\ 8 & 0 \end{bmatrix}$

(b) $\begin{bmatrix} 3 & 2 \\ -3 & -2 \end{bmatrix} \begin{bmatrix} 2 & -4 \\ -3 & 6 \end{bmatrix}$

3. Find A^2 and A^3 for each of the following matrices.

(a) $\begin{bmatrix} 1 & -1 \\ 3 & 2 \end{bmatrix}$

(b) $\begin{bmatrix} -3 & -1 \\ 2 & 1 \end{bmatrix}$

(c) $\begin{bmatrix} 1 & 0 \\ 2 & -1 \end{bmatrix}$

4. Find the values of x, y and z so that

$$\begin{bmatrix} 2 & 3 \\ -3 & 1 \end{bmatrix} \begin{bmatrix} x & 2 \\ -1 & y \end{bmatrix} = \begin{bmatrix} 1 & z \\ -7 & z \end{bmatrix}$$

5. Find the values of x, y and z so that

$$\begin{bmatrix} y & 3 \\ x & 1 \end{bmatrix} \begin{bmatrix} 2 & 3 \\ 1 & -2 \end{bmatrix} = \begin{bmatrix} -1 & -12 \\ z & z \end{bmatrix}$$

6. Find the values a, b, c and d so that

$$\begin{bmatrix} 2 & 3 \\ 3 & -2 \end{bmatrix} \begin{bmatrix} a & b \\ c & d \end{bmatrix} = \begin{bmatrix} -7 & 8 \\ -4 & -1 \end{bmatrix}$$

7. Find the values a, b, c and d so that

$$\begin{bmatrix} -3 & 2 \\ 1 & -2 \end{bmatrix} \begin{bmatrix} a & b \\ c & d \end{bmatrix} = \begin{bmatrix} 0 & 5 \\ -4 & 1 \end{bmatrix}$$

Identity Matrix (Unit Matrix)

A square matrix whose leading diagonals consist of 1's where all other elements are 0's is called an identity matrix (or unit matrix). The identity matrix is denoted by I.

The 2×2 identity matrix is $\begin{bmatrix} 1 & 0 \\ 0 & 1 \end{bmatrix}$,

and the 3×3 identity is $\begin{bmatrix} 1 & 0 & 0 \\ 0 & 1 & 0 \\ 0 & 0 & 1 \end{bmatrix}$

If A is a square matrix of order n and I the identity matrix of the same order, then $IA = AI = A$.

Example 5

Given that $A = \begin{bmatrix} -2 & 3 \\ 1 & 4 \end{bmatrix}$ find AI and IA

Solution $AI = \begin{bmatrix} -2 & 3 \\ 1 & 4 \end{bmatrix} \begin{bmatrix} 1 & 0 \\ 0 & 1 \end{bmatrix}$

$= \begin{bmatrix} -2 & 3 \\ 1 & 4 \end{bmatrix}$

$IA = \begin{bmatrix} 1 & 0 \\ 0 & 1 \end{bmatrix} \begin{bmatrix} -2 & 3 \\ 1 & 4 \end{bmatrix}$

$= \begin{bmatrix} -2 & 3 \\ 1 & 4 \end{bmatrix}$

Determinants

Associated with each square matrix is a real number, called the determinant. The determinant of a matrix A is denoted by $\det A$ or $|A|$, and is defined for 2×2 matrix $A = \begin{bmatrix} a & b \\ c & d \end{bmatrix}$ by

$$\begin{vmatrix} a & b \\ c & d \end{vmatrix} = ad - bc$$

Example 6

Find the determinant of the matrix $\begin{bmatrix} 3 & -4 \\ -2 & 2 \end{bmatrix}$

Solution $\begin{vmatrix} 3 & -4 \\ -2 & 2 \end{vmatrix} = (3)(2) - (-4)(-2)$

$$= -2$$

The Inverse of a Matrix

If A and B are two square matrices of the same order, then B is the inverse matrix of A, if $AB = BA = I$, where I is the identity matrix. The inverse matrix of A is denoted by A^{-1}, read "A inverse".

Not every square matrix has an inverse. A matrix which has an inverse is said to be non-singular (or invertible).

A matrix that has no inverse is said to be singular. The determinant of a singular matrix is zero.

If $A = \begin{bmatrix} a & b \\ c & d \end{bmatrix}$ and $ad - bc \neq 0$, then A^{-1} is obtained by interchanging the elements on the leading diagonal, changing the signs of the other two elements and finally multiplying by $\frac{1}{ad-bc}$, i.e. $A^{-1} = \frac{1}{ad-bc} \begin{bmatrix} d & -b \\ -c & a \end{bmatrix}$

Notice that $ad - bc$ is the determinant of the matrix A.

Example 7

Find the inverse of $A = \begin{bmatrix} 3 & 2 \\ 1 & 4 \end{bmatrix}$

Begin by finding the determinant of A

Solution $\begin{vmatrix} 3 & 2 \\ 1 & 4 \end{vmatrix} = (3)(4) - (2)(1) = 10$

Hence $A^{-1} = \frac{1}{10}\begin{bmatrix} 4 & -2 \\ -1 & 3 \end{bmatrix}$

$= \begin{bmatrix} 0.4 & -0.2 \\ -0.1 & 0.3 \end{bmatrix}$

Exercise 17.1(c)

1. Show that the second matrix of each pair of matrices is the inverse of the first matrix.

(a) $\begin{bmatrix} 1 & 2 \\ 2 & 3 \end{bmatrix}, \begin{bmatrix} -3 & 2 \\ 2 & 1 \end{bmatrix}$

(b) $\begin{bmatrix} 3 & 2 \\ 7 & 8 \end{bmatrix}, \begin{bmatrix} 5 & -2 \\ -7 & 3 \end{bmatrix}$

(c) $\begin{bmatrix} 5 & -2 \\ 3 & -1 \end{bmatrix}, \begin{bmatrix} 2 & -1 \\ 5 & -3 \end{bmatrix}$

(d) $\begin{bmatrix} 1 & -2 \\ -1 & 3 \end{bmatrix}, \begin{bmatrix} 3 & 2 \\ 1 & 1 \end{bmatrix}$

2. Evaluate:

(a) $\begin{vmatrix} 4 & 3 \\ 3 & 2 \end{vmatrix}$

(b) $\begin{vmatrix} 2 & 1 \\ 4 & 3 \end{vmatrix}$

(c) $\begin{vmatrix} -2 & 3 \\ -2 & 2 \end{vmatrix}$

(d) $\begin{vmatrix} -3 & -2 \\ 4 & 3 \end{vmatrix}$

(e) $\begin{vmatrix} 6 & 3 \\ 2 & 1 \end{vmatrix}$

(f) $\begin{vmatrix} 6 & 4 \\ 5 & 3 \end{vmatrix}$

3. If $A = \begin{bmatrix} -2 & 1 \\ 1 & 3 \end{bmatrix}$ and $B = \begin{bmatrix} -1 & 2 \\ 5 & 4 \end{bmatrix}$, find:

(a) $|A|$ (b) $|B|$ (c) $|AB|$ (d) $|BA|$ (e) $|A||B|$

4. Solve the following equations:

(a) $\begin{vmatrix} x & x \\ 3 & 4 \end{vmatrix} = -5$

(b) $\begin{vmatrix} 1 & x \\ x & 3 \end{vmatrix} = 2$

(c) $\begin{vmatrix} 4 & 5-x \\ 2-x & 1 \end{vmatrix} = 0$

5. Determine whether each of the following matrices has inverse or not:

(a) $\begin{bmatrix} 5 & 4 \\ 3 & 2 \end{bmatrix}$

(b) $\begin{bmatrix} 7 & 3 \\ -2 & 1 \end{bmatrix}$

(c) $\begin{bmatrix} 1 & 2 \\ 3 & 6 \end{bmatrix}$

(d) $\begin{bmatrix} 8 & 4 \\ -4 & -2 \end{bmatrix}$

(e) $\begin{bmatrix} -3 & 2 \\ -1 & 3 \end{bmatrix}$

(f) $\begin{bmatrix} 3 & -2 \\ 6 & 4 \end{bmatrix}$

6. Find the inverse of each of the following matrices:

(a) $\begin{bmatrix} -2 & 3 \\ 2 & 0 \end{bmatrix}$

(b) $\begin{bmatrix} 4 & 5 \\ 2 & 3 \end{bmatrix}$

(c) $\begin{bmatrix} -2 & -1 \\ 4 & 3 \end{bmatrix}$

(d) $\begin{bmatrix} 2 & 4 \\ 3 & 5 \end{bmatrix}$

(e) $\begin{bmatrix} 5 & -2 \\ 0 & 2 \end{bmatrix}$

(f) $\begin{bmatrix} 3 & 2 \\ -1 & 2 \end{bmatrix}$

17.2 Solving Simultaneous Linear Equations

Using Inverse Matrices

You can write the simultaneous equations in the form $AX = B$, where A is called the coefficient matrix and B the constant matrix.

If A is a square matrix and A^{-1} exists, then

$A^{-1}AX = A^{-1}B$

$IX = A^{-1}B$

$X = A^{-1}B$

Hence, the solution of the simultaneous linear equations is $X = A^{-1}B$.

Example 8

Solve the simultaneous equations

$x + 2y = 7$

$2x + 3y = 12$

Solution Rewrite the equations in matrix form

$$\begin{bmatrix} 1 & 2 \\ 2 & 3 \end{bmatrix} \begin{bmatrix} x \\ y \end{bmatrix} = \begin{bmatrix} 7 \\ 12 \end{bmatrix}$$

In this case

$$A = \begin{bmatrix} 1 & 2 \\ 2 & 3 \end{bmatrix}, X = \begin{bmatrix} x \\ y \end{bmatrix} \text{ and } B = \begin{bmatrix} 7 \\ 12 \end{bmatrix}$$

Determine the inverse of A

$$\begin{vmatrix} 1 & 2 \\ 2 & 3 \end{vmatrix} = 3 - 4 = -1$$

Hence $A^{-1} = \begin{bmatrix} -3 & 2 \\ 2 & -1 \end{bmatrix}$

Pre multiplying each side of the matrix equation by A^{-1} gives

$$\begin{bmatrix} x \\ y \end{bmatrix} = \begin{bmatrix} -3 & 2 \\ 2 & -1 \end{bmatrix} \begin{bmatrix} 7 \\ 12 \end{bmatrix}$$

$$= \begin{bmatrix} 3 \\ 2 \end{bmatrix}$$

Hence, the solutions are $x = 3$ and $y = 2$.

Using Determinants

Consider the simultaneous linear equations

$$a_{11}x + a_{12}y = b_1$$

$$a_{21}x + a_{22}y = b_2$$

Rewrite this in matrix form.

$$\begin{bmatrix} a_{11} & a_{12} \\ a_{21} & a_{22} \end{bmatrix} \begin{bmatrix} x \\ y \end{bmatrix} = \begin{bmatrix} b_1 \\ b_2 \end{bmatrix}$$

If $\begin{vmatrix} a_{11} & a_{12} \\ a_{21} & a_{22} \end{vmatrix} \neq 0$, then

$$\begin{bmatrix} x \\ y \end{bmatrix} = \frac{1}{a_{11}a_{22} - a_{21}a_{12}} \begin{bmatrix} a_{22} & -a_{12} \\ -a_{21} & a_{11} \end{bmatrix} \begin{bmatrix} b_1 \\ b_2 \end{bmatrix}$$

$$= \frac{1}{a_{11}a_{22} - a_{21}a_{12}} \begin{bmatrix} a_{22}b_1 - a_{12}b_2 \\ a_{11}b_2 - a_{21}b_1 \end{bmatrix}$$

So the solution of the simultaneous equations is given by

$$x = \frac{a_{22}b_1 - a_{12}b_2}{a_{11}a_{22} - a_{21}a_{12}} \quad \text{and} \quad y = \frac{a_{11}b_2 - a_{21}b_1}{a_{11}a_{22} - a_{21}a_{12}}$$

These can be written using determinants as:

$$x = \frac{\begin{vmatrix} b_1 & a_{12} \\ b_2 & a_{22} \end{vmatrix}}{\begin{vmatrix} a_{11} & a_{12} \\ a_{21} & a_{22} \end{vmatrix}} = \frac{D_x}{D} \quad \text{and} \quad y = \frac{\begin{vmatrix} a_{11} & b_1 \\ a_{21} & b_2 \end{vmatrix}}{\begin{vmatrix} a_{11} & a_{12} \\ a_{21} & a_{22} \end{vmatrix}} = \frac{D_y}{D} \quad \text{provided that } D \neq 0$$

Note that D is the determinant made up of the original coefficients of x and y. D_x is obtained by replacing the coefficients of the x-terms by the constants b_1 and b_2, D_y is found by replacing the coefficients of the y-terms by the constants b_1 and b_2.

Example 9

Solve the simultaneous equations

$$x + 2y = 7$$

$$2x + 3y = 12$$

Solution Begin by finding the determinants

$$D = \begin{vmatrix} 1 & 2 \\ 2 & 3 \end{vmatrix} = 3 - 4 = -1$$

$$D_x = \begin{vmatrix} 7 & 2 \\ 12 & 3 \end{vmatrix} = 21 - 24 = -3$$

$$D_y = \begin{vmatrix} 1 & 7 \\ 2 & 12 \end{vmatrix} = 12 - 14 = -2$$

So $x = \frac{D_x}{D} = \frac{-3}{-1} = 3$ and $y = \frac{D_y}{D} = \frac{-2}{-1} = 2$

Hence, the solutions are $x = 3$ and $y = 2$

Exercise 17.2

1. Solve the following simultaneous linear equations using matrix method

(a) $5x + 4y = 3$
   $3x + 2y = 1$

(b) $x + 2y = 1$
   $4x + y = 4$

(c) $-x + 2y = 4$
   $-2x + 3y = 5$

(d) $7x + 3y = -8$
   $5x + 2y = -6$

(e) $3x + 2y = 5$
   $-2x + 3y = -12$

(f) $2x - 3y = -5$
   $-x + 2y = 4$

2. Kojo has 22 coins with a total value of GH¢ 1.70. If the coins are all 10 Gp and 5 Gp, how many of each type of coin does he have?

3. 500 tickets were sold for a concert. The receipts from ticket sales were £3100 and the ticket prices were £5 and £8. How many of each price ticket were sold?

4. A man invests a part of £800 in bonds paying 12 percent interest. The remainder is in a savings account at 8 percent. If he receives £84 in interest for 1 year how much does he have invested at each rate?

5. A chemist has 15 % and 25 % acid solution. How much of each solution should be used to form 500 ml of a 21 % acid solution?

6. A plane flies 540 km with the wind in 3 hours. Flying back against the wind, the plane takes 9 hours to make the trip. What was the rate of the plane in still air? What was the rate of the wind?

17.3 Linear Transformations

Consider the equations

$x' = 2x + 3y$

$y' = x + 2y$

These equations can be used to transform any point (x,y) in the x - y plane into another point (x', y') in the plane. For example, the point (1, 2) is transformed into (8, 5). However, the point (0, 0) is mapped onto itself. The set of equation defines a linear transformation which will not change the origin. We can write the set of equations in matrix form as

$$\begin{bmatrix} x' \\ y' \end{bmatrix} = \begin{bmatrix} 2 & 3 \\ 1 & 2 \end{bmatrix} \begin{bmatrix} x \\ y \end{bmatrix}$$

The matrix $\begin{bmatrix} 2 & 3 \\ 1 & 2 \end{bmatrix}$ is called the transformation matrix.

In general the transformation matrix of the following linear transformation

$x' = ax + by$

$y' = cx + dy$

is $\begin{bmatrix} a & b \\ c & d \end{bmatrix}$, where a, b, c and d are constants.

Determining the Matrix of a Linear Transformation

The matrix of a linear transformation can be determined in two ways:

1. The matrix M of a linear transformation can be determined by expressing the position vector of the image of a point $P(x, y)$ in the form.

$$\begin{bmatrix} x' \\ y' \end{bmatrix} = M \begin{bmatrix} x \\ y \end{bmatrix}$$

For example the linear transformation under the reflection in the x-axis is

$x' = x$

$y' = -y$

We can write these equations in the equivalent forms shown below.

$x' = 1x + 0y$

$y' = 0x - 1y$

In matrix form, we write $\begin{bmatrix} x' \\ y' \end{bmatrix} = \begin{bmatrix} 1 & 0 \\ 0 & -1 \end{bmatrix} \begin{bmatrix} x \\ y \end{bmatrix}$

So the transformation matrix is $\begin{bmatrix} 1 & 0 \\ 0 & -1 \end{bmatrix}$.

2. If $\begin{bmatrix} a & b \\ c & d \end{bmatrix}$ is the matrix of a linear transformation T then under T

$$\begin{bmatrix} 1 \\ 0 \end{bmatrix} \to \begin{bmatrix} a & b \\ c & d \end{bmatrix}\begin{bmatrix} 1 \\ 0 \end{bmatrix} = \begin{bmatrix} a \\ c \end{bmatrix} \text{ and } \begin{bmatrix} 0 \\ 1 \end{bmatrix} \to \begin{bmatrix} a & b \\ c & d \end{bmatrix}\begin{bmatrix} 0 \\ 1 \end{bmatrix} = \begin{bmatrix} b \\ d \end{bmatrix}.$$

Notice that the first and second columns respectively of the transformation matrix are

$\begin{bmatrix} a \\ c \end{bmatrix}$ which is the image of $\begin{bmatrix} 1 \\ 0 \end{bmatrix}$ and $\begin{bmatrix} b \\ d \end{bmatrix}$ which is the image of $\begin{bmatrix} 0 \\ 1 \end{bmatrix}$.

You can find the matrix that represents a given transformation by using this property.

Example 10

(a) Reflection in the x-axis

Solution The mapping for this transformation is $\begin{bmatrix} x \\ y \end{bmatrix} \to \begin{bmatrix} x \\ -y \end{bmatrix}$

Under this transformation $\begin{bmatrix} 1 \\ 0 \end{bmatrix} \to \begin{bmatrix} 1 \\ 0 \end{bmatrix}$ and $\begin{bmatrix} 0 \\ 1 \end{bmatrix} \to \begin{bmatrix} 0 \\ -1 \end{bmatrix}$.

Therefore $\begin{bmatrix} 1 & 0 \\ 0 & -1 \end{bmatrix}$ is the matrix for this transformation.

(b) Reflection in the line $x = a$

Solution The mapping for this transformation is $\begin{bmatrix} x \\ y \end{bmatrix} \to \begin{bmatrix} 2a - x \\ y \end{bmatrix}$.

Under this transformation $\begin{bmatrix} 1 \\ 0 \end{bmatrix} \to \begin{bmatrix} 2a - 1 \\ 0 \end{bmatrix}$ and $\begin{bmatrix} 0 \\ 1 \end{bmatrix} \to \begin{bmatrix} 2a \\ 1 \end{bmatrix}$.

Therefore $\begin{bmatrix} 2a - 1 & 2a \\ 0 & 1 \end{bmatrix}$ is the matrix for this transformation.

(c) Anticlockwise rotation through 90° about the origin.

Solution The mapping for this transformation is $\begin{bmatrix} x \\ y \end{bmatrix} \to \begin{bmatrix} -y \\ x \end{bmatrix}$.

Under this transformation $\begin{bmatrix} 1 \\ 0 \end{bmatrix} \to \begin{bmatrix} 0 \\ 1 \end{bmatrix}$ and $\begin{bmatrix} 0 \\ 1 \end{bmatrix} \to \begin{bmatrix} -1 \\ 0 \end{bmatrix}$.

Therefore $\begin{bmatrix} 0 & -1 \\ 1 & 0 \end{bmatrix}$ is the matrix for this transformation.

(d) Enlargement from the origin with scale factor k.

Solution   The mapping for this transformation is $\begin{bmatrix} x \\ y \end{bmatrix} \rightarrow \begin{bmatrix} kx \\ ky \end{bmatrix}$.

Under this transformation $\begin{bmatrix} 1 \\ 0 \end{bmatrix} \rightarrow \begin{bmatrix} k \\ 0 \end{bmatrix}$ and $\begin{bmatrix} 0 \\ 1 \end{bmatrix} \rightarrow \begin{bmatrix} 0 \\ k \end{bmatrix}$.

Therefore $\begin{bmatrix} k & 0 \\ 0 & k \end{bmatrix}$ is the matrix for this transformation.

Rotation about the Origin

Any point in the x-y plane can be rotated about the origin O in an anticlockwise sense through an angle, θ, under a transformation whose matrix is given by $\begin{bmatrix} \cos\theta & -\sin\theta \\ \sin\theta & \cos\theta \end{bmatrix}$.

The matrix for clockwise rotation is given by

$$\begin{bmatrix} \cos(-\theta) & -\sin(-\theta) \\ \sin(-\theta) & \cos(-\theta) \end{bmatrix} = \begin{bmatrix} \cos\theta & \sin\theta \\ -\sin\theta & \cos\theta \end{bmatrix}$$

Combination of Transformations

Two transformations can be combined as follows; perform one transformation first and then perform the second transformation on the image of the first transformation. The final result is the combined transformation.

Suppose that a transformation with matrix $M = \begin{bmatrix} 0 & 1 \\ 1 & 0 \end{bmatrix}$ is followed by a transformation with matrix $N = \begin{bmatrix} 0 & 1 \\ -1 & 0 \end{bmatrix}$.

The matrix M transforms any point $P(x, y)$ to the point $P'(y, x)$ as you will see in the following matrix equation.

$$\begin{bmatrix} x' \\ y' \end{bmatrix} = \begin{bmatrix} 0 & 1 \\ 1 & 0 \end{bmatrix} \begin{bmatrix} x \\ y \end{bmatrix} = \begin{bmatrix} y \\ x \end{bmatrix}$$

The matrix N then transforms the image $P'(y, x)$ to give a final image $P''(x, -y)$ as the following matrix equation shows.

$$\begin{bmatrix} x'' \\ y'' \end{bmatrix} = \begin{bmatrix} 0 & 1 \\ -1 & 0 \end{bmatrix} \begin{bmatrix} y \\ x \end{bmatrix} = \begin{bmatrix} x \\ -y \end{bmatrix}$$

In brief the combined transformation is given by

$$\begin{bmatrix} x'' \\ y'' \end{bmatrix} = \begin{bmatrix} 0 & 1 \\ -1 & 0 \end{bmatrix} \begin{bmatrix} 0 & 1 \\ 1 & 0 \end{bmatrix} \begin{bmatrix} x \\ y \end{bmatrix} = \begin{bmatrix} 1 & 0 \\ 0 & -1 \end{bmatrix} \begin{bmatrix} x \\ y \end{bmatrix} = \begin{bmatrix} x \\ -y \end{bmatrix}$$

This result suggest that the combined transformation M followed by N is represented by a single transformation matrix $\begin{bmatrix} 1 & 0 \\ 0 & -1 \end{bmatrix}$. Notice that this matrix is obtained if we multiply the matrix N by matrix M. That is, the combined transformation matrix is NM.

The order of multiplication depends on the order of the transformations. Notice that M is the first transformation and N is the second transformation. Similarly, the transformation N followed by M has a combined transformation matrix MN.

Remember that the matrix MN is normally not equal to the matrix NM.

In general the transformation P followed by transformation Q, denoted by $Q \circ P$ and read "P followed by Q" or "Q follows P" has the combined matrix QP. Notice carefully the order in which the transformations Q and P are performed in $Q \circ P$.

Example 11

(a) Find the equation of the image of the line $y = 3x - 2$ under the transformation $\begin{bmatrix} 2 & -1 \\ 3 & 0 \end{bmatrix}$.

Solution Under this transformation the image of the point $P(x, y)$ is

$$\begin{bmatrix} x \\ y \end{bmatrix} \to \begin{bmatrix} 2 & -1 \\ 3 & 0 \end{bmatrix} \begin{bmatrix} x \\ y \end{bmatrix} = \begin{bmatrix} 2x - y \\ 3x \end{bmatrix}$$

If the transformation takes the point $P(x, y)$ to the image $P'(X, Y)$.

$X = 2x - y$ \quad (1)

$Y = 3x$ \quad (2)

Using (1) and (2) we eliminate x and y from $y = 3x - 2$ as follows. From (1) we have $y = 2x - X$.

So $2x - X = 3x - 2$

$x = -X + 2$

Also from (2) we have $x = \frac{1}{3}Y$

So $\frac{1}{3}Y = -X + 2$

$Y = -3X + 6$

Therefore the image of $y = 3x - 2$ is the line $y = -3x + 6$.

An alternative solution is given below.

Begin by finding two points on the given line

If $x = 2$ then $y = 3(2) - 2 = 4$

So (2, 4) is a point on the given line, and if

$x = -1$ then $y = 3(-1) - 2 = -5$

Also (-1, -5) is a point on the given line.

Use the matrix to find the images of these two points

$$\begin{bmatrix}2\\4\end{bmatrix} \to \begin{bmatrix}2 & -1\\3 & 0\end{bmatrix}\begin{bmatrix}2\\4\end{bmatrix} = \begin{bmatrix}0\\6\end{bmatrix} \text{ and } \begin{bmatrix}-1\\-5\end{bmatrix} \to \begin{bmatrix}2 & -1\\3 & 0\end{bmatrix}\begin{bmatrix}-1\\-5\end{bmatrix} = \begin{bmatrix}3\\-3\end{bmatrix}.$$

So (0, 6) and (3, -3) are two points on the image line.

Now the gradient of the image line is

$$\frac{-3 - 6}{3 - 0} = -3$$

Hence the equation of the image line is

$$\frac{y - 6}{x} = -3$$

$$y = -3x + 6$$

(b) Two linear transformations whose matrices are A and B are defined on the x-y plane by

$A: (x, y) \to (2x - y, 3x + 2y)$

$B: (x, y) \to (-2x + y, 2x - 3y)$

Find the image of the point (1, 2) under the transformation B followed by A.

Solution The matrices of transformation are $A = \begin{bmatrix}2 & -1\\3 & 2\end{bmatrix}$ and $B = \begin{bmatrix}-2 & 1\\2 & -3\end{bmatrix}$.

The combined matrix is

$$AB = \begin{bmatrix}2 & -1\\3 & 2\end{bmatrix}\begin{bmatrix}-2 & 1\\2 & -3\end{bmatrix} = \begin{bmatrix}-6 & 5\\-2 & -3\end{bmatrix}$$

Now $\begin{bmatrix} 1 \\ 2 \end{bmatrix} \to \begin{bmatrix} -6 & 5 \\ -2 & -3 \end{bmatrix} \begin{bmatrix} 1 \\ 2 \end{bmatrix} = \begin{bmatrix} 4 \\ -8 \end{bmatrix}$

Therefore the image of (1, 2) is (4, -8).

Exercise 17.3

1. Find the matrix of each of the following linear transformations:

(a) $x' = 3x + y$

$y' = 2x - 3y$

(b) $x' = 2x - 3y$

$y' = x + 2y$

(c) $x' = -3x + 2y$

$y' = 4x + 5y$

(d) $x' = -x + 4y$

$y' = x - 3y$

2. Write down the matrix of each of the following linear transformations:

(a) $A: (x, y) \to (4x - y, -3x + 2y)$

(b) $B: (x, y) \to (2x - y, 3x + y)$

(c) $P: (x, y) \to (5x - y, 2x + 3y)$

(d) $Q: (x, y) \to (-2x + 4y, -4x + y)$

3. Write down the matrix representing each of the following linear transformations:

(a) Anticlockwise rotation of 180° about the origin

(b) Reflection in the line $x = 0$

(c) Transformation by the vector $\begin{pmatrix} a \\ b \end{pmatrix}$

(d) Reflection in the line $y = x$

4. What transformations are represented by the following matrices?

(a) $\begin{bmatrix} 0 & 1 \\ -1 & 0 \end{bmatrix}$

(b) $\begin{bmatrix} -3 & 0 \\ 0 & -3 \end{bmatrix}$

(c) $\begin{bmatrix} 0 & -1 \\ -1 & 0 \end{bmatrix}$

5. A triangle have vertices A (1, 2), B (2, -1) and C (3, 0). What is the image under the transformation represented by the matrices?

(a) $\begin{bmatrix} 2 & -1 \\ 3 & -2 \end{bmatrix}$

(b) $\begin{bmatrix} 3 & 2 \\ 1 & 1 \end{bmatrix}$

(c) $\begin{bmatrix} 4 & -2 \\ -6 & 3 \end{bmatrix}$

6. What is the image of the point (-2, 3) under the matrix which moves

$\begin{bmatrix} 1 \\ 0 \end{bmatrix} \to \begin{bmatrix} 2 \\ 1 \end{bmatrix}$ and $\begin{bmatrix} 0 \\ 1 \end{bmatrix} \to \begin{bmatrix} -1 \\ 3 \end{bmatrix}$.

Matrices and Transformations   343

7. If $A = \begin{bmatrix} 0 & -1 \\ 1 & 0 \end{bmatrix}$, $B = \begin{bmatrix} -1 & 0 \\ 0 & 1 \end{bmatrix}$, $C = \begin{bmatrix} -1 & 0 \\ 0 & -1 \end{bmatrix}$ and $D = \begin{bmatrix} 2 & 0 \\ 0 & 2 \end{bmatrix}$ evaluate AB, BA and CD. Interpret the result geometrically.

8. Given the following transformations

A: Reflection in the line $y = 0$

B: Enlargement from the origin with scale factor 2

C: A clockwise rotation of $90°$ about the origin

D: A reflection in the line $y = x$

Find the combined matrix for:

(a) A followed by C

(b) C followed by B

(c) D followed by A

(d) D followed by B

9. Find the equation of the image of the line $y = 2x + 3$ under the transformation $\begin{bmatrix} -2 & 1 \\ 2 & 0 \end{bmatrix}$.

10. Find the equation of the image of the line $3x + 2y = 4$ under the transformation $\begin{bmatrix} 1 & 2 \\ -1 & 1 \end{bmatrix}$.

11. Find the image P' of the point $P\left(\sqrt{2}, \frac{1}{\sqrt{2}}\right)$ when it is rotated through $45°$ about the origin.

12. The point $A\left(-\frac{1}{\sqrt{3}}, 2\right)$ is rotated through $30°$ about the origin. Find its image A'.

Review Exercise 17

1. Use the following matrices to compute the given expressions

$A = \begin{bmatrix} 1 & 2 \\ 0 & 4 \end{bmatrix}$ $B = \begin{bmatrix} 3 & -1 \\ 4 & 2 \end{bmatrix}$ $C = \begin{bmatrix} 2 & 3 \\ 4 & -2 \end{bmatrix}$

(a) $A + B$ (b) $B + C$ (c) $2A - 3C$ (d) $A + C$ (e) $A - 2C$

(f) $3C - 4B$ (g) $(A + B) - 2C$ (h) $4C + (A - B)$ (i) $2A - 5(C + B)$

(j) $2(A - B) - C$ (k) $2(A - B) - C$ (l) $2(B - A) - C$

2. Use the following matrices to compute the given expression

$A = \begin{bmatrix} 3 & 2 \\ -1 & 4 \end{bmatrix}$ $B = \begin{bmatrix} 2 & 4 \\ 3 & -1 \end{bmatrix}$ $C = \begin{bmatrix} 2 & 0 \\ 4 & -2 \end{bmatrix}$ $D = \begin{bmatrix} 2 \\ 3 \end{bmatrix}$

(a) AB (b) CD (c) BC (d) $AD - BD$ (e) $(A + I)C$

(f) $(B - C)D$ (g) $B(A + C)$ (h) $A(B - C)$ (i) $2BD - AD$

3. Given that the matrix $A = \begin{bmatrix} -3 & 1 \\ 0 & 2 \end{bmatrix}$ find: (a) A^2 (b) A^3

4. Given that $A = \begin{bmatrix} 4 & -2 \\ 3 & -1 \end{bmatrix}$ and $B = \begin{bmatrix} 1 & 2 \\ -2 & 1 \end{bmatrix}$, find:

(a) A^2 (b) B^2 (a) $A^2 B$ (b) $B^2 A$

5. Find the values of a, b, c and d if $\begin{bmatrix} 1 & 3 \\ 1 & 4 \end{bmatrix}\begin{bmatrix} a & b \\ c & d \end{bmatrix} = \begin{bmatrix} 6 & -5 \\ 7 & -7 \end{bmatrix}$.

6. Find the values of x and y if $\begin{bmatrix} -1 & -1 \\ 4 & 2 \end{bmatrix}\begin{bmatrix} x \\ y \end{bmatrix} + \begin{bmatrix} -1 \\ 2 \end{bmatrix} = \begin{bmatrix} -5 \\ 0 \end{bmatrix}$.

7. If $A = \begin{bmatrix} 6 & 2 \\ 4 & y \end{bmatrix}$ and $B = \begin{bmatrix} 4 & 2 \\ x & 3 \end{bmatrix}$, find the values of x and y, given that A and B are commutative under matrix multiplication.

8. Find the value of each determinant:

(a) $\begin{vmatrix} 3 & 1 \\ 4 & 2 \end{vmatrix}$ (b) $\begin{vmatrix} 6 & 1 \\ 5 & 2 \end{vmatrix}$ (c) $\begin{vmatrix} 8 & -3 \\ 4 & 2 \end{vmatrix}$

(d) $\begin{vmatrix} 3 & 2 \\ -1 & 4 \end{vmatrix}$ (e) $\begin{vmatrix} 4 & -1 \\ 6 & -1 \end{vmatrix}$ (f) $\begin{vmatrix} 2 & 0 \\ 4 & -3 \end{vmatrix}$

9. Solve the following equations:

(a) $\begin{vmatrix} x & x \\ 4 & 3 \end{vmatrix} = 5$ (b) $\begin{vmatrix} x & 1 \\ 3 & x \end{vmatrix} = -2$

(c) $\begin{vmatrix} 5-x & 4 \\ 1 & 2-x \end{vmatrix} = 0$ (d) $\begin{vmatrix} 2 & x \\ -1 & x^2 - 3x - 6 \end{vmatrix} = 0$

10. Find the inverse of each of the following matrices:

(a) $\begin{bmatrix} 2 & 1 \\ 1 & 1 \end{bmatrix}$ (b) $\begin{bmatrix} 3 & -1 \\ -2 & 1 \end{bmatrix}$ (c) $\begin{bmatrix} 6 & 5 \\ 2 & 2 \end{bmatrix}$

(d) $\begin{bmatrix} -4 & 1 \\ 6 & -2 \end{bmatrix}$ (e) $\begin{bmatrix} -1 & -2 \\ 3 & 4 \end{bmatrix}$ (f) $\begin{bmatrix} 1 & 2 \\ 2 & 3 \end{bmatrix}$

11. Given that $A = \begin{bmatrix} -3 & 2 \\ -2 & 1 \end{bmatrix}$, find: (a) $A + 3I$ (b) $2I - A$

12. Given that $A = \begin{bmatrix} 3 & 2 \\ 1 & 4 \end{bmatrix}$, find the matrix B such that $A^2 B = I$.

13. Given that $A = \begin{bmatrix} 4 & 5 \\ -2 & -1 \end{bmatrix}$ and $B = \begin{bmatrix} -3 & 1 \\ 4 & -2 \end{bmatrix}$, find the matrix X such that:

(a) $(A + B)X = I$ \qquad (b) $(A - B)X = I$

14. Given a 2×2 matrix A such that $A^2 - 5A - 2I = 0$, show that $A^{-1} = \frac{1}{2}(A - 5I)$.

Hence find A^{-1} if $A = \begin{bmatrix} 1 & 2 \\ 3 & 4 \end{bmatrix}$.

15. Given that $A = \begin{bmatrix} 2 & 1 \\ -1 & 3 \end{bmatrix}$, evaluate A^{-1} and show that $A + 7A^{-1} = 5I$, where I is 2×2 identity matrix. Deduce that $A^2 - 5A + 7I = 0$ and use this equation to find A^2.

16. Solve each of the following simultaneous equations:

(a) $5x + 2y = 7$ \qquad (b) $5x - y = 13$ \qquad (c) $x + 3y = 5$

$3x - y = 13$ \qquad $2x + 3y = 12$ \qquad $2x - 3y = -8$

17. Given that the matrix $A = \begin{bmatrix} -3 & 4 \\ 3 & -2 \end{bmatrix}$, find a 2×2 matrix X such $AX = XA = I$, where I is the unit matrix. Hence solve the following simultaneous equations:

(a) $-3x + 4y = -1$ \qquad (b) $-3x + 4y = 9$

$3x - 2y = 5$ \qquad $3x - 2y = -7$

18. Given that $A = \begin{bmatrix} 4 & 3 \\ 3 & -2 \end{bmatrix}$, find the inverse matrix A^{-1} and hence solve the following simultaneous equations:

(a) $4x + 3y = 1$ \qquad (b) $4x + 3y = 11$

$3x - 2y = -12$ \qquad $3x - 2y = 4$

19. Two linear transformations whose matrices are A and B are defined on the x-y plane by:

$A: (x, y) \to (2x - 3y, 2x + y)$

$B: (x, y) \to (-2x + y, 3x - y)$

(a) Write down the matrices A and B. \qquad (b) Find the matrix AB.

(c) Find the image of the point (3, -2) under the transformation $A \circ B$, where $A \circ B$ means B followed by A.

20. Write down the matrices A and B representing the following transformation:

$A: (x, y) \to (-5x + 2y, 2x - 2y)$

$B: (x, y) \to (6x - 4y, -3x + 5y)$

Find the matrix X such that $(A + B)X = I$, where I is the 2×2 unit matrix.

21. The image $P'(1,8)$ of P is obtained by the transformation under the matrix $\begin{bmatrix} 4 & 2 \\ 3 & 5 \end{bmatrix}$ followed by a rotation through 90° about the origin. Find the coordinates of the point P.

22. Two linear transformations P and Q are defined by:

$P: (x, y) \to (4x + 3y, 2x + y)$

$Q: (x, y) \to (-x + 3y, 2x - 4y)$

(a) Write down the matrices P and Q. (b) Find the matrix PQ.

(c) Deduce the inverse of P.

23. Given $A = \begin{bmatrix} 3 & 4 \\ 1 & 2 \end{bmatrix}$ find the inverse A^{-1} of the matrix A. If the point Q is transformed by the matrix A^{-1} and the image is (-5, 4) find the coordinates of Q.

24. (a) Write down the matrix M of a linear transformation which rotates all points (x, y) through -60° about the origin.

(b) N is the matrix of the linear transformation which reflects the point of the plane in the line $y = x$. Find the image of the point A (1, -2) under the transformation $N \circ M$, correct to one decimal place.

25. Two linear transformations whose matrices are A and B are defined on the x-y plane by:

$A: (x, y) \to (-3x + 2y, x - y)$

$B: (x, y) \to (-2x + y, -3x + y)$

Find the image of a point (x, y) under a composite transformation $A \circ B$. Hence describe $A \circ B$ geometrically.

26. Under a linear transformation P of a plane the image of (1, 2) is (-4, 2) and that of (2, 3) is (-5, 5). Find the matrix P of the transformation.

Test covering Chapter 12 – 17

These tests are provided to help you in the process of reviewing the previous chapters. If you missed any answer go back and review the appropriate chapter section.

Take each test as you would take a test in class. Allow yourself 1 hour to take each test.

Test 3

1. The equation of a circle is $x^2 + y^2 + 8x - 10y + 16 = 0$. Find its centre and diameter.

2. The equation of a circle radius 3 is $x^2 + y^2 - 6x + 4y + c = 0$. Find the value of c.

3. The circle $x^2 + y^2 + 2x - 6y + 5 = 0$ touches the line whose points are (-5, 6) and (-1, -2) at its mid-point. If a diameter is drawn from the point of contact find the coordinates of its end points.

4. Using the expansion of $(1 + x)^5$ find the value of $(1.01)^5$.

5. Find the term containing a^6 in the expansion of $(a^2 - 2b)^7$.

6. Find the first five terms of the expansion of $(1 + x)^{-1/3}$.

7. Find $\sqrt{1.02}$ correct to five decimal places using the binomial theorem.

8. Find the value of $27^{1/3} + 9^{-1/3} \cdot 9^{4/3}$.

9. Solve $9^x - 3 \cdot 3^{x+1} = -8$.

10. Show that $2\log\left(\frac{2}{3}\right) + \log\left(\frac{3}{5}\right) - \log\left(\frac{4}{3}\right) = \log\left(\frac{1}{5}\right)$.

11. Given that $\log a = p + q$ and $\log b = p - q$ express in terms of p and q the value of $\log a^3 \cdot b^3$.

12. Find the values of x and y if

    $4^x \div 8^y = 32$

    $\log_y x = 2$

13. The second and fifth terms of a linear sequence are 7 and 19. What is the tenth term and the sum of the first ten terms?

14. Find the sum of the first seven terms of the sequence 8, 4, 2...

15. The second, fifth and fourteenth terms of an AP are in GP. What is the common ratio of the GP? $(d \neq 0)$

16. Find the sum of the terms between 21 and 81 divisible by 4.

17. Given that $tan\theta = \frac{3}{4}$, where $180° < \theta < 270°$, evaluate without using a calculator $sin2\theta$.

18. If $cosA = \frac{5}{13}$, such that $0° \leq A \leq 90°$, what is the value of $cos(90° - A)$?

19. Find the positive value of $cosx$ if $cos2x = \frac{1}{8}$.

20. If $sinx = \frac{1}{2}$ and x lies between $0°$ and $360°$ find the possible values of x and their cosines.

21. Find the values of x between $0°$ and $360°$ which satisfy $sec^2x - 5tanx + 5 = 0$.

22. Given that $\begin{bmatrix} 7 & 3x+1 \\ 2x+y & 3 \end{bmatrix} = \begin{bmatrix} 3x-y & 10 \\ 8 & 3 \end{bmatrix}$ find the values of x and y.

23. If $A = \begin{bmatrix} 1 & 2 \\ 2 & 1 \end{bmatrix}$ and $B = \begin{bmatrix} 3 & 2 \\ 1 & -1 \end{bmatrix}$ find AB.

24. Given the matrices $A = \begin{bmatrix} 1 & -1 \\ 0 & 2 \end{bmatrix}$ and $B = \begin{bmatrix} -3 & 2 \\ 1 & -2 \end{bmatrix}$ find the matrix C such that $C = A + 2B$.

25. A linear transformation T is defined by $T: (x, y) \to (2x - y, 3x + y)$. Find the image of $(2, -4)$.

Test 4

1. The line $2x + y - 4 = 0$ passes through the centre of a circle. If the circle passes through the x-intercept and the y-intercept of the line, find its equation.

2. Find the equation of the tangent to the circle $x^2 + y^2 - 6x - 8y + 20 = 0$ at the point $(5, 3)$.

3. The diameter of the circle $x^2 + y^2 - 6x - 4y = 0$ cuts the axes at A and B. If the gradient of the diameter is $-\frac{2}{3}$, find the coordinates of A and B. Hence find the area of triangle AOB.

4. Expand $(1 - 2x)^6$ and then use the expansion to evaluate $(0.98)^6$.

5. Find the term independent of y in the expansion of $\left(y^2 - \frac{2}{y}\right)^6$.

6. Find the first five terms of the expansion of $(1-x)^{-4}$.

7. Find $(0.99)^{-1/2}$ correct to five decimal places using the binomial theorem.

8. Simplify $\sqrt[3]{27x} \div \frac{1}{3} \cdot x^2$.

9. Find the value of $\frac{12^{1/2} \times 16^{1/8}}{27^{1/6} \times 18^{1/2}}$.

10. Find the value of $3\log 2 + \log \frac{5}{3} - \frac{1}{2}\log \frac{4}{25}$.

11. If $\log x = -2 + \log y = \frac{1}{2}\log 6\frac{1}{4}$ find the values of x and y if both are positive.

12. Solve $\log(x^2 + 1) - \log(x - 2) = 1$ for x.

13. Given that 27 is the fourth term of a GP whose first term is 8, find the common ratio.

14. The sum of the first and last terms of an AP is 42. The sum of all terms is 420. The second term is 4. What is the common difference?

15. The n th term of a GP is 2^{n-2}. Find the first two terms of the sequence and the sum of the first ten terms.

16. Find the number of terms which will make the GP $1 + \frac{1}{2} + \frac{1}{4} + \cdots$ just greater than 1.995.

17. A track changes direction by 20° when passing round a circular arc of length 500 metres. What is the radius of the arc? Take $\pi = 3.142$

18. Find the maximum value of the expression $3\cos\theta + 4\sin\theta$.

19. If $\tan A = \frac{4}{3}$ and $\cos B = \frac{5}{13}$ find the value of $\sin(A + B)$.

20. Find the values of x between 0° and 360° which satisfy $\sin 2x = 3\cos 2x$.

21. Prove the identity $\frac{1 + \cos 2\theta}{1 - \cos 2\theta} = \cot^2\theta$.

22. If $3X - 3\begin{bmatrix} -6 & 2 \\ 3 & -1 \end{bmatrix} = 4\begin{bmatrix} 3 & -3 \\ 0 & -3 \end{bmatrix}$ find X.

23. If $A = \begin{bmatrix} 6 & 9 \\ -4 & -6 \end{bmatrix}$ and $B = \begin{bmatrix} 1 & 2 \\ -1 & 0 \end{bmatrix}$ find $A^2 B$.

24. Find the values of x and y given that $\begin{bmatrix} 1 & 2 \\ -1 & 1 \end{bmatrix} \begin{bmatrix} x \\ y \end{bmatrix} = \begin{bmatrix} 4 \\ 5 \end{bmatrix}$.

25. A point A has image $A'(9,8)$ under the linear transformation matrix $P = \begin{bmatrix} 1 & 4 \\ 2 & 3 \end{bmatrix}$. Find A.

18 Differentiation and Integration

Differentiation

18.1 Limits of Functions

We will not give a formal definition of the limit of a function but we will give an intuitive description which should enable you calculate a few limits.

Consider the function $f(x)$ defined as follows:

$$f(x) = \begin{cases} x + 2 & x > 0 \\ 1 - x & x \leq 0 \end{cases}$$

The graph of the function is shown in Figure 18.1

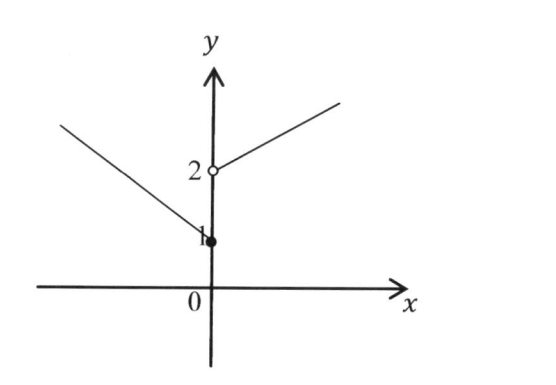

Figure 18.1

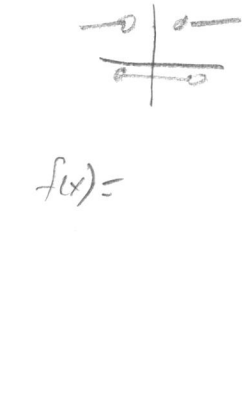

As x gets closer and closer to zero from the right $f(x)$ gets closer and closer to 2. By taking values of x close enough to 0, $f(x)$ can be made very close to 2. We say that the limit of $f(x)$ as x approaches 0 from the right is 2, which is written as $f(x) \to 2$ as $x \to 0^+$ or $\lim_{x \to 0^+} f(x) = 2$. 2 is called the right-hand limit. Notice that the function $f(x) = x + 2$ is undefined when $x = 2$. The definition of a limit describes what happens to a function $f(x)$ when x is near a given value a but not the value of the function at a.

However, by taking values x closer and closer to 0 from the left, we can make $f(x)$ very close to 1. We say that the limit of $f(x)$ from the left is 1, which is written as $f(x) \to 1$ as $x \to 0^-$ or $\lim_{x \to 0^-} it\ f(x) = 1$. 1 is called the left-hand limit.

These two limits are called one-sided limits.

Now let us consider another example. The graph of the function $f(x) = x^2 + 3$ is shown in Figure18.2

Figure 18.2

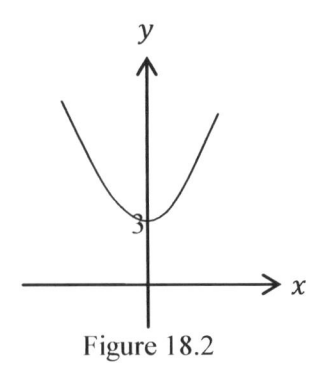

In this case, as x gets closer and closer to zero both from the right and from the left $f(x)$ gets closer and closer to 3. We say that $f(x)$ has a limit of 3 as x approaches 0, which is written as $f(x) \to 3$ as $x \to 0$ or $\lim_{x \to 0} f(x) = 3$. If both one-sided limits exist and are the same we call the common value the limit of the function. That is, a limit exists if both one-sided limits exist and are equal.

Remember that the limit of a function is an indication of the behaviour of the function at a point. The limit of a function, if it exists, as we approach a point is not necessary the same as the value of the function at that point.

The following properties of limit will be extremely useful in computing the limits of functions.

1. $\lim_{x \to a} c = c$, where c is a constant

If $f(x)$ and $g(x)$ are functions such that $\lim_{x \to a} f(x) = l$ and $\lim_{x \to a} g(x) = m$, then

2. $\lim_{x \to a}(f(x) \pm g(x)) = \lim_{x \to a} f(x) \pm \lim_{x \to a} g(x) = l \pm m$

3. $\lim_{x \to a}(f(x) \cdot g(x)) = \lim_{x \to a} f(x) \cdot \lim_{x \to a} g(x) = l \cdot m$

4. $\lim_{x \to a} \left(\dfrac{f(x)}{g(x)} \right) = \dfrac{\lim_{x \to a} f(x)}{\lim_{x \to a} g(x)} = \dfrac{l}{m}$ (provided that $m \neq 0$)

It is also useful to know that $\lim_{x \to a} x = a$, and

$\lim_{x \to a} x^n = \underbrace{\left(\lim_{x \to a} x\right)\left(\lim_{x \to a} x\right) \cdots \left(\lim_{x \to a} x\right)}_{n \text{ factors}} = a^n$

Example 1 Evaluate

(a) $\lim_{x \to 2}(2x - 1)$

Solution
$$\lim_{x \to 2}(2x-1) = \lim_{x \to 2} 2x - \lim_{x \to 2} 1$$
$$= 2 \lim_{x \to 2} x - 1$$
$$= 2(2) - 1$$
$$= 3$$

(b) $\lim_{x \to 1} \dfrac{x^2 - 1}{x - 1}$

Solution Because division by 0 is undefined you cannot substitute 1 for x.

Begin by factorising the numerator.

$$\lim_{x \to 1} \frac{x^2 - 1}{x - 1} = \lim_{x \to 1} \frac{(x-1)(x+1)}{x - 1}$$
$$= \lim_{x \to 1}(x + 1) \qquad \text{Simplify}$$
$$= \lim_{x \to 1} x + \lim_{x \to 1} 1$$
$$= 1 + 1 \qquad \text{Substitute 1 for } x$$
$$= 2$$

(c) $\lim_{x \to \infty} \dfrac{6x^2 + 3x + 2}{4x^2 - x - 7}$

Solution We divide the numerator and denominator by the highest power of x present in the denominator and use the result $\lim_{x \to \infty} \dfrac{1}{x} = 0$. In this case, the highest power of x is x^2.

$$\lim_{x \to \infty} \frac{6x^2 + 3x + 2}{4x^2 - x - 7} = \lim_{x \to \infty} \frac{6 + \dfrac{3}{x} + \dfrac{2}{x^2}}{4 - \dfrac{1}{x} - \dfrac{7}{x^2}}$$
$$= \frac{3}{2}$$

354 Further Mathematics

Exercise 18.1

1. Find the indicated limit:

(a) $\lim_{x \to 2}(2x+5)$ (b) $\lim_{x \to 3}(x+3)$ (c) $\lim_{x \to 0}(x^2+3)$ (d) $\lim_{x \to 4}(7-3x)$ (e) $\lim_{x \to 5} 6$

(f) $\lim_{x \to -1} 3x^2$ (g) $\lim_{x \to 3}(x^2-6x+15)$ (h) $\lim_{x \to -2}(x^2-x-2)$ (i) $\lim_{x \to 4}\sqrt{2x+1}$

2. Find the indicated limit:

(a) $\lim_{x \to 2} \dfrac{2x-1}{x+6}$ (b) $\lim_{x \to 3} \dfrac{x^2-9}{x-3}$ (c) $\lim_{x \to -2} \dfrac{x^2+x-2}{x+2}$

(d) $\lim_{x \to 1} \dfrac{x^2+6x-7}{x-1}$ (e) $\lim_{x \to 0} \dfrac{3x+5x^2}{x}$ (f) $\lim_{x \to 1} \dfrac{x^2+2x-3}{x^2+x-2}$

(g) $\lim_{x \to 0} \dfrac{x}{x+1}$ (h) $\lim_{x \to -1} \dfrac{x^3+1}{x+1}$ (i) $\lim_{x \to \frac{2}{3}} \dfrac{9x^2-4}{3x-2}$

(j) $\lim_{x \to -2} \dfrac{x^2+2x}{x^2+x-2}$ (k) $\lim_{x \to 1} \dfrac{x^3-1}{x-1}$ (l) $\lim_{x \to 9} \dfrac{x-9}{\sqrt{x}-3}$

3. Find the indicated limit:

(a) $\lim_{x \to \infty} \dfrac{2x-3}{4x+5}$ (b) $\lim_{x \to \infty} \dfrac{2x^2+3}{5x^2+2x}$ (c) $\lim_{x \to \infty} \dfrac{2x+3}{x^2}$

(d) $\lim_{x \to \infty} \dfrac{6x^2-8x}{7-3x^2}$ (e) $\lim_{x \to \infty} \dfrac{2x^2+5x}{1-x^3}$ (f) $\lim_{x \to \infty} \dfrac{2x^2+3x-4}{1-x^2}$

(g) $\lim_{x \to \infty} \dfrac{5x^2+4x+3}{5x^2-x-7}$ (h) $\lim_{x \to \infty} \dfrac{x(4-x^3)}{2x^4+3x^2}$ (i) $\lim_{x \to \infty} \dfrac{3x^2+x-1}{2x^2+3x+1}$

4. In each of the following, calculate $\lim_{h \to 0} \dfrac{f(a+h)-f(a)}{h}$

(a) $f(x)=2x$ (b) $f(x)=3x^2$ (c) $f(x)=2x+3$

(d) $f(x)=2x^2-3$ (e) $f(x)=x^3$ (f) $f(x)=\dfrac{1}{x}$

18.2 Differentiation of Polynomial Functions

The Gradient of a Curve

The gradient of a curve varies along its length. Because of this the gradient of a curve is not defined between two points as in the case of a straight line but at a single point. The gradient of a function at a given point is the same as the gradient of a line tangent to the graph of the function at the point, as illustrated in Figure 18.3.

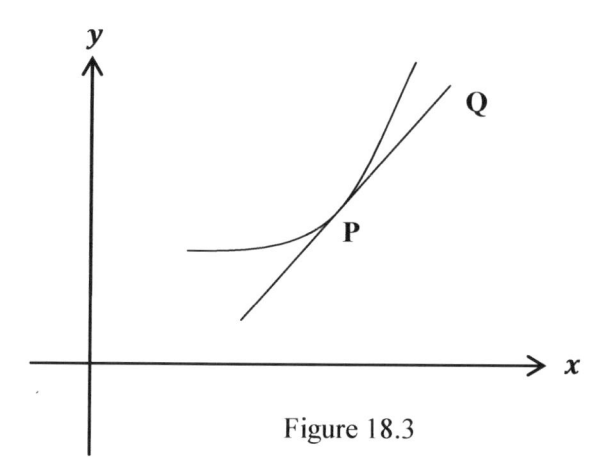

Figure 18.3

The gradient of the curve at P is equal to the gradient of the tangent PQ.

The Derivative

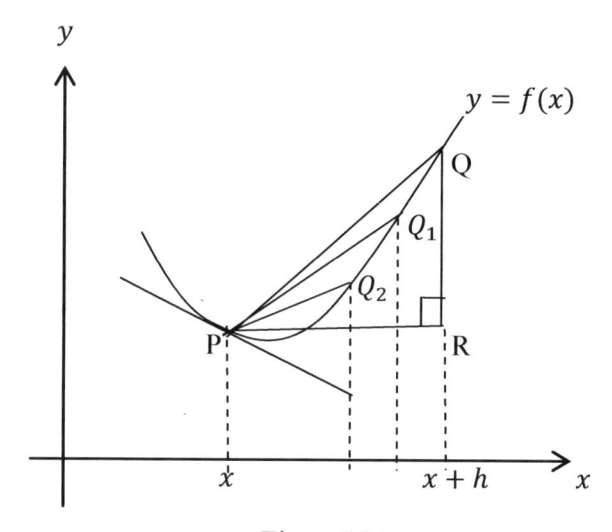

Figure 18.4

Figure 18.4 shows the graph of a function $y = f(x)$. The points P and Q have coordinates $(x, f(x))$ and $(x + h, f(x + h))$ respectively. The gradient of the chord PQ is given by

$$\frac{QR}{PR} = \frac{f(x+h) - f(x)}{h}$$

As Q moves closer and closer to P as shown in Figure 18.4, h approaches 0, and PQ becomes a tangent to the curve at P when Q coincides with P. That is, the gradient of the tangent at P is

$$\lim_{h \to 0} \frac{f(x+h) - f(x)}{h}$$

The gradient of the curve at P is the limiting value of the gradient of the chord as h approaches 0.

Consider the function $f(x) = x^2$. The gradient of the tangent to the curve at $(x, f(x))$ is given by

$$\lim_{x \to 0} \frac{f(x+h) - f(x)}{h} = \lim_{x \to 0} \frac{(x+h)^2 - x^2}{h}$$

$f(x) = x^2 + 2$

$$= \lim_{h \to 0} \frac{2xh + h^2}{h}$$

$$= \lim_{h \to 0} (2x + h)$$

$$= 2x$$

This new function is called the derivative of the function f with respect to x, denoted by $f'(x)$. The process that produces the derivative of a function is called differentiation.

The gradient of the tangent at any point on the graph of the function $f(x) = x^2$ is $2x$. To find the gradient of the tangent at a point $(x_0, f(x_0))$, we substitute x_0 into $2x$. For instance, the gradient of the tangent at the point (4, 16) is 2(4) i.e., 8.

The derivative of a function $f(x)$ with respect to x is denoted by

$f'(x)$, f', $D_x(x)$, $\frac{dy}{dx}$ or $\frac{d}{dx}[f(x)]$

Differentiation from first principles

The method used in computing the derivative of the function $f(x) = x^2$ is called differentiation from first principles.

Example 2 Find the derivative of the following functions from first principles.

(a) $f(x) = x^3$

Solution Let δx and δy represent small increment in x and y respectively.

$$\frac{\delta y}{\delta x} = \frac{f(x+\delta x)-f(x)}{\delta x}$$

$$= \frac{(x+\delta x)^3 - x^3}{\delta x}$$

$$= \frac{3x^2 \delta x + 3x\delta x^2 + \delta x^3}{\delta x}$$

$$= 3x^2 + 3x\delta x + \delta x^2$$

So, $\dfrac{dy}{dx} = \lim\limits_{\delta x \to 0} \dfrac{\delta y}{\delta x} = 3x^2$

The above solution can also be presented as follows:

$$y + \delta y = (x + \delta x)^3$$

$$\delta y = 3x^2 \delta x + 3x\delta x^2 + \delta x^3$$

$$\frac{\delta y}{\delta x} = 3x^2 + 3x\delta x + \delta x^2$$

So, $\dfrac{dy}{dx} = \lim\limits_{\delta x \to 0} \dfrac{\delta y}{\delta x} = 3x^2$

(b) $f(x) = 2x$

Solution

$$\frac{\delta y}{\delta x} = \frac{f(x+\delta x)-f(x)}{\delta x}$$

$$= \frac{2(x+\delta x) - 2x}{\delta x}$$

$$= 2$$

So, $\dfrac{dy}{dx} = \lim\limits_{\delta x \to 0} \dfrac{\delta y}{\delta x} = 2$

(c) $f(x) = -3$

Solution
$$\dfrac{\delta y}{\delta x} = \dfrac{f(x+\delta x) - f(x)}{\delta x}$$
$$= \dfrac{-3 - (-3)}{\delta x}$$
$$= 0$$

So, $\dfrac{dy}{dx} = \lim\limits_{\delta x \to 0} \dfrac{\delta y}{\delta x} = 0$

Exercise 18.2(a)

Find the derivative of each of the following functions from first principles:

1. $x \quad = 1$
2. $5x \quad = 5$
3. $3x^2 \quad 3(x+h)^2 - 3x^2/h$
4. $-2x^2 \quad = -4x$
5. 2
6. $-\dfrac{1}{2}x^2$
7. $\dfrac{2}{3}x^3$
8. $-3x^2 + 2$
9. $1 - 2x^3$
10. $3x^2 - 1$
11. $2x - 3x^2$
12. $x^2 + 3x - 2$

General Rule for Calculating Derivatives

To compute rapidly the derivative of any polynomial function we use the following rule, called the Power Rule.

For any real number $n > 0$, $\dfrac{d}{dx}(x^n) = nx^{n-1}$

This rule says that multiply x by the index n and then reduce the index on x by 1.

The symbol $\dfrac{d}{dx}$ is read " the derivative with respect to x of".

Example 3

Find the derivative of x^5

Solution $\dfrac{d}{dx}(x^5) = 5 \cdot x^{5-1} = 5x^4$

Exercise 18.2(b)

Differentiate each of the following functions with respect to x:

1. $-x$
2. 8
3. x^4
4. x^6
5. $-x^{-3}$
6. x^{-1}
7. x^{-8}
8. $x^{-3/4}$
9. $-x^{1/2}$
10. $x^{-3/2}$
11. $-x^{-2/3}$
12. $-x^{1/3}$

Rules for differentiation

You can compute the derivatives of multiples and combination of functions using the following rules of differentiation.

Derivative of a Constant Times a Function

If f and g are differentiable functions, then

$\dfrac{d}{dx}(cf(x)) = c\dfrac{d}{dx}f(x)$, where c is a constant

To differentiate leave the constant and differentiate the function and then multiply the result by the constant.

Example 4

(a) Differentiate the function $-3x^2$ with respect to x.

Solution $\dfrac{d}{dx}(-3x^2) = -3\dfrac{d}{dx}(x^2)$

$\phantom{\dfrac{d}{dx}(-3x^2)} = -3 \times 2x$

$\phantom{\dfrac{d}{dx}(-3x^2)} = -6x$

(b) Differentiate the function $\frac{1}{2}x^6$ with respect to x.

Solution $\dfrac{d}{dx}\left(\dfrac{1}{2}x^6\right) = \dfrac{1}{2} \times 6x^5$

$\phantom{\dfrac{d}{dx}\left(\dfrac{1}{2}x^6\right)} = 3x^5$

Exercise 18.2(c)

Find the derivative of each of the following functions:

1. $-2x$
2. $3x^4$
3. $-2x^5$
4. $\frac{1}{4}x^8$
5. $\frac{2}{3}x^6$
6. $-\frac{2}{5}x^{-5}$
7. $2x^{-3}$
8. $-4x^{-3}$
9. $3x^{-2}$
10. $3x^{1/3}$
11. $-2x^{3/2}$
12. $6x^{1/2}$
13. $-6x^{-2/3}$
14. $8x^{-3/4}$
15. $9x^{-2/3}$

Derivative of Sums of Functions

If f and g are differentiable functions, then $\frac{d}{dx}(f(x) \pm g(x)) = \frac{d}{dx}f(x) \pm \frac{d}{dx}g(x)$

To differentiate a sum or a difference of functions differentiate term by term

Example 5

Differentiate $x^3 + 3x^2 - 2x$ with respective to x

Solution   Let $y = x^3 + 3x^2 - 2x$

Then $\quad \frac{dy}{dx} = 3x^2 + 6x - 2$

Exercise 18.2(d)

Differentiate each of the following functions with respect to the variable.

1. $2x^3 + 3x^4$
2. $3x^5 - 2x$
3. $3 - t^4 + t^5$
4. $7 - 2t^{-1}$
5. $4t^3 + 3t$
6. $5x^4 - 3x^5$
7. $5 - x^{-3}$
8. $2x^3 - 3x^{-2}$
9. $3t^4 - 4t + 2$
10. $2t^2 - 3t$
11. $6x^2 + 5x + 2$
12. $\frac{1}{3}t^3 - \frac{1}{2}t^{-2}$
13. $3x^{1/3} - 2x^{-1/2}$
14. $4t^{-3/4} + \frac{1}{3}t^3$
15. $2 - 6x^{-2/3}$
16. $1 + \frac{1}{x^3}$
17. $3x - \frac{1}{\sqrt{x}} + \frac{1}{x}$
18. $\frac{5}{x^2} - \frac{1}{\sqrt{x^3}} + 2$
19. $t^2 + \frac{4}{t^3}$
20. $2x(3 - x)$
21. $3(x - 2)^2$
22. $\frac{3x^2 - 2x^3}{3x}$
23. $\frac{3x^2 + 2x - 1}{x}$
24. $\frac{(t+3)(2t-1)}{t^2}$

Derivatives of Products

If f and g are differentiable functions, then

$$\frac{d}{dx}(f(x) \cdot g(x)) = f(x)\frac{d}{dx}g(x) + g(x)\frac{d}{dx}f(x)$$

To differentiate a product of two functions, leave the first function and differentiate the second function + leave the second function and differentiate the first function

Example 6

Differentiate $x^3(2x - 3)$ with respect to x

Solution Let $y = x^3(2x - 3)$

then
$$\frac{dy}{dx} = x^3 \times 2 + (2x - 3) \times 3x^2$$

$$= 2x^3 + 6x^3 - 9x^2$$

$$= 8x^3 - 9x^2$$

The same result is obtained if you expand the bracket and then find the derivative.

$$y = 2x^4 - 3x^3$$

$$\frac{dy}{dx} = 8x^3 - 9x^2$$

Often it will be quite difficult or unnecessary to multiple out in order to differentiate term by term.

Exercise 18.2(e)

Differentiate the following functions with respect to x:

1. $x^3(4x^2 - 3)$
2. $(2x - 3)(x + 4)$
3. $3x^2(2x + 1)$
4. $(2x + 1)(3x - 2)$
5. $x(2x^3 - 3x^2)$
6. $x(3 - 2x)(2 + x)$
7. $\sqrt{x}(2x + 3)$
8. $(1 - x^2)(1 + 2x^2)$
9. $\sqrt[3]{x}(2 + x)$
10. $3x^{-2}(2x - 3)$
11. $(x^4 - 1)(x^3 + 1)$
12. $(2x - 5)(3x^2 + 2)$

Derivative of Quotients

If f and g are differentiable functions and $g(x) \neq 0$, then

$$\frac{d}{dx}\left(\frac{f(x)}{g(x)}\right) = \frac{g(x)\frac{d}{dx}f(x) - f(x)\frac{d}{dx}g(x)}{(g(x))^2}$$

To differentiate a quotient of two functions, leave the denominator and differentiate the numerator – leave the numerator and differentiate the denominator and divide all by the square of the denominator.

Example 7

Differentiate $\dfrac{2x}{3x-1}$ with respect to x

Solution Let $y = \dfrac{2x}{3x-1}$

Then $\dfrac{dy}{dx} = \dfrac{(3x-1)\times 2 - 2x \times 3}{(3x-1)^2}$

$= \dfrac{6x - 2 - 6x}{(3x-1)^2}$

$= -\dfrac{2}{(3x-1)^2}$

Exercise 18.2(f)

Differentiate the following functions with respect to x:

1. $\dfrac{x}{x+1}$
2. $\dfrac{x+1}{x-1}$
3. $\dfrac{1}{x^2-1}$
4. $\dfrac{2}{3x^2-1}$

5. $\dfrac{3x}{x^2-2}$
6. $\dfrac{3x^2-5}{2x-7}$
7. $\dfrac{1-\sqrt{x}}{1+\sqrt{x}}$
8. $\dfrac{1+x^2}{1-x^2}$

9. $\dfrac{x^2}{x+1}$
10. $\dfrac{x^2}{x^2-1}$
11. $\dfrac{x^2}{2x+3}$
12. $\dfrac{\sqrt{x}}{\sqrt{x}-1}$

Differentiation and Integration

The Chain Rule

If $y = f(u)$ where $u = g(x)$, then $\dfrac{dy}{dx} = \dfrac{dy}{du} \cdot \dfrac{du}{dx}$

The chain rule is used to compute the derivative of a function of a function of a variable.

Example 8

Differentiate $(2x - 3)^4$ with respect to x

Solution Let $y = (3x - 2)^4$ and $u = 3x - 2$

Then $\quad y = u^4$

$$\dfrac{dy}{du} = 4u^3$$

and $\quad \dfrac{du}{dx} = 3$

By the chain rule

$$\dfrac{dy}{dx} = \dfrac{dy}{du} \times \dfrac{du}{dx}$$

$$= 4u^3 \times 3$$

$$= 12\,u^3$$

$$= 12(3x - 2)^3 \qquad \text{Replace } u \text{ with } 3x - 2.$$

Exercise 18.2(g)

Differentiate the following function with respect to x:

1. $(x - 5)^3$
2. $(3x - 2)^5$
3. $(1 - 2x^2)^6$
4. $(x^2 + 1)^4$
5. $(x^3 - 2)^7$
6. $(2x + 1)^{-3}$
7. $(3 - 2x)^{-4}$
8. $(4x + 3)^{-2}$
9. $(1 - 2x^2)^{-1}$
10. $(x^2 - 1)^{1/2}$
11. $(3x + 2)^{-2/3}$
12. $(2x^2 - 3)^{3/4}$
13. $\dfrac{1}{(3x^2 + 2x)^2}$
14. $\dfrac{1}{\sqrt[3]{1 - x}}$
15. $\dfrac{1}{\sqrt{x^2 - 1}}$
16. $\dfrac{1}{\sqrt{2 + x^3}}$

The following rule is an alternative form of the Chain Rule which can be used to find the derivative of a function of a function of variable in one step without introducing another variable such as u.

$$\frac{d}{dx}[f(x)]^n = n[f(x)]^{n-1}\frac{d}{dx}f(x)$$

Example 9

Differentiate $(3 - 2x)^5$ with respect to x

Solution Let $y = (3 - 2x)^5$

then $\quad \dfrac{dy}{dx} = 5(3-2x)^4 \times -2$

$\qquad\qquad = -10(3 - 2x)^4$

Exercise 18.2(h)

Differentiate the following functions with respect to x:

1. $(x + 3)^4$
2. $(3x - 2)^3$
3. $(2x^2 + 3)^5$
4. $(x^2 + 1)^{1/2}$
5. $(2x^3 - 1)^{-1/3}$
6. $(3 + x^2)^{-2}$
7. $(1 - x^2)^3$
8. $(3x^2 - 1)^{-3}$
9. $(1 - x^2)^6$
10. $\dfrac{1}{(x^2 - 3)^4}$
11. $\dfrac{1}{\sqrt{2x^2 + 3}}$
12. $\dfrac{1}{\sqrt[3]{x^3 + 3x}}$

If $y = f(x)$, and $x = f^{-1}(y)$, then $\dfrac{dy}{dx}$ and $\dfrac{dx}{dy}$ are related by the following rule.

$$\frac{dy}{dx} = \frac{1}{dx/dy}$$

Example 10

Differentiate $x = y^2$

solution $\quad \dfrac{dx}{dy} = 2y$

$\therefore \dfrac{dy}{dx} = \dfrac{1}{dx/dy}$

$\qquad = \dfrac{1}{2y}$

But $y = \sqrt{x}$

Hence $\dfrac{dy}{dx} = \dfrac{1}{2\sqrt{x}}$

Exercise 18.2(i)

Differentiate the following functions with respect to x:

1. $y^3 = x$
2. $y^2 = 2x$
3. $\sqrt[3]{y} = x^2$
4. $1 = x^2 y^3$
5. $\dfrac{1}{3y^2} = x$
6. $\sqrt[3]{y^2} = x$

Sometimes a function is defined by expressing x and y separately in terms of a third independent variable say t, e.g. $x = 2t$, and $y = 3t^2$. The third variable, is called a parameter and the two expressions for x and y are called parametric equations. The following rule enables us to find the derivative of such functions with respect to x.

If $x = f(t)$ and $y = g(t)$, then $\dfrac{dy}{dx} = \dfrac{dy/dt}{dx/dt}$

Example 11

If $x = 2t^3$ and $y = 3t^2$ find $\dfrac{dy}{dx}$ in terms of t

solution $x = 2t^3$

$\dfrac{dx}{dt} = 6t^2$

$y = 3t^2$

$\dfrac{dy}{dt} = 6t$

$\therefore \dfrac{dy}{dx} = \dfrac{dy/dt}{dx/dt}$

$= \dfrac{6t}{6t^2}$

$$= \frac{1}{t}$$

Exercise 18.2(j)

In Problem 1 to 10, find $\dfrac{dy}{dx}$ in terms of t:

1. $x = 3t^2$, $y = 2t^3$
2. $x = 2at^2$, $y = 4at$
3. $x = t^{-3}$, $y = t^{-2}$
4. $x = (t+1)^2$, $y = 1 - t^2$
5. $x = (2t+1)^3$, $y = 3t^2 - 2$
6. $x = (1-t)^3$, $y = t^2 - t$
7. $x = \dfrac{t}{1-t}$, $y = \dfrac{t^2}{1-t}$
8. $x = \dfrac{2t}{2-t}$, $y = \dfrac{3t}{t+3}$
9. $x = \dfrac{2t}{1-t^2}$, $y = \dfrac{1+t^2}{1-t^2}$
10. $x = \dfrac{t}{1-t}$, $y = \dfrac{1-2t}{1-t}$

Exercise 18.2(k)

Differentiate the following functions with respect to x. You may have to use more than one rule to find the derivatives of the given functions.

1. $(x-3)^2(2x+1)^3$
2. $(x-1)^3(x-3)^2$
3. $(3x+2)^2(3x^2-1)$
4. $(x-1)^2(x^2+1)$
5. $\sqrt{(x-1)}\sqrt{(x+1)^3}$
6. $(x+1)\sqrt{(x^2-1)}$
7. $\dfrac{x^2}{\sqrt{(1+x^2)}}$
8. $\dfrac{3-2x}{1+x^2}$
9. $\dfrac{x^2-2}{(x-2)^2}$
10. $\sqrt{\dfrac{1-x}{1+x}}$
11. $\dfrac{\sqrt{x^3}}{1-x}$
12. $\dfrac{(1+x)^3}{1+x^3}$

Higher Order Derivatives

The differentiation of a function f produces a new function f', called the derivative of f. If f' is also a differentiable function, its derivative is denoted by f''. We call this the second derivative of f. The second derivative of $y = f(x)$ may be denoted by

$$\frac{d}{dx}\left(\frac{dy}{dx}\right) = \frac{d^2y}{dx^2},$$ read 'd two y by d x squared' or $D^2_x f(x)$. Similarly, the third derivative is denoted by f''', $\frac{d^3y}{dx^3}$ or $D^3_x f(x)$.

The n th order derivative of a function f, is obtained by differentiating n times.

However, if $y = f(t)$ and $y = g(t)$ then $\frac{d^2y}{dx^2} = \frac{\frac{d}{dt}(dy/dx)}{dx/dt}$

Example 12

(a) If $y = 3x^4 - 2x^3$ find $\frac{d^2y}{dx^2}$

Solution $y = 3x^4 - 2x^3$

$$\frac{dy}{dx} = 12x^3 - 6x^2$$

$$\frac{d^2y}{dx^2} = 36x^2 - 12x$$

(b) Find the second derivative of the function $f(x) = (x^2 + 1)^3$

Solution $f(x) = (x^2 + 1)^3$

$$f'(x) = 3(x^2 + 1)^2 \times 2x$$

$$= 6x(x^2 + 1)^2$$

$$f''(x) = 6x \cdot 2(x^2 + 1) \cdot 2x + (x^2 + 1)^2 \cdot 6$$

$$= 24x^2(x^2 + 1) + 6(x^2 + 1)^2$$

$$= 6(x^2 + 1)(5x^2 + 1)$$

(c) If $y = 3t^2$ and $x = 4t^3$ find $\frac{d^2y}{dx^2}$ in terms of t

Solution $y = 3t^2$

$$\frac{dy}{dt} = 6t$$

$$\frac{dx}{dt} = 12t^2$$

$$\frac{dy}{dx} = \frac{6t}{12t^2}$$

$$= \frac{1}{2t}$$

$$\therefore \frac{d^2y}{dx^2} = \frac{\dfrac{d}{dt}\left\{\dfrac{1}{2t}\right\}}{12t^2}$$

$$= -\frac{1}{24t^4}$$

Exercise 18.2(I)

Find the second derivative of the following functions:

1(a) $3x^2$ (b) $4x$ (c) $x^3 - 3x^2$

(d) $x^4 + x^2 - 2$ (e) $3x^2 + 2x$ (f) $3 - 2x + x^2 - x^3$

2(a) x^{-4} (b) $2x^{-3}$ (c) $2 + 3x^{-1}$ (d) $x^{-3} - 2x$

(e) $3x + \dfrac{2}{x^2}$ (f) $1 - \dfrac{4}{x^3}$ (g) $\dfrac{1}{2x^3} - \dfrac{1}{x^2} + \dfrac{2}{x}$

3(a) $x^{1/2}$ (b) $\sqrt[3]{x}$ (c) $\sqrt{x^3}$ (d) $\sqrt[4]{x^3}$

(e) $1 - \sqrt[3]{x}$ (f) $\dfrac{1}{\sqrt{x}}$ (g) $\dfrac{1}{\sqrt[3]{x}} - \dfrac{2}{\sqrt{x}} + \dfrac{3}{x}$

4(a) $(1 + x^3)^2$ (b) $(\sqrt{x} + 1)^2$ (c) $(2x - 3)^3$

(d) $x(x + 2)^2$ (e) $2x(x - 3)^3$ (f) $(x - 2)(x + 2)^2$

5(a) $\dfrac{x}{x+1}$ (b) $\dfrac{1+x}{1-x}$ (c) $\dfrac{x}{\sqrt{1+x^2}}$

(d) $\dfrac{1}{\sqrt{1+x}}$ (e) $\dfrac{1+x^2}{1-x^2}$ (f) $\dfrac{x^2}{\sqrt{x-1}}$

6. Given that $x = 2t^3$ and $y = 3t^2$ find $\dfrac{d^2y}{dx^2}$ in terms of t.

7. Given that $x = 3at^2$ and $y = 2at^3$ find $\dfrac{d^2y}{dx^2}$ in terms of t.

8. Given $x = (t^2 - 1)^2$ and $y = t^3$ find $\dfrac{d^2y}{dx^2}$ in terms of t.

Implicit Differentiation

The equation $y = 4x^3 + 5$ defines y as a function of x explicitly. However, not all equations in x and y can be expressed as explicit functions of x. For instance, it would be difficulty or even not possible to solve for y in the equation $y^3 - 2xy^2 + x^2 = 0$. In equations such as this, y is said to be given implicitly in terms of x. A function that cannot be expressed in the form $y = f(x)$, with y given explicitly in terms of x is called an implicit function.

You can calculate the derivatives of implicit functions by using a process called implicit differentiation.

Note that the chain rule gives $\dfrac{d}{dx}(y^2) = 2y\dfrac{dy}{dx}$ and $\dfrac{d}{dx}(y^3) = 3y^2\dfrac{dy}{dx}$.

Also using the product rule and chain rule we get $\dfrac{d}{dx}(xy) = x\dfrac{dy}{dx} + y$.

In the following example we show how to differentiate an implicit function.

Example 13

Find the derivative of $y^3 + 2xy - x^3 = 5$

Solution Differentiate both sides of the equation. Remember that y is a function of x.

$$y^3 + 2xy - x^3 = 5$$

$$3y^2\dfrac{dy}{dx} + 2x\dfrac{dy}{dx} + 2y - 3x^2 = 0 \qquad \text{Differentiate each term.}$$

$$(3y^2 + 2x)\dfrac{dy}{dx} = 3x^2 - 2y$$

$$\dfrac{dy}{dx} = \dfrac{3x^2 - 2y}{3y^2 + 2x}$$

Exercise 18.2(m)

Find $\dfrac{dy}{dx}$ for each of the following implicit functions:

1. $y^2 - x^2 = 8$
2. $2x - y^2 = 5$
3. $x^2 - 4xy + y^2 = 0$
4. $y^2 - 2xy = x^2$
5. $x^3 y^2 = 1$
6. $x^3 - y^3 = 2$
7. $3x^2 - 4xy = 0$
8. $x^2 + 3xy = y^2$
9. $x^2 - y^2 + 6y + 8x = 0$
10. $x^2 + y^2 + 2xy - 3y + 2x = 7$

18.3 Differentiation of Trigonometric Functions

The derivative of three trigonometric functions measured in radian is summarized in the table below.

y	$\dfrac{dy}{dx}$
$\sin x$	$\cos x$
$\cos x$	$-\sin x$
$\tan x$	$\sec^2 x$

You would use the same rules of differentiation stated above to compute derivatives of trigonometric functions as illustrated by the examples below.

Example 14

Find the derivatives of the following functions:

(a) $y = \sin 3x$ (b) $y = x\tan x + 1$ (c) $y = \cos^2 x$

Solution   Let $u = 3x$

$$\dfrac{du}{dx} = 3$$

and $y = \sin u$

$$\dfrac{dy}{du} = \cos u$$

The chain rule gives

$$\frac{dy}{dx} = \frac{dy}{du} \times \frac{du}{dx}$$

$$= \cos u \times 3$$

$$= 3\cos 3x$$

(b) $y = x\tan x + 1$

Solution Using the product rule and differentiating term by term we have

$$\frac{dy}{dx} = \tan x + x\sec^2 x$$

(c) $y = \cos^2 x$

solution Let $u = \cos x$

$$\frac{du}{dx} = -\sin x$$

and $y = u^2$

$$\frac{dy}{du} = 2u$$

By the chain rule

$$\frac{dy}{dx} = \frac{dy}{du} \times \frac{du}{dx}$$

$$= 2u \times -\sin x$$

$$= -2\sin x\cos x$$

$$= -\sin 2x$$

Alternatively, using the product rule we have

$$y = (\cos x)(\cos x)$$

$$\frac{dy}{dx} = \cos x \cdot -\sin x + (-\sin x \cdot \cos x)$$

$$= -\sin x\cos x - \sin x\cos x$$

$$= -2\sin x \cos x$$
$$= -\sin 2x$$

Exercise 18.3

1. Differentiate:

(a) $\cos 2x$ (b) $\sin 5x$ (c) $2\cos 3x$ (d) $-\frac{1}{2}\sin 6x$

(e) $\cos(3x - 2)$ (f) $\sin(2x + 3)$ (g) $\sin x^2$ (h) $6\cos\frac{1}{2}x$

(i) $\tan 3x^2$ (j) $\tan(2x - 5)$ (k) $\tan\frac{1}{3}x^2$ (l) $\tan(x^2 + 1)$

2. Differentiate:

(a) $\cos^2 2x$ (b) $3\sin^2 x$ (c) $\tan^2 2x$ (d) $\sin^2 3x$

3. Differentiate:

(a) $x\cos x$ (b) $x\sin 2x$ (c) $x^2 \sin x$ (d) $x^3 \tan x$

4. Differentiate:

(a) $2x\sin\frac{1}{2}x$ (b) $x\cos 2x$ (c) $x^2 \cos x$ (d) $x^2 \tan 2x$

(e) $\sin x \cos x$ (f) $\dfrac{\sin x}{x}$ (g) $\dfrac{\cos 2x}{x}$ (h) $\cos x \tan x$

5. Differentiate:

(a) $2\sin x + 3x$ (b) $1 - \cos 2x$ (c) $x^2 + 3\tan x$

(d) $2\cos x + 3$ (e) $2\sin x + x^2$ (f) $2x^3 - 3\cos x$

(g) $2x^2 + 5\tan x$ (h) $3\sin 2x - \cos x$ (i) $x\tan x - 3x^2$

(j) $x^2 \sin x - 1$ (k) $x^2 \tan x + \dfrac{1}{x}$ (l) $x^2 \sin x - \dfrac{1}{x^2}$

18.4 Tangents and Normals

Recall that the tangent to a curve at a point is the straight line that touches the curve at that point. The gradient of the tangent is the derivative of the function at this point. The perpendicular to a tangent is called the normal.

Remember that the gradients of two perpendiculars lines are negative reciprocal of each other. That is if the gradient of one line is m, then the gradient of the other line is $-\frac{1}{m}$.

Example 15

(a) Find the equations of the tangent and normal of the curve $y = 3x^2 - 2x - 5$ at the point where $x = 2$.

Solution First, find the y-coordinate of the point where $x = 2$

$$y = 3 \times 2^2 - 2 \times 2 - 5$$

$$= 3$$

Next, find the gradient of the tangent at the point $(2, 3)$

$$\frac{dy}{dx} = 6x - 2$$

$$= 6 \times 2 - 2$$

$$= 10$$

Finally, use the formula $y - y_1 = m(x - x_1)$ to find the equation of the tangent.

$$y - 3 = 10(x - 2)$$

$$y = 10x - 17$$

The equation of the tangent is $y = 10x - 17$

The gradient of the normal is $-\frac{1}{10}$, hence the equation of the normal is

$$y - 3 = -\frac{1}{10}(x - 2)$$

$$x + 10y - 32 = 0$$

The equation of the normal is $x + 10y - 32 = 0$

(b) At what point on the curve $y = 2x^2 - 3x + 1$ is the tangent parallel to the line $y = 5x - 7$.

Solution The gradient of the curve $y = 2x^2 - 3x + 1$ is given by $\frac{dy}{dx} = 4x - 3$

374 Further Mathematics

The gradient of the line $y = 5x - 7$ is 5

Because the tangent is parallel to the line

$4x - 3 = 5$

$\quad 4x = 8$

$\quad\ \ x = 2$

We use the equation $y = 2x^2 - 3x + 1$ to find the y-coordinate

$\quad y = 2 \times 2^2 - 3 \times 2 + 1$

$\quad\ \ = 8 - 6 + 1$

$\quad\ \ = 3$

Hence the point is (2, 3)

Exercise 18.4

1. Find the gradient of the tangent to each of the following curves at the given point:

(a) $y = x^2 - x + 1$, (2, 3)
(b) $y = 3 + 4x - 2x^2$, (3, -3)
(c) $y = 3x^2 - 2$, (0, -2)
(d) $y = x^3 - 3x$, (2, 2)
(e) $y = x^3 + 2x^2 + 1$, (-2, 1)
(f) $y = x^4 - 3x^2 + 2x - 1$, (1, -1)
(g) $y = 3x - x^3$, (2, 1)
(g) $y = x^2 - \frac{8}{x^3}$, (-2, 5)

2. Find the gradient to the following curves at the given points:

(a) $xy = 6$, (3, 2)
(b) $x^2 - y^2 = -8$, (1, 3)
(c) $x^3 + y^3 = 7$, (-1, 2)
(d) $3x^2 + 2y^2 = 14$, (2, 1)
(e) $x^2 + 3xy - y^2 = 9$, (2, 1)
(f) $x^3 - 2xy + y^3 = 0$, (1, 1)

3. Find the equation of the tangent to each of the following curves at the given points:

(a) $y = 2x^2 + 3$, (-1, 5)
(b) $y = 2x^2 - 3x - 5$, (3, 4)
(c) $y = 1 + 2x^2 - x^3$, (2, 1)
(d) $y = x^3 + 2x^2 - 3x - 1$, (-2, 5)
(e) $y = x^3(2 - x^2)$, (1, 1)
(f) $y = 3 - \frac{4}{x^2}$, (2, 2)

Differentiation and Integration 375

4. Find the equation of the normal to each of the following curves at the given point:

(a) $y = 3x + x^2$, $(1, 4)$

(b) $y = 2x^2 - 3$, $(-2, 5)$

(c) $y = 2 + 3x^2 - x^3$, $(2, 6)$

(d) $y = x^3 + 2x$, $(1, 3)$

(e) $y = 3x^3 + 2x^2 - 5x + 3$, $(-2, -3)$

(f) $y = (3x - 2)^4$, $(1, 1)$

5. Find the equation of the tangent and normal to each of the following curves at the given point:

(a) $x^2 - xy + y^2 = 7$, $(3, 2)$

(b) $x^2 + y^2 + 3xy - 11 = 0$, $(1, 2)$

(c) $y^2 - 3x + 2y + x^2 = 1$, $(2, 3)$

(d) $x^2y + x^3 + 2y = 2$, $(2, -1)$

6. At what point on the curve $y = 3x^2 + 4x - 5$ is the tangent parallel to the line $y = 7x - 19$.

7. At what point on the curve $y = x^3 - 3x^2$ is the normal parallel to the line $x + 4y - 17 = 0$.

8. At what point on the curve $y = 2x^2 - 3x + 4$ is the normal parallel to the line $3x - y + 14 = 0$.

18.5 Maximum and Minimum Values

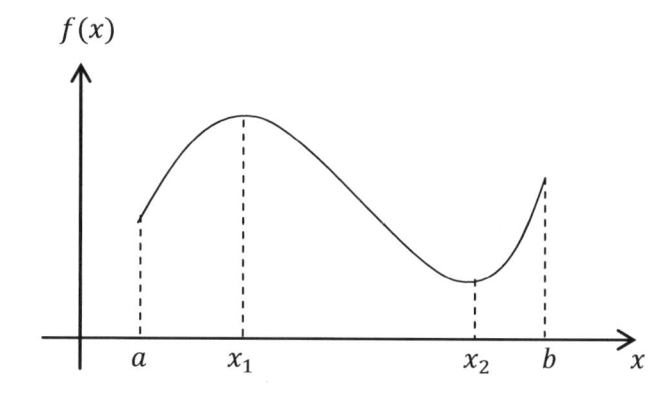

Figure 18.5

You can see from the graph in Figure 18.5 that as you move from left to right the function f increases for x-values from a to x_1, then decreases from x_1 to x_2, and finally increases from x_2 to b. You should notice that $f(x_1) \geq f(x)$ for all x on the interval $a \leq x \leq b$. So, a maximum value of f occurs at $x = x_1$. Also, notice that $f(x_2) \leq f(x)$ for all x on the interval $a \leq x \leq b$. So, f has a minimum value at $x = x_2$. The points $(x_1, f(x_1))$ and $(x_1, f(x_2))$ are called stationary points (or turning points) on the graph, or stationary values of f. The largest value of a function is called a maximum (plural maxima) and the smallest value is called a minimum (plural minima). A function has extremum (plural) at a point if it has either a maximum or a minimum there.

Local Maximum and Minimum

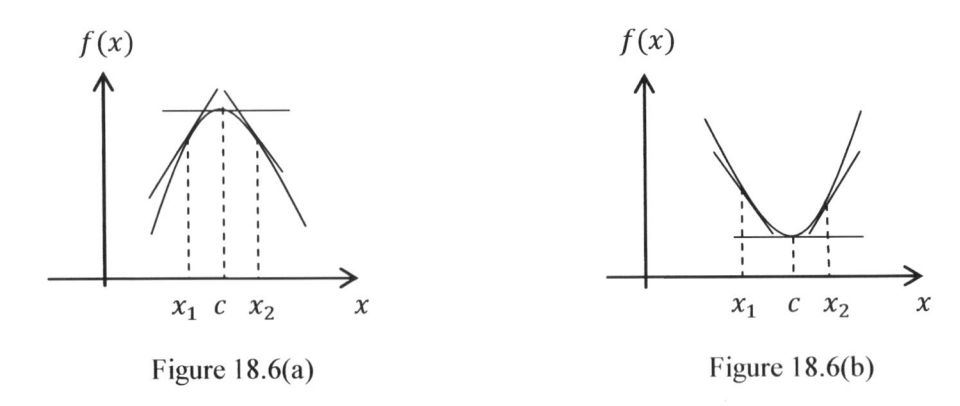

Figure 18.6(a)

Figure 18.6(b)

You can see from the graph of Figure 18.6(a) that the value of the function $f(x_1)$ at x_1 is less than the value of the function $f(c)$ at $x = c$. Again, you can see that $f(x_2)$ is less than $f(c)$. Notice that $f(c)$ is greater than the values of the function on either side of the point at $x = c$ and close to it. Hence, a maximum value of f occurs at $x = c$, called a local maximum, and c is called a critical number. A local maximum is the greatest value of a function in some region around a point in an open interval containing that point.

As you can see from the graph in Figure 18.6(b) both $f(x_1)$ and $f(x_2)$ are greater than $f(c)$. Also, notice that $f(c)$ is less than the values of the function on either side of the point at $x = c$ and close to it. The function has a local minimum at $x = c$. Similarly, a local minimum is the least value of a function in some region around the point.

As shown in Figure 18.6(a), the gradient of the tangent on the left hand side of the point at $x = c$ is positive, indicating that the function is increasing. At $x = c$ the tangent line is horizontal, so has gradient 0. The gradient of the tangent on the right hand side is negative, indicating that the function is decreasing. As shown by the tangent lines in Figure 18.6(b), the function is decreasing on the left hand side of the point at $x = c$, has a horizontal tangent at $x = c$, and is increasing on the right hand side.

Since the derivative of a function gives the gradient of a line tangent to the graph of the function, the preceding discussion suggest the following first derivative test.

First Derivative Test

Let c be a critical number for a function f.

1. If the sign of $f'(x)$ changes from positive to zero to negative as x increases through $x = c$, then $f(c)$ is a local maximum.

2. If the sign of $f'(x)$ changes from negative to zero to positive as x increases through $x = c$, then $f(c)$ is a local minimum.

Locating local Maxima and Minima

A function f has a local extrema at $x = c$ when $f'(c) = 0$. If a function f has a local extremum at c, then c is called a critical number.

To locate local extrema we find the critical number by solving the equation $f'(x) = 0$. Then identify local maxima and minima by using the first derivative test.

Example 16

Find all local extrema for the following functions and determine whether each extremum is a local maximum or minmum.

(a) $f(x) = x^2 - 6x + 5$

Solution Begin by finding all critical numbers. To do this set the derivative of the function equal to 0, and solve the resulting equation.

$$f(x) = x^2 - 6x + 5$$

$$f'(x) = 2x - 6 \qquad \text{Derivative of } f(x).$$

$$2x - 6 = 0$$

$$x = 3$$

The only critical number is 3.

Next determining where the function is increasing or decreasing. Choose a number just on left side of 3 and another number on the right side. In this case, we choose 2.5 on the left-hand side of 3 and 3.5 on the right hand-side. Evaluating $f'(2.5)$ gives $f'(2.5) = 2(2.5) - 6 = -1$ which is negative. Hence f is decreasing on the interval $-\infty < x < 3$.

Again, evaluating $f'(3.5)$ gives $f'(3.5) = 2(3.5) - 6 = 1$ which is positive. Hence f is increasing in the interval $3 < x < \infty$

These results are summarised in the following table.

	L	3	R
Sign of $\frac{dy}{dx}$	-	0	+
Gradient	\	—	/

The function has a local minimum of $f(3) = 3^2 - 6 \cdot 3 + 5 = -4$ when $x = 3$.

(b) $f(x) = 3 + 2x - x^2$

Solution Begin by finding the derivative of $f(x) = 3 + 2x - x^2$

$$f'(x) = 2 - 2x \qquad \text{Derivative of } f(x).$$

Setting the derivative equal to 0 gives

$$2 - 2x = 0$$
$$2x = 2$$
$$x = 1$$

The only critical number is 1

Now find the sign of f' just on the left side of 1 and the sign on the right side.

Choosing $x = 0.5$ and evaluating $f(0.5)$ gives

$$f'(0.5) = 2 - 2(0.5) = 1$$

which is positive, and therefore f is increasing on the interval $-\infty < x < 1$.

Choosing 1.5 on the right-hand side gives $f'(1.5) = 2 - 2(2.5) = -1$, so f is decreasing on the interval $1 < x < \infty$

Using the factorised form of the derivative, $f'(x) = 2(1 - x)$, makes it easier to see that $f'(x) > 0$ when $x < 1$, and $f'(x) < 0$ when $x > 1$.

These results are summarised in the following table.

	L	1	R
Sign of $\frac{dy}{dx}$	+	0	-
Gradient	/	—	\

The function has a local maximum of $f(1) = 3 + 2 - 1 = 4$ when $x = 1$.

Exercise 18.5(a)

Find the stationary points of each of the following functions. In each case investigate the nature of the stationary point

1. $y = x^2 - 4x + 3$
2. $y = 5 + 2x - x^2$
3. $y = 2x^2 - x^3$
4. $y = 3x^2 - x^3$
5. $y = x^3 - 3x - 7$
6. $y = x^4 - 4x^3$

7. $y = 2x^3 + 3x^2 - 12x + 6$ 8. $y = 2x^3 - x^2 - 8x + 3$ 9. $y = 15x - x^2 - \frac{1}{3}x^3$

10. $y = \frac{1}{3}x^3 - \frac{5}{2}x^2 + 6x$ 11. $y = x^2(x-2)$ 12. $y = x(2x-3)(x-4)$

Maxima and Minima and the second derivative

There is a connection between local extrema and the second derivative as you should notice from the following second derivative test for locating relative extrema.

Second Derivative Test

Suppose f is a continuous function such that $f'(c) = 0$ for some critical number c.

1. If $f''(c)$ is negative, then $f(c)$ is a local maximum'

2. If $f''(c)$ is positive, then $f(c)$ is a local minimum.

3. If $f''(c)$ is zero, then use the first derivative test to determine if $f(c)$ is a local maximum, a local minimum or neither.

Example 17

Find the stationary points of $y = \frac{1}{3}x^3 + \frac{1}{2}x^2 - 6x + 4$, distinguishing between them.

solution $y = \frac{1}{3}x^3 + \frac{1}{2}x^2 - 6x + 4$

$$\frac{dy}{dx} = x^2 + x - 6$$

At stationary point $\frac{dy}{dx} = 0$

i.e. $x^2 + x - 6 = 0$

$(x+3)(x-2) = 0$

$x + 3 = 0$ or $x - 2 = 0$

$x = -3$ $x = 2$

We have stationary points at $x = -3$ and $x = 2$.

Next we investigate the nature of each stationary point.

$$\frac{d^2y}{dx^2} = 2x + 1$$

When $x = -3$

$$\frac{d^2y}{dx^2} = 2(-3) + 1 = -5$$

So, y is maximum at $x = -3$ as indicated by the negative sign.

To find the maximum value substitute -3 for x in the equation for y.

$$y = \frac{1}{3}(-3)^3 + \frac{1}{2}(-3)^2 - 6(-3) + 4$$

$$= -9 + \frac{9}{2} + 18 + 4$$

$$= 17\frac{1}{2}$$

The maximum value of the function is $17\frac{1}{2}$.

Hence, the maximum point is $\left(-3, 17\frac{1}{2}\right)$.

When $x = 2$ $\quad \dfrac{d^2y}{dx^2} = 2(2) + 1 = 5$

So, y is minimum at $x = 2$ as indicated by the positive sign.

To find the minimum value substitute 2 for x in the equation for y.

$$y = \frac{1}{3}(2)^3 + \frac{1}{2}(2)^2 - 6(2) + 4$$

$$= \frac{8}{3} + 2 - 12 + 4$$

$$= -3\frac{1}{3}$$

The minimum value of the function is $-3\frac{1}{3}$

Hence the minimum point is $\left(2, -3\frac{1}{3}\right)$.

An alternative method of identifying local maxima and minima is to choose x_1 and x_2 on either side of a critical number c.

1. If both $f(x_1)$ and $f(x_2)$ are both less than $f(c)$, then $f(c)$ is a local maximum.

2. If both $f(x_1)$ and $f(x_2)$ are both greater than $f(c)$, then $f(c)$ is a local minimum.

Exercise 18.5(b)

1. Find and distinguish the stationary values of the following functions:

(a) $y = x^2 - 6x$ (b) $y = x - 4x^2$ (c) $y = x^2 - 4x + 7$

(d) $y = 5 + 6x - x^2$ (e) $y = 2x^3 - x^2 - 8x + 3$ (f) $y = x + \frac{1}{x}$

(g) $y = 2x^3 - 11x^2 + 12x - 5$ (h) $y = x - \frac{4}{x^2}$

2. Find and distinguish the stationary points on the following curve:

(a) $y = 8x + 5x^2$ (b) $y = x^3 - 3x + 7$ (c) $y = 3x - x^3$

(d) $y = x^3 - 6x^2 + 9x + 8$ (e) $y = x^3 - 3x^2 + 4$ (f) $y = 4x - 3x^3$

(g) $y = x^2(2 - x)$ (h) $y = 4x + \frac{1}{x}$ (i) $y = 2x^2 + \frac{1}{2x}$ (j) $y = x - \frac{4}{x^2}$

18.6 Problems Involving Maxima and Minima

The solution of many problems in life may require finding a maximum or a minimum. Your knowledge of stationary points and the method of determining maximum and minimum values may be useful tools in solving such problems.

Example 18

Thin metal is used to make cylindrical cans which are to hold 1,200 cm^3 of fruit juice. What should be the radius of the cans if they are to use the least amount of metal?

Solution

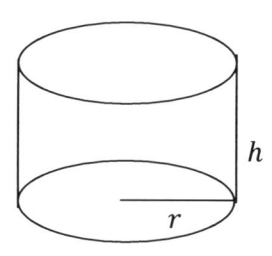

Let the height be h cm and the base radius r cm.

You are required to find the least surface area of the can. The total surface area, A cm^2, is given by $A = 2\pi r^2 + 2\pi rh$

You need to express A as a function of only one variable. You can do this by using the given information of the volume.

The volume, V cm³ is given by the formula $V = \pi r^2 h$

Therefore $\pi r^2 h = 1200$. Solving this for h we get

$$h = \frac{1200}{\pi r^2}$$

Substituting this expression for h into $A = 2\pi r^2 + 2\pi r h$ gives

$$A = 2\pi r^2 + \frac{2400}{r}$$

Differentiating A with respect to r gives

$$\frac{dA}{dr} = 4\pi r - \frac{2400}{r^2}$$

For maximum or minimum $\frac{dA}{dr} = 0$

$$0 = 4\pi r - \frac{2400}{r^2}$$

$$r^3 = \frac{600}{\pi}$$

$$r = \sqrt[3]{\frac{600}{\pi}}$$

$$r = 5.8$$

Hence, the radius of the can is 5.8 cm

The steps applied in solving the problem are summarized below:

1. Draw a diagram and label it, if possible.

2. Determine what the variables are and how they are related.

3. Decide what quantity needs to be maximized or minimized.

4. Write an expression for the quantity to be maximized or minimized in terms of only one variable.

5. Determine the minimum and maximum values.

6. Answer the question that is asked.

Exercise 18.6

1. A closed cylindrical can is made of a tin plate. If the volume of the can is 64 cm^3, find the radius of the can with the least possible surface area.

2. A manufacturer of tin cans wishes to produce a closed cylindrical can of volume 2 m^3. Find the dimensions of the can with the least possible surface area.

3. A metal can is made in the form of a cylinder with an open top. Its height is h cm and its radius r cm. It is to contain 10 cm^3 of liquid. What is the radius, if it is to consist of the smallest possible amount of metal?

4. A right-angled triangle has a hypotenuse of 9 metres. Find the maximum area, as the other two sides vary.

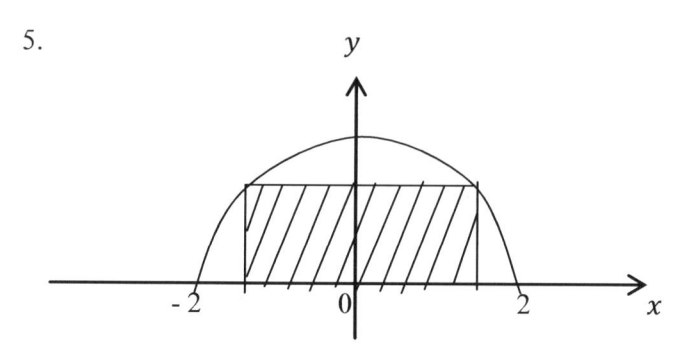

5.

The diagram shows a rectangle inscribed in a semicircle of radius 2 units. The base of the rectangle lies along the diameter of the semicircle. Find the largest area that can be inscribed in the semicircle.

6. Find the maximum possible area of a rectangle whose perimeter is 32 centimetres. What are the dimensions?

7. A three-sided fence is to be built by a farmer next to a straight section of a river, which forms the fourth side of a rectangular field. If there is 200 metres of fencing available, find the maximum enclosed area and the dimensions of the corresponding enclosure.

8. A rectangular fence is to be built by a farmer. If the enclosed area is to equal 450 m^2, find the minimum perimeter and the dimensions of the corresponding enclosure

9. An open box is made by cutting a square region from each corner of a sheet of metal 8 centimetres by 5 centimetres and turning up the sides. Find the length of the side of each square region that must be cut from each corner to produce a box of maximum volume.

384 Further Mathematics

10. An open box is made by cutting a square region from each corner of a sheet of metal 12 centimetres square and turning up the sides. Find the length of the side of each square region that must be cut from each corner to produce a box of maximum volume.

11. The bottom of a tank of height h centimetres is a square of side x metres and the tank is open at the top. It is designed to hold 40 m³ of liquid. Express in terms of x the total area of the bottom and the four sides of the tank. Find the value of x for which the area is minimum.

12. An open tank is to be constructed with a square horizontal base and vertical sides. The capacity of the tank is to be 400 m³. The cost of the material for the sides is $5 per square metre and the base is $3 per square metre. Find the minimum cost of the material, and give the corresponding dimensions of the tank.

18.7 Curve Sketching

The method of finding stationary points and determining their nature allows us to sketch curves.

To sketch the graph of a function f,

1. Find the intercept on the x- and y- axes;

   To find the y - intercept substitute $x = 0$ into $f(x)$.

   To find the x - intercept solve $f(x) = 0$.

2. Find the stationary points and determine their nature.

3. Find the behaviour (if necessary) as x becomes very large.

If x is large the sign of y will be determined by the term of the highest degree.

Note that If x is small, higher powers of x becomes negligible.

Example 19

Sketch the curve $y = x^3 - 3x^2$

Solution Begin by finding the x - and y - intercepts.

When $x = 0$, then $y = 0$

The y-intercept is (0, 0)

When $y = 0$, then

$x^3 - 3x^2 = 0$

$$x^2(x-3) = 0$$

$$x = 0 \text{ or } x = 3$$

The x-intercepts are $(0, 0)$ and $(3, 0)$

Next we determine the stationary points.

$$\frac{dy}{dx} = 3x^2 - 6x$$

For maximum or minimum $\frac{dy}{dx} = 0$

i.e. $3x^2 - 6x = 0$

$$3x(x-2) = 0$$

$$x = 0 \text{ or } x = 2$$

Now determine the nature of the stationary points.

$$\frac{d^2y}{dx^2} = 6x - 6$$

At $x = 0$, $\frac{d^2y}{dx^2} = 6(0) - 6 = -6$

So, y is maximum at $x = 0$ as indicated by the negative sign.

The function has a maximum point at $(0, 0)$.

Similarly, at $x = 2$ $\frac{d^2y}{dx^2} = 6(2) - 6 = 6$

So, y is minimum at $x = 2$ as indicated by the positive sign.

To find the minimum value substitute 2 for x in the equation for y.

$$y = 2^3 - 3 \times 2^2 = -4$$

The function has a minimum point at $(2, -4)$.

Finally, we examine the behaviour of the curve

As $x \to \pm\infty$, $y \approx x^3$

Hence, as $x \to +\infty$, $y \to +\infty$

and as $x \to -\infty$, $y \to -\infty$

From these results, we obtain the graph shown below.

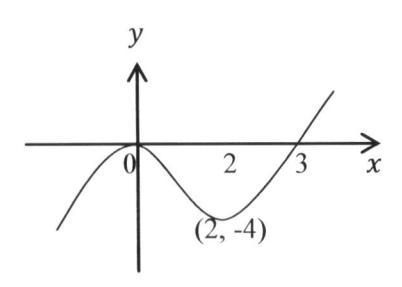

Exercise 18.7(a)

Sketch the graph of the following functions

1. $y = x^2 - x - 6$
2. $y = 3 + 2x - x^2$
3. $y = x^3 - 12x$
4. $y = x^3 - 3x$
5. $y = 3x^2 - x^3$
6. $x^3 - 6x^2$
7. $y = x^3 - 2x^2 + x$
8. $y = x^3 - 6x^2 + 9x$
9. $y = x^4 - 2x^3$
10. $y = x^2(x - 1)$
11. $y = 4x^5 - 5x^4$
12. $y = 4x^3 + 3x^4$

Sketching Graphs using Standard Curves

Standard Curves

The graphs shown below are standard graphs of some functions.

1. $y = x^2$

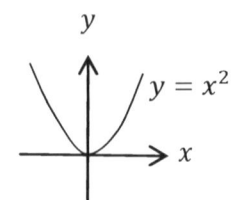

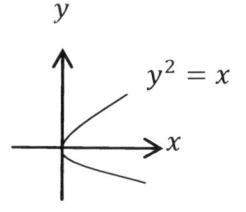

2. $y^2 = x$

3. $y = x^3$

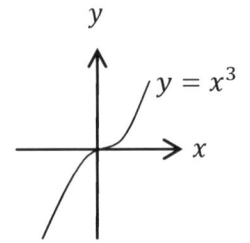

4. $y = \sqrt[3]{x}$

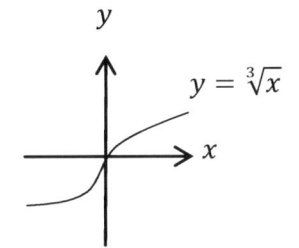

5.

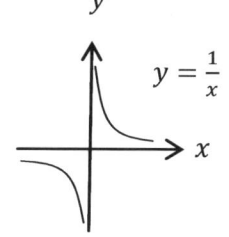

$y = \dfrac{1}{x}$

6.

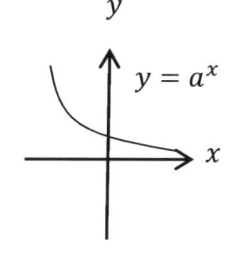

$y = a^x$

7.

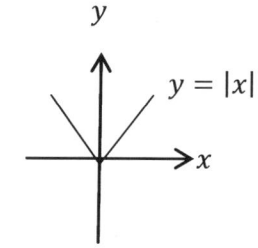

$y = |x|$

8.

$y = \dfrac{1}{x^2}$

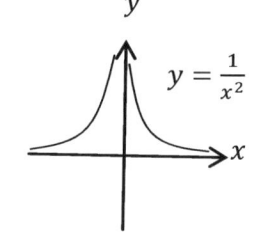

We can sketch the graphs of functions using as a starting point a standard curve. To do this you need to understand what happens to the shape of a graph when the following changes are made to the function.

1. Translation in the y- direction

$f(x) + a$ moves the graph a place up (down if a is negative).

Example 20

Sketch the graphs of $y = x^2$ and $y = x^2 + 3$

Solution

$y = x^2$

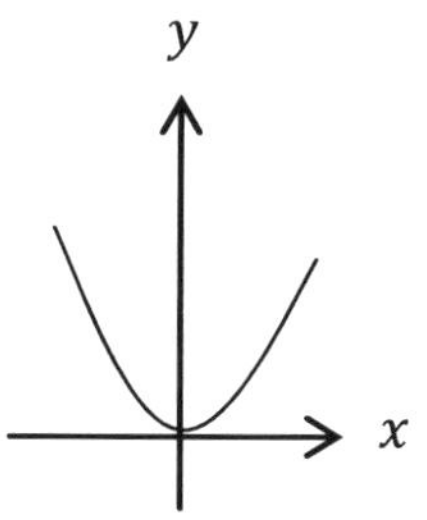

$y = x^2 + 3$

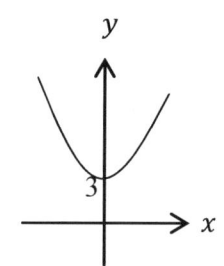

Figure 18.7(a)   Figure 18.7(b)

The two graphs are shown in Figure18.7(a) and Figure18.7(b) respectively. Notice that the graph of $y = x^2 + 3$ is the same as the graph of $y = x^2$ but has been moved up 3 units.

2. Translation in the x-axis

$f(x + a)$ moves the graph a places to the left (right if a is negative).

Example 21

Sketch the graphs of $y = x^2$ and $y = (x - 2)^2$

Solution

$y = x^2$

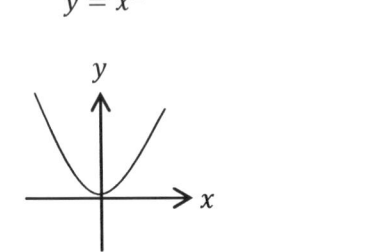

$y = (x - 2)^2$

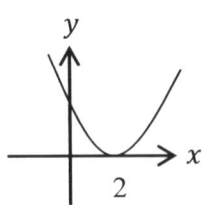

Figure 18.8(a)   Figure 18.8(b)

The graphs are shown in Figure 18.8(a) and Figure 18.8(b) respectively. Notice that the graph of $y = (x - 2)^2$ is the same as the graph of $y = x^2$ but has been moved 2 units to the right.

3. Reflection in the x-axis

The graph of $y = -f(x)$ is the graph of $y = f(x)$ reflected in the x-axis.

Example 22

Sketch the graphs of $y = x^2$ and $y = -x^2$

Solution

$y = x^2$

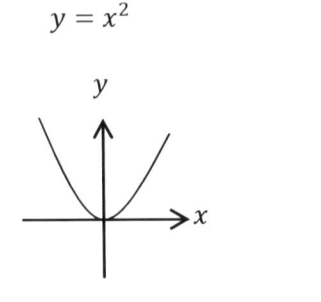

$y = -x^2$

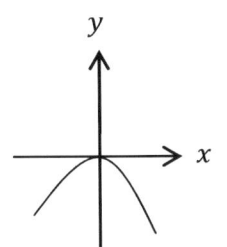

Figure 18.9(a)   Figure 18.9(b)

Notice that multiplying $f(x)$ by a negative number flips the graph of f upside down.

4. Reflection in the y-axis

The graph of $y = f(-x)$ is the graph of $y = f(x)$ reflected in the y-axis.

Example 23

Sketch the graphs of $f(x) = x^3$ and $f(-x) = -x^3$

Solution

$$f(x) = x^3 \qquad\qquad f(-x) = -x^3$$

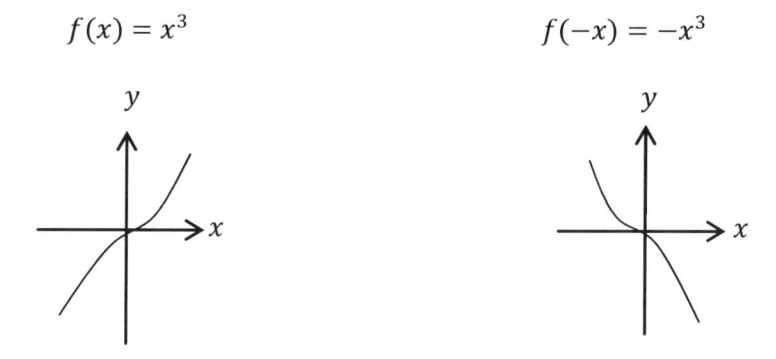

Figure 18.10(a)   Figure 18.10(b)

Other transformations are:

5. The graph of $y = af(x)$ is obtained by stretching the graph of $y = f(x)$ along the y-axis by a factor of a.

6. The graph of $y = f(ax)$ is obtained by a contraction of the graph of $y = f(x)$ along the x-axis by a factor of a.

7. The graph of $y = f^{-1}(x)$ is obtained by reflecting the graph of $y = f(x)$ in the line $y = x$.

You can use a combination of these transformations to sketch graphs of functions.

Example 24

Given the graph of $y = x^2$ sketch the graph of $y = x^2 - 6x + 8$.

Solution

First make $x^2 - 6x$ a perfect square by adding the square of one-half of the coefficient of x, which is $(-3)^2 = 9$. To make sure that the value of the function is not changed, we must subtract 9 from the result.

$$y = x^2 - 6x + 9 - 9 + 8$$
$$= (x-3)^2 - 1$$

The curve sketching can be done with steps shown in Figure 18.11.

$y = x^2$ $y = (x-3)^2$ $y = (x-3)^2 - 1$

(a)   (b)   (c)

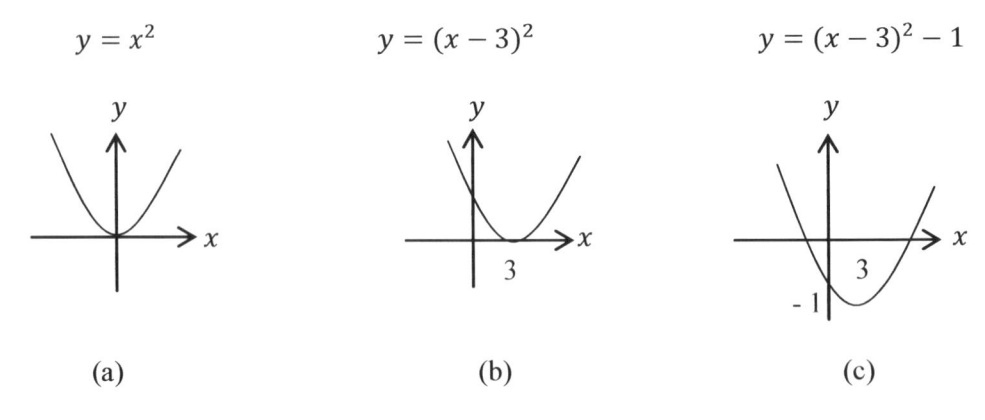

Figure 18.11

Exercise 18.7(b)

Sketch the graph of each of the following functions:

1. $y = -x^3 + 1$
2. $y = x^2 + 4$
3. $y = -x^2 + 3$
4. $y = x^3 - 4$
5. $y = (x - 3)^2$
6. $y = (x + 2)^2 - 4$
7. $y = 2 - (x + 3)^2$
8. $y = x^2 - x - 6$
9. $y = x^2 - 8x + 10$
10. $y = 20 - x - x^2$
11. $y = -x^2 + 3x - 2$
12. $y^2 = x + 4$
13. $x + y^3 = 0$
14. $y = -\frac{1}{x} + 1$
15. $y = -\sqrt[3]{x} - 1$

18.8 Small Changes

We can use the derivative of a function to estimate small changes. For any function $y = f(x)$, a small increment, δx, in x causes a change δy in y. The rate of change is $\frac{\delta y}{\delta x}$. As δx approaches zero $\frac{\delta y}{\delta x} \to \frac{dy}{dx}$.

Thus, if δx is made small

$$\frac{\delta y}{\delta x} \approx \frac{dy}{dx}$$

so, $\delta y \approx \frac{dy}{dx} \times \delta x$

Notice that the approximate increase in y is found by multiplying the increase in x by the derivative.

Example 25

(a) If the side of a cube is increased from 12 centimetres to 12.01 centimetres, what is the approximate increase in volume?

Solution Let the volume of a cube of side x cm by V. The volume of the cube is given by

$$V = x^3$$

Then $\quad \dfrac{dV}{dx} = 3x^2$

Now $\quad \delta V = \dfrac{dV}{dx} \times \delta x$

The change in length $\delta x = 12.01 - 12 = 0.01$

Therefore $\delta V = 3x^2 \times 0.01$

$$= 0.03x^2$$

But $x = 12$

So, $\delta V = 0.03 \times 12^2$

$$= 4.32$$

The approximate increase in volume is 4.32 cm

(b) Calculate $\sqrt[3]{125.1}$

Solution Let $y = \sqrt[3]{x}$

So, $\dfrac{dy}{dx} = \dfrac{1}{3} x^{-2/3}$

Now $\delta y = \dfrac{dy}{dx} \times \delta x$

If we let $x = 125$, then $\delta x = 0.1$

$$\delta y = \dfrac{1}{3\sqrt[3]{x^2}} \cdot \delta x$$

$$\delta y = \dfrac{1}{3\sqrt[3]{125^2}} \times 0.1$$

$$= \dfrac{1}{3} \times \dfrac{1}{25} \times 0.1$$

$$= 0.0013$$

$$\text{Now } y = \sqrt[3]{125} = 5$$

$$\text{So, } y + \delta y = 5.0013$$

Hence, $\sqrt[3]{125.1} = 5.0013$

Exercise 18.8

1. Find the approximate increase in area of a square when its side changes from 10 cm to 10.1 cm

2. A cube has side 8 cm. Find the approximate change in its volume if its side changes by 0.02 cm.

3. The radius of a sphere is measured as 6.5 cm, with possible error of 0.03 cm. What is the possible error in the volume?

4. The side of a square is measured with a possible error of 3 %, what is the approximate percentage error in the area?

5. The percentage error when measuring the area of a circle was 5 %. What was the approximate percentage error in its radius?

6. What is the error in the area of a circle and its circumference if its radius is 0.2 % greater than it correct radius of 1 m?

7. A $1\frac{1}{2}$ % error is made in measuring the radius of a sphere. Find the percentage error in its surface area.

8. An error of 2 % is made in measuring the radius of a sphere. What are the resulting errors in the calculation of its surface area and volume?

9. The height of a cylinder is 6 cm and its radius is 3 cm. Find the approximate increase in volume when the radius increases to 3.02 cm.

10. The volume of a sphere increases by 2 %. Find the corresponding percentage increase in its surface area

11. Find without using a calculator:

(a) $\sqrt{5.3}$ (b) $\sqrt[3]{27.1}$ (c) $\sqrt[5]{32.01}$ (d) $\sqrt{17.2}$ (e) $\sqrt[4]{81.02}$

12. Find without using a calculator:

(a) $\sqrt[3]{29}$ (b) $\sqrt{147}$ (c) $\sqrt[5]{36}$ (d) $\sqrt{627}$ (e) $\sqrt[3]{1003}$

18.9 Rate of Change

If a cylindrical container collects water from a leaking tap, the volume, V, and depth, h, of water increases with respect to time t. In your daily life you may have noticed that an increase in volume of water in a container causes a rise in depth of water.

Because V is a function of h, and V and h are functions of t, the chain rule gives

$$\frac{dV}{dt} = \frac{dV}{dh} \cdot \frac{dh}{dt}$$

Using this equation you can calculate the rate of change of volume over time when the value of the rate of depth over time is known or vice versa.

Rate problems usually involve three steps:

1. Find the functional relationship between the variables in the problem.

2. Using the chain rule write an equation connecting the rate of change of the variables.

3. Using the data of the problem, find one rate in terms of the other.

Example 26

(a) The side of a cube is increasing at 5 cm s^{-1}, what is the rate of increase of the volume when $x = 3$ cm

Solution Let the side of the cube be x cm.

Then the volume of the cube is given by

$$V = x^3$$

So, $\dfrac{dV}{dx} = 3x^2$

The chain rule gives

$$\frac{dV}{dt} = \frac{dV}{dx} \cdot \frac{dx}{dt}$$

$$= 3x^2 \times 5$$

$$= 15x^2$$

When $x = 3$, we have

$$\frac{dV}{dt} = 15 \times 3^2$$

$$= 135$$

The rate of increase of the volume is 135 cm^3 s^{-1}

(b) Air is pumped into a spherical balloon at 12 cm^3 s^{-1}. How fast is the radius increasing when it is 3 cm?

Solution The volume, V, of the balloon is given by

$$V = \frac{4}{3}\pi r^3$$

So, $\frac{dy}{dr} = 4\pi r^2$

The chain rule gives

$$\frac{dV}{dt} = \frac{dV}{dr} \times \frac{dr}{dt}$$

$$12 = 4\pi r^2 \times \frac{dr}{dt}$$

$$\frac{dr}{dt} = \frac{3}{\pi r^2} \qquad \text{Isolate } \frac{dr}{dt}$$

$$\frac{dr}{dt} = \frac{3}{\pi(3)^2} \qquad \text{Substitute 3 for } r.$$

$$\frac{dr}{dt} = \frac{1}{3\pi}$$

The rate of increase of the radius is $\frac{1}{3\pi}$ cm s^{-1}

Exercise 18.9

1. The side of a square is increasing at 0.8 cm s^{-1}. At what rate is the area increasing at a time when the side is 10 cm?

2. The radius of a sphere is increasing at 0.2 cm s^{-1}. At what rate, is the surface area increasing when the radius is 3 cm? At what rate is the volume increasing?

3. The area of a circle is increasing at 12 cm^2 s^{-1}. At what rate is the radius increasing, when it is 20 cm?

4. Water is poured into a cone of vertical angle 90^0 at 10 cm^3 s^{-1}. When the height of water is 15 cm, at what rate is it increasing?

5. The volume of a cube is decreasing at 6 cm^3 s^{-1}. When the side is 4 cm, what is the rate of decrease of (a) the side and (b) the surface area?

6. A pump is inflating a spherical balloon. If the radius at a certain instant is 5 m and it is increasing at a rate of 10 cm s^{-1}, at what rate is the pump working?

7. If air is pumped into a balloon at the rate of 0.50 m³ s⁻¹ at what rate will the radius be increasing when it is 5 m?

8. A spherical balloon is losing air at a rate of 18 m³ s⁻¹. When its radius is 3 m, at what rate is it diminishing?

9. Liquid is dropping through a conical funnel at a rate of 5 cm³ s⁻¹. When the depth of liquid in the funnel is x cm, its volume is $\frac{1}{3}\pi x^3$. Find the rate at which the level of liquid is falling when $x = 10$.

10. A funnel is made in the shape of a right circular cone of height 8 cm and base radius 6 cm. Water drains from the vertex of the funnel at the rate of 1.8 cm³ s⁻¹. Find the rate at which the water level is dropping when the water has receded 4 cm from the top.

18.10 Kinematics

Kinematics is the study of motion. The position of an object relative to some fixed location at time t is called displacement and it is normally denoted by the letter s.

Velocity and Acceleration

If a point P moves so that a small change in displacement δs takes place over time δt, the average speed is given by $\frac{\delta s}{\delta t}$. The limit of $\frac{\delta s}{\delta t}$ as $\delta t \to 0$ is $\frac{ds}{dt}$, called the instantaneous velocity. If v is the instantaneous velocity at a time t then $v = \frac{ds}{dt}$.

The change in velocity over time, t, is called the acceleration. For a small change in velocity, δv, in time δt, the average acceleration is $\frac{\delta v}{\delta t}$. The limit of $\frac{\delta v}{\delta t}$ as $\delta t \to 0$ is $\frac{dv}{dt}$, called the instantaneous acceleration. If a is the acceleration at the instant t, then $a = \frac{dv}{dt}$. Since acceleration is the derivative of velocity then $\frac{d}{dt}(v) = \frac{d}{dt}\left(\frac{ds}{dt}\right) = \frac{d^2s}{dt^2}$.

Unless otherwise stated, distance and time are measured in metres (m) and seconds (s) respectively. The velocity of an object is measured in metres per second (ms⁻¹) and acceleration in metres per second squared (ms⁻²).

Example 27

(a) A particle moves along a straight line so that its distance in metres after t s is $s = 3t^2 + 2t$. Find its distance, velocity and acceleration after 5 seconds.

Solution The distance is given by

$$s = 3t^2 + 2t$$

Substitute $t = 5$ into the equation to get

$$s = 3 \times 5^2 + 2 \times 5$$

$$= 75 + 10$$

$$= 85$$

The distance moved is 85 m

To find the velocity differentiate the equation for s.

$$v = \frac{ds}{dt} = 6t + 2$$

Substitute $t = 5$ into this equation to get

$$v = 6 \times 5 + 2 = 30 + 2 = 32$$

The velocity after 5 seconds is 32 ms^{-1}

To find the acceleration differentiate the equation for v.

$$a = \frac{d^2s}{dt^2} = 6$$

The acceleration after 5 seconds is 6 ms^{-2}

(b) A ball is thrown vertically upwards and its height after t s is s m where $s = 29.4t - 4.9t^2$. Find when the ball is momentary at rest and the greatest height reached.

Solution $\quad s = 29.4t - 4.9t^2$

$$\frac{ds}{dt} = 29.4 - 9.8t$$

The ball is momentary at rest when $\frac{ds}{dt} = 0$

So, $\quad 29.4 - 9.8t = 0$

$$t = 3$$

The ball is momentary at rest when $t = 3$ s.

The ball will reach it's greatest height at $t = 3$ s.

So, $s = 29.4 \times 3 - 4.9 \times 3^2$

$$= 88.2 - 44.1$$

= 44.1

The greatest height reached is 44.1 m.

(c) A particle is moving along a straight line so that its distance from the starting point is s metres after time t second is given by $s = 5t^3 - t^4$. Find the distance moved in the 3rd second and the maximum velocity attained.

Solution $s = 5t^3 - t^4$

When $t = 2$ we have

$s = 5 \times 2^3 - 2^4 = 40 - 16 = 24$

When $t = 3$ we have

$s = 5 \times 3^3 - 3^4 = 135 - 81 = 54$

The distance moved in the 3rd second is $54 - 24 = 30$ m

To find the velocity find the derivative of $s = 5t^3 - t^4$.

$v = \frac{ds}{dt} = 15t^2 - 4t^3$

Now, $\frac{d^2s}{dt^2} = 30t - 12t^2$

The particle will attain the maximum or minimum velocity when $\frac{d^2s}{dt^2} = 0$.

So, $30t - 12t^2 = 0$

$6t(5 - 2t) = 0$

$t = 0$ or $t = 2.5$

The maximum velocity $= 15 \times 2.5^2 - 4 \times 2.5^3$

$= 93.75 - 62.5$

$= 31.25$

The maximum velocity is 31.25 ms^{-2}

Exercise 18.10

1. A particle moves along a straight line so that after t s, its distance from a fixed point on the line is s m, where $s = 2t^3 - 3t^2 + 4$. Find the distance, velocity and acceleration after 3 s.

2. A body moves so that its distance from the starting point s in metres is $s = t^3 + 3t$ after t seconds. What are the distance, velocity and acceleration after 2 s?

3. The distance s m of a moving point A at time t s is given by $s = 6t^2 - t^3$. Find the total distance moved during the 2nd second and the maximum distance moved.

4. The distance s m of a moving particle at time t s is given by $s = t^4 - 3t^2 + 4t$. Find the velocity and acceleration of the particle at the instant $t = 2$ s.

5. If $s = \frac{1}{3}t^3 - \frac{1}{2}t^2 - 2t$, when t is in seconds and s in metres, find when the velocity becomes zero.

6. If $s = 5t^3 - 3t^2 + 2t + 1$, where t is in seconds and s in metres, find the velocity when the acceleration is zero.

7. If $v = s^2 + 6s + 5$, where s is in metres, and v in metres per seconds, find the acceleration after 3 m.

8. If $v = s^2 + 2s + 3$, where s is in metres, and v in metres per second, find s when $v = 38$ m and the acceleration after 3 m.

9. A ball is thrown vertically upwards and its height after t s is s m, where $s = 19.6t - 4.9t^2$. Find when the ball is momentarily at rest. What is the greatest height reached?

10. A particle moves from O towards A. It is s m from O after t s where $s = t(t - 3)$. When is it again at O? What is the particle's greatest displacement from O?

11. A particle moves along a straight line so that its distance from O, a fixed point on the line, is s m where $s = t^3 - 5t^2 + 6t$. When is the particle at O? What is its velocity and acceleration at these times?

12. A particle P is travelling along a straight line so that its distance from the starting point s m after time t s is given by the equation $s = 2t^2 - \frac{1}{3}t^3$. Calculate the velocity and acceleration of P after 3 s, and the distance travelled by P when it first comes to rest.

13. A particle is moving in a straight line and its distance s m from a fixed point in the line after t s is given by $s = 18t - 21t^2 + 4t^3$. Find:

(a) the velocity and acceleration of the particle after 3 s

(b) the distance travelled between the two times when the velocity is instantaneously zero.

14. A bus which runs from P to Q stops at two adjacent rest stops. The distance s km, travelled by the bus t hours after passing a railway bridge between the two rest stops is $s = 9t + 3t^2 - t^3$. Find the distance of each rest stop from the railway bridge.

15. A bus which stops at two adjacent bus stops passes a police check point at 8 am. If the bus is s km past the check point t hours past 8 am, where $s = 6t + 3t^2 - 4t^3$, find the time of departure from the first stop and the time of arrival at the second stop.

18.11 Indefinite Integrals

Finding an integral is the reverse of finding a derivative. That is, to integrate a function $f(x)$ is to find a function $F(x)$ whose derivative is $f(x)$. For example, the derivative of x^3 is $3x^2$, so an integral of $3x^2$ is x^3. Because the derivative of a constant is zero $f(x) = x^3$ is not the only function whose derivative is $3x^2$. For example, $x^3 + 2$ and $x^3 - \frac{1}{3}$, are all integrals of $3x^2$. So a constant is added to all integrals. The constant is represented by the letter c, and is called the constant of integration (or arbitrary constant). The symbol $\int$ is the integral sign and the integral of $f(x)$ is written $\int f(x)\, dx$, called an indefinite integral. Hence, $\int 3x^2\, dx = x^3 + c$.

Similarly, $\int x^4\, dx = \frac{1}{5}x^5 + c$ because $\frac{d}{dx}\left(\frac{1}{5}x^5 + c\right) = x^4$ and $\int \frac{1}{2}\, dx = \frac{1}{2}x + c$, because $\frac{d}{dx}\left(\frac{1}{2}x + c\right) = \frac{1}{2}$. The process of computing an integral is called integration.

Rule for integrating powers of x

For any real number $n \neq -1$, $\int x^n\, dx = \frac{x^{n+1}}{n+1} + c$.

That is, to integrate x^n increase the index by 1, and divide by the new index.

If $n = -1$, the expression in the denominator is 0, and the above rule cannot be used.

Example 28 Evaluate

(a) $\int x^7\, dx$

Solution $\int x^7\, dx = \frac{x^{7+1}}{7+1} + c = \frac{1}{8}x^8 + c$

(b) $\int x^{-4}\, dx$

Solution $\int x^{-4}\, dx = \frac{x^{-4+1}}{-4+1} + c = -\frac{1}{3}x^{-3} + c$

(c) $\int x^{1/2}\, dx$

Solution $\int x\, dx = \dfrac{x^{\frac{1}{2}+1}}{\frac{1}{2}+1} + c = \dfrac{2}{3}x^{3/2} + c$

The Sum Rule

If f and g are functions, then

$$\int (f(x) \pm g(x))\, dx = \int f(x)\, dx \pm \int g(x)\, dx$$

To integrate the sum of functions, integrate term by term

Multiplication by Constant Rule

$$\int Af(x)\, dx = A\int f(x)\, dx$$

To integrate a function multiplied by a constant, integrate the function and then multiply the integral by the constant.

Example 29 Evaluate:

(a) $\int (x^3 + x^2 + 1)\, dx$

Solution $\int (x^3 + x^2 + 1)\, dx = \dfrac{1}{4}x^4 + \dfrac{1}{3}x^3 + x + c$

(b) $\int 2x^3\, dx$

Solution $\int 2x^3\, dx = 2 \cdot \dfrac{1}{4}x^4 + c = \dfrac{1}{2}x^4 + c$

(c) $\int (3x^2 + 2x + 4)\, dx$

Solution $\int (3x^2 + 2x + 4)\, dx = x^3 + x^2 + 4x + c$

(d) If $\dfrac{dy}{dx} = 4x^3 + 3x^2$, find y in terms of x given that $y = 3$ when $x = -2$.

Solution $y = \int (4x^3 + 3x^2)\, dx = x^4 + x^3 + c$

Substituting $x = -2$ and $y = 3$ into the equation for y gives

$3 = 16 - 8 + c$

$c = -5$

So, $y = x^4 + x^3 - 5$

$\int x^n dx = \dfrac{x^{n+1}}{n+1} + C$, $n \neq -1$

$\dfrac{1}{x^4} = x^{-4}$

$\dfrac{3}{2}+1 = \dfrac{3+2}{2}$

$\dfrac{2}{3}+1 = \dfrac{2+3}{3} = \dfrac{5}{3}$

Exercise 18.11

Evaluate the following integrals:

1(a) $\int x^2\, dx = \dfrac{1}{3}x^3 + C$ (b) $\int x^4\, dx = \dfrac{1}{5}x^5 + C$ (c) $\int x^9\, dx = \dfrac{1}{10}x^{10} + C$

(d) $\int x^{11}\, dx = \dfrac{1}{12}x^{12} + C$ (e) $\int 5\, dx = 5x + C$ (f) $\int \dfrac{1}{3} dx = \dfrac{1}{3}x + C$

2(a) $\int 6x\, dx = 3x^2 + C$ (b) $\int 4x^5\, dx = \dfrac{2}{3}x^6 + C$ (c) $\int 8x^3\, dx = 2x^4 + C$

(d) $\int \dfrac{3}{4}x^2\, dx = \dfrac{1}{4}x^3 + C$ (e) $\int \dfrac{5}{4}x^4\, dx = \dfrac{1}{4}x^5 + C$ (f) $\int \dfrac{2}{3}x^7\, dx = \dfrac{1}{12}x^8 + C$

3(a) $\int x^{-3}\, dx = -\dfrac{1}{2}x^{-2} + C$ (b) $\int x^{-2}\, dx = -x^{-1} + C$ (c) $\int x^{-5}\, dx = -\dfrac{1}{4}x^{-4} + C$

(d) $\int \dfrac{1}{x^4} dx = -\dfrac{1}{3x^3} + C$ (e) $\int \dfrac{1}{x^7} dx = -\dfrac{1}{6x^6} + C$ (f) $\int \dfrac{1}{x^6} dx = -\dfrac{1}{5x^5} + C$

4(a) $\int x^{\frac{1}{3}} dx = \dfrac{3}{4}x^{4/3} + C$ (b) $\int x^{\frac{3}{4}} dx = \dfrac{4}{7}x^{7/4} + C$ (c) $\int x^{\frac{3}{2}} dx = \dfrac{2}{5}x^{5/2} + C$

(d) $\int \sqrt[5]{x}\, dx = \int x^{1/5} dx = \dfrac{5}{6}x^{6/5} + C$ (e) $\int \sqrt[3]{x^2}\, dx = \dfrac{3}{5}x^{5/3} + C$ (f) $\int \sqrt[4]{x^5}\, dx$

5(a) $\int x^{-\frac{1}{4}} dx = \int \dfrac{1}{x^{1/4}}$ (b) $\int x^{-\frac{2}{3}} dx = 3x^{1/3} + C$ (c) $\int x^{-\frac{4}{3}} dx$

(d) $\int \dfrac{1}{\sqrt[5]{x^4}} dx$ (e) $\int \dfrac{1}{\sqrt{x^3}} dx$ (f) $\int \dfrac{1}{\sqrt[4]{x^3}} dx$

6(a) $\int (2x^2 + x^3)\, dx = \int 2x^2 + \int x^3 = \dfrac{2}{3}x^3 + \dfrac{1}{4}x^4 + C$ (b) $\int (2x^3 - 5x^4)\, dx = \int 2x^3 - \int 5x^4$

(c) $\int (3x^2 - 2x + 1)\, dx = x^3 - x^2 + x + C$ (d) $\int (3 - 2x + 4x^2)\, dx = 3x - x^2 + \dfrac{4}{3}x^3 + C$

(e) $\int (1 - 2x + 6x^2)\, dx = x - x^2 + 2x^3 + C$ (f) $\int (5x^4 - 3x^2 + 2x)\, dx = \dfrac{5x^5}{5} - \dfrac{3x^3}{3} + \dfrac{2x^2}{2} + C$

7(a) $\int x(x^2 - 2)\, dx = \int x^3 - 2x\, dx = \dfrac{x^4}{4} - x^2 + C$ (b) $\int x(4x^2 + 3x)\, dx = \int 4x^3 + 3x^2\, dx = x^4 + x^3 + C$

(c) $\int x^2(2 + 3x^2)\, dx = \int 2x^2 + 3x^4\, dx = \dfrac{2x^3}{3} + \dfrac{3x^5}{5} + C$ (d) $\int x(x^4 - 3x)\, dx = \int x^5 - 3x^2\, dx = \dfrac{x^6}{6} - x^3 + C$

(e) $\int 2x^2(2x - x^2)\, dx = \int 4x^3 - 2x^4\, dx = x^4 - \dfrac{2}{5}x^5 + C$ (f) $\int 3x^2(x^2 + 1)\, dx = \int 3x^4 + 3x^2\, dx = \dfrac{3}{5}x^5 + x^3 + C$

8(a) $\int (x+1)(x-3)\, dx = x^3 - \dfrac{2}{5}x^5 + C$ (b) $\int (2x+1)(x+3)\, dx = \dfrac{2}{3}x^5 + \dfrac{3}{3}x^3 + C$

(c) $\int (3 - 2x)^2\, dx$ (d) $\int x(x-2)(x-2)\, dx$

(e) $\int (3x + 2)^2\, dx$ (f) $\int (x^3 - 2x^2)^2\, dx$

402 Further Mathematics

9(a) $\int \frac{x^3+1}{x^2}dx = \int x+1\,dx = \frac{1}{2}x^2+x+C$

(b) $\int \frac{5x-3x^2}{x}dx = \int (5-3x)\,dx = 5x - \frac{3}{2}x^2 + C$

(c) $\int \frac{x^2+2x}{x^5}dx = \int \frac{x(x+2)}{x^5}\,dx = \int \frac{x+2}{x^4}\,dx = \frac{\frac{1}{2}x^2+2x}{\frac{1}{5}x^5}$

(d) $\int \frac{3x^2-8x^5}{x^4}dx = \int x^{-2}(3-8x^3)\,dx$

(e) $\int \frac{4-3x^2+2x^3}{x^2}dx$

(f) $\int \frac{(1+x^2)(1-2x^2)}{x^2}dx =$

10(a) $\int \frac{\sqrt{x}-5}{x^2}dx$

(b) $\int \frac{1+x^2}{\sqrt{x}}dx$

(c) $\int \frac{(x-2)^2}{\sqrt[3]{x}}dx$

(d) $\int (3\sqrt{x}-2)(1+\sqrt{x})\,dx$

(e) $\int (\sqrt{x}-\sqrt[3]{x})^2\,dx$

(f) $\int \left(\sqrt{x}-\frac{1}{\sqrt{x}}\right)(x+3\sqrt{x})\,dx$

In Problem 11 – 16, find y as a function of x

11. $\frac{dy}{dx}=2x$, $y=2$ for $x=-1$

12. $\frac{dy}{dx}=4x-3$, $y=2$ for $x=1$

13. $\frac{dy}{dx}=3x^2+x$, $y=1$ for $x=1$

14. $\frac{dy}{dx}=2x+3x^2$, $y=3$ for $x=2$

15. $\frac{dy}{dx}=4x-x^3$, $y=0$ for $x=2$

16. $\frac{dy}{dx}=x(x^2+1)$, $y=2$ for $x=-1$

18.12 Definite Integrals

A definite integral can be obtained by substituting values into the indefinite integral. The definite integral of a function f, denoted $\int_a^b f(x)\,dx$ is given by

$$\int_a^b f(x)\,dx = F(b) - F(a), \text{ where } \int f(x)\,dx = F(x) + c$$

The b above the integral sign is called the upper limit of integration, and the a is the lower limit of integration.

Note: The symbol $[f(x)]_a^b$ is used to represent $F(b) - F(a)$.

Rules of Definite Integrals

If f and g are functions then

1. $\int_a^b (f(x) \pm g(x))\, dx = \int_a^b f(x)\, dx \pm \int_a^b g(x)\, dx$

2. $\int_a^b Af(x)\, dx = A \int_a^b f(x)\, dx$

3. $\int_a^b f(x)\, dx = \int_a^c f(x)\, dx + \int_c^b f(x)\, dx$ for any real number c.

4. $\int_a^b f(x)\, dx = - \int_b^a f(x)\, dx$

Example 30 Evaluate:

(a) $\int_1^2 x^3\, dx$

Solution $\int_1^2 x^3\, dx = \left[\dfrac{1}{4}x^4 + c\right]_1^2$

$\qquad\qquad\qquad = \left(\dfrac{1}{4} \times 2^4 + c\right) - \left(\dfrac{1}{4} \times 1^4 + c\right)$ Substitute the limits for x

$\qquad\qquad\qquad = 3\dfrac{3}{4}$

This example shows that the constant c is not required in this case, because it would be eliminated in the final answer.

(b) $\int_0^2 (3x^2 - 4x^3)\, dx$

Solution $\int_0^2 (3x^2 - 4x^3)\, dx = [x^3 - x^4]_0^2$

$\qquad\qquad\qquad\qquad = (2^3 - 2^4) - 0$ Substitute the limits for x

$\qquad\qquad\qquad\qquad = -8$

Exercise 18.12

Evaluate the following definite integrals:

1(a) $\int_{-1}^{2} x\, dx$ 　　　　　(b) $\int_1^3 x^2\, dx$ 　　　　　(c) $\int_2^4 x^3\, dx$

(d) $\int_0^3 3x^2\, dx$ 　　　　　(e) $\int_{-2}^{-1} x^4\, dx$ 　　　　　(f) $\int_{-1\frac{1}{2}}^{1\frac{3}{2}} x^2\, dx$

2(a) $\int_{-2}^{2}(x^2 + x)\,dx$ (b) $\int_{0}^{2}(4 - x^2)\,dx$ (c) $\int_{2}^{4}(3x^2 - 2x)\,dx$

(d) $\int_{1}^{2}(x^3 - x)\,dx$ (e) $\int_{-1}^{2}(3 - x^2)\,dx$ (f) $\int_{1}^{3}(x^2 - 4x + 3)\,dx$

3(a) $\int_{1}^{2} x(1 + 4x)\,dx$ (b) $\int_{1}^{4} x(x + 3)\,dx$ (c) $\int_{-1}^{2}(x^3 + 2)^2\,dx$

(d) $\int_{-1}^{1}(2x + 1)^3\,dx$ (e) $\int_{-1}^{1} x(x + 1)(x + 2)\,dx$ (f) $\int_{0}^{1} \sqrt{x}(x + 2)\,dx$

4(a) $\int_{1}^{2} \dfrac{4}{x^3}\,dx$ (b) $\int_{1}^{9} \dfrac{dx}{\sqrt{x}}$ (c) $\int_{4}^{9} \dfrac{dx}{\sqrt{x^3}}$

(d) $\int_{1}^{2} \dfrac{dx}{x^3}$ (e) $\int_{2}^{3} \dfrac{2}{3x^2}\,dx$ (f) $\int_{1}^{4} \sqrt{x}\,dx$

5(a) $\int_{1}^{3} \dfrac{2x^2 + 1}{x^2}\,dx$ (b) $\int_{1}^{2} \dfrac{x^2 + 2x}{x^3}\,dx$ (c) $\int_{1}^{2}\left(1 - \dfrac{8}{x^3}\right)dx$

(d) $\int_{-1}^{2} \dfrac{x^3 + 2x^2 - 3x}{x}\,dx$ (e) $\int_{1}^{4} \dfrac{x^2 - 1}{\sqrt{x}}\,dx$ (f) $\int_{1}^{9}\left(\sqrt{x} + \dfrac{1}{\sqrt{x}}\right)dx$

18.13 Integration by Substitution

Occasionally we evaluate $\int f(x)\,dx$ by making the substation $x = g(u)$ and using the following rule.

$$\int f(x)\,dx = \int f[g(u)]\,g'(u)\,du$$

After obtaining the indefinite integral on the right we replace u with the original expression in x.

The corresponding rule for definite integrals is

$$\int_{a}^{b} f(x)\,dx = \int_{\alpha}^{\beta} f[g(u)]\,g'(u)\,du, \text{ where } \alpha = g^{-1}(a) \text{ and } \beta = g^{-1}(b)$$

Observe that the limits of integration had been changed to correspond to the new variable.

Example 31 Evaluate

(a) $\int (3x - 2)^3\,dx$

Solution Let $u = 3x - 2$

then $du = 3dx$

With substitution the integral becomes

$$\int (3x-2)^3 \, dx = \int u^3 \times \frac{1}{3} du$$

$$= \frac{1}{3} \int u^3 \, dx$$

$$= \frac{1}{12} u^4 + c$$

Substitute $3x - 2$ for u in the integral to get

$$\int (3x-2)^3 \, dx = \frac{1}{12}(3x-2)^4 + c$$

(b) $\int_{-1}^{1} x^2(1-x^3) \, dx$

Solution Let $u = 1 - x^3$

then $du = -3x^2 dx$

The new limits can be found as follows.

If $x = 1$, then $u = 1 - 1 = 0$

If $x = -1$, then $u = 1 + 1 = 2$

Hence $\int_{-1}^{1} x^2(1-x^3) \, dx = \int_{2}^{0} x^2 \times \frac{u}{-3x^2} du$ Use new limits

$$= -\frac{1}{3} \int_{2}^{0} u \, du$$

$$= \left[-\frac{1}{6} u^2 \right]_{2}^{0}$$

$$= \left(-\frac{1}{6} \times 0^2 \right) - \left(-\frac{1}{6} \times 2^2 \right)$$

$$= \frac{2}{3}$$

Note: You can use the original limits when you substitute $1 - x^3$ for u.

Exercise 18.13

Find by substitution:

1. $\int (3x+2)^3 \, dx$
2. $\int (2x-1)^{-2} \, dx$
3. $\int \sqrt[3]{x-2} \, dx$
4. $\int x(x^2+3)^2 \, dx$
5. $\int x(x^2+1)^4 \, dx$
6. $\int x^2(2x^3+3)^4 \, dx$
7. $\int x\sqrt{1-x^2} \, dx$
8. $\int x^2\sqrt{x^3-2} \, dx$
9. $\int x\sqrt{1-x^2} \, dx$
10. $\int x^2\sqrt{x^3+4} \, dx$
11. $\int x^3\sqrt{2-x^4} \, dx$
12. $\int x^2\sqrt{2x^3-3} \, dx$
13. $\int \dfrac{1}{(2x-5)^4} \, dx$
14. $\int \dfrac{1}{\sqrt[3]{3x+1}} \, dx$
15. $\int \dfrac{2x}{\sqrt{1+3x^2}} \, dx$
16. $\int \dfrac{3x}{(x^2-1)^4} \, dx$

Evaluate

17. $\int_0^1 (1-x)^4 \, dx$
18. $\int_2^7 \sqrt{x+2} \, dx$
19. $\int_1^2 x\sqrt{x^2+1} \, dx$
20. $\int_{-1}^{2} \dfrac{dx}{(2x+1)^2}$
21. $\int_1^2 \dfrac{dx}{(3x-2)}$
22. $\int_0^1 \dfrac{x^3}{(x^4+1)^3} \, dx$

18.14 Integration of Trigonometric functions

The standard integrals of three trigonometric functions are listed in the table below:

y	$\int y \, dx$
$\sin x$	$-\cos x$
$\cos x$	$\sin x$
$\sec^2 x$	$\tan x$

Example 32 Integrate:

(a) $\int \sin 2x \, dx$

Solution $\int \sin 2x \, dx$

Let $u = 2x$

then $du = 2dx$

With substitution, we get

$$\int \sin 2x \, dx = \int \sin u \times \frac{1}{2} du$$

$$= \frac{1}{2} \int \sin u \, dx$$

$$= -\frac{1}{2} \cos u + c$$

$$= -\frac{1}{2} \cos 2x + c \qquad \text{Substitute } 2x \text{ for } u.$$

(b) $\int \cos(3x - 2) \, dx$

Solution $\int \cos(3x - 2) \, dx$

Let $u = 3x - 2$

$du = 3dx$

With substitution, we get

$$\int \cos(3x - 2) \, dx = \int \cos u \times \frac{1}{3} du$$

$$= \frac{1}{3} \int \cos u \, dx$$

$$= \frac{1}{3} \sin u + c$$

$$= \frac{1}{3} \sin(3x - 2) + c$$

In general, $\int \sin ax \, dx = -\frac{1}{a} \cos ax + c$ and $\int \cos ax \, dx = \frac{1}{a} \sin ax + c$

Exercise 18.14

1. Integrate the following functions:

(a) $\cos 2x$ (b) $\sin 3x$ (c) $2 \sec^2 4x$ (d) $3 \sin 2x$

(e) $3 \cos 6x$ (f) $-6 \sin 4x$ (g) $\frac{3}{4} \sin \frac{1}{3} x$ (h) $\sec^2 \frac{1}{2} x$

2. Integrate the following functions:

(a) $\cos(2x+1)$ (b) $3\sin(3-2x)$ (c) $\cos(5x+2)$

(d) $\sec^2(3x-2)$ (e) $\sin(3-\frac{1}{2}x)$ (f) $6\sec^2(\frac{3}{2}x+2)$

18.15 The Trapezium Rule

Not all functions can be evaluated by any of the technique discuss so far. An approximate numerical value of the definite integral of such functions could be obtained, using the Trapezium Rule (called Trapezoidal Rule in the Americans).

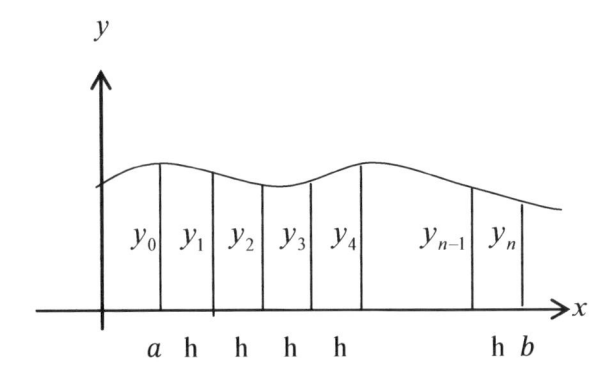

Figure 18.12

Suppose that we wish to find the area under the curve from $x = a$ to $x = b$, as shown in Figure 18.12. We divide this area into n equal parallel strips, each of width h. The ordinates of the curve at successive point are $y_0, y_1, y_2, y_3, \ldots, y_{n-1}, y_n$. The area under the curve is approximated with n trapezia. From geometry, the area of a trapezium is half the product of the sum of the bases and the height.

Hence, the area under the curve

$$\approx h\frac{y_0+y_1}{2} + h\frac{y_1+y_2}{2} + h\frac{y_2+y_3}{2} + \cdots + h\frac{y_{n-1}+y_n}{2}$$

$$\approx h\left\{\frac{1}{2}y_0 + y_1 + y_2 + y_3 + \cdots + y_{n-1} + \frac{1}{2}y_n\right\}$$

$$\approx h\left\{\frac{1}{2}(y_0 + y_n) + y_1 + y_2 + \cdots + y_{n-1}\right\}$$

So Area $\approx$ (width of interval)[$\frac{1}{2}$ (first + last ordinate) + sum of remaining ordinate]

Example 33

(a) Evaluate $\int_1^6 \dfrac{dx}{x^2+1}$, using the trapezium rule with 5 equal intervals.

Solution We divide the interval [1, 6] into 5 equal intervals of width, $h = \dfrac{6-1}{5} = 1$

The rest of the calculation is set out as follows:

x	x^2+1	$\dfrac{1}{x^2+1}$	First and last ordinate	Remaining ordinate
1	2	0.5000	0.5000	
2	5	0.2000		0.2000
3	10	0.1000		0.1000
4	17	0.0588		0.0588
5	26	0.0385		0.0385
6	37	0.0270	0.0270	
		Total	0.5270	0.3973

$$\int_1^6 \dfrac{dx}{x^2+1} = 1 \times \left[\dfrac{0.5270}{2} + 0.3973\right]$$

$$= [0.2635 + 0.3973]$$

$$= 0.6608$$

Note: To get a better approximation you can divide the interval into more subintervals.

(b) y is a function of x: the table below gives six values for x and y.

x	0	5	10	15	20	25
y	2.3	2.7	3.3	4.0	4.9	5.9

Use the trapezium rule to find an approximation for $\int_0^{25} y\,dx$

Solution Here h = 5, hence

$$\int_0^{25} y\,dx = 5 \times \left[\frac{2.3+5.9}{2} + 2.7 + 3.3 + 4.0 + 4.9\right]$$

$$= 5 \times [4.1 + 14.9]$$

$$= 95$$

Exercise 18.15

1. Evaluate $\int_1^3 \frac{1}{x}dx$, using the trapezium rule with 4 equal intervals.

2. Evaluate $\int_0^6 \sqrt{1+x^2}\,dx$, using the trapezium rule with 6 equal intervals.

3. Evaluate $\int_0^2 \frac{dx}{\sqrt{x^3+1}}$, using the trapeziums rule with 4 equal intervals.

4. Evaluate to 3 decimal places, the value of $\int_0^{2.5} \frac{4}{x+5}dx$ for values of x from 0 to 2.5 at intervals of 0.5.

5. y is a function of x, with values given by the following table:

x	1	2	3	4	5	6
y	75	64	56	49	41	37

Use the trapezium rule to approximate $\int_1^6 y\,dx$.

6. y is given in terms of x by the following table:

x	0	0.1	0.2	0.3	0.4	0.5	0.6
y	0.35	0.47	0.57	0.65	0.79	0.86	1.03

Use the trapezium rule to approximate $\int_0^{0.6} y\,dx$

7. The depth of a river of width 40 m is measured at 5 m intervals of its cross-section as follows

Distance (m)	0	5	10	15	20	25	30	35	40
Depth (m)	0	0.31	2.17	5.24	2.79	0.16	4.25	2.76	1.48

Using the trapezium rule, find approximately the area of the cross-section of the river, correct to one decimal place.

8. The speed of a particle is recorded at half-second intervals as follows:

t/s	0	0.5	1.0	1.5	2.0	2.5	3.0
v/ms^{-1}	0	1.2	4.9	11.0	19.6	30.6	44.1

Using the trapezium rule, find approximately the distance it travels in 3 s.

18.16 Area under a Curve

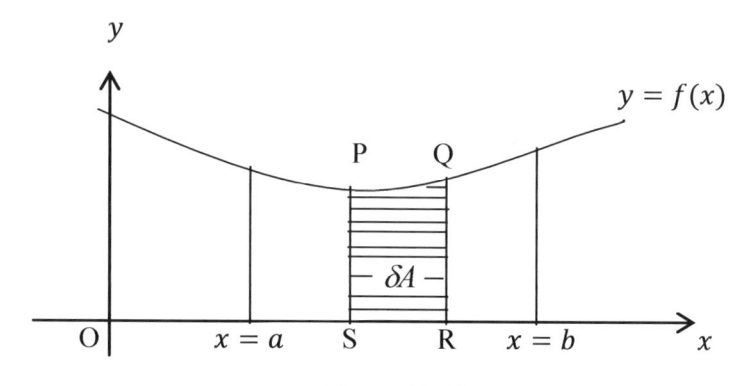

Figure 18.13

Consider the area bounded by the graph of $y = f(x)$, the line $x = a$, the line $x = b$ and the x-axis, as shown in Figure 18.13. Let A represents the area under the curve up to the point P on the curve with coordinates (x, y). If we move the point P to the point Q with coordinates $(x + \delta x, y + \delta y)$ the right-hand boundary of A moves from PS to QR and the area is increased by a small area δA. If we make the width δx very small, the area of PQRS can be approximated with a rectangle having width δx and height y. So, $\delta A \approx y \delta x$. Dividing both side by δx gives $\frac{\delta A}{\delta x} \approx y$. Taking the limit as $\delta x \to 0$ gives $\lim_{\partial x \to 0} \frac{\delta A}{\delta x} = \frac{dA}{dx} = y$. Hence $A = \int y \, dx$. This represents the area under the curve up to the point P. The area between the graph and the x-axis from $x = a$ to $x = b$ is equal to the area up to $x = b$ minus the area up to $x = a$. This is written briefly as $A = \int_a^b y \, dx$.

You can see that the definite integral $\int_a^b f(x)\,dx$ gives the area below the graph of the function $y = f(x)$ above the x-axis and between the lines $x = a$ and $x = b$.

Example 34

(a) Find the area of the region bounded by the curve $y = 3x$, the x-axis and the vertical lies $x = 1$ and $x = 2$

Solution It is usually necessary to sketch the curve in order to avoid misleading result when calculating areas.

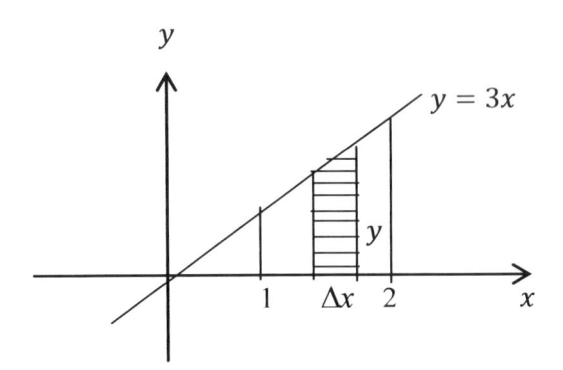

Area $= \int_1^2 y\,dx$

$= \int_1^2 3x\,dx$

$= \left[\dfrac{3}{2}x^2\right]_1^2$

$= 6 - \dfrac{3}{2}$

$= 4\dfrac{1}{2}$

The area is $4\dfrac{1}{2}$ square units

You may confirm this result by using the area formula $A = \dfrac{1}{2}bh$.

The area of the larger triangle − area of the smaller triangle gives the area between $x = 2$ and $x = 1$ That is the area is $\dfrac{1}{2} \times 2 \times 6 - \dfrac{1}{2} \times 1 \times 3 = 6 - \dfrac{3}{2} = 4\dfrac{1}{2}$ square unit.

(b) Find the area enclosed by the curve $y = x^2 - 5x + 6$ and the x-axis.

Solution Begin by finding the point(s) where the graph crosses the x-axis. You can do this by solving the equation

$$x^2 - 5x + 6 = 0$$

$$(x - 2)(x - 3) = 0$$

$$x = 2 \text{ or } x = 3$$

The solutions of this equation are 2 and 3. The area required is the area bounded by the x-axis and the curve between $x = 2$ and $x = 3$, as shown in the graph below.

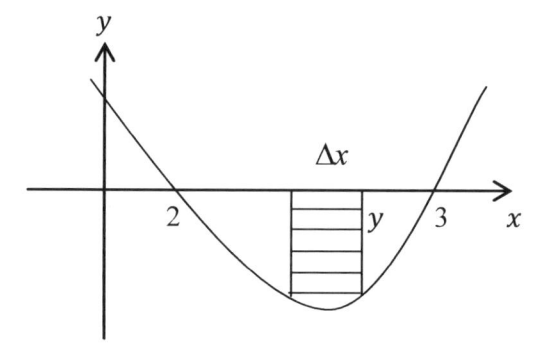

Area $= \int_2^3 y \, dx$

$= \int_2^3 (x^2 - 5x + 6) \, dx$

$= \left[\frac{1}{3} x^3 - \frac{5}{2} x^2 + 6x \right]_2^3$

$= \left(9 - \frac{45}{2} + 18 \right) - \left(\frac{8}{3} - 10 + 12 \right)$

$= \frac{9}{2} - \frac{14}{3}$

$= -\frac{1}{6}$

The result is a negative number because the value of the function, $f(x)$ is negative for values of x between $x = 2$ and $x = 3$. Since δx is always positive $f(x) \cdot \delta x$ is negative.

Generally, the definite integral will be negative for areas below the x-axis.

Since area is nonnegative, the required area is given by $\frac{1}{6}$ square units.

(c) Find the area enclosed by the curve $y = x^3 - 6x^2 + 8x$ and the x-axis.

Solution Begin by finding the point(s) where the graph crosses the x-axis.

$$x^3 - 6x^2 + 8 = x(x-2)(x-4) = 0$$

The solutions are 0, 2 and 4.

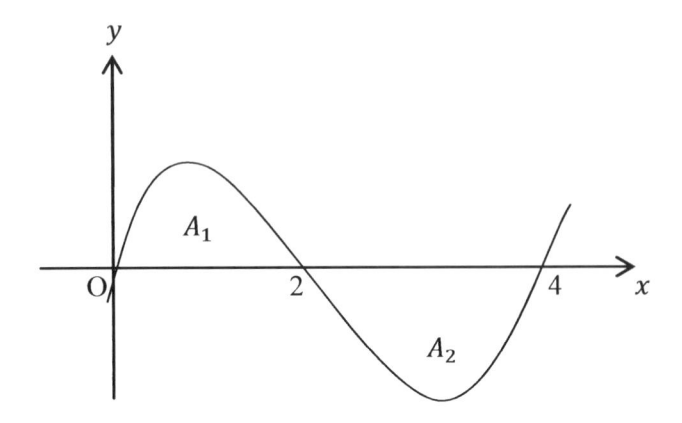

The area A_1 is above the x-axis whereas the area A_2 is below. To find the area, integrate A_1 and A_2 separately and then add the absolute values of the results to get the total area A.

$$A_1 = \int_0^2 (x^3 - 6x^2 + 8x)\, dx$$

$$= \left[\frac{1}{4}x^4 - 2x^3 + 4x^2\right]_0^2$$

$$= (4 - 16 + 16) - 0$$

$$= 4$$

$$A_2 = \int_2^4 (x^3 - 6x^2 + 8x)\, dx$$

$$= \left[\frac{1}{4}x^4 - 2x^3 + 4x^2\right]_2^4$$

$$= (64 - 128 + 64) - (4 - 16 + 16)$$

$$= 0 - 4$$

$$= -4$$

Now $A = A_1 + A_2 = |4| + |-4| = 4 + 4 = 8$

Hence, the total area is 8 square units.

Differentiation and Integration

If we work out the integral over the entire interval, i.e. from $x = 0$ to $x = 4$, we get

$A = \int_0^4 (x^3 - 6x^2 + 8x)\, dx$

$= 0$

The values of two areas that are numerically equal and opposite have been added together in the process of integration, so they have cancelled each other out.

Exercise 18.16(a)

1. Find the areas enclosed by the following curves and the x-axis, between the limits shown:

(a) $y = 5x$, $x = 1$, $x = 4$
(b) $y = x^2$, $x = 1$, $x = 3$
(c) $y = 2x^2 - x + 1$, $x = -1$, $x = 2$
(d) $y = x^2 + 3$, $x = -1$, $x = 2$
(e) $y = (x - 1)(x - 3)$, $x = 0$, $x = 3$
(f) $y = \frac{1}{x^2}$, $x = 1$, $x = 8$

2. Sketch the following curves and find the area enclosed by them, and by the x-axis, and the given straight lines:

(a) $y = x(3 - x)$, $x = 4$
(b) $y = -x^3$, $x = -2$
(c) $y = x^2(x - 1)$, $x = 2$
(d) $y = x^3 - 4x$, $x = 3$

3. Find the area below the curve $y = 6 - x - x^2$ and above the x-axis

4. Find the area above the curve $y = x^2 - 4x + 3$ but below the x-axis

5. Find area below the curve $y = 1 - x^2$ and above the x-axis

6. Sketch the curve $y = x^2 - 3x + 2$ and find the area cut off below the x-axis

7. Sketch the curve $y = x(x + 2)(x - 2)$, and find the area of each of the two segments cut off by the x-axis

8. Find the area below the curve $y = 2x^3 + 4x^2$, from $x = -2$ to $x = 0$

9. Find the area between the y-axis, the curve $y = x^3 - 4$ and the line $y = 4$

10. Find the area enclosed by the y-axis and the following curves and straight lines:

(a) $y^2 = x$, $y = 3$
(b) $y^2 - x + 3 = 0$, $y = -1$, $y = 2$
(c) $y = 4x^2$, $y = 1$, $y = 4$
(d) $y = x^3$, $y = 1$, $y = 27$

Area enclosed by two curves

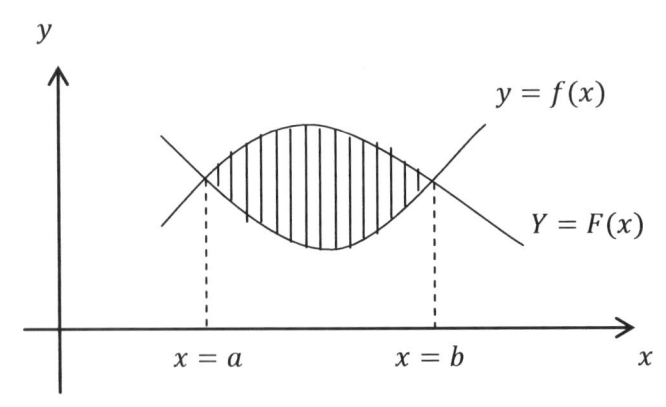

Figure 18.14

Figure 18.14 shows the graphs of the functions $y = f(x)$ and $Y = F(x)$. The area between the curves from $x = a$ to $x = b$ is the same as the area under the graph of $Y = F(x)$ minus the area under the graph of $y = f(x)$. That is, the area between the curves is given by

$$\int_a^b F(x)\,dx - \int_a^b f(x)\,dx.$$

This can be written as $\int_a^b [F(x) - f(x)]\,dx$.

Example 35

(a) Find the area enclose by the curves $y = x^2$ and $y = 2x$

Solution

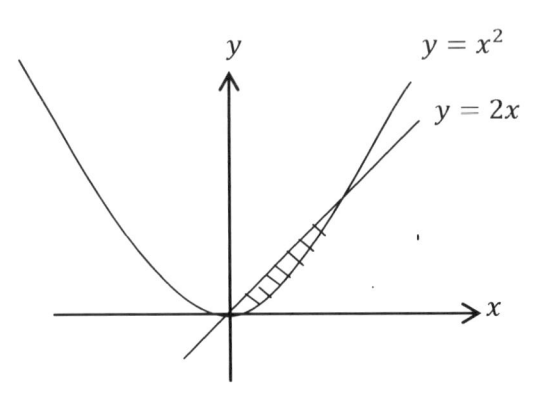

Begin by finding the limits of the integration. This is done by solving the following equation.

$$x^2 = 2x$$

$$x(x-2) = 0$$

$$x = 0 \text{ or } x = 2$$

The required area = $\int_0^2 (2x - x^2)\, dx$

$$= \left[x^2 - \frac{1}{3}x^3 \right]_0^2$$

$$= \left(4 - \frac{8}{3} \right) - 0$$

$$= 1\frac{1}{3}$$

The area enclosed between the curves is $1\frac{1}{3}$ square units.

(b) Find the area enclosed between the curves $y = x(x-2)$ and $y = x(4-x)$

Solution

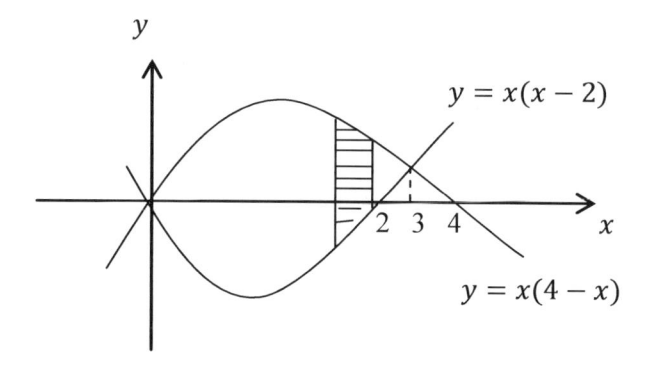

Begin by finding the limit of integration

$$x(x-2) = x(4-x)$$

$$2x^2 - 6x = 0$$

$$2x(x-3) = 0$$

$$x = 0 \text{ or } x = 3$$

The required area = $\int_0^3 \{x(4-x) - x(x-2)\}\, dx$

$$= \int_0^3 (6x - 2x^2)\, dx$$

$$= \left[3x^2 - \frac{2}{3}x^3 \right]_0^3$$

$$= (27 - 18) - 0$$

$$= 9$$

The area enclosed between the curves is 9 square units

Exercise 18.16(b)

1. Find the area of the segment cut off from each of the following curves by the given straight lines:

(a) $y = x^2 - 3x - 4$, $y = 6$

(b) $y = x^2 + 2x - 3$, $y = 5$

(c) $y = -x^2 + x + 6$, $y = -6$

(d) $y = x(x - 1)$, $y = x$

(e) $y^2 = x$, $y = x$

(f) $y = 8 - 2x - x^2$, $y + x - 2 = 0$

(g) $y = x^2 - x - 6$, $y = 2x - 2$

(h) $y = x^2 + 1$, 5

2. Find the area enclosed by each of the following pairs of curves:

(a) $y = 4 - x^2$, $y = x(x - 2)$

(b) $y = x(x - 2)$, $y = x(3 - x)$

(c) $y = 2x^2 - 4x$, $y = x^2 - 3x$

(d) $y^2 = x$, $y = x^3$

18.17 Solids of Revolution

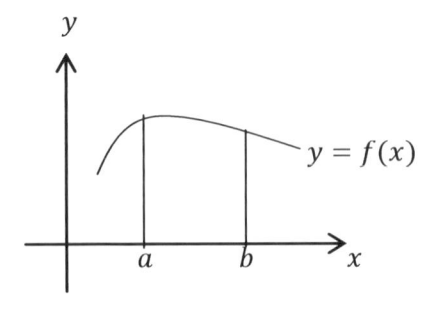

Figure 18.15(a)

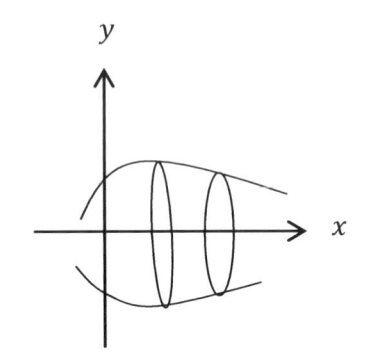

Figure 18.15(b)

Many three dimensional solids can be generated by rotating a plane figure bounded by a curve, the x-axis and the ordinates through a complete revolution about the x–axis or the y–axis. Figure 18.15(a) shows the region below the graph of a function $y = f(x)$, above the x-axis and between $x = a$ and $x = b$. Figure 18.15(b) shows the solid generated when this region is revolved about the x-axis. The solid generated is called a solid of revolution. The volume of a solid of revolution can be found by integration.

Volume of solid of revolution

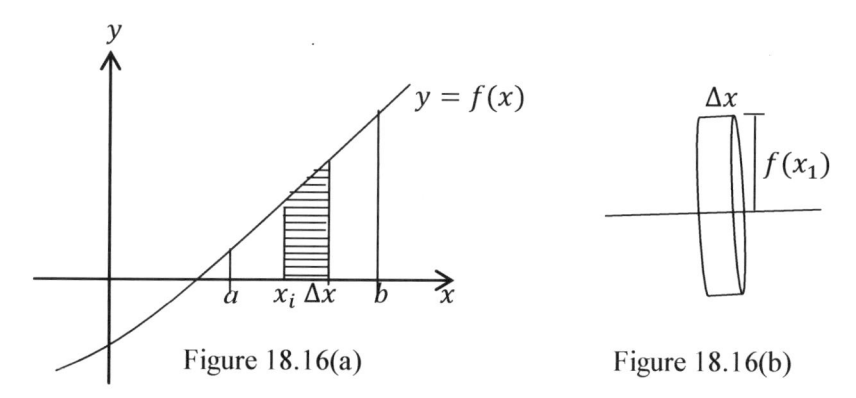

Figure 18.16(a)

Figure 18.16(b)

We start by dividing the region between $x = a$ and $x = b$ into n subintervals of equal width Δx by the points $a = x_0, x_1, x_2, \ldots, x_i, \ldots, x_n = b$. Consider a small element of length $f(x_i)$ and thickness Δx, as shown in Figure 18.16(a). Revolving this area about the x-axis we generate a right circular cylinder of height Δx and radius $f(x_i)$, as shown in Figure 18.16(b). The formula for the volume of a right circular cylinder is $\pi r^2 h$, where r is the radius of the circular base and h is the height of the cylinder. Thus, the volume of the cylinder generated by this element is approximately $\pi [f(x_i)]^2 \Delta x$. The sum of the volumes of all the cylinders is an approximation to the volume of the solid of revolution, or

$V \approx \sum_{i=1}^{n} \pi [f(x_i)]^2 \Delta x$. The volume of the solid of revolution is the limit of this sum as Δx approaches 0, i.e. $V = \lim_{\Delta x \to 0} \sum_{i=0}^{n} \pi [f(x_1)]^2 \Delta x$. This limit may be evaluated by the definite integral $V = \int_a^b \pi [f(x)]^2 \, dx$.

Example 36

(a) Find the volume of the solid generated by rotating about the x-axis the area under $y = x^2$ from $x = 1$ to $x = 2$.

Solution

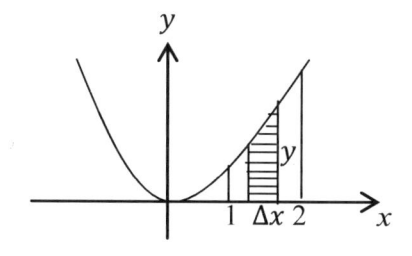

Volume $= \int_1^2 \pi y^2 \, dx$

$= \int_1^2 \pi x^4 \, dx$

$= \left[\frac{1}{5}\pi x^5\right]_1^2$

$= \left(\frac{1}{5} \cdot \pi \cdot 2^5\right) - \left(\frac{1}{5} \cdot \pi \cdot 1^5\right)$

$= \frac{32}{5}\pi - \frac{1}{5}\pi$

$= \frac{31}{5}\pi$

The volume of the solid is $\frac{31}{5}\pi$ cubic unit

(b) Find the volume of the solid generated by rotating about the y-axis the area in the first quadrant enclosed by $y = x^3$, $y = 1$, $y = 8$ and the y-axis.

Solution

Volume $= \int_1^8 \pi x^2 \, dy$

$= \int_1^8 \pi y^{2/3} \, dy$

$= \left[\frac{3}{5}\pi y^{5/3}\right]_1^8$

$= \left(\frac{3}{5} \cdot \pi \cdot 8^{5/3}\right) - \left(\frac{3}{5} \cdot \pi \cdot 1\right)$

$= \frac{96}{5}\pi - \frac{3}{5}\pi$

$= \frac{93}{5}\pi$

The volume of the solid is $\frac{93}{5}\pi$ cubic unit

Exercise 18.17

1. Find the volumes enclosed when the following curves between the limits shown are rotated about the x-axis:

(a) $y = x$, $x = 1$, $x = 2$

(b) $y = 3x$, $x = 0$, $x = 4$

(c) $y = x^2 + 1$, $x = 0$, $x = 1$

(d) $y = \frac{1}{x}$, $x = 2$, $x = 3$

(e) $y = \sqrt{x}$, $x = 0$, $x = 3$

(f) $y^2 = x$, $x = 0$, $x = 1$

(g) $y = 2x - 3$, $x = 3$, $x = 4$

(h) $y = x(x - 2)$, $y = 0$

2. Find the volumes enclosed when the following curves between the limits shown are rotated about the y-axis:

(a) $y = x^2 + 1$, $x = 0$, $y = 4$

(b) $y = 1 - x^3$, $x = 0$, $y = 0$

(c) $y = 2x - 4$, $y = 2$, $x = 0$

(d) $y = \sqrt{x}$, $y = 2$, $x = 0$

(e) $y = x^2$, $y = 1$, $y = 4$, $x = 0$

(f) $y^2 = x - 2$, $x = 0$, $y = 0$, $y = 3$

3. By considering the revolution of the line $y = rx/h$ about the x-axis between the limits $x = 0$ and $x = h$ (where r and h are constants), find the volume of a cone radius r and height h.

4. The equation of a circle radius r, is $x^2 + y^2 = r^2$. By finding the volume of revolution between $x = -r$ and $x = +r$, find the volume of a sphere.

5. Find the volume of the solid formed when the area enclosed between the straight line $y = x$ and the curve $y = x(2 - x)$ makes a complete revolution about the x-axis.

6. Find the volume generated when the area between the curves $y^2 = 4x$ and $x^2 = 4y$ is rotated through four right angles about the x-axis.

18.18 Kinematics

Recall that if the displacement of a particle at time t seconds is s metres, then its velocity and its acceleration are given by $v = \frac{ds}{dt}$ and $a = \frac{dv}{dt} = \frac{d^2s}{dt^2}$. You can use integrals to find the displacement or velocity of a particle when the acceleration is given.

Example 37

(a) A particle starting from rest at O moves along a straight line OA so that its acceleration after t s is $(3t^2 - 4t)$ ms^{-2}. Find its displacement after 3 seconds

Solution   Let a = the acceleration of the particle at time t

$$a = 3t^2 - 4t$$

$$v = \int (3t^2 - 4t)\, dx$$

$$= t^3 - 2t^2 + c \quad \text{for some constant } c$$

Since the particle starts from rest $v = 0$ when $t = 0$

$$\therefore c = 0$$

So $v = t^3 - 2t^2$

$$s = \int (t^3 - 2t^2)\, dt$$

$$= \frac{1}{4}t^4 - \frac{2}{3}t^3 + c$$

Since the particle starts from rest, $s = 0$ when $t = 0$

$$\therefore c = 0$$

So $s = \frac{1}{4}t^4 - \frac{2}{3}t^3$

Substituting $t = 3$ into the equation for s we get

$$s = \frac{1}{4} \times 3^4 - \frac{2}{3} \times 3^3$$

$$= \frac{9}{4}$$

$$= 2.25$$

The displacement of the particle is 2.25 m

(b) A particle is thrown vertically upwards from the ground level with a velocity of 24.5 ms^{-1}. If the acceleration due to gravity is 9.8 ms^{-2}, find the highest height reached.

Solution Let a = the acceleration of the particle

So $\quad a = -9.8$

$\therefore \quad v = \int -9.8 \, dt = -9.8t + c$

When $t = 0$, $v = 24.5$

$\therefore \quad 24.5 = -9.8(0) + c$

$\quad\quad c = 24.5$

So $\quad v = -9.8t + 24.5$

Now integrate v to find the displacement.

$$s = \int(-9.8t + 24.5) \, dt$$

$$= -4.9t^2 + 24.5t + c$$

When $t = 0$, $s = 0$ and $c = 0$

So $\quad s = -4.9t^2 + 24.5t$

At the highest height $v = 0$

i.e. $\quad -9.8t + 24.5 = 0$

$$t = 2.5 \text{ s}$$

Substitute $t = 2.5$ into the equation for s to get

$$s = -4.9 \times 2.5^2 + 24.5 \times 2.5$$

$$= 30.6$$

The highest height reached is 30.6 m

Exercise 18.18

1. Find the displacement s m of a particle at time t s if its velocity is $(3t^2 + 4t)$ m s^{-1} and $s = 2$ when $t = 0$.

2. Find the velocity v ms^{-1} and displacement s m of a particle at time t s, if its acceleration a is $(6t - 8)$ ms^{-2} and $s = 4$, $v = 6$ when $t = 0$.

3. A particle moves with a velocity equal to $(2t^2 - 3t)$ ms^{-1}, where t is the time in seconds. If it starts at the point P, find the distance from P after 6 seconds.

4. A particle has an acceleration of $(18t + 3)$ ms^{-2} and starts from rest. What is its velocity after 2 s and how far has it moved?

5. A particle has an acceleration of $6t$ ms^{-2}, where t is the time in seconds. It starts with velocity of 2 ms^{-1} from O. What is its velocity and distance from O after 3 seconds?

6. If the velocity of a particle is $v = 64 - t^3$ in ms^{-1} and it starts so that $s = 0$ when $t = 0$, find the distance from the starting point when the velocity is zero for an instant.

7. The velocity of a particle is $(3t^2 + 2t)$ ms^{-1} and it starts $+6$ m from the point O when $t = 0$. Calculate its acceleration and distance from the point O after 5 s.

8. A particle moves so that its velocity is given by $(12t - t^2)$ ms^{-1} where t is the time in seconds. If $s = 0$ when $t = 0$, find the distance from the starting point where the acceleration is zero.

9. A particle starts from rest at O and moves along a straight line. After t s its acceleration is a ms^{-2}, where $a = 3t^2 - t^3$. When was the particle momentary at rest?

10. A bus starts at a terminal and stops at a rest stop t hours after it leaves the terminal, with a speed of $(240t - 150t^2)$ km h^{-1}. Find the distance from the terminal to the rest stop.

11. A ball is thrown vertically downwards at 20 ms^{-1} from the fifth floor of a building. If the acceleration due to gravity is 9.8 ms^{-2}, find its velocity and position after 2 s.

12. A ball is thrown vertically upwards from ground level with a velocity of 25 ms^{-1}. If the acceleration due to gravity is 10 ms^{-2}, find the greatest height reached.

Review Exercise 18

1. Find the indicated limits:

(a) $\lim\limits_{x \to 1}(4x - 3)$

(b) $\lim\limits_{x \to 3}(x^2 - 6x + 8)$

(c) $\lim\limits_{x \to -2} \dfrac{1}{x+4}$

(d) $\lim\limits_{x \to -1} \dfrac{x^2 - 1}{x+1}$

(e) $\lim\limits_{x \to \infty} \dfrac{x(4 - x^3)}{3x^4 + 2x^2}$

(f) $\lim\limits_{x \to \infty} \dfrac{3x^2 + 7x - 4}{1 + x^2}$

2. Differentiate the following with respect to x:

(a) $y = 3x^2 + 2x$

(b) $y = 2 + 2x - 4x^2$

(c) $y = x^4 + x$

(d) $y = x^4 - x^2 + 2$

(e) $y = 1 + \frac{2}{\sqrt{x}} - \frac{3}{x}$

(f) $y = 2x^5 - \frac{4}{x^4} + \frac{3}{\sqrt[3]{x}}$

3. Differentiate the following with respective to x:

(a) $y = x(x - 2)$

(b) $y = (3 - 2x^2)(2 + x^3)$

(c) $y = (3 - 2x^2)^2$

(d) $y = \frac{3x^2 + 2x + 1}{x}$

(e) $y = \frac{x-8}{\sqrt{x}}$

(f) $y = \frac{\sqrt[3]{x} + \sqrt[4]{x} - 3}{\sqrt{x}}$

4. Differentiate the following with respective to x:

(a) $y = (3x - 7)^{1/2}$

(b) $y = (2 + x^2)^5$

(c) $y = \sqrt{1 - 2x^2}$

(d) $y = x(1 - 2x)^{1/2}$

(e) $y = \frac{x-4}{\sqrt{x^2-1}}$

(f) $y = \frac{\sqrt{x}}{\sqrt{x-1}}$

5. Differentiate the following with respect to x:

(a) $\sin 2x$

(b) $\cos 5x$

(c) $2 \sin 3x$

(d) $-\frac{1}{2} \cos 6x$

(e) $\sin (3x - 2)$

(f) $6 \sin \frac{1}{2} x$

(g) $\cos(2x + 3)$

(h) $\tan (3x + 5)$

(i) $\cos x^2$

(j) $\tan(x^2 - 1)$

(k) $\sin (x^2 + 3)$

(l) $\tan 3x^2$

6. Find $\frac{dy}{dx}$ in terms of the parameter t for the following functions:

(a) $x = t^2$, $y = 2t^3 - 1$

(b) $x = \frac{1}{1-t}$, $y = t^2$

(c) $x = t - 2t^2$, $y = t^2 + t^3$ 	(d) $x = \frac{1}{1-t}$, $y = \frac{1}{1-t^2}$

7. For each of the following functions express $\frac{dy}{dx}$ in terms of x and y

(a) $x^2 - y^2 = 4$ 	(b) $x^2 + y^2 + 3x - y = 0$

(c) $x^2 - 4xy + y^2 = 5$ 	(d) $x^2 - 3x^3y - 4y = 0$

8. Find the second derivative of the following functions:

(a) $y = 2x^2 + 3x - 1$ 	(b) $y = 6x^2 - 3x + 5$

(c) $y = -2x^3 + 4x + 7$ 	(d) $y = (x^3 - 1)(x - 4)$

(e) $y = 3\sqrt{x} + 2x^2$ 	(f) $y = (x + 1)^3$

9. Find the differential coefficient of $y = 2x^3 + 3x^2 - 4x - 1$ and determine the gradient of the curve at $x = 2$.

10. The gradient of a curve is given by $2x - 3x^2$. Find the equation of the curve if (1, 2) lies on it.

11. Find the gradients of the following functions at the points shown:

(a) $x^2 - y^2 = 5$, (2,1) 	(b) $x^2 + y^3 = 1$, (−3,2)

(c) $x^2 + y^2 - 3x + 2y = 1$, (2,1) 	(d) $x^2 - 2xy = 5$, (−1,2)

(e) $x^2y + x^3 + 2y = 2$, (2,−1) 	(f) $x^2y^2 + 2x - 3y = 0$, (1,1)

12. Find the equations of the tangents and normals to the following curves at the points given:

(a) $y = 2x^2 - 3x + 1$, (−1,6) 	(b) $y = x^3 + 2x$, (1,3)

(c) $y = x^3 + 2x^2$, (−2,0) 	(d) $y = 1 + \frac{2}{x}$, (2,2)

13. Find the equation of the tangent of the curve $y = 6 - x - x^2$ at the point where $x = 2$.

14. Find the equation of the tangent to the curve $y = x^3 - 11x$ at the point (1, -10) and also the equation of the tangents which are parallel to the line $x - y = 0$.

15. Find the equation of the tangent at the point (2, 2) on the curve $y = x^3 - 3x$ and the coordinates of the point at which this tangent meets the curve again.

16. Find any maximum or minimum points on the following curves:

(a) $y = x^2 + 2x - 2$ (b) $y = x(x^2 - 12)$

(c) $y = x^3 - 2x^2 - 4x$ (d) $y = x^3 + x^2 - x$

17. Sketch the curve of the following functions:

(a) $y = x^2 - 2x$ (b) $y = 3 - 2x - x^2$ (c) $y = x^3 - x$

(d) $y = x^2(x - 3)$ (e) $y = x^2(3 - 2x)$ (f) $y = x^2(x^2 - 8)$

18. A metal basket is made in the form of a cylinder with an open top. Its height is h cm and its radius r cm. It is to contain 2,000 cm² of liquid. What is the radius, if it is to consist of the smallest possible amount of mental?

19. A cardboard box is to be made in the form of a cuboid with square cross-section. Its volume must be 8,000 cm². Let x cm be the side of the square and x cm be the length. Whet value of x will use the least cardboard?

20. An open box is to be made by cutting a square from each corner of a piece of metal 16 cm by 10 cm and then folding up the sides. Find the maximum possible volume of the box.

21. 100 cm of wire is cut into two pieces, one of which is made into a square and the other into a circle. Find the radius of the circle if the sum of the two areas is to be as small as possible

22. A three-sided fence is to be built by a farmer next to a straight section of a river which forms the fourth side of a rectangular field. The side of the fence parallel to the river will cost $ 5 per metre to build, whereas the sides perpendicular to the river will cost $ 3 per metre. If the enclosed area is equal to 450 m², find the minimum perimeter and the cost of fencing the field.

23. A cube has side 3 cm. Find the approximate change in its volume if its side changes by 0.1 cm

24. The side of a square is measured with a possible error of 2 %. What is the approximate error in areas?

25. The percentage error when measuring the area of a circle was 5 %. What was the approximate percentage error in its radius?

26. The side of a square is increasing at 0.5 cm s^{-1}. At what rate is the area increasing at a time when the side is 25 cm

27. The radius of a sphere is increasing at 0.02 cm s^{-1}. At what rate is the surface area increasing when the radius is 8 cm?

28. Water is poured into a cone, of semi-vertical angle 45°, at 10 cm^3 s^{-1}. When the height of the water is 15 cm, at what rate is it increasing?

29. For each of the following find y as a function of x:

(a) $\frac{dy}{dx} = 2x, y = 1$ for $x = 0$

(b) $\frac{dy}{dx} = 3x^2 + x, y = 1$ for $x = 1$

(c) $\frac{dy}{dx} = x^2 + x + 1, y = 3$ for $x = 1$

(d) $\frac{dy}{dx} = x(x^2 + 1), y = 0$ for $x = 1$

30. Find the following integrals:

(a) $\int 2x \, dx$
(b) $\int (3x^2 + 2x + 1) \, dx$
(c) $\int (5 + x^2 + 2x^3) \, dx$
(d) $\int x(3x^2 + 7) \, dx$
(e) $\int (x - 3)(x^2 + 2x) \, dx$
(f) $\int x(4x^2 - 2x^3) \, dx$

31. Integrate the following functions:

(a) $\cos 3x$
(b) $\sin 2x$
(c) $4\sec^2 2x$
(d) $\sin(2x + 1)$
(e) $\sec^2(4x - 3)$
(f) $4\cos(\frac{1}{2}x + 3)$

32. Evaluate the following integrals:

(a) $\int_0^3 (2x + 3x^2) \, dx$
(b) $\int_1^9 (x^2 - \frac{1}{6}x) \, dx$
(c) $\int_1^9 (\sqrt{x} - \frac{3}{x^2}) \, dx$
(d) $\int_1^4 x(x + 1) \, dx$
(e) $\int_{-1}^2 (x^3 + 2)^2 \, dx$
(f) $\int_1^4 \frac{x^2 - 1}{\sqrt{x}} \, dx$

33. Determine each of the following integrals:

(a) $\int (2x + 3)^2 \, dx$
(b) $\int (3x - 2)^{-2} \, dx$
(c) $\int \sqrt{2x + 1} \, dx$

(d) $\int \sqrt[3]{2-3x}\, dx$ (e) $\int x^2(x^3-1)\, dx$ (f) $\int x(3x^2-4)\, dx$

(g) $\int x\sqrt{2x^2-3}\, dx$ (h) $\int x^2\sqrt{x^3+2}\, dx$ (i) $\int \frac{2x}{(x^2-3)^3}\, dx$

34. Evaluate each of the following integrals:

(a) $\int_0^2 x(3-x^2)\, dx$ (b) $\int_2^7 \sqrt{x-2}\, dx$ (c) $\int_1^3 x\sqrt[3]{1-x^2}\, dx$

(d) $\int_{-1}^2 \frac{dx}{(2x-1)^2}$ (e) $\int_0^1 \frac{dx}{(3x+2)^3}$ (f) $\int_{-1}^0 \frac{x}{(x^2-1)^3}\, dx$

35. Evaluate $\int_1^2 \sqrt{x^3+1}\, dx$, using the trapezium rule with four intervals

36. Evaluate $\int_1^3 (x^2+1)\, dx$, using the trapezium rule with 8 intervals

37. y is a function of x, with values given by the following tables

x	1	2	3	4	5	6
y	85	74	66	59	51	47

Use the trapezium rule to approximate $\int_1^6 y\, dx$

38. Find the areas enclosed by the following curves and the x-axis between the limits indicated:

(a) $y = x$, $x = 1$ and $x = 3$ (b) $y = \frac{1}{x^2}$, $x = 1$ and $x = 20$

(c) $y = x^{-1/2}$, $x = 0$ and $x = 9$ (d) $y = x^3 - x$, $x = 0$ and $x = 1$

39. Find the volumes enclosed when the following curves between the limits indicated are rotated about the x – axis

(a) $y = 3x^2$, $x = 1$ and $x = 2$ (b) $y = \sqrt{x}$, $x = 0$ and $x = 3$

(c) $y = \frac{1}{x}$, $x = 1$ and $x = 10$ (d) $y = 2x - 3$, $x = 2$ and $x = 4$

(e) $y = 3x - \frac{2}{x}$, $x = 1$ and $x = 4$ (f) $y = \sqrt{1+x^2}$, $x = 0$ and $x = 3$

40. Find the displacement s m of a particle at time t s if its velocity is $(3t^2 + 4t)$ ms^{-1} and $s = 2$ when $t = 0$

41. Find the velocity v ms^{-1} and displacement s m of a particle at time t s if its acceleration a is $(6t - 8)$ ms^{-2} and $s = 4$, $v = 6$ when $t = 0$.

42. Find the velocity v ms^{-1} and displacement s m of a particle at time t s if we know that its acceleration a ms^{-2} is given by:

(a) $a = 3t - 4$, and when $t = 0, s = 5$ and $v = 6$

(b) $a = -10$, and when $t = 1, s = 6$ and $v = 3$

43. A particle starts from rest at O and moves along a straight line. After t s its velocity is v ms^{-1}, where $v = 2t - t^2$. Show that the particle is momentarily at rest after 2 s and find its distance from O at this time. Find also the maximum velocity of the particle during the first 2 s of the motion

44. A particle starts from rest at a point 6 m from O and moves in a straight line away from O with a velocity v ms^{-1} at time t s given by $v = t - \frac{1}{12}t^2$. Find:

(a) its acceleration and distance from O, each in terms of t

(b) the time at which it begins to return, and the time at which it again reaches its starting point

19 Vectors

19.1 Vector Algebra

Quantities such as displacement, force and velocity can be described by stating both a magnitude (size or length) and a direction. Such quantities are called vectors. Other quantities, such as area, time and temperature can be represented by a single real number. Such quantities have only magnitude, and are called scalar quantities.

Vectors can be represented diagrammatically as directed line segments.

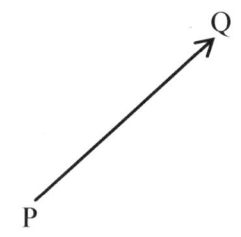

Figure 19.2

Figure 19.2 shows a vector which has its initial point at P and terminal point at Q. This vector is written as $\overrightarrow{PQ}$. The arrow indicates the direction of the vector, and the length of the line segment represents the magnitude of the vector, denoted by $|\overrightarrow{PQ}|$.

In print, we usually denote vectors by boldface characters such as **PQ** or **a**. In handwriting you can indicate a vector as a character with an arrow on it or a character with a line under it, such as $\vec{A}$, $\vec{a}$ or $\underline{a}$.

Writing vectors in magnitude-bearing form

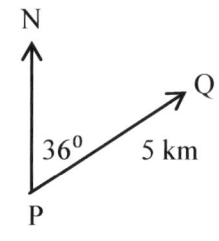

Figure 19.3

The vector shown in Figure 19.3 has magnitude 5 kilometres and direction 036°. This can be written briefly as $\overrightarrow{PQ} = (5\ km, 036°)$. This representation of vectors is called the magnitude-bearing form.

Writing vectors as column vectors

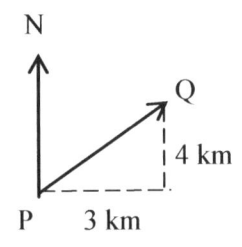

Figure 19.4

The vector, shown in Figure 19.4, can be written in terms of its movement in the x- and y-directions as $\overrightarrow{PQ} = \binom{3}{4}$. In this form the vector is said to be written in component form or the vector is expressed as a column vector.

The movement in the x- direction is called the horizontal component, and the movement in the y-direction is called the vertical component. In the x-direction, movement to the right is indicated as positive and movement to the left is indicated as negative. In the y-direction, movement up is positive and movement down is negative.

Equal Vectors

Two vectors $\overrightarrow{AB}$ and $\overrightarrow{PQ}$ are equal if they have the same magnitude (length) and have the same direction. We write

$\overrightarrow{AB} = \overrightarrow{PQ}$.

Addition of Vectors

The Triangle Law

Figure 19.5   Figure 19.6

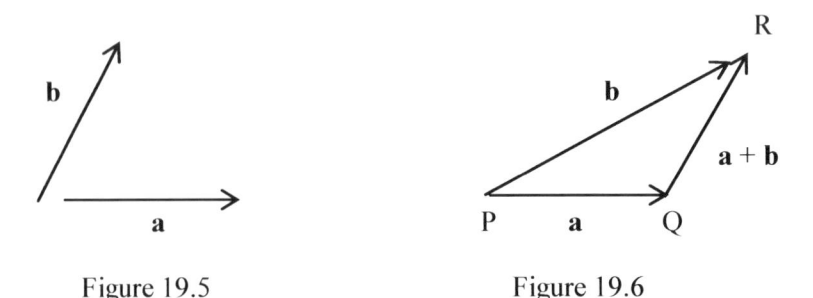

You can add vectors **a** and **b** by putting the initial point of **b,** represented by QR in magnitude and direction, and the terminal point of **a**, represented by PQ in magnitude and direction, together (see Figure 19.6). The sum of the vectors is the vector from the initial point of **a** to the terminal point of **b**. That is $\overrightarrow{PR} = \boldsymbol{a} + \boldsymbol{b}$. $\overrightarrow{PR}$ is called the resultant vector of **a** and **b**. This definition of addition of vector is known as the triangle law.

The Parallelogram Law

To add the vectors shown in Figure 19,7, we translate the initial point of vector **b** to the terminal point of vector **a** and then complete the parallelogram as shown in Figure 19.8.

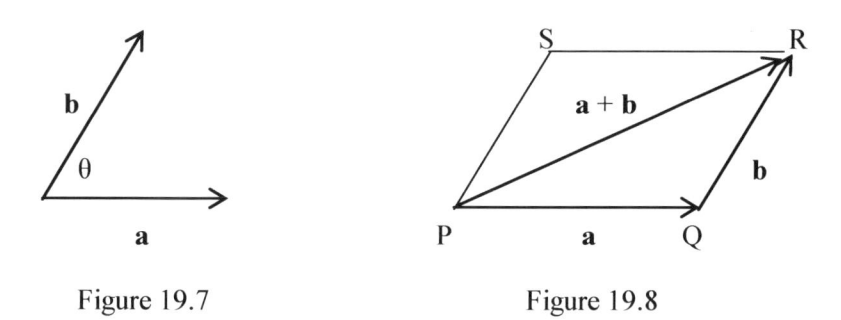

Figure 19.7   Figure 19.8

The vector $\overrightarrow{PQ}$ represents vector **a** in magnitude and direction. The vector $\overrightarrow{QR}$ is drawn parallel to vector **b**. The vector lying along the diagonal PR is called the resultant vector of **a** and **b**. That is $\overrightarrow{PR} = a + b$, This definition of addition of vectors is known as the parallelogram law.

Changing vectors in magnitude-bearing form to column vectors

Example 1

Express $\overrightarrow{PQ} = (15\ units, 120°)$ as a column vector

Solution

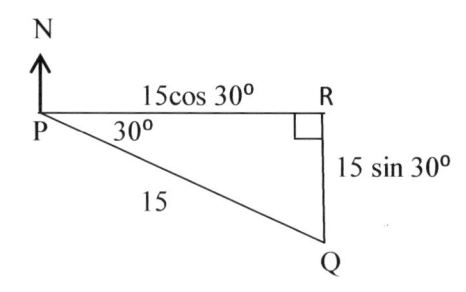

Figure 19.9

By trigonometry $PR = 15\ cos\ 30°$ and $QR = 15\ sin\ 30°$. The vector $\overrightarrow{PQ}$ can be written as

$$\overrightarrow{PQ} = \begin{pmatrix} 15cos30° \\ -15sin30° \end{pmatrix} = \begin{pmatrix} 13.0 \\ -7.5 \end{pmatrix}$$

Exercise 19.1(a)

Express each of the following vectors as column vectors. Obtain components correct to one decimal place

1. (8 units, $030°$)
2. (12 units, $120°$)
3. (10 units, $215°$)
4. (6 units, $350°$)

Adding Column Vectors

The sum of two vectors expressed as column vectors is obtained by adding their corresponding components. If $a = \begin{pmatrix} a_1 \\ a_2 \end{pmatrix}$ and $b = \begin{pmatrix} b_1 \\ b_2 \end{pmatrix}$, then $a + b = \begin{pmatrix} a_1 + b_1 \\ a_2 + b_2 \end{pmatrix}$.

Example 2

Find $\begin{pmatrix} -3 \\ 2 \end{pmatrix} + \begin{pmatrix} 5 \\ -3 \end{pmatrix}$

solution $\begin{pmatrix} -3 \\ 2 \end{pmatrix} + \begin{pmatrix} 5 \\ -3 \end{pmatrix} = \begin{pmatrix} 2 \\ -1 \end{pmatrix}$

Exercise 19.1(b)

1. Find:

(a) $\begin{pmatrix} 2 \\ 3 \end{pmatrix} + \begin{pmatrix} 1 \\ -4 \end{pmatrix}$
(b) $\begin{pmatrix} 3 \\ -1 \end{pmatrix} + \begin{pmatrix} -4 \\ 3 \end{pmatrix}$
(c) $\begin{pmatrix} 3 \\ -4 \end{pmatrix} + \begin{pmatrix} -3 \\ 5 \end{pmatrix}$

(d) $\begin{pmatrix} 2 \\ 4 \end{pmatrix} + \begin{pmatrix} -3 \\ -5 \end{pmatrix}$
(e) $\begin{pmatrix} -7 \\ 2 \end{pmatrix} + \begin{pmatrix} 1 \\ 5 \end{pmatrix}$
(f) $\begin{pmatrix} -2 \\ -3 \end{pmatrix} + \begin{pmatrix} 5 \\ 1 \end{pmatrix}$

2. Find the value of x and y from the following vector equations

(a) $\begin{pmatrix} 3 \\ x \end{pmatrix} + \begin{pmatrix} y \\ 2 \end{pmatrix} = \begin{pmatrix} 8 \\ 7 \end{pmatrix}$
(b) $\begin{pmatrix} 1 \\ 2 \end{pmatrix} + \begin{pmatrix} y \\ -5 \end{pmatrix} = \begin{pmatrix} 3 \\ x \end{pmatrix}$

(c) $\begin{pmatrix} 2x + 3 \\ x \end{pmatrix} + \begin{pmatrix} -3 \\ 3y - 2 \end{pmatrix} = \begin{pmatrix} y \\ 5 \end{pmatrix}$
(d) $\begin{pmatrix} 3x \\ 2x - y \end{pmatrix} + \begin{pmatrix} y \\ -7 \end{pmatrix} = \begin{pmatrix} 2 - 3y \\ 2y \end{pmatrix}$

3. Given that $a = \begin{pmatrix} 2 \\ -1 \end{pmatrix}$, $b = \begin{pmatrix} x \\ y \end{pmatrix}$, $c = \begin{pmatrix} -1 \\ 3 \end{pmatrix}$ and $a + b = c$, find x and y.

Negative Vectors

The vector $\overrightarrow{BA}$ has the same magnitude as $\overrightarrow{AB}$ but has a direction opposite to that of $\overrightarrow{AB}$. The vector $\overrightarrow{BA}$ is called the negative vector of $\overrightarrow{AB}$. That is $\overrightarrow{BA} = -\overrightarrow{AB}$.

If $\overrightarrow{AB} = \begin{pmatrix} a \\ b \end{pmatrix}$, then $\overrightarrow{BA} = -\begin{pmatrix} a \\ b \end{pmatrix} = \begin{pmatrix} -a \\ -b \end{pmatrix}$

Example 3

Find $-\begin{pmatrix} -3 \\ 4 \end{pmatrix}$

Solution $-\begin{pmatrix} -3 \\ 4 \end{pmatrix} = \begin{pmatrix} 3 \\ -4 \end{pmatrix}$

Exercise 19.1(c)

1. Find

(a) $-\begin{pmatrix} 1 \\ 3 \end{pmatrix}$ (b) $-\begin{pmatrix} 3 \\ -4 \end{pmatrix}$ (c) $-\begin{pmatrix} -2 \\ 5 \end{pmatrix}$ (d) $-\begin{pmatrix} -2 \\ 0 \end{pmatrix}$

2. Given that the vectors $a = \begin{pmatrix} 2 \\ 3 \end{pmatrix}$, $b = \begin{pmatrix} -2 \\ 4 \end{pmatrix}$ and $c = \begin{pmatrix} 4 \\ 1 \end{pmatrix}$, find

(a) $-a$ (b) $-b$ (c) $-c + b$ (d) $-b + a$

Subtraction of Vectors

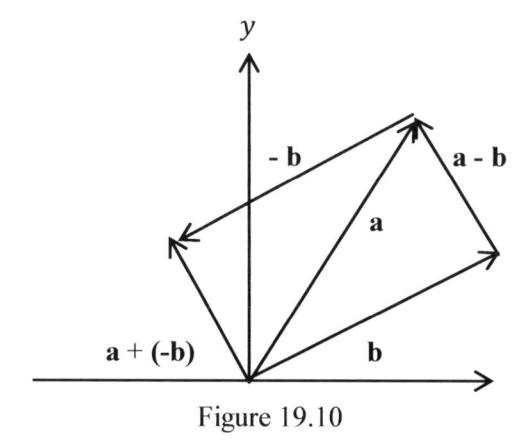

Figure 19.10

The difference $a - b$ is the vector from the terminal point of b to the terminal point of a. This result is equal to $a + (-b)$, as shown in Figure 19.10.

If a and b are column vectors, $a - b$ is obtained by subtracting the corresponding components. For example, if $a = \begin{pmatrix} a_1 \\ a_2 \end{pmatrix}$ and $b = \begin{pmatrix} b_1 \\ b_2 \end{pmatrix}$ then $a - b = \begin{pmatrix} a_1 - b_1 \\ a_2 - b_2 \end{pmatrix}$.

Example 4

Find $\begin{pmatrix} -3 \\ 2 \end{pmatrix} - \begin{pmatrix} -5 \\ 6 \end{pmatrix}$

Solution $\begin{pmatrix} -3 \\ 2 \end{pmatrix} - \begin{pmatrix} -5 \\ 6 \end{pmatrix} = \begin{pmatrix} 2 \\ -4 \end{pmatrix}$

Exercise 19.1(d)

1. Find

(a) $\begin{pmatrix} 3 \\ 5 \end{pmatrix} - \begin{pmatrix} 1 \\ 2 \end{pmatrix}$
(b) $\begin{pmatrix} 5 \\ 2 \end{pmatrix} - \begin{pmatrix} -2 \\ 3 \end{pmatrix}$
(c) $\begin{pmatrix} 1 \\ 2 \end{pmatrix} - \begin{pmatrix} 1 \\ -2 \end{pmatrix}$

(d) $\begin{pmatrix} -7 \\ -6 \end{pmatrix} - \begin{pmatrix} -5 \\ 3 \end{pmatrix}$
(e) $\begin{pmatrix} -2 \\ 1 \end{pmatrix} - \begin{pmatrix} 1 \\ 2 \end{pmatrix}$
(f) $\begin{pmatrix} -2 \\ 1 \end{pmatrix} - \begin{pmatrix} -8 \\ -4 \end{pmatrix}$

2. Given that $a = \begin{pmatrix} 2 \\ -3 \end{pmatrix}$, $b = \begin{pmatrix} 3 \\ -4 \end{pmatrix}$ and $c = \begin{pmatrix} 4 \\ 1 \end{pmatrix}$, find

(a) $b - a$
(b) $c - a$
(c) $b - c$

Multiplying a Vector by a Scalar

If we multiply a vector **a** by a scalar (a real number) k > 0, the result is a vector in the same direction as **a** with magnitude $k|a|$ (see Figure 19.11).

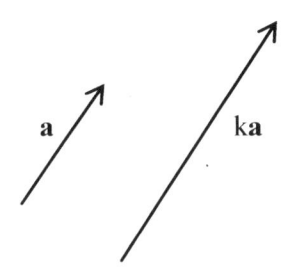

Figure 19.11

If we multiply a vector **a** by scalar k < 0, the result is a vector in the opposite direction as **a** with magnitude $|k||a|$ (see Figure 20.6).

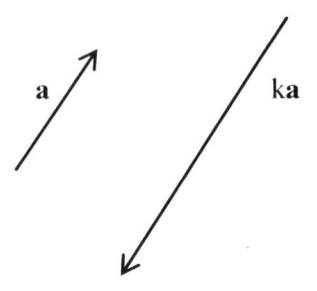

Figure 19.12

To multiply the vector $a = \begin{pmatrix} a_1 \\ a_2 \end{pmatrix}$ by the scalar k, multiply both components by k. That is $ka = \begin{pmatrix} ka_1 \\ ka_2 \end{pmatrix}$.

Example 5

Find $-3 \begin{pmatrix} 1 \\ -2 \end{pmatrix}$

Solution $-3 \begin{pmatrix} 1 \\ -2 \end{pmatrix} = \begin{pmatrix} -3 \\ 6 \end{pmatrix}$

Exercise 19.1(e)

1. Find:

(a) $2 \begin{pmatrix} 3 \\ -1 \end{pmatrix}$ (b) $-3 \begin{pmatrix} -1 \\ 2 \end{pmatrix}$ (c) $\frac{1}{2} \begin{pmatrix} -2 \\ 6 \end{pmatrix}$ (d) $-\frac{2}{3} \begin{pmatrix} 3 \\ -6 \end{pmatrix}$ (e) $-2 \begin{pmatrix} -3 \\ 0 \end{pmatrix}$

2. Given that $a = \begin{pmatrix} 1 \\ 2 \end{pmatrix}$, $b = \begin{pmatrix} 2 \\ -3 \end{pmatrix}$ and $c = \begin{pmatrix} 3 \\ 4 \end{pmatrix}$, find

(a) $-2a$ (b) $3b$ (c) $3a + 2b$ (d) $2b - c$

3. Given that $a = \begin{pmatrix} 2 \\ 3 \end{pmatrix}$ and $b = \begin{pmatrix} 3 \\ -1 \end{pmatrix}$, solve the equation $xa + yb = \begin{pmatrix} 12 \\ 7 \end{pmatrix}$

4. Given that $a = \begin{pmatrix} 3 \\ 4 \end{pmatrix}$, $b = \begin{pmatrix} 2 \\ -3 \end{pmatrix}$ and $c = \begin{pmatrix} 5 \\ 1 \end{pmatrix}$, solve the equation $xa + yb = c$

5. Given that $a = \begin{pmatrix} 3 \\ 2 \end{pmatrix}$, $b = \begin{pmatrix} 2 \\ -3 \end{pmatrix}$ and $c = \begin{pmatrix} 12 \\ -5 \end{pmatrix}$, find k and m such that $ka + mb = c$

Parallel Vectors

Two vectors **a** and **b** are parallel if $a = kb$, where k is any scalar. Parallel vectors may have the same direction or opposite direction.

Example 6

Determine whether or not $a = \begin{pmatrix} 3 \\ 2 \end{pmatrix}$ and $b = \begin{pmatrix} -9 \\ -6 \end{pmatrix}$ are parallel

Solution We have $\begin{pmatrix} -9 \\ -6 \end{pmatrix} = -3 \begin{pmatrix} 3 \\ 2 \end{pmatrix}$, so **a** and **b** are parallel

Exercise 19.1(f)

Determine whether the vectors **a** and **b** are parallel

1. $a = \begin{pmatrix} -2 \\ -4 \end{pmatrix}$, $b = \begin{pmatrix} 1 \\ 2 \end{pmatrix}$
2. $a = \begin{pmatrix} 1 \\ 2 \end{pmatrix}$, $b = \begin{pmatrix} -2 \\ 1 \end{pmatrix}$
3. $a = \begin{pmatrix} 6 \\ 4 \end{pmatrix}$, $b = \begin{pmatrix} 3 \\ -2 \end{pmatrix}$
4. $a = \begin{pmatrix} -2 \\ 1 \end{pmatrix}$, $b = \begin{pmatrix} 8 \\ -4 \end{pmatrix}$

Magnitude of Vectors

The magnitude of a vector is represented geometrically by the length of a line segment. The magnitude of vector $\vec{AB}$, written $|\vec{AB}|$ or simply AB, is given by the length of line AB.

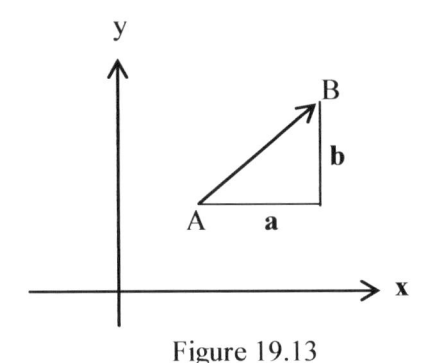

Figure 19.13

Using the Pythagoras' theorem, we get

$$|\vec{AB}| = \sqrt{a^2 + b^2}$$

Example 7

Find the magnitude of the vector $a = \begin{pmatrix} -8 \\ 15 \end{pmatrix}$

Solution $|a| = \sqrt{(-8)^2 + 15^2}$

$= \sqrt{289}$

$= 17$ units

Exercise 19.1(g)

1. Find the magnitude of each of the following vectors

(a) $\begin{pmatrix} 4 \\ 3 \end{pmatrix}$ (b) $\begin{pmatrix} -2 \\ 4 \end{pmatrix}$ (c) $\begin{pmatrix} -3 \\ 5 \end{pmatrix}$ (d) $\begin{pmatrix} -5 \\ -12 \end{pmatrix}$

2. Given that $\mathbf{a} = \begin{pmatrix} -7 \\ 2 \end{pmatrix}$ and $\mathbf{b} = \begin{pmatrix} 1 \\ 5 \end{pmatrix}$, evaluate:

(a) $|\mathbf{a} + \mathbf{b}|$ (b) $|\mathbf{b} - \mathbf{a}|$ (c) $|2\mathbf{a} + \mathbf{b}|$ (d) $|\mathbf{a} - 2\mathbf{b}|$

Unit Vectors

A vector with magnitude 1 is called a unit vector. If vector **a** has magnitude $|\mathbf{a}| > 0$, then the unit vector in the direction of **a**, denoted by $\hat{a}$ is given by

$$\hat{a} = \frac{1}{|\mathbf{a}|} \mathbf{a}$$

Example 8

Find the unit vector in the direction of $\mathbf{a} = \begin{pmatrix} 3 \\ -4 \end{pmatrix}$

Solution $|\mathbf{a}| = \sqrt{3^2 + (-4)^2}$

$$= \sqrt{25}$$

$$= 5$$

$$\hat{a} = \frac{1}{5}\begin{pmatrix} 3 \\ -4 \end{pmatrix} = \begin{pmatrix} 3/5 \\ -4/5 \end{pmatrix}$$

Exercise 19.1(h)

1. Find the unit vector of each of the following vectors

(a) $\begin{pmatrix} -8 \\ 6 \end{pmatrix}$ (b) $\begin{pmatrix} 15 \\ 8 \end{pmatrix}$ (c) $\begin{pmatrix} -12 \\ -5 \end{pmatrix}$ (d) $\begin{pmatrix} 7 \\ -24 \end{pmatrix}$

2. Given that $\mathbf{a} = \begin{pmatrix} 4 \\ -3 \end{pmatrix}$ and $\mathbf{b} = \begin{pmatrix} -1 \\ 2 \end{pmatrix}$, find the unit vector in the direction of **c** in each case:

(a) $\mathbf{c} = \mathbf{a} + \mathbf{b}$ (b) $\mathbf{c} = \mathbf{b} - \mathbf{a}$ (c) $\mathbf{c} = \mathbf{a} + 2\mathbf{b}$ (d) $\mathbf{c} = 2\mathbf{a} + 3\mathbf{b}$

Unit Vectors along the axes

Sometimes, it may be convenient to write vectors in terms of two standard unit vectors in the x and y directions, denoted by **i** and **j** respectively (see Figure 19.14).

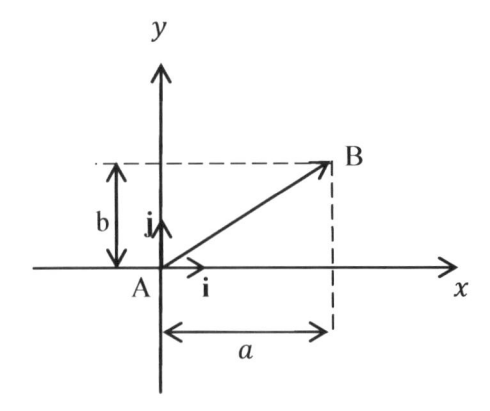

Figure 19.14

Using the unit vectors **i** and **j** we can write vector $\vec{AB}$ as

$$\vec{AB} = a\mathbf{i} + b\mathbf{j}$$

$a\mathbf{i}$ and $b\mathbf{j}$ are called the components of vector $\vec{AB}$

Notice that $a\mathbf{i} + b\mathbf{j}$ can also be written as $\begin{pmatrix} a \\ b \end{pmatrix}$.

Using vectors in the form $a\mathbf{i} + b\mathbf{j}$

Example 9

(a) Given that $\mathbf{a} = 3\mathbf{i} + 2\mathbf{j}$ and $\mathbf{b} = -5\mathbf{i} + \mathbf{j}$, find $\mathbf{a} + \mathbf{b}$

Solution $\mathbf{a} + \mathbf{b} = (3\mathbf{i} + 2\mathbf{j}) + (-5\mathbf{i} + \mathbf{j})$

$$= (3 - 5)\mathbf{i} + (2 + 1)\mathbf{j}$$

$$= -2\mathbf{i} + 3\mathbf{j}$$

(b) Given that $\mathbf{a} = 3\mathbf{i} - 4\mathbf{j}$, find $|-3\mathbf{a}|$

Solution $-3\mathbf{a} = -3(3\mathbf{i} - 4\mathbf{j})$

$$= -9\mathbf{i} + 12\mathbf{j}$$

$$|-3\mathbf{a}| = \sqrt{(-9)^2 + 12^2}$$

$$= \sqrt{225}$$

= 15 units

(c) Find the unit vector in the direction of $a = -5i + 12j$

Solution $|a| = \sqrt{(-5)^2 + 12^2}$

$= \sqrt{169}$

$= 13$ units

$\hat{a} = -\frac{5}{13}i + \frac{12}{13}j$

(d) Given that $a = -i + 3j$ and $b = -2i + j$, find $|2a - 3b|$

Solution $2a - 3b = 2(-i + 3j) - 3(-2i + j)$

$= -2i + 6j + 6i - 3j$

$= 4i + 3j$

$|4i + 3j| = \sqrt{4^2 + 3^2}$

$= \sqrt{16 + 9}$

$= \sqrt{25}$

$= 5$

Exercise 19.1(i)

1. Express each of the following vectors in terms of i and j

(a) $\binom{4}{3}$ (b) $\binom{3}{-2}$ (c) $\binom{-5}{-2}$ (d) $\binom{0}{5}$ (e) $\binom{-7}{0}$

2. Given that $a = -3i + 2j$, $b = 2i - 3j$ and $c = 5i + 4j$, find

(a) $a + b$ (b) $b - c$ (c) $2c + b$ (d) $3a - 2b$

3. Find the magnitude of each of the following vectors

(a) $-3i + j$ (b) $4i - 3j$ (c) $-12j$ (d) $-15i - 8j$

4. Find the unit vector in the direction of the following vectors

(a) $i + j$ (b) $-3i + 2j$ (c) $2i - j$ (d) $-i - 7j$

Properties of Vectors

If a, b and c are vectors, and λ and μ are scalars then the following holds:

1. $a + b = b + a$ Commutative property

2. $(a + b) + c = a + (b + c)$ Associative property

3. $\lambda(\mu a) = (\lambda \mu)a$ Associative property

4. $(\lambda + \mu)a = \lambda a + \mu a$ Distributive property

5. $\lambda(a + b) = \lambda a + \lambda b$ Distributive property

Position Vectors

A vector with its initial point located at the origin is called a position vector. A position vector is uniquely represented by the coordinates of its terminal point.

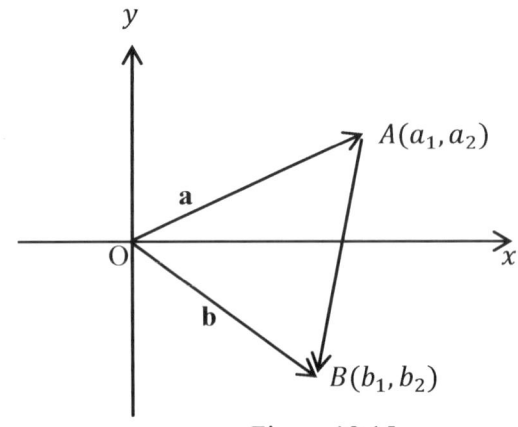

Figure 19.15

The position vector with initial point at the origin and terminal point at the point $A(a_1, a_2)$ (see Figure 19.15) is denoted by $a = \overrightarrow{OA} = \begin{pmatrix} a_1 \\ a_2 \end{pmatrix}$. Similarly, the position vector of B is denoted by $b = \overrightarrow{OB} = \begin{pmatrix} b_1 \\ b_2 \end{pmatrix}$.

Note: We customary denote the position vector by the letter used for its terminal point.

We can express vector $\vec{AB}$ in terms of the position vectors of A and B.

$\vec{AB} = \vec{AO} + \vec{OB}$

$\quad\quad = \vec{OB} + (-\vec{OA})$

$\quad\quad = \mathbf{b} - \mathbf{a}$

Example 10

(a) Given that A (-7, 5) and B (- 2, 3), find $\vec{AB}$

Solution $\vec{AB} = \vec{OB} - \vec{OA}$

$$= \begin{pmatrix} -2 \\ 3 \end{pmatrix} - \begin{pmatrix} -7 \\ 5 \end{pmatrix}$$

$$= \begin{pmatrix} 5 \\ -2 \end{pmatrix}$$

(b) Given that P (- 3, 2) and $\vec{PQ} = \begin{pmatrix} 5 \\ 3 \end{pmatrix}$, find the coordinates of Q

Solution $\vec{PQ} = \vec{OQ} - \vec{OP}$

$\vec{OQ} = \vec{PQ} + \vec{OP}$

$$= \begin{pmatrix} 5 \\ 3 \end{pmatrix} + \begin{pmatrix} -3 \\ 2 \end{pmatrix}$$

$$= \begin{pmatrix} 2 \\ 5 \end{pmatrix}$$

The coordinates of Q is (2, 5)

Exercise 19.1(j)

1. Express as column vectors the position vectors of each of the following points:

(a) (4, 2)   (b) (-5, 3)   (c) (0, 4)   (d) (- 3, -2)   (e) (-5, 0)

2. Express the position vectors of each of the following points in terms of **i** and **j**:

(a) (3, 2)   (b) (1, 3)   (c) (-2, 0)   (d) (4, -5)   (e) (0, 2)

3. Find the vector $\vec{AB}$ in each case:

(a) A(7, 3), B(9, 4)   (b) A(1, - 2), B(3, - 5)   (c) A(3, 4), B(2, -1)   (d) A(2, 3), B(-2, 4)

4. Given $A(x, y)$, $B(3, 4)$ and $\overrightarrow{AB} = \begin{pmatrix} -2 \\ 1 \end{pmatrix}$, find x and y.

5. Given $A(4, -5)$ and $\overrightarrow{AB} = \begin{pmatrix} 2 \\ 3 \end{pmatrix}$, find the coordinates of B.

6. Given $Q(-2, -3)$ and $\overrightarrow{PQ} = \begin{pmatrix} -3 \\ 4 \end{pmatrix}$, find the coordinates of P.

7. The position vectors relative to the origin O, of three points A, B and C are $2i + 3j$, $4i + 7j$ and $i + 2j$ respectively. Given that $\overrightarrow{OB} = m\overrightarrow{OA} + n\overrightarrow{OC}$, where m and n are scalar constants, find the values of m and n.

8. The position vectors of A and B are $-3i + 2j$ and $-i - j$. Given that $\overrightarrow{AC} = 5i + j$, find $\overrightarrow{BC}$.

9. A and B have position vectors $5i + 2j$ and $-i + 4j$ respectively. Find the position vector of C if $3\overrightarrow{OA} = 2\overrightarrow{OB} + \overrightarrow{OC}$.

10. The coordinates of A and B are $(2, 3)$ and $(-2, 5)$ respectively. Find the position vector of C if $2\overrightarrow{OA} = 2\overrightarrow{OB} + \overrightarrow{BC}$.

Using Vectors in Geometry

You can use vector algebra to prove some theorem in geometry.

Example 11

(a) In the triangle ABC, M and N are the midpoint of the sides AB and AC respectively. Show that MN is parallel to BC and that $MN = \frac{1}{2} BC$.

Solution

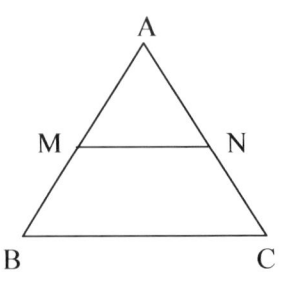

Let $\overrightarrow{AB} = b$ and $\overrightarrow{AC} = a$

Then $\overrightarrow{AM} = \frac{1}{2} b$ and $\overrightarrow{AN} = \frac{1}{2} a$

$\overrightarrow{MN} = \overrightarrow{MA} + \overrightarrow{AN}$

$$= \overrightarrow{AN} - \overrightarrow{AM}$$

$$= \frac{1}{2}(a - b)$$

$$= \frac{1}{2}\overrightarrow{BC}$$

∴ MN is parallel to BC and $MN = \frac{1}{2}BC$

(b) OAB is a triangle with $\overrightarrow{OA} = a$ and $\overrightarrow{OB} = b$, X and Y are such that $\overrightarrow{OX} = \lambda a$ and $\overrightarrow{OY} = \mu b$, if XY is parallel to AB. Show that $\lambda = \mu$.

Solution

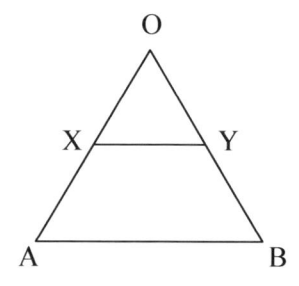

$$\overrightarrow{XY} = \overrightarrow{OY} - \overrightarrow{OX}$$
$$= \mu b - \lambda a$$

$$\overrightarrow{AB} = b - a$$

Because $\overrightarrow{XY}$ is parallel to $\overrightarrow{AB}$,

$$\overrightarrow{AB} = k\overrightarrow{XY}, \text{ where } k \text{ is a scalar}$$

$$b - a = k(\mu b - \lambda a)$$

$$(1 - k\mu)b - (1 - k\lambda)a = 0$$

$$\therefore 1 - k\mu = 0$$

$$\mu = \frac{1}{k}$$

Also $1 - k\lambda = 0$

$$\lambda = \frac{1}{k}$$

So $\lambda = \mu$

Exercise 19.1(k)

1. PQRS is a parallelogram in which PR is perpendicular to QS. Prove that PQRS is a rhombus.

2. ABC is a triangle and $\overrightarrow{AB} = \mathbf{b}$, $\overrightarrow{AC} = \mathbf{c}$, X and Y are the midpoint of AB and AC respectively. Express in terms of **b** and **c**:

(a) $\overrightarrow{BC}$ (b) $\overrightarrow{AX}$ (c) $\overrightarrow{AY}$ (d) $\overrightarrow{XY}$

What can you conclude about BC and XY?

If $\triangle ABC$ has area 6, what is the area of $\triangle AXY$?

3. M and N divides the sides PQ and PR respectively of triangle PQR in the ratio 1 : 2. Show that $\overrightarrow{MN}$ is parallel to $\overrightarrow{QR}$

4. ABCD is any quadrilateral where E, F, G, H are the midpoints of the sides

(a) Express $\overrightarrow{EF}$ in terms of $\overrightarrow{AC}$

(b) Express $\overrightarrow{HG}$ in terms of $\overrightarrow{AC}$

(c) Prove that EFGH is a parallelogram

5. In a given quadrilateral ABCD, k is the midpoint of AC and also the midpoint of BD. Prove that ABCD is a parallelogram

6. Using vector method prove that a quadrilateral with two sides equal and parallel is a parallelogram

7. Prove that the diagonals of a parallelogram bisect each other

8. Using vector method prove that if ABCD is a parallelogram and X and Y are the midpoint of AB and CD respectively, then CXAY is a parallelogram

9. Given a trapezium ABCD with AD//BC, if X and Y are the mid-point of AB and DC respectively prove that $\overrightarrow{XY} = (\overrightarrow{AD} + \overrightarrow{BC})$

[Hint: Let $\overrightarrow{XA} = \mathbf{a}, \overrightarrow{DY} = \mathbf{b}$]

10. OABC is a parallelogram with the position vectors of A and C relative to O being **a** and **c**. X is the midpoint of OA and OB, and XC meet at Y. Find the position vector of Y.

19.2 The Ratio Theorem

Internal Division of a Line Segment

A line segment AB is said to be divided internally in the ratio $a : b$ if the dividing point P is between A and B, as shown in Figure 19.16, such that $\frac{AP}{PB} = \frac{a}{b}$.

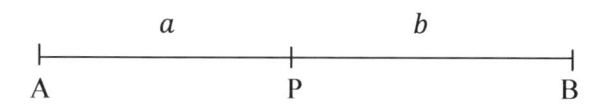

Figure 19.16

As shown in Figure 19.17, $\boldsymbol{a}$ and $\boldsymbol{b}$ are the position vectors of the points A and B with respect to a fixed origin O.

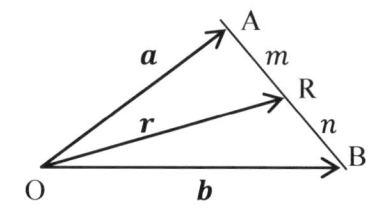

Figure 19.17

Let $\boldsymbol{r}$ be the position vector of the point R which divides AB internally in the ratio m to n.

$\overrightarrow{OR} = \overrightarrow{OA} + \overrightarrow{AR}$

$\overrightarrow{OR} - \overrightarrow{OA} = \overrightarrow{AR}$

$\boldsymbol{r} - \boldsymbol{a} = \frac{m}{m+n}(\boldsymbol{b} - \boldsymbol{a})$

$m\boldsymbol{r} + n\boldsymbol{r} = m\boldsymbol{b} + n\boldsymbol{a}$

$\boldsymbol{r} = \frac{m\boldsymbol{b} + n\boldsymbol{a}}{m+n}$

Example 12

A point A has coordinates (3, 1) and B has coordinates (2, 3). If O has coordinates (0, 0) find the coordinates of the point P which divides AB internally in the ratio 1 : 2.

Solution

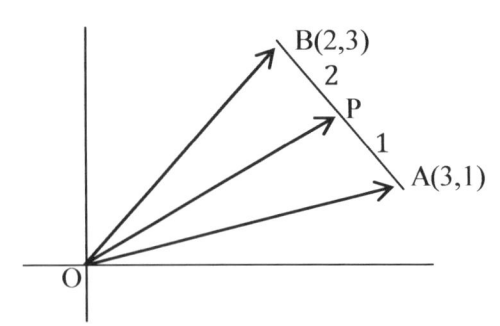

If P is the point dividing AB internally in the ratio 1: 2, then AP : PB = 1: 2.

So $\overrightarrow{OP} = \dfrac{2\overrightarrow{OA}+\overrightarrow{OB}}{2+1}$

$= \dfrac{2\binom{3}{1}+\binom{2}{3}}{2+1}$

$= \begin{pmatrix}\frac{8}{3}\\ \frac{5}{3}\end{pmatrix}$

Thus the coordinates of P are $\left(\dfrac{8}{3}, \dfrac{5}{3}\right)$.

External Division of a Line Segment

A point P divides a line segment AB externally in the ratio $a : b$ if P is in the line containing the line segment AB but not between A and B, as shown in Figure 19.18.

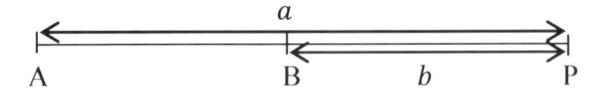

Figure 19.18

Note the following:

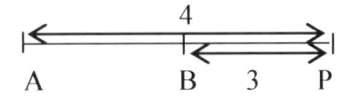

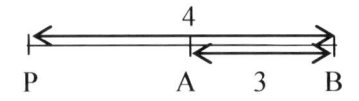

If a point P divides AB externally in the ratio 4 : 3 then P lies in AB produced.

If a point P divides AB externally in the ratio 3 : 4 then P lies in BA produced.

As shown in Figure 19.19 a and b are the position vectors of the points A and B with respect to a fixed origin O.

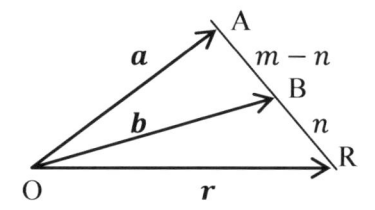

Figure 19.19

Let r be the position vector of the point R which divides AB externally in the ratio m to n.

$\overrightarrow{OB} = \overrightarrow{OA} + \overrightarrow{AB}$

$\overrightarrow{OB} - \overrightarrow{OA} = \overrightarrow{AB}$

$b - a = \dfrac{m-n}{m}(r - a)$

$mr - nr = mb - na$

$r = \dfrac{mb - na}{m - n}$

Example 13

The coordinates of the point A and B are (1, 2) and (3, 2) respectively. If O has coordinates (0, 0) find the coordinates of the point P which divides AB externally in the ratio 3 : 1.

Solution

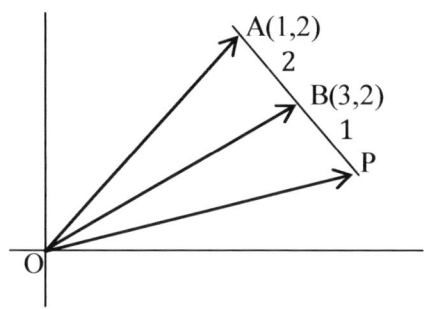

If P is the point dividing AB internally in the ratio 3: 1, then AP : BP = 3: 1.

So $\overrightarrow{OP} = \dfrac{3OB - \overrightarrow{OA}}{3 - 1}$

$= \dfrac{3\binom{3}{2} - \binom{1}{2}}{3 - 1}$

$= \binom{4}{2}$

Thus the coordinates of P are (4, 2).

Exercise 19.2

1. Given O(0, 0), A (4, -3), B (-2, 4) find the coordinates of the point P which divides AB internally in each of the given ratio:

(a) 4 : 1   (b) 2 : 1   (c) 3 : 2

2. Given O(0, 0), A (4, 1), B (1, 5) find the coordinates of the point P which divides AB externally in each of the given ratio.

(a) 3 : 2   (b) 1 : 4   (c) 5 : 2

3. OABC is a parallelogram. Q is the midpoint of OC. P and X are points on $\overrightarrow{OB}$ and $\overrightarrow{AQ}$ respectively. If $\overrightarrow{OA} = \boldsymbol{a}$ and $\overrightarrow{OC} = \boldsymbol{b}$ find the following in terms of $\boldsymbol{a}$ and $\boldsymbol{b}$:

(a) $\overrightarrow{OQ}$

(b) $\overrightarrow{OX}$ such that X divides $\overrightarrow{AQ}$ internally in the ratio 2 : 1.

(c) $\overrightarrow{OB}$

(d) $\overrightarrow{OP}$ such that P divides $\overrightarrow{OB}$ in the ratio 1 : 2.

(e) Draw conclusions from your results.

19.3 Scalar Product

The scalar (or dot) product $\boldsymbol{a} \cdot \boldsymbol{b}$ of two vectors **a** and **b** is a real number, defined by

$\boldsymbol{a} \cdot \boldsymbol{b} = |a||b|\cos\theta$, where θ is the angle between **a** and **b**.

Since $\cos(\pi/2) = 0$, $\boldsymbol{a} \cdot \boldsymbol{b} = 0$ when $\theta = \pi/2$. Hence, two vectors **a** and **b** are orthogonal or perpendicular vectors if and only if $\boldsymbol{a} \cdot \boldsymbol{b} = 0$.

If **a** and **b** are parallel but in the same direction, then $\theta = 0°$, and $\boldsymbol{a} \cdot \boldsymbol{b} = |a||b|$.

If **a** and **b** are parallel but have opposite direction then $\theta = \pi$, and $\boldsymbol{a} \cdot \boldsymbol{b} = -|a||b|$

The scalar product is defined for vectors in component form as follows:

If $\boldsymbol{a} = \begin{pmatrix} a_1 \\ a_2 \end{pmatrix}$ and $\boldsymbol{b} = \begin{pmatrix} b_1 \\ b_2 \end{pmatrix}$ are two vectors then the scalar product is defined by

$\boldsymbol{a} \cdot \boldsymbol{b} = a_1 b_1 + a_2 b_2$

Example 14

(a) Find the scalar product of $a = \begin{pmatrix} 3 \\ 2 \end{pmatrix}$ and $b = \begin{pmatrix} -2 \\ 4 \end{pmatrix}$.

Solution We have $a \cdot b = 3(-2) + 2(4) = 2$

(b) Find the scalar product $a \cdot b$ for $a = 3i - 6j$ and $b = 2i + 5j$.

Solution We have $a \cdot b = 3(2) + (-6)(5) = 6 - 30 = -24$

(c) Determine whether the two vectors $a = 3i - 4j$ and $b = 4i + 3j$ are perpendicular.

Solution We have $a \cdot b = 3(4) + (-4)(3) = 0$

Hence **a** and **b** are perpendicular vectors

Properties of Scalar Product

For vectors **a**, **b**, **c** and any scalar k, the following hold:

1. $a \cdot b = b \cdot a$ $\qquad\qquad\qquad\qquad$ Commutative property

2. $a \cdot (b + c) = a \cdot b + a \cdot c$ $\qquad\qquad$ Distributive property

3. $(ka) \cdot b = k(a \cdot b) = a \cdot (kb)$

4. $a \cdot a = |a|^2$

Angles between Vectors

Recall, that the scalar product of two vectors **a** and **b** is defined by $a \cdot b = |a||b|\cos\theta$.

From this definition we get

$$\cos\theta = \frac{a \cdot b}{|a||b|}$$

which allows us to find the angle between any two vectors.

Given $a = \begin{pmatrix} a_1 \\ a_2 \end{pmatrix}$ and $b = \begin{pmatrix} b_1 \\ b_2 \end{pmatrix}$, we have

$$\cos\theta = \frac{a_1 b_1 + a_2 b_2}{\left(\sqrt{a_1^2 + a_2^2}\right)\left(\sqrt{b_1^2 + b_2^2}\right)}$$

Example 15

Find the acute angle between $a = \begin{pmatrix} 2 \\ 3 \end{pmatrix}$ and $b = \begin{pmatrix} -1 \\ 5 \end{pmatrix}$.

Solution We have $\cos\theta = \dfrac{2(-1)+3(5)}{(\sqrt{1^2+3^2})(\sqrt{(-1)^2+5^2})}$

$$= \dfrac{13}{(\sqrt{13})(\sqrt{26})}$$

$$= \dfrac{1}{\sqrt{2}}$$

$$\theta = 45°$$

Exercise 19.3(a)

1. Find $a \cdot b$

(a) $a = \begin{pmatrix} 3 \\ 1 \end{pmatrix}, b = \begin{pmatrix} 2 \\ 5 \end{pmatrix}$ (b) $a = \begin{pmatrix} 2 \\ 5 \end{pmatrix}, b = \begin{pmatrix} 2 \\ 1 \end{pmatrix}$ (c) $a = \begin{pmatrix} 0 \\ -2 \end{pmatrix}, b = \begin{pmatrix} -2 \\ 4 \end{pmatrix}$

(d) $a = 3i + j$ (e) $a = -2i - 3j$ (f) $a = 4i + 3j$

    $b = -2i + 3j$ $b = 2i - 3j$ $b = -3i + 4j$

2. Determine if the vectors are perpendicular

(a) $a = \begin{pmatrix} 1 \\ 4 \end{pmatrix}, b = \begin{pmatrix} 8 \\ -2 \end{pmatrix}$ (b) $a = \begin{pmatrix} 2 \\ 3 \end{pmatrix}, b = \begin{pmatrix} -1 \\ 4 \end{pmatrix}$ (c) $a = \begin{pmatrix} 2 \\ -1 \end{pmatrix}, b = \begin{pmatrix} 2 \\ 4 \end{pmatrix}$

(d) $a = 6i + 2j$ (e) $a = 5i - 3j$ (f) $a = -2i + 7j$

    $b = -i + 3j$ $b = 2i + 3j$ $b = 7i + 2j$

3. Find the angle between the vectors

(a) $a = \begin{pmatrix} 1 \\ 4 \end{pmatrix}, b = \begin{pmatrix} -2 \\ 3 \end{pmatrix}$ (b) $a = \begin{pmatrix} 3 \\ -4 \end{pmatrix}, b = \begin{pmatrix} 12 \\ 5 \end{pmatrix}$ (c) $a = \begin{pmatrix} -3 \\ 2 \end{pmatrix}, b = \begin{pmatrix} 2 \\ 4 \end{pmatrix}$

(d) $a = 3i - 2j$ (e) $a = -5i + 3j$ (f) $a = 4i - j$

    $b = i + j$ $b = 2i + 7j$ $b = 2i + 3j$

The Cosine and Sine Formulas

We can derive the cosine and sine formulas by using the properties of the scalar product.

The Cosine Formula

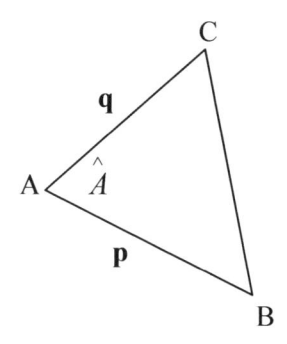

Figure 19.20

Consider the triangle ABC shown in Figure 19.20. Let $\vec{AB} = \boldsymbol{P}$ and $\vec{AC} = \boldsymbol{q}$.

Then $\vec{BC} = \boldsymbol{q} - \boldsymbol{p}$

Now $\vec{BC} \cdot \vec{BC} = (\boldsymbol{q} - \boldsymbol{p}) \cdot (\boldsymbol{q} - \boldsymbol{p})$

$$= \boldsymbol{q} \cdot \boldsymbol{q} - \boldsymbol{q} \cdot \boldsymbol{p} - \boldsymbol{p} \cdot \boldsymbol{q} + \boldsymbol{p} \cdot \boldsymbol{p}$$

$$= |q|^2 + |p|^2 - 2(\boldsymbol{q} \cdot \boldsymbol{p})$$

$$= b^2 + c^2 - 2bc\cos \hat{A}$$

But $\vec{BC} \cdot \vec{BC} = a^2$ so $a^2 = b^2 + c^2 - 2bc\cos \hat{A}$

Similarly $b^2 = a^2 + c^2 - 2ac\cos \hat{B}$ and $c^2 = a^2 + b^2 - 2ab\cos \hat{C}$

The Sine Formula

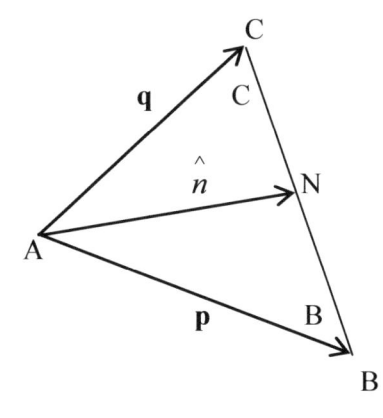

Figure 19.21

Consider the triangle AB shown in Figure 19.21. Let $\hat{n}$ be a unit vector drawn through A perpendicular to BC at N. Then AN is the component of **p** in the direction of $\hat{n}$, but it also the component of **q** in the direction $\hat{n}$. So

$$AN = p \cdot \hat{n} = c \sin \hat{B}$$

and $AN = q \cdot \hat{n} = b \sin \hat{C}$

Then $c \sin \hat{B} = b \sin \hat{C}$

So $\dfrac{b}{\sin \hat{B}} = \dfrac{c}{\sin \hat{C}}$

Similarly, $\dfrac{a}{\sin \hat{A}} = \dfrac{b}{\sin \hat{B}}$

The relations can be combined as follows:

$$\frac{a}{\sin \hat{A}} = \frac{b}{\sin \hat{B}} = \frac{c}{\sin \hat{C}}$$

Vectors and Trigonometry

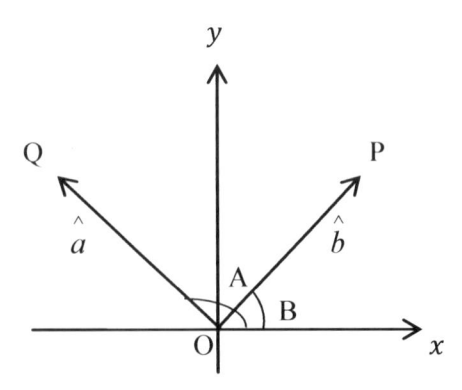

Figure 19.22

In Figure 19.22, $\hat{a}$ is a unit vector making an angle A with Ox and $\hat{b}$ is a unit vector making an angle B with Ox. Then, angle POQ is $A - B$. Now

$\hat{a}$ and $\hat{b}$ can be written in component form as

$\hat{a} = -i\cos(180° - A) + j\sin(180° - A)$

$\phantom{\hat{a}} = i\cos A + j\sin A$

and $\hat{b} = i\cos B + j\sin B$

$\hat{a} \cdot \hat{b} = |\hat{a}||\hat{b}|\cos(A - B) = \cos(A - B)$

Also $\hat{a} \cdot \hat{b} = (i\cos A + j\sin A) \cdot (i\cos B + j\sin B)$

$\phantom{\text{Also } \hat{a} \cdot \hat{b}} = \cos A \cos B + \sin A \sin B$

Hence $\cos(A - B) = \cos A \cos B + \sin A \sin B$

Putting $-B$ for B, we get $\cos(A + B) = \cos A \cos B - \sin A \sin B$

Using the results above together with the relation $\cos(90° - \theta) = \sin\theta$, we can show that $\sin(A - B) = \sin A \cos B - \cos A \sin B$.

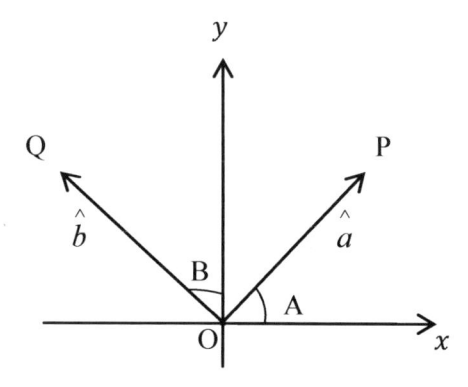

Figure 19.23

In Figure 19.23, $\hat{a}$ is a unit vector making an angle A with Ox and $\hat{b}$ is a unit vector making an angle B with Oy. The angle POQ is then $(90° - A) + B = 90° - (A - B)$.

Now $\cos[90° - (A - B)] = \sin(A - B)$.

So $\hat{a} \cdot \hat{b} = |\hat{a}||\hat{b}|\sin(A - B) = \sin(A - B)$

Now $\hat{a}$ and $\hat{b}$ can be written in component form as $\hat{a} = i\cos A + j\sin A$ and
$\hat{b} = -i\cos(90° - B) + j\sin(90° - B) = -i\sin B + j\cos B$

Also $\hat{a} \cdot \hat{b} = (i\cos A + j\sin A) \cdot (-i\sin B + j\cos B)$

$\qquad = -\cos A \sin B + \sin A \cos B$

Hence $\sin(A - B) = \sin A \cos B - \sin A \cos B$

Putting $-B$ for B we get $\sin(A + B) = \sin A \cos B + \cos A \sin B$

Exercise 19.3(b)

1. Evaluate the following:

(a) $\begin{pmatrix} 3 \\ 4 \end{pmatrix} \cdot \begin{pmatrix} 5 \\ 2 \end{pmatrix}$
(b) $\begin{pmatrix} -2 \\ -5 \end{pmatrix} \cdot \begin{pmatrix} 3 \\ 1 \end{pmatrix}$
(c) $\begin{pmatrix} 2 \\ -4 \end{pmatrix} \cdot \begin{pmatrix} 7 \\ -1 \end{pmatrix}$

2. Evaluate the following:

(a) $(3i + 2j) \cdot (i - 5j)$ (b) $(i + 2j) \cdot (3i + 4j)$ (c) $(-i + j) \cdot (2i + 3j)$

3. Show that $a = \begin{pmatrix} 3 \\ 4 \end{pmatrix}$ and $b = \begin{pmatrix} -4 \\ 3 \end{pmatrix}$ are perpendicular vectors.

4. If $a = \begin{pmatrix} 3 \\ 5 \end{pmatrix}$ and $b = \begin{pmatrix} 10 \\ x \end{pmatrix}$ are perpendicular solve for x.

5. Find the value of x if $\begin{pmatrix} x \\ 3 \end{pmatrix}$ is perpendicular to $\begin{pmatrix} 3 \\ 1 \end{pmatrix}$.

6. Find the angle between the pairs of vectors in Question 1.

7. Find the angles between the pairs of vectors in Question 2

8. Prove that $(c - a) \cdot (c + a) = |c|^2 - |a|^2$

9. Given that $a = a_1 i + a_2 j$ and $b = b_1 i + b_2 j$ show that $a \cdot b = a_1 b_1 + a_2 b_2$

10. ABCD is a rhombus in which $\overrightarrow{AB} = a$ and $\overrightarrow{AD} = b$.

(a) Express $\overrightarrow{AC}$ in terms of **a** and **b**.

(b) Express $\overrightarrow{BD}$ in terms of **a** and **b**.

(c) How are $|a|$ and $|b|$ related?

(d) Show that the diagonals of a rhombus are perpendicular.

11. OABC is a parallelogram with $\overrightarrow{OA} = \boldsymbol{a}$ and $\overrightarrow{OC} = \boldsymbol{c}$. By considering $\overrightarrow{OB}, \overrightarrow{CA}$ show that OABC is a rhombus if and only if its diagonals are perpendicular.

12. Given A (5, 11), B (2, 7) and C (7, 7)

(a) express $\overrightarrow{BA}$ as a column vector

(b) express $\overrightarrow{BC}$ as a column vector

(c) Find $\overrightarrow{BA} \cdot \overrightarrow{BC}$, $|\overrightarrow{BA}|$ and $|\overrightarrow{BC}|$

(d) Find $\angle ABC$

13. Given A (5, -6), B (-1, -2) and C (4, -1)

(a) express $\overrightarrow{BA}$ and $\overrightarrow{BC}$ as column vectors.

(b) find $\overrightarrow{BA} \cdot \overrightarrow{BC}$, $|\overrightarrow{BA}|$ and $|\overrightarrow{BC}|$

(c) Find $\angle ABC$

14. ABC is a right-angled triangle in which $\overrightarrow{AB} = \boldsymbol{c}, \overrightarrow{CA} = \boldsymbol{b}$ and $\angle BAC = 90°$.

(a) Express $\boldsymbol{a}$ in terms of $\boldsymbol{b}$ and $\boldsymbol{c}$

(b) Evaluate $\boldsymbol{b} \cdot \boldsymbol{c}$

(c) Show that $|\boldsymbol{a}|^2 = |\boldsymbol{b}|^2 + |\boldsymbol{c}|^2$

15. Find the angles and the lengths of the sides of the triangle whose vertices are A (-5, -5), B (-2, -8) and C (-10, -10).

16. OAB is a triangle with the sides OA and AB represented by the vectors $\boldsymbol{p} = \begin{pmatrix} 5 \\ 1 \end{pmatrix}$ and $\boldsymbol{q} = \begin{pmatrix} -3 \\ 3 \end{pmatrix}$ respectively.

(a) Find the coordinates of A

(b) Find the coordinates of B

(c) Express $\overrightarrow{OA}$ as a column vector

(d) Find $\angle OBA$.

Review Exercise 19

1. Given that $a = 3i + 2j$, $b = -i + 3j$ and $c = 2i - 3j$, express in terms of i and j.

   (a) $a + 2b$ (b) $2c - 3b$ (c) $3a + 2c$

2. Given $p = 3i + 2j$ and $q = 2i - 3j$, find numbers x and y such that $xp + yq = 3i - 11j$

3. If $a = i - 3j$, $b = 2i + 3j$ and $c = 2i + 4j$, find the values of the constants m and n such that $2a = 5mb - 3nc$.

4. Given that $a = -12i + 4j$ and that $b = i + pj$, find the values of the constants p and q such that $a + qb = -27i + 19j$.

5. Given that $a = \binom{3}{1}$, $b = \binom{-4}{3}$ and $c = \binom{17}{-3}$, find m and n such that $ma + nb = c$.

6. Given $a = -2i + j$, $b = 3i + 2j$ and $c = -i - 2j$, find:

   (a) $|a + 2b|$ (b) $|2c - 3b|$ (c) $|3a - 2c|$

7. Find the magnitude and the angle made with the x-axis of the vectors.

   (a) $\binom{-3}{4}$ (b) $\binom{5}{-12}$ (c) $-2i + 3i$ (d) $4i - 5j$

8. Find the unit vectors in the direction of the following vectors:

   (a) $a = -3i + 4j$ (b) $b = 3i - j$

   (c) $a = 2i + j$ (d) $b = -5i + 12j$

9. A(-3, 2), B(1, -2) and C(2, 3) are points in the $x - y$ plane. Find $\overrightarrow{BA}$ and $\overrightarrow{BC}$ in the form $\binom{x}{y}$.

10. A, B and C are points with position vectors $3i - 2j$, $2i + j$ and $-i + 4j$ respectively. Find in terms of i and j, the vectors $\overrightarrow{AB}$, $\overrightarrow{BC}$ and $\overrightarrow{CA}$.

11. The coordinates of A are (2, -3) and the position vector of B is $4i + 2j$. Find the vector $\overrightarrow{BA}$.

12. If the coordinates of A are (4, 2) and $\overrightarrow{AB} = 2i + j$, find the position vector of B.

13. A and B have position vectors $2i + 5j$ and $4i - j$ respectively. Find the position vector of C if $3\overrightarrow{OA} = 2\overrightarrow{OB} + \overrightarrow{OC}$.

14. The coordinates of A and B are (3, 2) and (5, - 2) respectively. Find the position vector of C if $2\overrightarrow{OA} = 2\overrightarrow{OB} + \overrightarrow{BC}$.

15. The position vectors relative to an origin O, of three points A, B and C are $3i + 2j$, $7i + 4j$ and $2i + j$ respectively. Given that $\overrightarrow{OB} = m\overrightarrow{OA} + n\overrightarrow{OC}$, where m and n are scalar constants, find the value of m and n.

16. The position vector of A and B are $2i - 3j$ and $-i - j$. Given that $\overrightarrow{AC} = i + 5j$, find the position vector of C. Find $|\overrightarrow{BC}|$ and the angle $\overrightarrow{BC}$ makes with the x-axis.

17. A quadrilateral has vertices A (4, 0), B (7, - 3), C (- 2, - 2) and D (- 5, 1). Show that ABCD is a parallelogram.

18. Show that the triangle whose vertices have position vectors $4i + 2j$, $2i + 5j$ and $5i + 3j$ is isosceles.

19. A (1, 2), B (3, 5), C (3, - 6) and $D(x, y)$ are the vertices of the parallelogram ABCD. Find the values of x and y.

20. P and Q divides the sides BC and AC respectively of triangle, ABC, in the ratio 2 : 1. If $\overrightarrow{AB} = a$ and $\overrightarrow{AC} = b$, find (a) $\overrightarrow{QP}$ and (b) show that $\overrightarrow{QP}$ is parallel to $\overrightarrow{AB}$ and one-third its length.

21. OABC is a parallelogram in which the position vectors of A, B and C relative to O are **a**, **b** and **c** respectively. Let X and Y be the midpoints of the diagonals OB and AC respectively.

(a) Express **b** in terms of **a** and **c**

(b) Find expressions for $\overrightarrow{AC}, \overrightarrow{AY}, \overrightarrow{OX}, \overrightarrow{OY}$

(c) What can you say about the points X and Y?

22. Given that $a = 3i + 2j$, $b = 2i - j$ and $c = -2i + 3j$ find the scalar product.

(a) $a \cdot c$ (b) $a \cdot b$ (c) $b \cdot c$

23. Given that $a = 3i + 2j$ and $b = 2i - 3j$ show that **a** is perpendicular to **b**.

24. Given that $a = i + 3j$ and $b = 6i + 2j$ find the angle between **a** and **b**.

25. A (2, 3), B (-1, 4) and C (5, -2) are three points. Evaluate $\overrightarrow{BA} \cdot \overrightarrow{BC}$ and hence find $\angle ABC$.

26. Given $a = -2i + 3j$, $b = i + 4j$ and $c = 7i + 2j$ verify that $a \cdot (b + c) = a \cdot b + a \cdot c$

27. If $p = 4i + 3j$, $q = -3i + 2j$ and $r = i + 5j$ show that $p \cdot (q - r) = p \cdot q - p \cdot r$ and find the angle between **p** and **q**.

28. The position vectors of A and B are $3i + 4j$ and $-i + 7j$ respectively. Show that OA is perpendicular to AB and find $\angle AOB$.

29. The vertices of the triangle ABC are A (1, 1), B (3, 1) and C (3, 4). Find $C\hat{A}B$, correct to two significant figures.

30. PQRS is a quadrilateral with vertices P (-2, 1), Q (1, 5), R (4, 1) and S (1, -3). Using the dot (scalar) products of vectors

(a) Show that $\overrightarrow{PR}$ is perpendicular to $\overrightarrow{QS}$.

(b) Find, correct to the nearest degree, angle SPQ.

20 Mechanics

20.1 Kinematics

Kinematics is the branch of mathematics that study how bodies move.

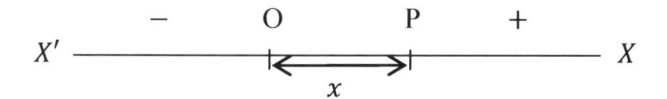

Figure 20.1

Consider a particle P moving along a straight line $X'OX$. The position of the particle at time, t, is given by its displacement x from a fixed point O as shown in Figure 20.1. The displacement in the direction OX is taken as positive and the displacement in the direction OX' is taken as negative.

The rate of change of the displacement of the particle is called velocity and the rate of change of velocity is called the acceleration. Similarly, the velocity and acceleration of the particle would be considered positive in the direction OX and negative in the direction OX'.

Unless otherwise stated, displacement and time are measured in metres, m, and seconds, s, respectively. Velocity is measured in metres per second, ms^{-1}, and acceleration in metres per second squared, ms^{-2}.

Uniform Motion in a Straight Line

We consider a motion in a straight line with constant acceleration (or retardation). Suppose a body moves from rest with initial velocity u and then moves with a constant acceleration, a, until it reaches a final velocity, v, after time, t. Figure 20.2 shows the velocity-time graph.

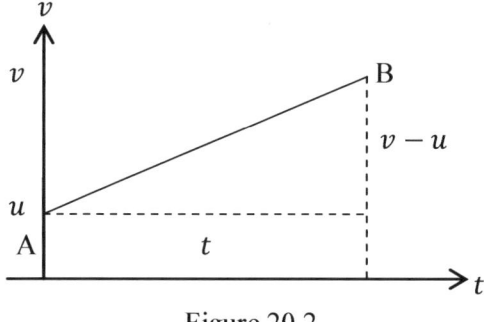

Figure 20.2

The gradient of AB is $\frac{v-u}{t}$

The gradient is the rate of the change of the velocity

Hence $a = \frac{v-u}{t}$

and $at = v - u$

Rearranging this equation we get

$$v = u + at \qquad (1)$$

The area under the graph represents the displacement. If we treat the area under the graph as a trapezium we get

$$s = \frac{1}{2}(u + v)t \qquad (2)$$

Substituting $v = u + at$ into Equation (2) gives $s = \frac{1}{2}(u + u + at)t$, so

$$s = ut + \frac{1}{2}at^2 \qquad (3)$$

From (1) $t = \frac{v-u}{a}$

Substitute this expression for t into Equation 2 to get

$$s = \left(\frac{u+v}{2}\right)\left(\frac{v-u}{a}\right)$$

and $2as = v^2 - u^2$

Rearranging this equation we get

$$v^2 = u^2 + 2as \qquad (4)$$

The above results apply to motion in a straight line with constant acceleration (or retardation). The velocity of a particle which has moved a distance s in time t can be measured in m s^{-1} or km h^{-1} and a useful relationship is 18 km h^{-1} = 5 m s^{-1}.

A knot is a nautical mile per hour where a nautical mile can be taken as 1850 m. The nautical mile is the international unit of length for sea and air navigation.

Example 1

(a) A particle increases its velocity from 15 m s^{-1} to 25 m s^{-1} in 100 m. Find the acceleration and how long will it take to travel.

Solution Using the equation $s = \frac{1}{2}(u + v)t$, with $u = 15$, $v = 25$ and $s = 100$ we have

$$100 = \frac{1}{2}(15 + 25)t$$

$$100 = 20t$$

$$t = 5$$

Hence the time taken is 5 s.

Using the equation $v = u + at$ we have

$$25 = 15 + 5a$$

$$a = 2$$

Hence the acceleration of the particle is 2 m s^{-2}.

Alternatively, using $v^2 = u^2 + 2as$ we have

$$25^2 = 15^2 + 2a \times 100$$

$$400 = 200a$$

$$a = 2$$

Hence the acceleration of the particle is 2 m s^{-2}.

(b) A particle is accelerated at 4 m s^{-1} over 100 m in 5 s. Find its initial and final velocities.

Solution Using the equation $s = ut + \frac{1}{2}at^2$ with $a = 4$, $s = 100$ and $t = 5$ we have

$$100 = u \times 5 + \frac{1}{2} \times 4 \times 5^2$$

Simplifying we have $u = 10$

Hence the initial velocity is 10 m s^{-1}.

Using the equation $v = u + at$ we have

$$v = 10 + 4 \times 5$$

$$v = 30$$

Hence the final velocity is 30 m s^{-1}.

(c) A body decelerating at 2 m s^{-2} passes a certain point with a speed m s^{-1}. Find its velocity after 5 s and the distance covered in that time.

Solution Using the equation $v = u + at$ we have $v = 25 + (-2) \times 5 = 15$

Hence the velocity after 5 s is 15 m s^{-1}.

Using the equation $s = \frac{1}{2}(u+v)t$ we have

$$s = \frac{1}{2}(25+15) \times 5$$

$$= 100$$

Hence the distance covered in 5 s is 100 m.

Exercise 20.1(a)

1. A particle increases its velocity from 20 m s^{-1} to 30 m s^{-1} in 4 s. What is its acceleration and the distance it travelled?

2. A particle accelerates at 6 m s^{-2} from 25 m s^{-1} to 97 m s^{-1}. What distance has it travelled and how long does it take?

3. A particle accelerates at 5 m s^{-2} over 200 m to finish with a speed of 60 m s^{-1}. What was the initial velocity and how long does it take?

4. A particle starts with a velocity of 15 m s^{-1} and travels 126 m in 6 seconds. What is the acceleration and the final velocity?

5. A particle is accelerated uniformly from rest so that after 12 seconds it has achieved the speed of 30 m s^{-1}. Find its acceleration and the distance it has covered.

6. A car accelerates uniformly from rest, achieving a speed of 30 m s^{-1} after covering 150 m. How long did it take?

7. A car accelerates uniformly from rest and after 12 seconds it has covered 40 m. What is its acceleration and its final velocity?

8. A train slows down from 20 m s^{-1} to 15 m s^{-1} covering 200 m. How long did it take?

9. A car accelerates from rest with constant acceleration 2.5 m s^{-1}. How long does it take to cover 15 m?

10. A particle is accelerated from 4 m s^{-1} to 6 m s^{-1} over a distance of 25 m. Find the acceleration and the time taken.

11. A cyclist travelling at 30 m s^{-1} brakes uniformly and stops in 45 m. What is its rate of deceleration and how long does this take?

12. A car travelling at 20 m s^{-1} accelerates uniformly so that in the next 2 s it covers 46 m. What is its acceleration and its final speed?

Velocity Time Graph

The velocity-time graph relates the velocity to the time.

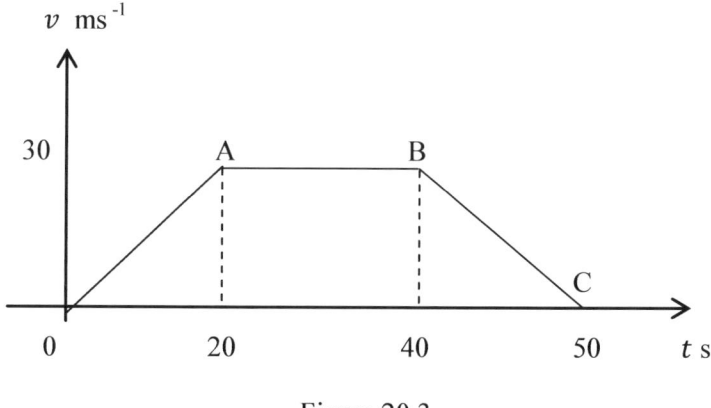

Figure 20.3

Figure 20.3 shows the velocity-time graph of a particle. The graph shows that the particle starts from rest, moving with a steady increasing velocity to 30 m s^{-1} in 20 seconds. Then moves with a constant velocity between 20 and 40 seconds, and finally has a steadily decreasing speed between 40 and 50 seconds.

The gradient of a line on a velocity-time graph represents the acceleration of the particle. You can see that the acceleration is uniform between 0 and 20 seconds and between 40 and 50 seconds.

The gradient of the line OA gives the acceleration as $\frac{30}{20}$, i.e. 1.5 ms^{-2}. The line BC slopes downwards with a negative gradient. This represents an object that is steadily slowing down. The acceleration is negative. A negative acceleration is called a deceleration (or retardation). In this case deceleration is $\frac{30}{10}$, i.e. 3 ms^{-2}.

The area under velocity-time graph represents the distance travelled by the particle. The graph and t- axis formed a trapezium. The area of the trapezium gives the total distance travelled by the particle.

Note that the distance moved in each phase of the journey is equal to the area between the corresponding graph and the time-axis.

Using the formula $A = \frac{1}{2}h(a + b)$ with $a = 20$, $b = 50$ and $h = 30$ we have

Total distance $= \frac{1}{2} \times 30 \times (20 + 50) = 1050$ m

The same result is obtained by considering the area under the graph as the sum of the areas of two triangles and the area of a rectangle.

Total distance = $\left(\frac{1}{2} \times 30 \times 20\right) + (20 \times 30) + \left(\frac{1}{2} \times 10 \times 30\right)$

$= 300 + 600 + 150$

$= 1050$ m

Example 2

(a) A car travels on a horizontal road from village A to another village B. The car starts from rest and accelerates uniformly for 20 seconds reaching a speed of 16 ms^{-1}. It maintains this speed for 15 seconds and then decelerates uniformly coming to rest in 15 seconds. Find:

(i) the distance AB (ii) the average speed of the journey

(iii) the acceleration during the first 20 seconds

Solution The motion of the car is shown in the velocity-time graph.

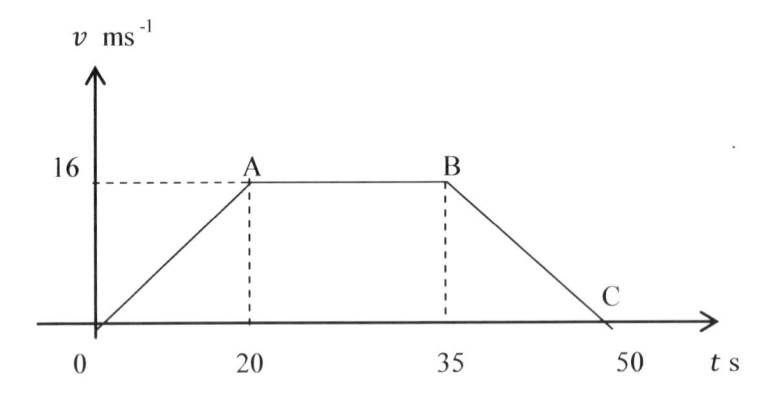

(i) Area of the trapezium = $\frac{1}{2} \times 16 \times (15 + 50)$

$= 520$ m

The distance AB is 520 m

(ii) Average speed = $\frac{total\ distanc}{time}$

$= \frac{520}{50}$

= 10.4

The average speed is 10.4 ms^{-1}

(iii) The acceleration is given by the gradient of the line OA

$$\text{Gradient} = \frac{16}{20}$$

$$= 0.8$$

The acceleration during the first 20 seconds is 0.8 ms^{-2}

(b) A train at a station P accelerates uniformly from rest until it attains a speed of 150 km h^{-1}. It then continues at this speed for some time and decelerates uniformly until it comes to a stop at a station Q 120 km from P. The total time taken for the journey is 1 hour. If the rate of deceleration is twice that of acceleration calculate the time taken during which the constant speed is maintained and the acceleration of the train.

Solution

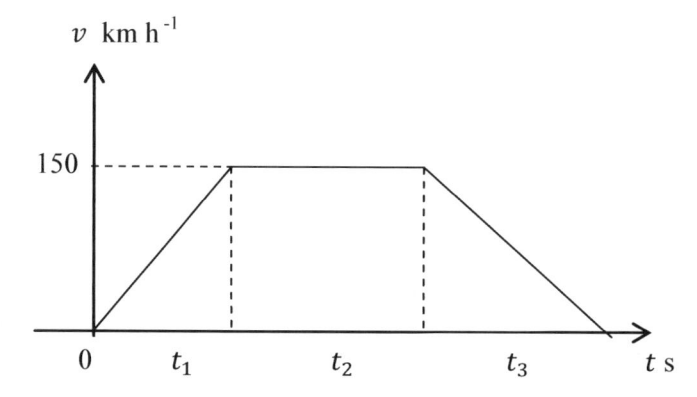

Suppose the train takes t_1, t_2 and t_3 hours for the three parts of the journey as indicated in the diagram above. The train travels at constant speed for t_2 hours. The area under the graph represents the distance travelled.

Therefore $\frac{1}{2}(t_2 + 1) \times 150 = 120$

$$t_2 = \frac{3}{5}$$

Hence the train travels at constant speed for $\frac{3}{5}$ hours or 36 minutes.

The time taken to decelerates is $\frac{1}{2}$ the time it takes to accelerates so $t_3 = \frac{1}{2}t_1$. We know that the journey takes 1 hour so

$$t_1 + \frac{3}{5} + \frac{1}{2}t_1 = 1$$

which simplifies to
$$\frac{3}{2}t_1 = \frac{2}{5}$$

giving
$$t_1 = \frac{4}{15}$$

Hence the time it takes to accelerates is $\frac{4}{15}$ hours or 16 minutes.

Therefore $a = \dfrac{150}{\frac{4}{15}}$

$= 562.5$

Hence the acceleration is 562.5 km h^{-2}.

Alternatively, using the equation $v = u + at$ the acceleration a m s^{-2} is given by

$$150 = 0 + a \times \frac{4}{15}$$

giving $a = 562.5$ km h^{-2}

Exercise 20.1(b)

1. A vehicle accelerates uniformly for 6 seconds, travels at a constant speed for 15 seconds, and then decelerates uniformly to rest in 4 seconds. If the total distance travelled is 800 metres, find the maximum speed.

2. A car accelerates at 3 ms^{-2} for 5 seconds, then travels at a constant speed for 16 seconds, and then decelerates uniformly to rest. The total distance travelled is 300 m. Draw a velocity-time graph and hence find the deceleration.

3. A cyclist starts from rest, and accelerates uniformly to a maximum speed of 15 ms^{-1}. This speed is maintained for 20 seconds, and then the cyclist decelerates uniformly to rest in 2 s. The total distance covered is 375 metres. Sketch a velocity-time graph. Find the total time taken and the acceleration.

4. A train starts from rest at a station P and accelerates uniformly until it reaches a velocity of 36 km h^{-1} after 3 hours. It maintains this velocity for some time and then retards uniformly for 2 hours to come to rest at station Q. If the distance between P and Q is 270 kilometres, find:

(a) the time the train travelled a maximum velocity

(b) the acceleration of the train

(c) the retardation of the train

5. A train starts from rest from station P and accelerates uniformly for 3 minutes reaching a speed of 60 km h^{-1}. It maintains this speed for 5 minutes and then retards uniformly for 2 minutes to come to rest at station Q. Find:

(a) the distance PQ in kilometres   (b) the average speed of the train

(c) the acceleration in ms^{-2}

6. A train starts from station X and accelerates uniformly to a speed of 20 ms^{-1}. It maintains this speed and then retards uniformly until it comes to rest at a station Y. The distance between the stations is 2 kilometres. The total time taken is 2 minutes. If the retardation is twice the acceleration in magnitude, find:

(a) the time for which the train is travelling at constant speed

(b) the acceleration of the train

Vertical Motion under Gravity

A special case of motion in a straight line with uniform acceleration is a body moving in space under the influence of gravitational force.

The acceleration of the body is then the acceleration due to gravity, approximately 9.8 m s^{-2}, but this differs slightly at different places on the earth's surface. The acceleration due to gravity is represented by g.

Example 3

A particle is projected vertically upwards with a velocity of 30 m s^{-1} from a point O, find:

(a) the maximum height reached

Solution   The positive direction is taken to be vertically upwards, making the distance, velocity and acceleration downwards negative. The initial velocity u is 30 m s^{-1} and the acceleration a is -10 m s^{-2}.

At the maximum height h m, the velocity v of the particle is 0.

Using the equation $v^2 = u^2 + 2as$ we have

$$0 = 30^2 + 2 \times (-10) \times h$$

giving $h = 45$

Hence the maximum height reached is 45 m

(b) the time taken to reach the maximum height

Solution Using the equation $v = u + at$ we have

$$0 = 30 + (-10) \times t$$

giving $\quad t = 3$

Hence the time taken to reach the maximum height is 3 s.

(c) the time taken for it to return to O

Solution When the particle returns to O, then the displacement s is 0.

Using the equation $s = ut + \frac{1}{2}at^2$ we have

$$0 = 30 \times t + \frac{1}{2} \times (-10) \times t^2$$

$$5t^2 - 30t = 0$$

Solving this gives $t = 0$ or $t = 6$

Hence the time taken for it to return to O is 6 s.

You can see from this result that the time taken to return to O is twice the time taken to reach the greatest height. That is, the time spent moving upwards is the same as the time spent moving downwards.

(d) its velocity when it returns to O

Solution Using the equation $v = u + at$ we have

$$v = 30 + (-10) \times 6$$

giving $\quad v = -30$

Hence the velocity when it returns to O is 30 m s^{-1}.

Notice that the magnitude of the velocity when it returns to O is the same as the magnitude of the velocity of projection.

Example 4

A stone is dropped from the top of a building of height 45 m. Find the time it takes to reach the ground and the velocity with which it hits the ground. [Take g = 10 m s^{-2}]

Solution Here the positive direction is taken to be vertically downwards, making upwards distance, velocity and acceleration negative.

initial velocity is $u = 0$ and the acceleration is $a = 10$.

Using the equation $s = ut + \frac{1}{2}at^2$ we have

$$45 = 0 + \frac{1}{2} \times 10 \times t^2$$

giving $\quad t^2 = 9$

$\quad\quad\quad t = 3 \quad\quad\quad$ Take only the positive value of t.

Hence the time taken is 3 s.

Using the equation $v = u + at$ we have

$$v = 0 + 10 \times 3 = 30$$

Hence the velocity is 30 m s^{-1} when it hit the ground.

Exercise 20.1(c)

In Problem 1 – 10, take $g = 10$ m s^{-2}

1. A body is dropped from the top of a building. How far has it fallen in metres at the end of 1 s, 2 s and 5 s?

2. A particle is projected downwards with a velocity of 25 m s^{-1}. How far has it travelled after 4 seconds? How far does it travel in the next 4 seconds?

3. An object is projected upwards with a velocity of 180 m s^{-1}. Calculate its position after 10 seconds. Find the total time that has elapsed when it has returned to that position again.

4. A body falls from the top of a 80 m cliff. Find the time taken and the speed at the bottom.

5. A body is thrown down from a 120 metre-cliff at 10 m s^{-1}. Find the time it takes to reach the bottom and its speed at the bottom.

6. A cliff is 100 m high. A particle is projected upwards from the top with a velocity of 40 m s^{-1}. How long does it take to reach the sea?

7. A particle is projected vertically upwards with a velocity of 30 m s^{-1} from a point O, find:

(a) the maximum height reached

(b) the time taken for it to return to O

(c) the time taken for it to be 35 m below O

8. From the top of a 50 m cliff a body is thrown up at 15 m s^{-1}, so that it then falls over the edge of the cliff. Find the greatest height reached and the time it take to reach the bottom of the cliff. Find its speed at the bottom of the cliff.

9. A particle P is projected vertically upwards from O with velocity 50 m s^{-1}. One second later another particle Q is projected from O with the same vertical velocity. After what time and at what height will the two particles collide?

10. A body is thrown up in the air at 20 m s^{-1}, and 2 seconds later a second body is thrown up at 50 m s^{-1}. Find how high the bodies are when they collide.

20.2 Composition of Velocities

We can compose two or more velocities into a single resultant velocity. Consider two particles moving with velocities as illustrated in Figure 20.4 (a). The resultant of the two velocities is the vector sum of the two velocities. This is illustrated in Figure 20.4(b).

Figure 20.4(a)

Figure 20.4(b)

Generally, the resultant of two or more velocities can be found by drawing lines representing the velocities in magnitude and direction to form a polygon. The line which completes the polygon represents the resultant velocity.

Resolution of Velocities

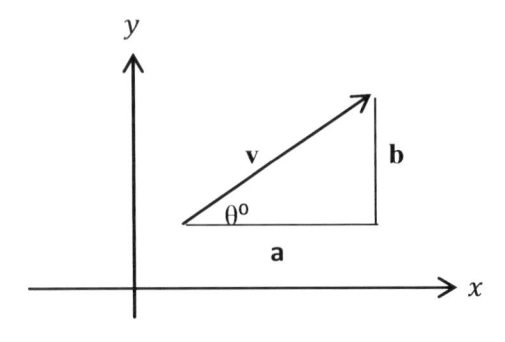

Figure 20.5

Figure 20.5 shows a velocity vector **v** resolved into two perpendicular components **a** and **b** along the x- direction and the y- direction respectively.

Mechanics

The vector **v** makes an angle θ with vector **a**. By trigonometry $|a| = |v|\cos\theta$ and $|b| = |v|\sin\theta$, so **v** can be written as $v = \begin{pmatrix} v\cos\theta \\ v\sin\theta \end{pmatrix}$. The magnitude of **v** is given by $|v| = \sqrt{a^2 + b^2}$. From the triangle $\tan\theta = \frac{b}{a}$. Hence, the angle which **v** makes with horizontal direction is $\tan^{-1}\left(\frac{b}{a}\right)$.

The resultant of two or more velocities can be obtained by resolving each velocity and then adding the corresponding components. The resultant velocity can also be obtained by scale drawing.

Example 5

Two particles travel with velocities **u** = (5 ms⁻¹, 050°) and **v** = (12 ms⁻¹, 150°). Calculate the magnitude and direction of the resultant velocity.

Solution The diagram below illustrates the velocity vectors. The resultant is **u** + **v**

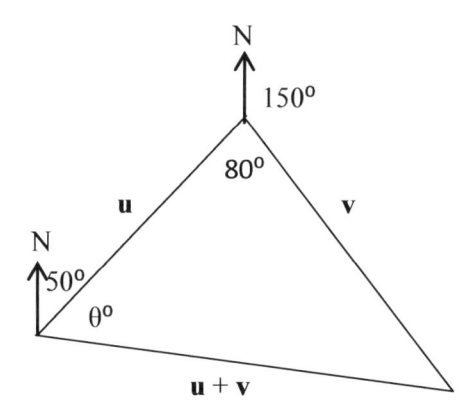

Figure 20.6

Using vectors

Resolving both **u** and **v**, we have

$$u + v = \begin{pmatrix} 5\cos 40° \\ 5\sin 40° \end{pmatrix} + \begin{pmatrix} 12\cos 60° \\ -12\sin 60° \end{pmatrix}$$

$$= \begin{pmatrix} 3.83 \\ 3.21 \end{pmatrix} + \begin{pmatrix} 6 \\ -10.39 \end{pmatrix}$$

$$= \begin{pmatrix} 9.83 \\ -7.18 \end{pmatrix}$$

$$|u + v| = \sqrt{9.83^2 + (-7.18)^2}$$

$$= \sqrt{148.1813}$$

$= 12.2 \text{ ms}^{-1}$

The direction of the resultant velocity can be determined from the diagram below.

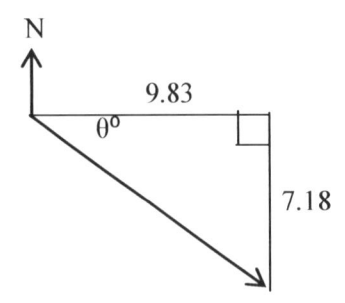

From the triangle

$$\tan \theta = \frac{7.18}{9.83}$$

$$= 0.7304$$

$$\theta = 36.1°$$

The direction of the resultant velocity is 126^0

Using Trigonometry

$$|u + v|^2 = 5^2 + 12^2 - 2(5)(12) \cos 80°$$

$$= 148.16$$

$$|u + v| = \sqrt{148.16}$$

$$= 12.2 \text{ ms}^{-1}$$

From Figure 13.18

$$\frac{\sin \theta}{12} = \frac{\sin 80°}{12.2}$$

$$\sin \theta = \frac{12 \sin 80°}{12.2}$$

$$= 0.9687$$

$$\theta = 75.6°$$

The direction of the resultant velocity is 126^0

Relative Velocity

There are many situations where a velocity is made up of two other velocities. Relative velocity is the velocity of one body as seen from another.

Suppose an airplane travelling with velocity **a** north encounter a wind blowing from the west with velocity **b** as illustrated in Figure 20.7(a)

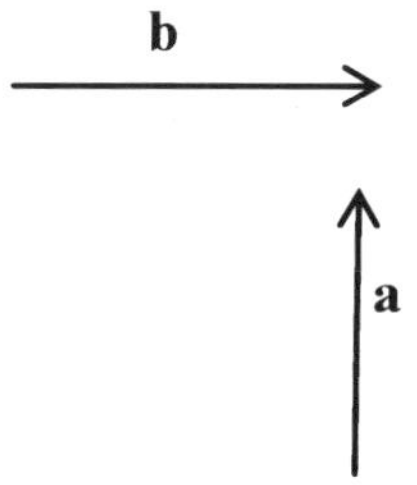

Figure 20.7(a) 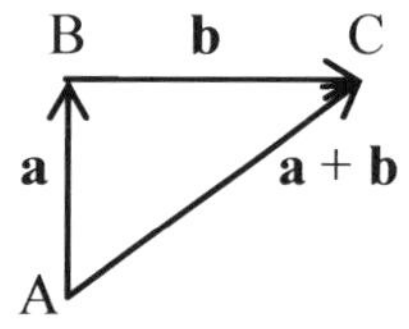   Figure 20.7(b)

An observer on the ground would see the airplane as moving along $\overrightarrow{AC}$, as shown in Figure 20.7 (b). The vector $\overrightarrow{AC}$ represents the velocity relative to the ground, called the ground speed, and the direction of the plane as seen from the ground, called the track. Figure 20.7(b) shows that the velocity of the airplane relative to the ground is the vector sum of the velocity of the airplane relative to the air, called the airspeed and the velocity of the wind relative to the ground, called the wind speed. The direction in which the pilot steers an airplane is called its course. The various velocities and directions are summarised in the following table.

Velocity	Air speed	Wind speed	Ground speed
Direction	Course	Wind direction	Track

The effect of the river current upon a boat is similar to the effect of the wind upon a plane. The velocity of a boat as seen by an observer on the ground is the vector sum of the velocity of the boat in the water and the velocity of the current.

Notice that a current is described as moving towards a certain direction whereas a wind is described as coming from a certain direction.

Example 6

(a) An airplane moves on a bearing of $315°$ at 125 km h^{-1} relative to the ground due to the fact that there is a wind blowing from the west of 50 km h^{-1} relative to the ground. How fast and in what direction will the plane have travelled in still air?

Solution The diagram below illustrates the velocity vectors for the airplane and wind.

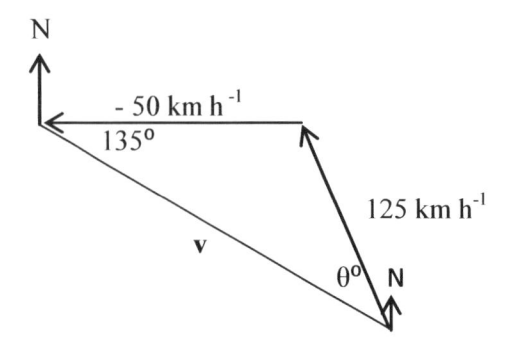

Figure 20.8

Let the airplane's velocity in still air be **v**.

Using vectors

$$v = \begin{pmatrix} -125 \cos 45° \\ 125 \sin 45° \end{pmatrix} + \begin{pmatrix} -50 \\ 0 \end{pmatrix}$$

$$= \begin{pmatrix} -138.39 \\ 88.39 \end{pmatrix}$$

$$|v| = \sqrt{(-138.39)^2 + 88.39^2}$$

$$= \sqrt{26964.12}$$

$$= 164.2 \text{ km h}^{-1}$$

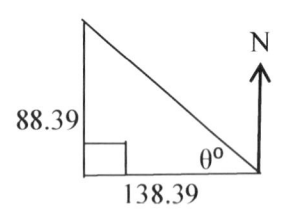

From the triangle

$$\tan \theta = \frac{88.39}{138.39}$$

$$= 0.6387$$

$$\theta = 32.6°$$

Direction of plane in still air is $303°$

Using Trigonometry

$$v^2 = 50^2 + 125^2 - 2(50)(125) \cos 135°$$

$$= 26963.83$$

$$v = \sqrt{26963.83}$$

$$= 164.2 \text{ km h}^{-1}$$

From Figure 20.8

$$\frac{\sin \theta}{50} = \frac{\sin 135°}{164.2}$$

$$\sin \theta = 0.2153$$

$$\theta = 12.4°$$

Direction of the airplane in still air is $303°$

(b) A motor boat travelling east from P on one bank of a river with a speed of 8 ms^{-1} encounters a current flowing to the north with speed of 5 ms^{-1}.

(i) What is the velocity of the boat relative to an observer on the shore?

Solution The vector triangle of velocities is shown below. Let **v** be the velocity of the boat relative to the observer.

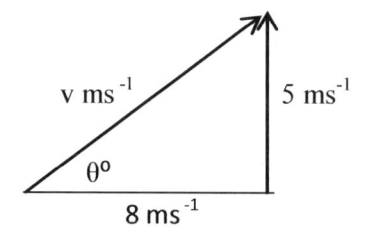

The resultant velocity is the vector sum of the two individual velocities. The magnitude of the resultant can be determined using Pythagoras' theorem.
$$v^2 = 8^3 + 5^2$$

$$= 89$$

$$v = 9.4$$

The velocity of the boat relative to the observer is 9.4 ms^{-1}

(ii) A point Q on the other bank is directly opposite P. If the width of the river is 120 metres wide, how long does it take the boat to travel from P to Q.

The time to travel from P to Q can be determined using the average speed equation $s = \frac{d}{t}$.
Rearranging this equation we get $t = \frac{d}{s}$.

Substitute $d = 120$ and $v = 8$ into the equation for t to get

$t = \frac{120}{8} = 15$

The time it takes the boat to travel from P to Q is 15 s.

Exercise 20.2

1. Two boats leave a harbour at 8:00 am with velocities **u** = (20 km h^{-1}, 180°) and **v** = (15 km h^{-1}, 060°) respectively. Calculate

(a) the resultant of their velocities

(b) the distance between the boats at 9:00 am.

2. A pilot is steering an aircraft due east and its airspeed is 600 km h^{-1}. There is a wind blowing from the south at a speed of 50 km h^{-1}. Find the direction in which the aircraft travels and its speed over the ground.

3. A pilot was steering his aircraft on a bearing of 136° and his airspeed indicator showed 200 km h^{-1}. However there was a wind blowing at 50 km h^{-1} from the west. What was its speed over the ground and its direction?

4. A helicopter leaves an airfield A to fly to B 500 km away on a bearing of 140°. There is a steady wind of 30 km h^{-1} from NE. The helicopter has airspeed of 150 km h^{-1}. Find the course the pilot must take and the time taken for him to reach B.

5. A current flows at 5 km h^{-1} on a bearing of 150°. A boat which can travel at 12 km h^{-1} is to sail 60 km due east. In what direction should it be steered, and how long will the journey take?

6. A pilot steered an aircraft on a bearing 060°. There was a 45 km h^{-1} wind blowing from the south-east. If the airspeed of the aircraft was 235 km h^{-1}, find the aircraft's track and its ground speed

7. A river is flowing at 2 ms^{-1}. A man can row at 2.5 ms^{-1}. In which direction should he row if he is to cross the river directly? If the river is 60 metres wide how long will it take him?

8. A ship sets off on a bearing of 147° at a speed of 30 km h^{-1} through the water. The current flows at 5 km h^{-1} in the direction 083°. Find the direction the ship travels and the distance of the ship after 3 hours?

9. A motor boat can travel at 8 ms^{-1} in still water. The boat travels from P on one bank of a river flowing north at 5 ms^{-1}.

(a) Find the speed of the boat in the water

(b) Find the angle to the bank that the boat should travel in order to reach a point Q which is directly opposite P

10. A river with parallel banks is 50 m wide and is flowing east at 1.2 ms^{-1}. A point Q on the other bank is directly opposite a point P. If a boat takes 10 seconds to travel in a straight line from P to Q, calculate the speed of the boat in still water and the angle to the bank at which the boat should be steered

11. A canoe travelling 12 ms^{-1} East encounters a current travelling 5 ms^{-1} North. The river is 90 metres wide.

(a) What is the resultant velocity of the canoe?

(b) How long does it take the canoe to travel to a point which is directly opposite its starting point?

(c) What distance downstream does the boat reach the opposite shoe?

12. An airplane whose speed in still air is 450 km h^{-1} travels directly from A to B, a distance of 1200 km. The bearing of B from A is $215°$ and there is a wind of 60 km h^{-1} from the east

(a) Find the bearing on which the plane was steered.

(b) Find the time taken for the journey.

20.3 Forces

Force and Acceleration

We shall consider some problems of forces acting on a particle and the motion they produce. We begin by stating the following Newton's Laws.

Newton's Laws

Newton's First Law states that:

Everybody remains at rest or moves with uniform velocity unless it is made to change this state by external forces.

Newton's Second Law states that:

If a force acts on a body and produces a constant acceleration, then the force is proportional to the product of the mass of the body and the acceleration. Also the acceleration takes place in the direction of the force.

Newton's Second Law can be summarised by the formula

$F = ma$

F like acceleration is a vector. The unit of force is the newton, N. 1 N is that force which gives a mass of 1 kg an acceleration of 1 m s^{-2}.

Weight

A mass of m kg if released from rest would fall vertically with a constant acceleration g m s^{-2}. The force that produces this acceleration is the gravitational pull of the earth on the particle. Using Newton's Second law, $F = ma$, the force acting on the body is thus mg N, and this is known as the weight of the body. For example, the weight of a particle of mass 1 kg is approximately 9.8 N.

The weight of a particle varies with locality, depending on the value of g which varies slightly over the surface of the earth.

Newton's Third Law states that:

To every action there is an equal and opposite reaction.

Whenever a box rests on the ground, it is acted upon by at least one force, its own weight, W. The box is kept in equilibrium by a force N, exerted by the ground on the box, which acts vertically upwards and it is equal to the weight of the box, i.e. $W = N$. The force N, is called the normal reaction.

Generally, whenever a particle is in contact with a smooth plane, the particle will exert a force on the plane at right angle to the plane and the plane will exert on the particle a force, known as the normal reaction, of equal magnitude but opposite in direction.

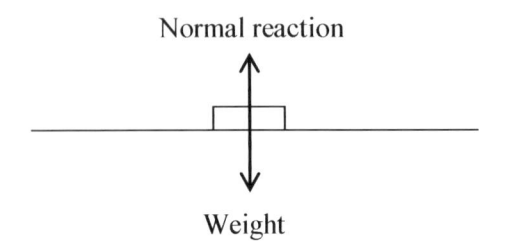

Figure 20.9

Example 7

(a) A particle of mass 10 kg is acted on by a force of 15 N for 4 s. If it's initial velocity is 2 m s^{-1}, what is it's velocity at the end of the 4 s period.

Solution Using $F = ma$, gives

$$15 = 10a$$

$$a = 1.5 \text{ m s}^{-2}$$

Since F is constant, the acceleration is constant and so we can use the equations for uniformly accelerated motion. Using $v = u + at$, with $u = 2$ m s^{-1} and $t = 4$ s gives

$$v = 2 + 1.5 \times 4$$

$$= 8$$

Therefore it's velocity after 4 s is 8 m s^{-1}.

(b) The engine of a car of mass 1.5 tonnes travelling on a straight level road exerts a constant pull of magnitude P newtons. The car accelerates from a speed of 36 km h^{-1} to a speed of 108 km h^{-1} in 12 s. If there is a constant resistance to the motion of 200 newtons, find the value of P.

Solution We first convert the speed from km h^{-1} to m s^{-1}. 36 km h^{-1} = 10 m s^{-1} and 108 km h^{-1} = 30 m s^{-1}

Using $v = u + at$ with $u = 10$, $v = 30$ and $t = 12$ gives

$$30 = 10 + 12a$$

giving $\qquad a = \dfrac{5}{3} \text{ m s}^{-2}$

Now the resultant force is $P - 200$.

Hence $\qquad P - 200 = 1500a$

$$P = 1500 \left(\dfrac{5}{3}\right) + 200$$

$$= 2700 \text{ N}$$

(c) A box of mass 50 kg is placed on the floor of a lift. Find the reaction of the floor of the lift on the box when the:

(i) lift is ascending with constant speed

Solution There are only two forces acting on the box, its own weight, mg N and the normal reaction of magnitude N newtons. As long as the box is in contact with the floor of the lift, the box will have the speed and acceleration of the lift. From $F = ma$ we get

$$N - mg = ma$$

Since the speed is constant $a = 0$. Taking g to be 10 m s^{-2} we get

$$N - 50 \times 10 = 50 \times 0$$

$$N - 500 = 0$$

$$N = 500 \text{ N}$$

(ii) moving with acceleration 4 m s^{-2} upwards

Solution Taking the upward direction as positive we have

$$N - 500 = 50 \times 4$$

$$N = 700 \text{ N}$$

(iii) moving with acceleration 4 m s^{-2} downwards

Solution The downward acceleration is negative

Thus $\quad N - 500 = 50 \times -4$

$$N = 300 \text{ N}$$

Exercise 20.3(a)

1. A body of mass 4 kg is accelerated at 15 m s^{-2}. Find the force applied to it.

2. A body experience a force of 30 N, and accelerates at 20 m s^{-2}. Find the mass of the body.

3. A mass of 3 kg receives a force of 12 N. Find its acceleration.

4. A stone of mass 5 kg is skimmed across the ice at 10 m s^{-1} and comes to rest in 100 m. Find the force acting upon it, assuming that it is constant.

5. A body of mass 65 kg is dragged across the floor at a constant speed of $\frac{1}{2}$ m s^{-1}. If the force dragging it is 30 N, find the resistance force.

6. A train of mass 24,000 kg comes to rest from 5 m s^{-1} in 30 seconds. Find the breaking force, assuming that it is constant.

7. A car of mass 1,500 kg accelerates from 3 m s^{-2} to 15 m s^{-2} in 20 seconds. Find the tractive force, assuming that it is constant.

8. A sprinter of mass 80 kg accelerates constantly over 50 metres reaching a speed of 15 m s^{-1} at the end. Find the force that he exerts.

9. A train of mass 50,000 kg accelerates constantly from rest until it reaches 20 m s^{-1} over a distance of 400 m, find the tractive force of the train.

10. A train of mass 81,000 kg is brought to rest from a speed of $\frac{1}{3}$ m s^{-1} by buffers which contract through 30 cm. Find the force, assumed constant, exerted by the buffers.

11. A parachutist of mass 200 kg falls at a steady speed of 1.5 m s^{-1}. What is the force of air resistance?

12. A lift cage weighs 1,000 kg. The tension in the supporting cable is 9,300 N. Find the deceleration of the lift.

13. A parachutist has mass 100 kg and is falling at 2 m s^{-1}. Air resistance is equal to 600 N at this speed. Find his acceleration at this speed.

14. A man weighs out diamonds in a lift which is accelerating upwards at 2 m s^{-2}. The balance registers 25 grams. What is the true weight of the diamonds? [Take g to be 9.8 m s^{-2}]

15. A 80 kg man weighs himself on a weighing machine which is in a lift accelerating up at 1.5 m s^{-2}. What does he seem to weigh?

16. A 60 kg woman weighs herself in a lift and the dial registers 45 kg. What is the deceleration of the lift?

17. When a man weighs himself in a lift accelerating up at 2 m s^{-2} the dial registers 75 kg. What is his true weight?

18. A body of mass 2 kg experiences a force of $(4\mathbf{i} + 3\mathbf{j})$ N. Find its acceleration in vector form.

19. A bucket has a mass of 25 kg when it is full of water. Find its acceleration upwards when the tension in the rope is 300 N.

20. A boy lifted a box off the ground with an initial acceleration of 2.5 m s^{-2}. If the box had a mass of 12 kg, what vertical force did the boy applied?

Combination of Forces

Forces like vectors can be combined to produce a single resultant force.

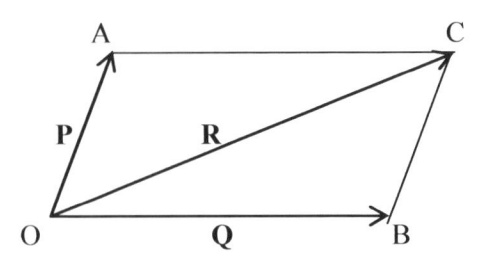

Figure 20.10

Consider the two forces **P** and **Q** represented by the adjacent sides OA and OB of the parallelogram OBCA. OC represents the magnitude and direction of the resultant force **R** of the forces **P** and **Q**. Thus **R** = **P** + **Q**. In general, the resultant of two or more forces acting on an object is simply the sum of the force vectors.

Determination of Resultant Forces by Scale Diagram

Scale diagram can be used to compose two or more forces to obtain a single force called the resultant force.

Example 8

A force of 4 N is inclined at an angle of 45° to a second force of 7 N, both forces acting at a point. Find the magnitude of the resultant of these two forces and the direction of the resultant with respect to the 7 N force.

Solution

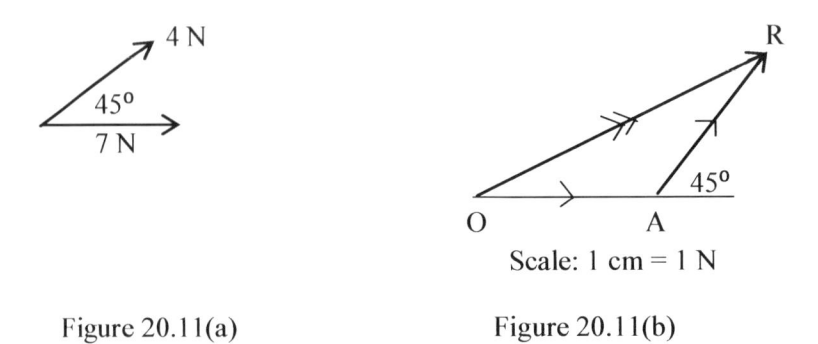

Figure 20.11(a)   Figure 20.11(b)

Scale: 1 cm = 1 N

We draw a force diagram to scale, using a scale of 1 N to 1 cm. $\overrightarrow{OA}$ represents the 7 N force in both magnitude and direction and $\overrightarrow{AR}$ represents the 4 N force in both magnitude and direction.

The line OR represents the resultant force, and by measurement $|OR| = 10.2$ cm long and makes an angle of 16° with OA. Hence, the resultant force is 10.2 N making an angle of 16° with the 7 N force.

(b) Forces 2 N and 5 N are at an angle of 67°. A force of 7 N makes an angle of 120° with the second force on the opposite side to the first. Find the resultant force.

Solution

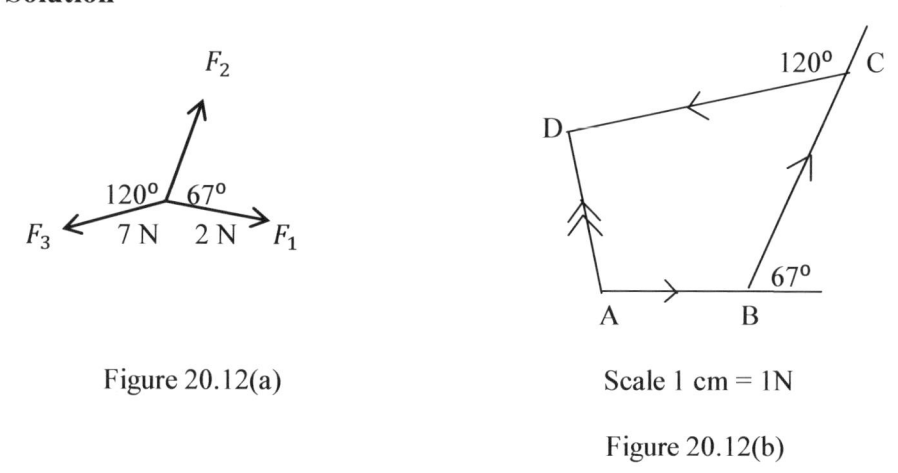

Figure 20.12(a)

Scale 1 cm = 1N

Figure 20.12(b)

$\vec{AB}$ is drawn to represent F_1, as shown in Figure 20.12(b). F_2 is added to F_1 by drawing $\vec{BC}$. F_1 and F_2 are at an angle of 67°. Finally F_3 is added to $F_1 + F_2$. F_3 makes an angle of 120° with F_2. The resultant of the vector addition is represented by $\vec{AD}$. By measurement $|AD|$ is 4.8 cm long and makes an angle of 128° with F_1. Hence the resultant force is 4.8 N making an angle 128° with the force F_1.

Note that a different order can be taken to obtain the same result.

Exercise 20.3(b)

1. Find, by drawing, the resultant of each pair of forces acting at a point and the angle it makes with the first force in each of the following cases:

Forces	Angle between forces
(a) 4 N, 7 N	45°
(b) 3 N, 4 N	60°
(c) 18 N, 15 N	150°
(d) 15 N, 20 N	120°
(e) 20 N, 25 N	70°

2. Forces of 30 N and 10 N act at a point and are inclined at 40° to each other. Find, by drawing, the resultant force and its direction relative to 10 N force.

3. Find, by drawing, the resultant of two forces $F_1 = (40\ N, 045°)$ and $F_2 = (30\ N, 125°)$ which act at a point.

4. A force 2 N makes an angle of 50° with a force of 5 N. A third force of 3 N makes an angle of 90° with the first on the side as the second. Determine graphically the resultant force.

5. Forces A, B and C are coplanar and act at a point. Force A is 15 N at 90°, B is 8 N at 180° and C is 17 N at 293°. Determine graphically the resultant force.

Resolution of Forces

We have just learned to compose two or more forces into a single resultant force. We shall now reserve the process and split a force into two components which are at right angles to each other.

Figure 20.13(a)

Figure 20.13(b)

The vector **F** can be resolved into two component vectors, one along OA and the other along OB as shown in Figure 20.13(b). By trigonometry the component along OA is $F\cos\theta$ and the one along OB is $F\sin\theta$, where θ is the angle between the force **F** and the line OA. Using these results, we show that the horizontal component of a force of 8 N making an angle of 30° with the line OA (see Figure 20.14(a)) is $8\cos 30°$ N, i.e. $4\sqrt{3}$ N, and the perpendicular component is $8\sin 30°$ N, i.e. 4 N.

Figure 20.14(a)

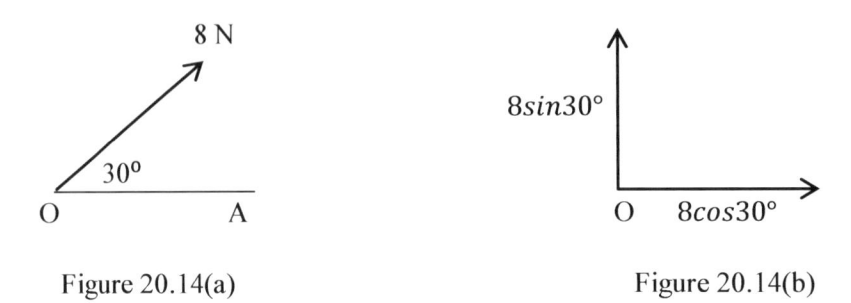

Figure 20.14(b)

Mechanics 487

The vectors F_1 and F_2, shown in Figure 20.15(a) makes angles θ_1 and θ_2 respectively with the horizontal.

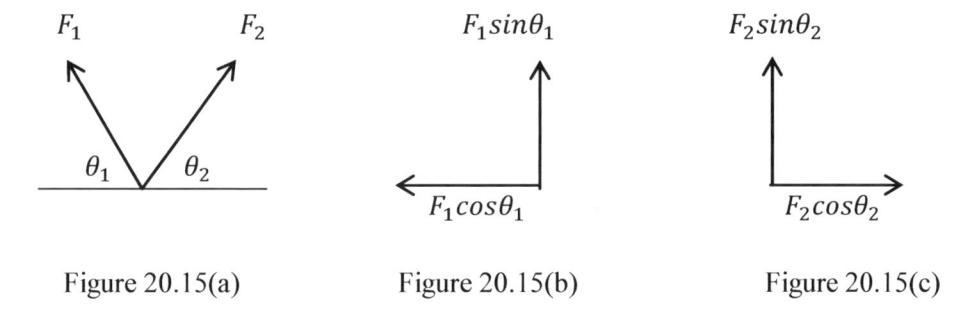

Figure 20.15(a) Figure 20.15(b) Figure 20.15(c)

The horizontal and the vertical components of F_1 and F_2 are shown separately in Figure 20.15(b) and Figure 20.15(c) respectively. The net horizontal force $H = F_2 cos\theta_2 - F_1 cos\theta_1$ and the net vertical force $V = F_1 sin\theta_1 + F_2 sin\theta_2$.

The magnitude of the resultant vector **R** is given by $|R| = \sqrt{H^2 + V^2}$ and its angle to the horizontal is given by $tan^{-1}\left(\frac{V}{H}\right)$. As shown in Example 9, these results can be used to find the resultant of forces arithmetically.

Example 9

A force of 10 N is inclined at an angle of 60° to a second force of 8 N, both forces acting at a point. Find the magnitude of the resultant of these forces and the direction of the resultant force with respect to the 8 N force.

Solution The two forces are resolved as shown in Figure 20.16(b).

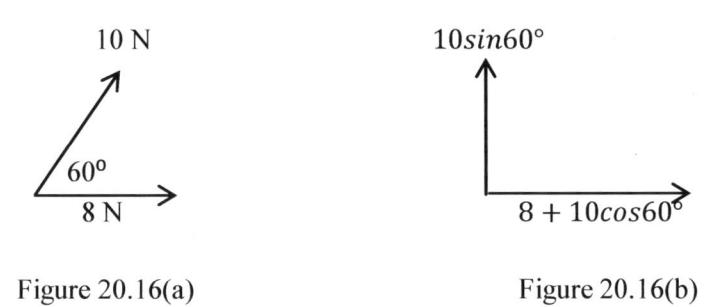

Figure 20.16(a) Figure 20.16(b)

Notice that the net horizontal force is $H = 8 + 10cos60° = 13$ N, and the net vertical force is $V = 10sin60° = 8.66$ N.

Hence the magnitude of the resultant force $R = \sqrt{13^2 + 8.66^2} = 15.6$ N.

The direction of the resultant force is $tan^{-1}\left(\frac{8.66}{13}\right) = 33.7°$. Thus, the resultant of the two forces is a single force of 15.6 N at 33.7° to the 8 N force.

Exercise 20.3(c)

1. For each pair of forces, determine by calculation the resultant force and its direction.

Forces	Angle between forces
(a) 4 N, 7 N	45°
(b) 3 N, 4 N	60°
(c) 18 N, 15 N	150°
(d) 15 N, 20 N	120°
(e) 20 N, 25 N	70°

2. Force, $F_1 = 1.5$ N at $90°$ and $F_2 = 2.6$ N at $45°$ act at a point. Determine by calculation the resultant force.

3. Find, by calculation the resultant of the two forces $F_1 = (4\,N.045°)$ and $F_2 = (3\,N, 125°)$ which act at a point.

4. Three forces, $F_1 = (4N, 030°)$, $F_2 = (3N, 090°)$ and $F_3 = (2N, 150°)$ act at a point. Calculate the magnitude of the resultant force and its direction relative to the 2 N force.

5. Forces A, B and C are coplanar and act at a point. Force A is 12 N at $90°$, B is 5 N at $180°$ and C is 13 N at $300°$. Calculate the result force.

Forces in Equilibrium

A particle at rest is said to be in a state of equilibrium. Intuition tells us that if a particle under the action of two forces is in equilibrium, then the two forces must be equal in magnitude but opposite in direction.

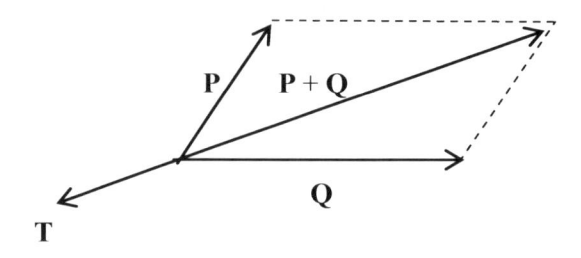

Figure 20.17

Consider two forces **P** and **Q** acting on a particle in equilibrium as shown in Figure 20.17. The force **T** that keeps the particle in equilibrium and the resultant force (**P** + **Q**) are equal in magnitude and opposite in direction, and so **T** = - (**P** + **Q**) or **T** + **P** + **Q** = 0.

In particular, a particle under the action of two or more forces would be in equilibrium if the vector sum of all the forces is zero.

The resultant of any number of coplanar forces acting on a particle can be found by resolving each of the forces into two components at right angles to each other and then adding the corresponding components to obtain the resultant force.

Example 10

(a) A body of mass 6 kg resting on a horizontal floor is acted upon by three forces $F_1 = (12\,N, 030°)$, $F_2 = (20\,N, 300°)$ and $F_3 = (15\,N, 180°)$. Find, the resultant force, correct to one decimal place and the initial acceleration of the body.

Solution Resolving the forces into components we have

$$F_1 = \begin{pmatrix} 12\cos 60° \\ 12\sin 60° \end{pmatrix} = \begin{pmatrix} 6 \\ 10.39 \end{pmatrix}$$

$$F_2 = \begin{pmatrix} -20\cos 30° \\ 20\sin 30° \end{pmatrix} = \begin{pmatrix} -17.32 \\ 10 \end{pmatrix}$$

and $\quad F_3 = \begin{pmatrix} 0 \\ -15 \end{pmatrix}$

If **R** is the resultant force then

$$R = F_1 + F_2 + F_3$$

So $\quad R = \begin{pmatrix} 6 \\ 10.39 \end{pmatrix} + \begin{pmatrix} -17.32 \\ 10 \end{pmatrix} + \begin{pmatrix} 0 \\ -15 \end{pmatrix} = \begin{pmatrix} -11.32 \\ 5.39 \end{pmatrix}$

The magnitude of the resultant force is given by

$$|R| = \sqrt{(-11.32)^2 + 5.39^2}$$

$$= 12.5\,N$$

Now we will find the direction of the force.

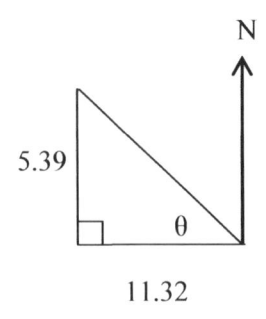

Figure 20.18

From Figure 20.18 we have

$$\tan\theta = \frac{5.39}{11.32}$$

$$= 0.4761$$

$$\theta = 25.5°$$

Direction of resultant force is 296°

Therefore the resultant force is $R = (12.5\ N, 296°)$

(b) Three forces $F_1 = (6N, 180°)$, $F_2 = (8N, 330°)$ and $F_3 = (10N, 060°)$ act on a body. Find the magnitude and direction of the single force needed to keep the body in equilibrium.

Solution Resolving the forces we have

$$F_1 = \begin{pmatrix} 0 \\ -6 \end{pmatrix}$$

$$F_2 = \begin{pmatrix} -8\cos 60° \\ 8\sin 60° \end{pmatrix} = \begin{pmatrix} -4 \\ 6.928 \end{pmatrix}$$

$$F_3 = \begin{pmatrix} 10\cos 30° \\ 10\sin 30° \end{pmatrix} = \begin{pmatrix} 8.66 \\ 5 \end{pmatrix}$$

So $\quad F_1 + F_2 + F_3 = \begin{pmatrix} 4.66 \\ 5.928 \end{pmatrix}$

Let the force needed to keep the body in equilibrium be F_4. Then

$$F_1 + F_2 + F_3 + F_4 = 0$$

Hence $\quad F_4 = \begin{pmatrix} -4.66 \\ -5.928 \end{pmatrix}$

The magnitude of F_4 is

$$|F_4| = \sqrt{(-4.66)^2 + (-5.928)^2}$$

$$= 7.5 \text{ N}$$

Now we will find the direction of the force.

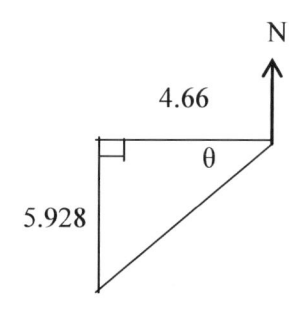

Figure 20.19

From Figure 20.19 we have

$$tan\theta = \frac{5.928}{4.66}$$

$$\theta = 51.8°$$

Direction of the force is 218°

The force needed is 7.5 N, and the direction of the force is 218°.

Exercise 20.3(d)

1. Find the resultant of the following set of forces

(a) (14 N, 060°), (22 N, 315°)   (b) (2 N, 048°), (3 N, 185°)

(c) (5 N, 065°), (16 N, 060°), (6 N, 215°)

2. Find the magnitude and direction of the single force needed to keep the following sets of forces in equilibrium.

(a) (18 N, 000°), (17 N, 270°)

(b) (3 N, 120°), (4 N, 210°)

(c) (12 N, 180°), (15 N, 060°), (20 N, 330°)

(d) (5 N, 090°), (16 N, 180°), (8 N, 030°), (12 N, 300°)

3. A body of mass 8 kg resting on a horizontal floor is acted upon by three forces $F_1 = (18\ N, 330°)$, $F_2 = (10\ N, 090°)$ and $F_3 = (25\ N, 180°)$. Find:

(a) the resultant force, correct to one decimal place

(b) it's initial acceleration

4. Four forces $P = (5\ N, 090°)$, $Q = (4\ N, 000°)$, $R = (3\sqrt{2}\ N, 045°)$ and $S = (3\sqrt{2}\ N, 135°)$ act on a body of mass 6 kg. Calculate to the nearest whole number:

(a) the magnitude and direction of the resultant force

(b) the acceleration with which the body begins to move

Triangle of Forces

Consider a particle O, in equilibrium, acted upon by three forces **P**, **Q** and **R** as shown in Figure 20.20(a).

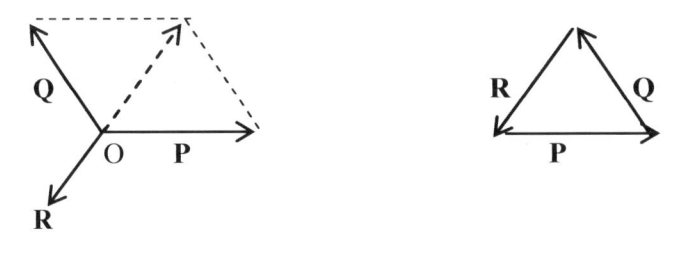

Figure 20.20(a)   Figure 20.20(b)

Since the forces are in equilibrium, the force **R** and the resultant force (**P** + **Q**) are equal in magnitude and opposite in direction. The three forces form the triangle shown in Figure 20.20(b), where the three forces **P**, **Q** and **R** are represented in magnitude and direction by the sides of the triangle taken in order. This triangle is called the triangle of forces.

The result of the preceding discussion can be summarized as follows.

If three forces act at a point and are in equilibrium then they can be represented in magnitude and direction by the three sides of a triangle taken in order.

Lami's Theorem

Figure 20.21(a) shows three forces **P**, **Q** and **R** in equilibrium with angle α, β and γ between them. They form the triangle of forces shown in Figure 20.21(b).

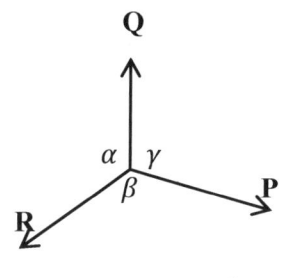

Figure 20.21(a)

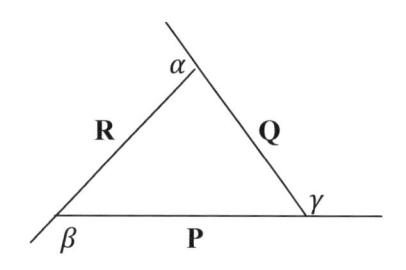

Figure 20.21(b)

Using Figure 20.21(b) and the sine rule we have

$$\frac{P}{\sin(180°-\alpha)} = \frac{Q}{\sin(180°-\beta)} = \frac{R}{\sin(180°-\gamma)}$$

So $\dfrac{P}{\sin\alpha} = \dfrac{Q}{\sin\beta} = \dfrac{R}{\sin\gamma}$

where $|P| = P$, $|Q| = Q$ and $|R| = R$.

This result is known as Lami's Theorem and may be started as follows:

If three forces acting at a point are in equilibrium then the magnitude of each force is proportional to the sine of the angle between the other two forces.

The triangle of forces and Lami's Theorem can be useful in solving problems involving particles in equilibrium under the action of three forces. If the magnitude of one of the forces and the direction of all three forces are known, then the magnitude of the other two forces can be found using either the triangle of forces or the Lami's Theorem.

Tensions

Consider a particle suspended from a ceiling by a string, as shown in Figure 20.22

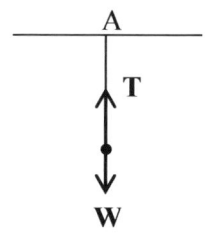

Figure 20.22

If the particle hangs at rest there will be no resultant force on the particle. However, the particle is acted upon by at least one force its own weight, **W**. This must be balanced by a force, **T** of equal magnitude acting in the opposite direction. The force **T** is the tension in the string, which acts vertically upwards.

Example 11

A body P of mass 10 kg is suspended by two light inextensible strings which are inclined at 40° and 60° respectively. Find the tensions in both strings? [Take $g = 10$ m s^{-2}].

Solution

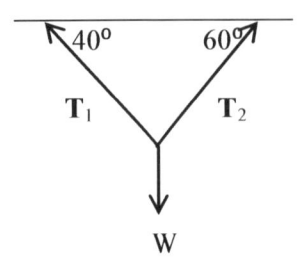

Figure 20.23(a)

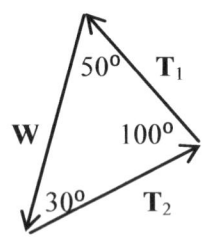

Figure 20.23(b)

Resolving the forces

Resolving the forces into components gives

$$\begin{pmatrix} -T_1 \cos 40° \\ T_1 \sin 40° \end{pmatrix} + \begin{pmatrix} T_2 \cos 60° \\ T_2 \sin 60° \end{pmatrix} + \begin{pmatrix} 0 \\ -100 \end{pmatrix} = \begin{pmatrix} 0 \\ 0 \end{pmatrix}$$

So $-0.766T_1 + 0.5T_2 = 0$

$0.5428T_1 + 0.866T_2 = 100$

Solving the simultaneous equations gives

$T_1 = 50.8$ N and $T_2 = 77.8$ N

Using the triangle of forces

Using the triangle of forces shown in Figure 20.23(b) we have

$$\frac{T_1}{\sin 30°} = \frac{100}{\sin 100°}$$

$$T_1 = \frac{100 \sin 30°}{\sin 100°}$$

$$= 50.8 \text{ N}$$

Also $\dfrac{T_2}{\sin 50°} = \dfrac{100}{\sin 100°}$

$$T_2 = \frac{100 \sin 50°}{\sin 100°}$$

$$= 77.8 \text{ N}$$

Using Lami's Theorem

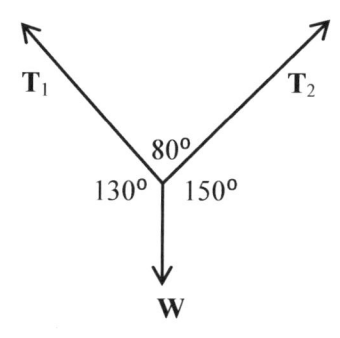

Figure 20.24

By the Lami's Theorem we have

$$\frac{T_1}{sin150°} = \frac{T_2}{sin130°} = \frac{100}{sin80°}$$

So $T_1 = \frac{100 sin150°}{sin80°}$

$= 50.8$ N

Also $T_2 = \frac{100 sin130°}{sin80°}$

$= 77.8$ N

Exercise 20.3(e)

1. A mass of 6 kg hangs at the end of a string and is pulled aside by a horizontal force of 25 N. Find the tension in the string and the angle it makes with the vertical.

2. A mass of 12 kg hangs at the end of a string and is pulled aside by a horizontal force so that the string makes 15° with the vertical. Find the force.

3. A mass of 8 kg hangs at the end of a string and is pulled aside by a horizontal force, so that the tension in the string is 100 N. Find the force.

4. A mass of 6 kg is supported by two strings, which makes 30° and 40° with the vertical. Find the tension in the strings.

5. A body of mass 5 kg is suspended by two light inextensible strings inclined at 40° and 50°. Find the tensions in the strings.

6. A mass of 6 kg is supported by two strings each of which exerts a tension of 36 N. Find the angles that strings make with the vertical.

7. A mass of 8 kg is supported by two strings. One string exerts 50 N and is at 25° to the vertical. Find the tension in the other string and the angle it makes with the vertical.

Friction

When two surfaces rub together there is a force which opposes the motion called friction.

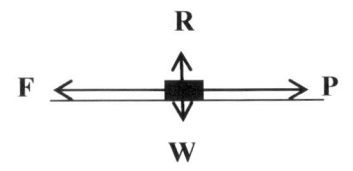

Figure 20.25

Consider a particle of weight **W** at rest on a rough horizontal plane, as shown in Figure 20.25. The normal reaction **R** of the plane on the particle has a magnitude equal to the magnitude of **W**. When a force **P** is applied horizontally to the particle, slowly increasing its magnitude from zero, the frictional force **F** always acts to oppose the movement of the particle. Initially, the particle does not move because the frictional force is equal (but opposite) to the applied force. As the force **P** is increased, the frictional force also increases to equal the applied force. Eventually, however, the frictional force reaches its maximum value, and a slight increase in **P** will make the particle move in the direction of the force **P**. The maximum value of the frictional force **F** is called the limiting friction.

For any given surfaces the value $\dfrac{limiting\ frictional\ force}{normal\ reaction}$

is a constant. This constant is known as coefficient of friction and is denoted by μ. The value of μ depends only on the nature of the surfaces in contact and is independent of the area in contact or the forces present.

Consider a particle of mass m kg at rest on a rough plane inclined at an angle of $\theta°$ to the horizontal.

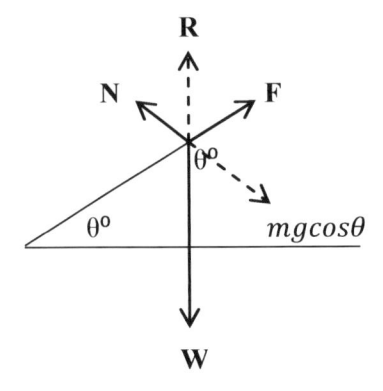

Figure 20.26

The particle has a weight mg N vertically downwards so the plane must be supporting it with a force **R** acting vertically upwards, called the reaction force. By Newton's third law, **R** must be equal in magnitude to the weight i.e. R = mg. The force **R** can be resolved into two components, the normal reaction, **N**, which is perpendicular to the plane and the force of friction, **F**, which acts up the plane, as shown in Figure 20.26.

When the particle is at rest, resolving perpendicular to the plane

$N = mg\cos\theta$

and resolving along the plane

$F = mg\sin\theta$

As the plane is raised, initially **F** will increase to oppose the movement of the particle. However, eventually **F** reaches its maximum value at an angle α to the horizontal, such that the particle is about to slip down the plane. At this point the force is said to be limiting. If the plane is raised further, **F** will exceed the force of limiting friction, and the particle will slide down the plane. The frictional force (still at its maximum value) will continue to act up.

When in limiting equilibrium, $F = mg\sin\alpha$ and $N = mg\cos\alpha$ and

$$\mu = \frac{force\ of\ limiting\ friction}{normal\ reaction}$$

$$= \frac{mg\sin\alpha}{mg\cos\alpha}$$

$$= \tan\alpha$$

Thus, the angle of the plane at the point of slipping depends on the coefficient of friction.

When the particle is in limiting equilibrium or when the particle is in motion, the frictional force **F** has magnitude μN i.e. $F = \mu N$. Note that until sliding takes place, the frictional force is less than μN.

After the limiting friction has been reached and the particle starts to slide, the frictional force will continue to act at this maximum value, μN, to oppose the motion of the particle.

Example 12

(a) A body of mass 5 kg rests on a smooth plane inclined at 30° to the horizontal. Find the least value of the force required to keep it in equilibrium and the resultant reaction of the plane. [Take g = 9.8 m s^{-2}]

Solution

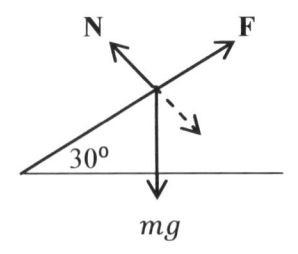

Resolving horizontally we have

$$F = mg\sin 30°$$
$$= 5 \times 9.8 \times 0.5$$
$$= 24.5$$

The least force required is 24.5 N

Resolving vertically we have

$$N = mg\cos 30°$$
$$= 5 \times 9.8 \times 0.866$$
$$= 42.4$$

The resultant reaction is 42.4 N

(b) A body of mass 4 kg lies on a slope of angle 20°, where $\mu = \frac{1}{4}$. Find the acceleration of the body down the slope.

Solution

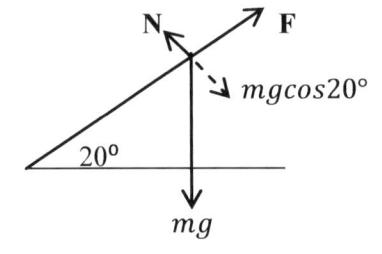

Resolving perpendicular to the slope we have

$$N = mg\cos 20°$$
$$= 4 \times 9.8 \times \cos 20° = 36.8 \text{ N}$$

Resolving parallel to the slope we have

$$mg\sin 20° - F = ma$$

$$4 \times 9.8 \times \sin 20° - F = 4a$$

Since the body is sliding the friction is limiting, so $F = \mu N = \frac{1}{4}N$.

Thus $\qquad 4 \times 9.8 \times \sin 20° - \frac{1}{4} \times 36.8 = 4a$

Simplifying this gives

$$4a = 4.2$$

$$a = 1.05$$

The acceleration is 1.05 m s^{-2}

(c) From a point P a particle is projected with a speed 30 m s^{-1} up the slope of a rough plane inclined at an angle 25° to the horizontal. The coefficient of friction between the particles and the plane is $\frac{1}{4}$. If the particle comes to instantaneous rest at Q, find the distance PQ and the time the particle takes to go from P to Q.

Solution

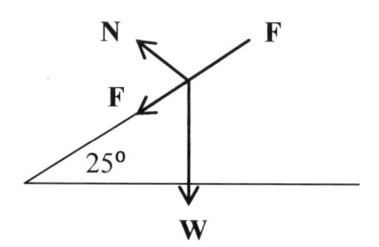

Resolving perpendicular to the slope we have

$$N = mg\cos 25°$$

Since the particle is moving up the plane, the frictional force will act down the plane.
Resolving parallel to the slope

$$F + mg\sin 25° = ma$$

Since the particle is in motion

$$F = \mu N = \frac{1}{4}N$$

Hence $\frac{1}{4} \times m \times 9.8 \times \cos 25° + m \times 9.8 \times \sin 25° = ma$

Simplifying this gives $a = 6.4$ m s^{-2}.

Because the acceleration a is constant, you can use the equations for uniformly accelerated motion to find the distance.

Using the equation $v^2 = u^2 + 2as$, we have

$$0^2 = 30^2 + 2 \times -6.4 \times s$$

$$s = 70.3$$

So the distance PQ is 70.3 m

Using the equation $v = u + at$ we have

$$0 = 30 - 6.4t$$

So $\quad t = 4.7$

Hence the time taken to go from P to Q is 4.7 s.

Exercise 20.3(f)

1. A mass of 7 kg lies on a rough horizontal table and a horizontal force of 25 N is applied. If the body does not move find the frictional force.

2. A mass of 9 kg on a rough floor can just be moved by a horizontal force of 50 N. Find the coefficient of friction.

3. A mass of 5 kg lies on a rough table where the coefficient of friction μ is 0.6. Find the least horizontal force needed to move it.

4. A body of mass 4 kg slides on a rough horizontal table where the coefficient of friction is $\frac{1}{2}$. If it starts with a speed of 3 m s^{-1}, find how far it skids.

5. A body of mass 8 kg slides along a rough horizontal table coming to rest from 2 m s^{-1} in 1 minute. Find the coefficient of friction.

6. A body of mass 25 kg lies on a floor where the coefficient of friction is $\frac{1}{2}$. A rope is attached to it at an of angle 20° to the horizontal. Find the least tension in the rope which will pull the body.

7. A body of mass 60 kg lies on a floor where the coefficient of friction is $\frac{1}{4}$. It is pushed by a rod which makes an angle of 15° with the horizontal. Find the least force in the rod which will move the body.

8. A body of 6 kg rests on a smooth plane inclined at 30° to the horizontal. Find the frictional force and the normal reaction.

9. A man of mass 65 kg is standing on a roof which slopes at 25° to the horizontal. If the man remains stationary, find the frictional force and the normal reaction.

10. A box of mass 50 kg placed on a sloping ramp which is inclined at an angle of 20° to the horizontal. Find the frictional force if:

(a) the box does not slide

(b) the box slips with an initial acceleration of 1.2 m s^{-2}

11. A car of mass 600 kg is packed on a hill which slopes at an angle of 8°. If the brakes exert a frictional force of 500 N, find the acceleration in which the car starts to move.

12. A particle of mass 0.6 kg is resting on a plane inclined at an angle of 15° to the horizontal. What are the normal reaction and friction? Find the coefficient of friction if the particle is about to slip down the plane when its angle with the horizontal is 27°.

13. A body of mass 2 kg lies on a slope of angle 30°, where the coefficient of friction is $\frac{1}{4}$. Find the least force parallel to the slope which will prevent it from sliding down. Find the least force parallel to the slope which will push it up the slope.

14. A body lies on a slope of angle 20°, where the coefficient of friction is $\frac{1}{4}$. Find its acceleration down the slope.

15. A body lies on a slope of angle 42°, and accelerates at 6 m s^{-2}. Find the coefficient of friction.

16. A body of mass 4 kg is resting on a plane inclined at 36° to the horizontal and has a coefficient of friction of $\frac{1}{2}$. Find the force required to act up the plane, if the body is about to slip:

(a) down the plane

(b) up the plane

17. A particle slides down a rough plane inclined at an angle of 30° to the horizontal. The coefficient of friction between the plane and the particle is $\frac{1}{2}$. Find the time taken for the speed of the particle to increase from 5 m s^{-1} to 30 m s^{-1} and the distanced travelled during this time.

18. At a given instant a particle is moving with speed 8 m s^{-1} down a rough plane inclined at an angle of 60° to the horizontal. The coefficient of friction between the particle and the plane is $\frac{3}{4}$. Find how long it will take the particle to reach a speed of 25 m s^{-1} and how far it will move in this time.

Moments

In many situations in our daily life, we see cases in which forces exert a turning effect. Examples include when a door is open, when we use a screw driver or a spanner. The turning effect of a force about a point is measured by its moment. The moment of a force about a point is defined as the product of the force and the perpendicular distance from the point to the line of action of the force. The unit of moment is the newton-metre (N m).

Figure 20.27(a)   Figure 20.27(b)

In Figure 20.27(a) the moment of the force about A is $F \times d$. The force tends to move the body round A in anticlockwise direction. When the turning effect is anticlockwise, the moment is taken to be positive. When the turning effect is clockwise as in Figure 20.27(b), the moment is taken to be negative.

The Principle of Moments

If a body is in equilibrium the sum of the moments in a clockwise direction about any point is equal to the sum of the moments in an anticlockwise direction about the same points.

Taking Moments about a Point

The process of equating the clockwise and anticlockwise moments about a point is called taking moments about that point.

Consider a rod supported by a fulcrum. If a mass m kg is hung from the rod, it will exert a force mg N vertically downwards. This is the weight of the particle, and will be shown in a diagram as an arrow. Generally, all forces acting on the rod will be represented by arrows. When the weight of the rod is negligible compared with the other forces acting on the rod, the rod is called a light rod. The weight of an object can be taken to act at point called the centre of gravity. A uniform rod will have its centre of gravity in the middle of the rod. The support given by a fulcrum is called the reaction or thrust.

Example 13

A uniform beam AB of length 12 m and weight 6 N rest on a fulcrum at its mid-point P. A force of 4 N acts vertically downwards at A, a force of 3 N acts vertically downward at Q, 4 m from A, and a force of magnitude x N acts vertically downward at B. If the beam is in equilibrium, what is the value of x? Find the force exerted by the fulcrum on the beam.

Solution Begin by drawing a diagram showing all forces acting on the beam.

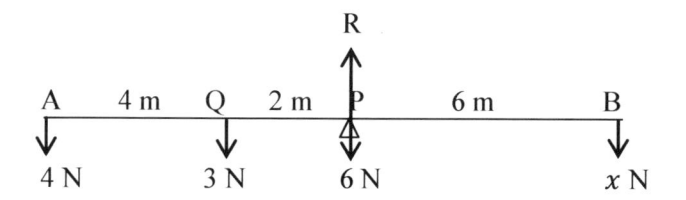

Figure 20.28

Let R newtons be the magnitude of the force exerted by the fulcrum.

Taking moments about P, we have

$$x \times 6 = 4 \times 6 + 3 \times 2$$

$$6x = 30$$

$$x = 5$$

We can choose any point about which to take moments. The work is simplified by choosing a point which gives one of the unknowns zero moment.

Taking moments about B, we have

$$R \times 6 = 4 \times 12 + 3 \times 8 + 6 \times 6$$

$$6R = 108$$

$$R = 18$$

Therefore the reaction force is 18 N

Because the beam is in equilibrium, you can obtained the same result by equating the sum of the upward forces and the sum of the downward forces to get

$$R = 4 + 3 + 6 + 5$$

$$= 18 \text{ N}$$

Exercise 20.3(g)

1. A light rod of length 2 m has weights of 8 kg and 12 kg at each end. Where should it be pivoted if it is to balance?

2. A light rod of length 3 m is pivoted 1.8 m from one end. 5 kg is placed at the longer end; how much weight should be put at the shorter end to balance it?

3. A uniform beam of length 2 m and mass 15 kg is pivoted 1.6 m from one end. What weight should be put at the shorter end in order to balance it?

4. A non-uniform rod of length 3 m and mass 50 kg is such that its weight acts through a point 60 cm from the centre. If it is pivoted at its centre, what weight should be put at an end in order to balance it?

5. A uniform rod of mass 15 kg and length 1.8 m is smoothly hinged at one end. It is held horizontal by a force acting vertically at its free end. What is the force?

6. A uniform rod of mass 20 kg and length 2.4 m is smoothly hinged at one end. It is held horizontal by a force of 12 kg vertically. Where does the force act?

7. A non-uniform beam of length 4 m is supported at both ends. The up thrusts from the supports are 8 g N and 12 g N. Where does the weight of the beam act?

8. A light rod 2.4 m long is pivoted 0.9 m from the end P. A 5 kg mass is hung 2.1 m from P and a 2.5 kg mass is hung 1.5 m from P. What mass must be placed at P and what is the reaction?

9. A uniform metre rule has a fulcrum under the 0.3 m mark. It is balanced by a 5 kg mass hung under the 0.15 m mark and a 0.5 kg mass under the 0.6 m mark. What is the mass of the rule?

10. A uniform beam PQ of length 3.6 m and mass 15 kg is balanced by fulcrums 2 m and 3 m from P. A mass of 8 kg is hung 2.4 m from P and a mass of 1 kg is hung 3.2 m from P. What are the reactions?

20.4 Momentum

If a particle of mass m moves with velocity v, the product of the mass and the velocity is known as the momentum of the particle, i.e.,

momentum $= mv$

Since momentum is the product of a scalar and a vector, momentum is a vector in the same direction as the velocity, v, along a straight line passing through the particle. Since mass is measured in kg and velocity in m s^{-1}, we measure momentum in kg m s^{-1}.

Impulse

Consider a constant force **F** acting on a particle of mass m kg for a time t. Suppose the velocity of the particle changes from u to v in time t, the force will give the particle a constant acceleration $\frac{v-u}{t}$ in the direction of the force.

From Newton's Second Law

$$F = ma$$

we obtain

$$F = \frac{m(v-u)}{t}$$

or
$$Ft = mv - mu \qquad (1)$$

The product Ft is called the impulse of the force **F** acting for the time t. The equation (1) can be restated as

impulse of a force = change in momentum

Since force is measured in newtons and time in seconds, the unit of impulse is the newton-second (N s).

Conservation of Linear Momentum

A particle P of mass m_1, moving in a straight line with a velocity u_1 collides with a particle Q of mass m_2 moving in the same direction with velocity u_2. Suppose P and Q move with velocities v_1 and v_2 after collision respectively. By Newton's third law the force F_1 exerted by P on Q when the particle collide is equal and opposite to the force F_2 exerted by Q on P. Hence,

$$F_1 + F_2 = 0$$

assuming no external forces act on the particles.

Since these forces last for the same time t, and *force = mass × acceleration*, we have

$$\frac{m_1(v_1-u_1)}{t} + \frac{m_2(v_2-u_2)}{t} = 0$$

or $\quad m_1 u_1 + m_2 u_2 = m_1 v_1 + m_2 v_2 \qquad (1)$

The left hand side of (1) gives the total momentum of the two particles before the collision and the right hand side gives the total momentum after collision. Equation (1) suggests that if no external forces act on the particles, the total momentum before the collision equals the total momentum after collision. This result illustrates the following principle of mechanics.

The Principle of Conservation of Momentum

If there is no resultant external force acting on a system of objects, its total momentum is constant.

Example 14

(a) The velocity of a particle moving in a straight line is reduced from 15 m s^{-1} to 5 m s^{-1} by a constant force acting for 2 s. If the mass of the particle is 12 kg, find the magnitude of force.

Solution Initial momentum; $mu = 12 \times 5 = 60$

Final momentum; $mv = 12 \times 15 = 180$

Use the equation $Ft = mv - mu$ to get

$$F \times 2 = 180 - 60$$

$$F = 60$$

The force is 60 N

(b) Two particles P and Q are moving in the same straight line towards one another. The mass of P is 30 kg and its speed is 6 m s^{-1} while the mass of Q is 20 kg and its speed is 3 m s^{-1}. Given that the particles coalesce on impact, find their speed after collision.

Solution Let m_1 be the mass of P and m_2 the mass of Q, and u_1 and u_2 the velocities before collision of P and Q respectively. Let v be their common velocity after the collision.

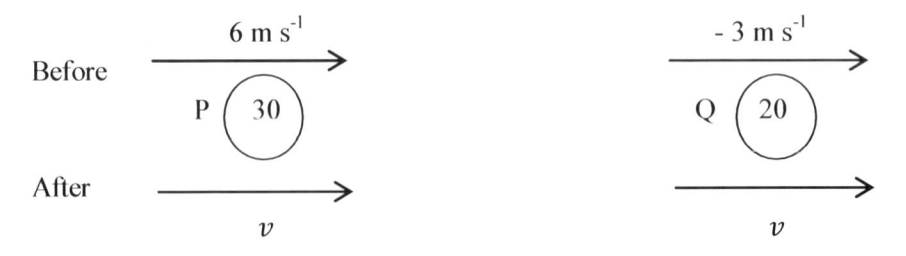

Before
6 m s^{-1} → P (30)   -3 m s^{-1} → Q (20)

After
→ v → v

Taking the direction of P as positive the velocity of Q will be -3 m s^{-1}.
Momentum before collision $= m_1 u_1 + m_2 u_2$

$$= 30 \times 6 + 20 \times (-3)$$

$$= 120$$

Momentum after collision $= (m_1 + m_2)v$

$$= (30 + 20)v$$

$$= 50v$$

Momentum after collision = Momentum before collision

Hence $\quad 50v = 120$

$$v = 2.4$$

The common velocity after collision is 2.4 m s^{-1}.

Exercise 20.4

1. What is the momentum of a mass 350 g moving with a velocity of 6 ms^{-1}?

2. What is the impulse that has acted to move a body from rest to 12 m s^{-1} if the body has mass 5 kg?

3. An impulse of 15 N s is applied to a stationary mass 3 kg. Find its velocity afterwards.

4. An impulse of 25 N s is applied to a mass 5 kg moving at 3 m s^{-1} in the direction of the impulse. Find the new velocity of the mass.

5. A body of mass 4 kg moving at 6 m s^{-1} is struck by an impulse acting in the same direction as the motion. If it now moves at 8 m s^{-1} find the impulse.

6. A rifle of mass 5 kg fires a bullet of mass 12 g at 600 m s^{-1}. Find the initial speed of recoil of the rifle.

7. A particle of mass 5 kg moves at 10 m s^{-1} and collides with a stationary particle of mass 3 kg. If they move on together find their common velocity.

8. A bullet of mass 0.04 kg is fired into a stationary block of wood of mass 0.76 kg and remains embedded in the wood. The speed of the bullet was 240 m s^{-1}. Find the final speed of the block and bullet.

9. A gun of mass 600 kg fires a shell of 3 kg. If the gun moves back with an initial velocity 8 m s^{-1}, find the initial velocity of the bullet and the constant force necessary to stop the gun in 2 seconds.

10. An arrow of mass 60 g flying horizontally at 50 m s^{-1} strikes a hanging target of mass 3 kg and sticks in it. Find the speed with which they move after impact.

11. A car of mass 400 kg moving at 30 m s^{-1} bumps into a car of mass 500 kg which is moving in the same direction at 25 m s^{-1}. If they stay fixed together find their speed.

12. A car of mass 300 kg travelling at 12 m s^{-1} has a head-on collision with another car travelling in the opposite direction at 20 m s^{-1}. If they are both brought to a halt find the mass of the second car.

13. A particle of mass 500 g moving with speed 12 m s^{-1} collides directly with a particle moving in the opposite direction at 8 m s^{-1}. Both particles remain at rest after the impact. Find the mass of the second particle.

14. A particle of mass 15 kg moving with speed 18 m s^{-1} collides with a particle of mass 30 kg moving with speed 3 m s^{-1} in the same direction. The lighter particle is brought to rest by the impact. Find the speed of the heavier particle after the collision.

15. A mass of 1.5 kg moving with velocity 4 m s^{-1} collides with a mass of 1 kg moving with velocity 2 m s^{-1}. The first mass moves in its initial direction with speed 1.5 m s^{-1}. Find the velocity of the other mass if they were moving in:

(a) the same direction

(b) opposite direction

16. A particle of mass 5 kg moves with velocity $3\mathbf{i} + 5\mathbf{j}$ m s^{-1}. After a blow it moves with velocity $4\mathbf{i} + 7\mathbf{j}$ m s^{-1}. Find the magnitude of the impulse it has received.

17. Two particles of mass 6 kg and 3 kg and velocities $2\mathbf{i} + \mathbf{j}$ m s^{-1} and $-3\mathbf{i} - 2\mathbf{j}$ collide and coalesce. Find their new velocity.

18. A particle A of mass 200 g moving with velocity $3\mathbf{i} + 4\mathbf{j}$ m s^{-1} collides with another particle B of mass 100 g which is at rest. If particle A now moves with velocity $2\mathbf{i} - 3\mathbf{j}$ m s^{-1}, find the velocity with which particle B moves.

19. A body, A of mass 12 kg moving with a velocity of (10 m s^{-1}, 270°) collides with another body, B, of mass 8 kg moving with velocity (8 m s^{-1}, 060°). If the velocity of A immediately after the collision is (4 m s^{-1}, 330°), find the magnitude of the velocity of B immediately after collision.

20. A particle of mass 4 kg moving with velocity $5\mathbf{i} + 3\mathbf{j}$ m s^{-1} collides with another particle of mass 6 kg moving with velocity $5\mathbf{i} + 12\mathbf{j}$ m s^{-1}. They move together after collision. Find the magnitude of their common velocity if they were moving in opposite direction.

Review Exercise 20

1. A particle increases its velocity from 2.5 m s^{-1} to 10 m s^{-1} in 5 seconds. Find its acceleration and the distance covered.

2. A particle is accelerated at 4 m s^{-2} over a distance of 100 m in 5 seconds. Find its initial and final velocities.

Mechanics 509

3. A particle accelerates at 3 m s^{-1} from 15 m s^{-2} to 45 m s^{-1}. Find the distance covered and the time it takes to cover this distance.

4. A particle accelerates at 5 m s^{-2} over 50 m to finish with a speed of 30 m s^{-1}. What was the initial velocity and how long does it take?

5. A particle starts with a velocity of 15 m s^{-1} and travels 50 m in 4 seconds. What is the acceleration and the final velocity?

6. A particle is projected vertically upwards with velocity of 60 m s^{-1} from a point O. Find the maximum height reached by the particle.

7. A particle is projected vertically upwards with velocity of 30 m s^{-1} from a point O. Find the time it takes to return to the point of projection.

8. A stone is dropped from the top of a building 125 m high. Find the time it takes to reach the ground.

9. A stone is dropped from the top of a building 45 m high. Find the velocity with which it hits the ground.

10. A particle is projected vertically upwards with a velocity of 25 m s^{-1} from a point O. Find the time taken for it to be 30 m below O.

11. A car starts from rest and accelerates uniformly for 2 minutes at 0.15 m s^{-2}, then travels at a constant speed for 4 minutes and retards uniformly to rest after covering a distance of 720 m. Find the total distance covered and the magnitude of the uniform retardation.

12. A particle accelerates uniformly for 2 seconds, travels at a constant speed for 12 seconds, then decelerates uniformly to rest in 4 seconds. If the total distance travelled is 750 m, find the maxinum speed.

13. A car accelerates at 3 m s^{-2} for 4 seconds, then travels at a constant speed for 20 seconds, and then decelerates uniformly to rest. The total distance travelled is 300 m. Draw a velocity-time graph and hence find the deceleration.

14. A cyclist starts from rest, and accelerates uniformly to a maximum speed of 15 m s^{-1}. The speed is maintained for 30 seconds, and then the cyclist decelerates uniformly to rest. The total distance covered is 525 m. Sketch a velocity-time graph. Find the total time taken.

15. A train accelerates uniformly from rest at station P until it attains a speed of 100 km h^{-1}. This speed is maintained for some time and then the train decelerates uniformly until it comes to a stop at station Q. The distance between the stations is 60 km and the time taken for the journey is 1 hour. If the rate of deceleration is twice that of acceleration, calculate the time for which the speed is constant and the acceleration of the train.

16. A light aircraft has an airspeed of 120 knots and there is a north-westerly wind (i.e. blowing from the north-west) of 50 knots. The aircraft flies in a north-easterly direction. Find the velocity of the aircraft relative to the ground.

17. A light aircraft has an airspeed of 120 km h^{-1} and flies on a course of S40°E. There is a 35 km h^{-1} wind blowing from north-west. Find the ground speed and the direction of the aircraft.

18. An airplane moves on a bearing of 315° at 130 km h^{-1} relative to the ground, due to the fact that there is a wind blowing from the west at 45 km h^{-1}. Find the speed and the direction the plane will have travelled if there was no wind.

19. A force acting on a body of mass 2 kg changes its speed from 20 m s^{-1} to 50 m s^{-1} in 6 s. Find the magnitude of the force.

20. A force of magnitude 12 N acting on a body of mass 3 kg changes its speed from 10 m s^{-1} to 30 m s^{-1}. Find the time during which the force acts on the body.

21. A constant force acts on a toy car of mass 5 kg and increases its velocity from 5 m s^{-1} to 9 m s^{-1} in 2 seconds. Find the velocity of the car 3 seconds after attaining the velocity of 9 m s^{-1}.

22. A train moves under a net forward force of 25 k N. It starts from rest and reaches a speed of 10 m s^{-1} after 50 s. Find the acceleration and hence the total mass of the train.

23. The engine of a car of mass 1,000 kg produces a forward force of 2,700 N. Find the acceleration of the car if the resistance to motion is 1,200 N.

24. Find by drawing the magnitude and direction of the resultant force of the forces, 3 N at 18° and 7 N at 115° when acting simultaneously at a point.

25. A body is acted upon by three forces **P** (10 N, 030°), **Q** (15 N, 120°) and **R** (15 N, 240°). Find by drawing the magnitude and direction of the resultant force.

26. Three forces **P** (2 N, 060°), **Q** (4 N, 180°) and **R** (5 N, 300°) act on a body of mass 2 kg at rest. Find the magnitude and direction of the resultant force.

27. A body of mass 500 g lying on a horizontal floor is acted upon by the force **P** (10 N, 090°), **Q** (16 N, 180°), **R** (8 N, 030°) and **S** (12 N, 300°). Find the magnitude and direction of the resulting acceleration.

28. Three forces $3\mathbf{i} - 8\mathbf{j}$, $3\mathbf{i} + 3\mathbf{j}$ and $4\mathbf{i} - \mathbf{j}$ act on a body of mass 4 kg. Find the acceleration of the body.

29. A bucket of water of mass 20 kg is being pulled out of a well. Find the magnitude of the tension in the rope when it has an acceleration of 0.3 ms^{-2}.

30. A body of mass 12 kg hangs from a string and is pulled aside by a horizontal force of 30 N. Find the tension in the string and the angle it makes with the vertical.

31. A particle of mass 0.3 kg is suspended in equilibrium by two light strings inclined at 30° and 60° respectively to the downward vertical. Calculate the tensions in the strings.

32. A particle of mass 5 kg is suspended in equilibrium by two light strings inclined to the horizontal at 40° and 50° respectively. Calculate the tension in each string.

33. Two particles A and B with mass 3 kg and 4 kg moving with velocities 5 m s^{-1} and 8 m s^{-1} respectively collide. After collision the particle A moves with velocity 6 m s^{-1}. Determine the velocity of the particle B after collision.

34. A body of mass 5 kg moving along a track at 10 m s^{-1} collides with a body of mass 8 kg moving at 6 m s^{-1} on the same track. If the bodies coalesce after collision, find their common speed, if before collision they were moving:

(a) in the same direction

(b) in opposite direction

35. Two objects P and Q are of masses 12 kg and 8 kg respectively. They move towards each other in a straight line with velocities 3 m s^{-1} and 4 m s^{-1} respectively and collide. After collision P moves with a velocity of 2 m s^{-1} in the opposite direction. Calculate the magnitude of the velocity of Q.

36. A particle of mass 4 kg moving with velocity $5\mathbf{i} + 3\mathbf{j}$ m s^{-1} collides with another particle of mass 6 kg moving with velocity $5\mathbf{i} + 18\mathbf{j}$ m s^{-1}. If the particles coalesce after collision, find the magnitude and the direction of their common velocity.

37. A body of mass 2 kg lies on a slope of a rough plane inclined at an angle $tan^{-1}\left(\frac{1}{\sqrt{3}}\right)$ to the horizontal, where the coefficient of friction is $\frac{1}{4}$. Find the acceleration of the body down the slope.

38. A body of 5 kg lies on a slope of 30° to the horizontal. If the coefficient of friction is $\frac{1}{2}$, find the least force parallel to the slope which will move the body up the slope.

39. A uniform plane XY of length 30 m and mass 40 kg rest horizontally on two supports P and Q at 8 m and 18 m from X respectively. An object of mass 10 kg is hung 10 m from X and an object of mass 5 kg is hung 6 m from Y. If the system remains in equilibrium under the action of these forces, find the forces exerted by the supports.

40. A uniform beam XY of length 6 m rests on two supports P and Q at 0.8 m from X and 1.8 m from Y respectively. Masses of 12 kg and 15 kg are suspended at 1.7 m from X and 0.4 m from Y respectively. If the reaction at P is 150 N and the system remains in equilibrium under the action of these forces, find the mass of the beam, and the reaction at Q.

21 Permutations and Combinations

21.1 Permutation

The Multiplication Principles

If a student drives to school, he can take either one of the two roads to an intersection, which we shall label 1 and 2, and then take any one of the three roads that leads to the school, label A, B and C as shown Figure 20.1.

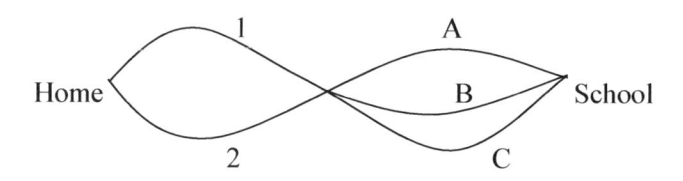

Figure 20.1

He has two choices of roads from his home to the intersection and three choices from the intersection to the school, giving the number of possible routes as 2×3 i.e. 6. Notice that the same result is obtained if we list the possible routes: 1A, 1B, 1C, 2A, 2B and 2C, giving six possible routes. If you determine the total number of possible choices by multiplying the number of each independent choice you are using the Counting Principles.

The Counting Principles

If there are m possible ways of making a first choice and for each of the m possible ways there are n possible ways of making a second choice, then there are $m \times n$ possible ways of making the two choices.

Example 1

(a) How many different 4-digit numbers can be produced if the first digit may not be 0?

Solution We have 9 choices for the first digit 10 choices for the second, third and fourth digits. Hence, the number of different numbers formed is

$9 \cdot 10 \cdot 10 \cdot 10 = 9,000$

(b) A class has 15 boys and 12 girls. In how many possible ways can one boy and one girl be selected to participate in a quiz?

Solution A boy can be selected in 15 different ways and a girl in 12 ways.

Hence the number of possible selection is $15 \times 12 = 180$

Exercise 21.1(a)

1. A woman has 4 blouses and 5 skirts. How many different outfits can she wear?

2. A car company has 3 different car model and 6 colour schemes. How many cars must a dealer display to show each possibility?

3. A man has 3 pairs of shoes, 8 pairs of socks, 4 pairs of trousers, and 9 shirts. How many outfits can he wear?

4. A school board has 12 members. The board must select from its members a chairman, vice-chairman and a secretary. In how many ways can this be done?

5. How many 4-letter code words are possible from the first 6 letters of the alphabet with no letters repeated?

6. How many ways are there to rank 7 candidates who apply for a job?

7. In how many ways can 3 boys and 3 girls be seated in a row of 6 seats if the seating must alternate boys and girls, starting with a girl?

8. How many7-digit telephone numbers can be produced if the first digit may not be 0 and the digits are not repeated?

9. How many 7-digit telephone numbers can be assigned by a telephone company in a city if the first three digits are 446 and 447?

10. Using only the digits 0 and 1, how many different numbers consisting of 8 digits can be formed?

11. From a group of 5 people we are required to select a different person to participate in each of 3 different tests. In how many ways can the selections be made?

12. In how ways can10 blue flowers and 5 red flowers be planted in a row, if we do not distinguish between flowers of the same colour?

Permutation

Suppose four boys, A, B, C and D who took part in a contest are to be ranked. The first position can be taken by any one of the four boys.

$$A \quad B \quad C \quad D$$

For each of the four choices for first position, the second position can be taken by any one of the three remaining boys. For example, if A is ranked first B, C or D may be ranked second. The possible choices for the first and second positions are listed below.

$$AB \qquad AC \qquad AD$$

BA	BC	BD
CA	CB	CD
DA	DB	DC

You can see that the number of choices for the first and second positions is 4×3 i.e. 12. Now, for each of the twelve choices, the third position can be taken by any one of the two remaining boys. For example, if A and B are ranked first and second respectively C or D can be ranked third, that is, ABC or ABD. Hence, the number of choices for the first, second and third positions is $4 \times 3 \times 2$. Finally, we have only one choice for fourth position. Hence, the four boys can be ranked in $4 \times 3 \times 2 \times 1$ ways written as 4! ways.

Each arrangement of a number of objects in a particular order is called permutation. In general, the number of distinct permutations or arrangements of n objects is denoted by the symbol ^nP_n and is given by $^nP_n = n!$.

In how many ways can we arrange 5 objects taking two objects at a time?

Since the first object can be chosen in 5 ways and the second object in 4 ways, the total number of arrangements is 5×4 i.e. 20. We can write 5×4 as

$$\frac{5 \cdot 4 \cdot 3 \cdot 2 \cdot 1}{3 \cdot 2 \cdot 1} = \frac{5!}{3!} = \frac{5!}{(5-2)!}$$

So $^5P_2 = \dfrac{5!}{(5-2)!}$

In general, the number of permutation of n different objects taken r at a time, denoted in symbol by ^nP_r is given by $^nP_r = \dfrac{n!}{(n-r)!}$.

Example 2

(a) Find the number of arrangement of the word FIGURE.

Solution The word could be arranged in $^6P_6 = 6! = 720$ ways

(b) In how many ways can three prizes be awarded to a class of 12 boys, if no boy wins more than one prize?

Solution The number of ways in which the prizes could be awarded is

$$^{12}P_3 = 1320$$

(c) How many four digit numbers can be formed with the digits 0, 1, 2, 5 and 8?

Solution The number cannot begin with 0, so the first digit can be chosen in 4 ways. The second digit can be 0 or any of the three remaining digits, so the second digit can be chosen in 4 ways. We have 3 choices for the third digit and 2 choices for fourth digit.

Hence, the number of four digit numbers that can be formed is $4 \cdot 4 \cdot 3 \cdot 2 = 96$

Permutation with conditions

Example 3

(a) Find the number of arrangements of the letters in the word ABSENT that begin with a vowel.

Solution There are two choices for the first letter as it must be filled by either A or E. To fill the second position, we have a choice of five letters, one vowel together with four consonants. The third place can be filled in 4 ways, and so on.

This gives $2 \cdot 5 \cdot 4 \cdot 3 \cdot 2 \cdot 1 = 240$ arrangements.

(b) How many arrangements of the letters of the word OBJECT will end with a consonant?

Solution We can fill the places in any order, and so we start by filling the final position. There are 4 consonants that can fill this position. The first position can be filled with any of the five remaining letters; there are 4 ways to fill the second position, 3 ways to fill the third and so on. This gives $5 \cdot 4 \cdot 3 \cdot 2 \cdot 1 \cdot 4 = 480$ arrangements.

(c) How many arrangements of the word DETAILS have their odd position filled by consonant and even position by vowels.

Solution We start by filling first, third, fifth and seventh positions with four consonants in $4 \cdot 3 \cdot 2 \cdot 1 = 24$ ways. The second, fourth and sixth positions can be filled by three vowels in $3 \cdot 2 \cdot 1 = 6$ ways. Any set of consonants can be placed with any set of vowels to give the total number of arrangements as $24 \times 6 = 144$.

(d) How many five-figure even numbers can be formed from the digits 0, 1, 2, 3, 4, 6 if no digit is repeated?

Solution We start by filling the final position. Because the number is even it must end with the number 0, 2, 4 or 6, so the last number can be chosen in four ways. The number cannot begin with 0, so the first number can be chosen in 4 ways. The second number can be chosen from any of the remaining numbers, including 0, in 4 ways. There will be 3 choices for the third number and 2 choices for the fourth number. The total number of even numbers that can be formed is $4 \cdot 4 \cdot 3 \cdot 2 \cdot 4 = 384$.

(e) In how many ways can 5 people sit at a round table?

Solution   We consider one person fixed and then arrange the rest about him. The second person has four choices; the third person has three choices and so. This gives a total of $1 \times 4 \times 3 \times 2 \times 1 = 24$ arrangements.

Permutation with Repetition

Our previous discussion of permutations required that the objects we were arranging should be distinct. We now examine permutations that may have an object repeated.

In how many ways can the word SEE be rearranged?

If the E's are treated as different, the 3 letters can be arranged in 3!. In every distinct arrangement the 2 E's can be arranged in 2! ways. Then the number of distinct arrangements of the word SEE is 3!/2!, i.e. 3.

In general the number of permutations of n objects of which r are identical is $n!/r!$. If r and s of the n objects are the same, the number of different arrangements is $n!/r!\,s!$.

Example 4

(a) In how many ways can the letters in the word THERE be arranged?

Solution   The total number of arrangement is $5!/2! = 5 \cdot 4 \cdot 3 = 60$

(b) Find the number of arrangement of the word SUCCESS.

Solution   The number of arrangements is $\frac{7!}{3!2!} = 420$

(c) Find the number of permutations of the letters of the word ALLOW in which the two L's are not adjacent.

Solution   The total number of permutations of the five letters is $\frac{5!}{2!}$ i.e. 60. When the two L's are next to one another, they can be regarded as a single letter. The number of permutations of A, O, W and double L is 4!, that is the number of permutations in which the L's are adjacent is 24. The number of permutations in which the two L's are not adjacent will be $60 - 24 = 36$.

Exercise 21.1(b)

1. Ten athletes run in a race. Find the number of ways in which the first three places can be filled.

2. How many different ways are there to arrange the 6 letters in the word FRIDAY?

3. How many ways are there to rank 8 candidates who apply for a job?

4. A headmaster, a first assistant and a second assistant for a school are chosen from a group of 20 teachers. In how many ways can the 3 teachers be chosen for the 3 position?

5. A form captain and an assistant for a class are to be chosen from a group of 10 students. In how many ways can the student be chosen for the two positions?

6. From a pool of 10 job applicants a list ranking the top 4 must be made. How many such list are possible?

7. A school bus has 16 seats. In how many different ways can 5 students be seated in it?

8. How many ways are there to arrange 7 people in a line?

9. How many ways can you arrange the letters from the word TYPES?

10. A telephone number contains 6 digits. How many possible telephone numbers are there in which no digit is repeated?

11. A school inspector must visit 6 schools. In how many different ways can he plan his visit?

12. In how many ways can the second runner-up, the first runner-up, and the winner of a beauty pageant be chosen from 25 contestants?

13. The digits 0 through 9 are written on 10 cards. Four different cards are drawn and a 4-digit number is formed. How many different 4-digit numbers can be formed in this way?

14. A science class contains 30 students. In how many ways can prizes be awarded in Mathematics, Physics and Chemistry, if no student can win more than one prize?

15. In how many ways can 6 different books be distributed to 8 children if no child gets more than one book?

16. A computer must assign each of 5 outputs to one of 8 different printers. In how many ways can it do this provided no printer gets more than one output?

17. Find the number of three-figure integers that can be formed from the numbers 2, 3, 4, 5, 6

(a) if no number is used twice

(b) if any number may be used more than once

18. How many numbers between 100 and 1000 can be formed which the digit 0 does not appear?

19. How many numbers between 5000 and 10,000 can be formed, using only the digits 2, 3, 4, 6, 8?

20. How many odd numbers between 500 and 1,000 can be formed, using only the digits 1, 3, 6, 7, 8?

21. How many five-figure odd numbers can be made from the digits 1, 2, 3, 4, 5 if no digits is repeated?

22. How many even numbers, greater than 2,000 can be formed with the digits 1, 2, 4, 8, if each digit may be used only once in each number?

23. How many numbers between 1,000 and 10,000 can be formed using only odd digits, if no digit is repeated?

24. Find the number of arrangements of the letters of the word MONDAY that begin either with A or O.

25. Find the number of arrangements of the letters of the word AROUND in which the vowels and the consonants come alternately.

26. Find the number of arrangements of the letters of the word MINIMUM which begin with M and end in I.

27. Find the number of arrangements of the letters of the word ARREARS. Find also the number of these arrangements in which no two R's are adjacent.

28. How many different 11-letter words (real or imaginary) can be formed from the letters in the word MATHEMATICS?

29. How many different 9-letter words (real or imaginary) can be formed from the letters in the word ECONOMICS?

30. In how many ways can 12 children be placed in 3 distinct teams of 3, 5 and 4 members?

21.2 Combination

Two letters can be chosen out of the letters A, B, C in the following ways:

        AB   BA   AC   CA   BC   CB

This gives six different permutations of the letter A, B and C. If the order does not matter, only three different selections can be made:

        AB   AC   BC

Notice that AB and BA are different permutations that contain the same letters. However, if the order of the letters does not matter then AB is the same selection as BA.

The selection of objects without regard to order is referred to as combination. Hence, the number of ways in selecting two letters out of three letters, denoted by $^3C_2$, is 3.

Each combination of two letters can be arranged in 2! ways. Hence, the number of permutation of the 3 letters taken 2 at a time equals the number of combinations multiplied by 2!, i.e. $^3C_2 \times 2! = {^3P_2}$

$$\therefore {^3C_2} = \frac{^3P_2}{2!} = \frac{3!}{2!(3-2)!} = 3$$

In general, the number of combination of n distinct objects taken r at a time is denoted by ^nC_r, where $^nC_r = \frac{n!}{(n-r)!r!}$

By definition $0! = 1$ and $1! = 1$.

Some calculators have in built functions for calculating ^nC_r and ^nP_r.

Example 5

(a) A class is to select 3 students from 10 for a prize. In how many ways can the selection be made?

Solution 3 students can be selected in $^{10}C_3 = \frac{10!}{(10-3)!3!} = 120$ ways

(b) In how many ways can a committee of 3 be formed from 8 male and 7 female teachers?

Solution The 3 members are to be selected from 15 teachers. The number of ways of forming the committee is

$$^{15}C_3 = \frac{15!}{12!3!} = \frac{15 \cdot 14 \cdot 13}{3 \cdot 2 \cdot 1} = 455$$

(c) A book shop has in stock 6 different mathematics books and 5 different science books. If 2 mathematics and 3 science books are to be selected, in how many ways can this be done?

Solution The mathematics book can be selected in $^6C_2$ ways and the science book can be selected in $^5C_3$ ways. Each selection of two mathematics books can be taken with each selection of three science books. This gives $^6C_2 \times {^5C_3} = 15 \times 10 = 150$ different selections

(d) A mathematics test has two sections. There are 5 questions in Section A, and 4 questions in Section B. A student must answer 3 questions in all, with at least one question from each section. In how many different ways can a student complete the test?

Solution A student could answer 2 questions from section A and 1 question from section B or 1 question from section A and 2 questions from section B.

Hence, the student can complete the test in $^5C_2 \cdot {}^4C_1 + {}^5C_1 \cdot {}^4C_2 = 40 + 30 = 70$ ways

Exercise 21.2

1. In how many ways can a committee of 3 parliamentarians be selected from a group of 8 parliamentarians?

2. In how many ways can a committee of 4 teachers be selected from a group of 9 teachers?

3. A school is to select 4 of 10 eligible students for a scholarship. In how many ways can the selection be made?

4. There are 25 members in a club. In how many ways can a subcommittee of 3 members be formed?

5. How many different relay teams of 4 persons can be chosen from a group of 12 runners?

6. A disciplinary committee has 3 teachers, 2 administration members and 3 students in it. In how many ways can a subcommittee of 2 teachers, 1 administrator and 1 student be formed?

7. A test has 2 parts. In part 1 a student must do 3 of 5 questions and in part 2 a student must choose 2 of 4 questions. In how many different ways can a student complete the test?

8. A sample of 8 persons is selected for a test from a group containing 10 students and 5 teachers. In how many ways can the 8 persons be selected?

9. A box contains 20 light builds. The quality control engineer will pick a sample of 4 light bulbs for inspection. How many different samples are there?

10. Nine points are marked on a circle. How many chords do they form?

11. There are 12 points on a plane, no three of them being in a straight line. How many triangles can be drawn using the points as vertices?

12. How many triangles can be formed by joining any three of the vertices of a heptagon?

13. Find the number of ways in which a group of eight boys can be split into two groups of four.

14. Three out of ten houses are to be painted white and the rest green. Find the number of ways in which the choice can be made.

15. A team of 11 players is to be chosen from 15 boys. Find the number of ways in which this can be done if there are two boys who refuse to play in the same team.

16. In how many ways can six girls be divided into two teams of three?

17. A school library contains 10 Mathematics and 8 Physics textbooks. In how many ways can a borrower select 2 of each?

18. A committee must contain 3 men and 4 women. In how many ways can the committee be chosen from 10 men and 6 women?

19. From 6 teachers and 5 students a committee of five is to be chosen that includes three students and two teachers. In how many ways can this be done?

20. How many committee of three can be formed from six girls and six boys, if the girls are in majority?

21. From 7 women and 5 men a committee of three is to be formed. The committee must include at least two women. In how many ways can this be done?

22. In how many ways can two taxis, each of which will take at most 4 passengers, take a party of 7 people, if 2 of them refuse to be in the same taxis?

23. There are five male teachers and seven female teachers. How many ways can they form a committee of four with a female as chairperson?

24. A team of 6 people is to be selected from 8 men and 5 women. Find the number of different teams that can be selected if:

(a) there are no restriction

(b) the team contains 4 women

(c) the team contains at least 4 men

25. A committee of 7 members is to be selected from 6 teachers and 9 students. Find the number of different committees that may be selected if:

(a) there are no restrictions

(b) the committee must consist of 2 teachers and 5 students

(c) the committee must contains at most 1 teacher

Review Exercise 21

1. How many different ways can the letters in the word STUDY be arranged?

2. Six athletes run in a race, assuming there are no ties, how many different results are possible?

3. How many ways can a captain and vice-captain be chosen from a football team of 11?

4. How many ways can three examination questions be chosen from seven?

5. If 6 couples go to a party, in how many ways can they pair off to dance?

6. Ama invited 4 friends to go with her to the school speech day. There are 120 different ways in which they can sit together in a row. In how many of those ways is Ama sitting in the middle?

7. Ten children are present at a party. In how many ways can four children be chosen to a game if two children refuse to play the game together?

8. How many 3-digit positive integers are odd and do not contain the digit " 5 "?

9. From a box of 10 identical balls, 4 are to be removed. How many different sets of 4 balls could be removed?

10. A talent contest has 8 contestants. Judges must award prices for first, second and third places. If there are no ties, in how many different ways can the 3 prizes be awarded, and how many different groups of 3 people can get prizes?

11. In how many ways can a student answer 8 out of 10 questions, if he must answer the first five questions?

12. How many 5-digits numbers can be formed from the nine digits 1, 2, 3, $\cdots$, 9:

(a) if no digit is repeated?

(b) if repetitions are allowed?

13. (a) Find how many different 4-digit numbers can be formed from the digits 0. 3, 4, 5, 6 and 7, if each digit may be used only once.

(b) Find how many of these 4-digit numbers are even.

14. How many different 4-digit numbers greater than 5000 can be formed using the digits 1, 3, 5, 6, 7 and 8, if no digit can be used more than once?

15. A 4-digit number is formed by using four of the six digits 2, 3, 4, 5, 6 and 8; no digit may be used more than once in any number. How many different 4-digit numbers can be formed if:

(a) there are no restriction

(b) the number is odd and more than 4000.

16. A committee of 5 members is to be selected from 6 women and 9 men. Find the number of different committees that may be selected if:

(a) there are no restriction

(b) the committee must consist of 2 women and 3 men

(c) the committee must contain at least 3 women

17. A committee of 6 members is to be selected from 8 teachers and 4 students. Find the number of different committees that can be selected if:

(a) there are no restriction

(b) the committee contains all 4 students

(c) the committee contains at least 4 teachers

22 Probabilities

22.1 Probability

Probability of an event is a measure of how likely it is that the event will occur. Probability can assume a value between 0 and 1, inclusive. Some events will never occur, and others will always occur. The probability of an event that is impossible is 0 and the probability of an event that is certain is 1. Any other event will have a probability lying within the range 0 to 1. If an event is very unlikely to occur, the probability would be close to 0. If an event is very likely to occur, the probability would be close to 1.

The probability of an event can be expressed as a fraction, percent or decimal.

Basic Probability Terms

Experiments

We can determine the probability of an event by collecting data from an experiment. An experiment is any process that leads to a set of results. Examples of experiments include tossing a coin, rolling a die and drawing a card from a pack of cards.

Trials and Outcomes

Suppose a boy draws a ball at random from a bag containing 10 balls. Each individual draw is considered a trial. The possible results of each trials are its outcomes. For instance, the outcomes of tossing a die are 1, 2, 3, 4, 5 and 6, and the outcomes of tossing a coin are a head and a tail.

Sample Space

The set of all possible outcomes of an experiment is called the sample space. The sample space when a coin is tossed is {head, tail}. The sample space when a die is rolled is {1, 2, 3, 4, 5, 6}.

Events

Any subset of a sample space is called an event. For example, the event {1, 3, 5} is satisfied by rolling either a 1, 3 or 5, and the event {2} is satisfied when a 2 is rolled. If an event has exactly one element, it is called a simple event.

Assignment of Probabilities

In any particular toss of a coin, there is no way to know whether a head or a tail will occur. However, if the coin is tossed many times about $\frac{1}{2}$ of the tosses will be heads.

Thus, when a coin is tossed a head is as likely as a tail to occur. Hence, it is reasonable to assign a probability of $\frac{1}{2}$ for obtaining a head (or tail) in any particular toss of a coin.

Each roll of a die results in one of the numbers 1, 2, 3, 4, 5 or 6. If we roll the die many times, any face is as equally likely to occur as any other, provided the die is fair. Here we assign a probability of $\frac{1}{6}$ for obtaining a particular face.

In general, if a sample space S has n equally likely outcomes, the probability assigned to each outcome is $\frac{1}{n}$.

Probability of an Event

If an experiment has n equally likely outcomes, among which an event E occurs r times, then the probability that the event E occurs, denoted P (E) is given by

$$P(E) = \frac{\text{number of times the event E occurs}}{\text{total number of possible outcomes}} = \frac{r}{n}$$

For example, if a bag contains 15 red balls and 5 blue balls, the probability that a ball selected at random is red is the number of red balls divided by the total number of balls, i.e. $\frac{15}{20} = \frac{3}{5}$.

The above definition is called the classical definition of probability.

Relative Frequency

The classical definition of probability covers cases where the outcomes are equally likely to occur. If the possible outcomes are not equally likely to occur, the probability that the event will occur can be estimated by collecting data from an experiment.

If an event E occurs in a number of independent random trials the ratio

$$\frac{\text{number of times the event occurs}}{\text{total number of trials}}$$

is called the relative frequency. The relative frequency is used to estimate the probability of an event whose outcomes are not equally likely. If you take a large sample or carry out a larger number of trials you get a better estimate of the probability. The relative frequency get closer to the actual probability as the number of trails (or the size of the sample) increases. For example, if 36 of 300 boys are 16 years old the probability of selecting at random a 16 year old is 36/300, i.e., 0.12.

The definition of probability based on previous known results is called an empirical definition of probability.

Complement of an Event

The complement of an event E is the set of outcomes in the sample space not in E. The complement of an event E is denoted by E', and read not E.

Since an event and its complement together make all the possible outcomes,

$$P(E) + P(E') = 1 \text{ or } P(E') = 1 - P(E).$$

Sometimes it would be easier to find the probability of an event E by subtracting $P(E')$ from 1. For example, if the probability that it will rain tomorrow is 0.8 then the probability that it will not rain tomorrow is 1 − 0.8 or 0.2.

Example 1

(a) If a fair six-sided die is rolled, what is the probability of rolling a 5 or a 6?

Solution Here the sample space S = {1, 2, 3, 4, 5, 6} and the event E = {5, 6}. So $P(E) = \frac{2}{6} = \frac{1}{3}$.

(b) In a class of 40 students 15 students regularly complete their homework. If a student is picked at random, what is the probability that he did not complete his homework?

Solution If A = {students who complete their homework}, then $P(A') = 1 - \frac{15}{40} = \frac{5}{8}$

Exercise 22.1

1. Two coins are tossed. List the elements of the sample space. What is the probability that one coin falls head.

2. What is the probability that at least one head occur in the toss of two coins.

3. A bent coin was tossed 60 times, and Heads came up 20 times. What is the probability of obtaining Heads with this coin?

4. If a letter is chosen at random from the word, "friend", find the probability of getting a vowel.

5. In a survey of 500 students, 200 said they read the Graphic news paper. If a student is picked at random, what is the probability that he or she read the Graphic?

6. A card is drawn from a pack. Find the probability that it is red.

7. A die is thrown, what is the probability of throwing an even number.

8. What is the probability that a number greater than 2 results from a single throw of a die?

9. A boy chooses a number at random from 1 to 20 inclusive. What is the probability that the number is a multiple of 3?

10. A bag contains 30 balls of which 16 are red and the rest blue. If a ball is drawn at random, find the probability that it is blue?

11. Calculate the probability of selecting at random a winning horse in a race in which 10 horses are running.

12. A ball is drawn at random from a box containing 15 red balls and 20 black balls. Find the probability of drawing a red ball.

13. The probability that a boy does not hit a target is $\frac{2}{3}$. Find the probability that the boy will hit the target.

14. The weatherman reports that the probability that it will rain during the week end is $\frac{3}{10}$. What is the probability that it will not rain?

15. The probability of a boy passing a test is 0.7. Find the probability that he will fail the test.

16. In a batch of 45 lamps there are 10 faulty lamps. If one lamp is drawn at random, find the probability that it is not faulty.

17. Determine the probability of selecting at random a man from a group of 50 people containing 30 women.

18. A card is drawn from a pack. What is the probability that it is not an Ace?

19. In a raffle there is one prize of GH¢ 10 and four prizes of GH¢ 5. Thousand tickets are sold. Find the probability that a man won a prize.

20. The probability that a football club win is 0.6 and draw or a defeat are equally likely. Find the probability that the team will draw.

22.2 Laws of Probability

There are two basic laws of probability, the Addition Law and the Multiplication Law. Before we consider these laws, we need to define the following terms.

Mutually Exclusive Events

Two events are mutually exclusive if the two events cannot happen at the same time. For example, when you toss a coin the events "tossing a head" and "tossing a tail" are mutually exclusive events because it is not possible for both events to occur simultaneously.

Similarly, when you roll a die the event " rolling a 3" and " rolling a 2 " are mutually exclusive events. However, the events, "rolling a 2" and "rolling an even number" are not mutually exclusive because it is possible for both to occur simultaneously. If a 2 is rolled then, both events would be satisfied.

Independent Events

Two events are independent if the probability of one event is not affected by the occurrence of the previous event. For example, if you throw a die on two occasions, the outcome of the first throw does not affect the outcome of the second throw. Similarly, outcomes of two successive tosses of a coin are independent, since the second toss cannot be influenced by the outcome of the first toss.

The probability of drawing a white ball from a bag containing 3 white balls and 2 black balls is $\frac{3}{5}$. If the ball is replaced before a second ball is drawn the probability of drawing a white ball on the second occasion is also $\frac{3}{5}$. Because the ball is replaced between draws the second selection is said to be with replacement. Since the probability of the second draw is the same as that of the first draw the two events are independent.

Dependent Events

Two events are dependent if the probability of one event is affected by the occurrence of the previous event. For example, if a ball is drawn from a bag containing 3 white balls and 2 black balls the probability of drawing a white ball is $\frac{3}{5}$. If the first ball drawn is not replaced, the probability of selecting a white ball on the second occasion is now $\frac{2}{4}$ i.e. $\frac{1}{2}$, because one ball has been removed, leaving four balls in the bag of which two are white. The two events are dependent because the occurrence of one event affects the probability of the other. Since the ball is not replaced between draws, the second selection is said to be without replacement.

The Addition Law of Probability

If a coin is tossed, the probability it would fall head is $\frac{1}{2}$. The probability that it would fall tail is also $\frac{1}{2}$. It is certain that the coin will fall on one side or the other.

Thus, the probability that it will fall either head or tail is 1. Notice that $\frac{1}{2} + \frac{1}{2} = 1$. This example is an illustration of the addition law of probability.

The Addition Law

If two events are mutually exclusive then the probability of any one event occurring is the sum of the individual probabilities. For instance, if A and B are two mutually exclusive events, then

$P(A \text{ or } B) = P(A) + P(B)$

The symbol ∪ can be used for 'or'.

Example 2

What is the probability of throwing a 5 or a 6 with a throw of a die?

Solution $P(throwing\ a\ 5\ or\ a\ 6) = \frac{1}{6} + \frac{1}{6} = \frac{1}{3}$

Exercise 22.2(a)

1. If P (A) = 0.25 and P (B) = 0.30, find $P(A \cup B)$. If A, B are mutually exclusive.

2. If A and B represent two mutually exclusive events such that P (A) = 0.35 and P (B) = 0.50, find $P(A \cup B)$.

3. If A and B represent mutually exclusive events $P(A) = 0.75$ and $P(B) = 0.55$, find $P(A' \cup B')$.

4. $P(A \cup B) = 0.60$ and $P(A) = 0.20$, find $P(B)$ given that the events A and B are mutually exclusive.

5. The events A and B are such that $P(A) = \frac{1}{6}$, $P(B) = \frac{1}{5}$, $P(A \cup B) = \frac{11}{30}$. Show that the events A and B are mutually exclusive.

6. A card is drawn at random from a standard deck of 52 cards, find the probability of drawing a card that is either a queen or king of any suit.

7. A ball is picked at random from a box containing 3 white, 5 red and 7 blue balls. Find the probability that a white or red ball is picked.

8. A ball is picked at random from a box containing 5 red, 8 blue and 7 green balls. Find the probability that neither red nor green ball is picked.

9. A ball is picked at random from a box containing 5 blue, 4 green and 3 brown balls. Find the probability a blue is not picked.

10. Ama has probability $\frac{1}{10}$ of becoming the next girls' prefect, and that Esi's chance is $\frac{1}{15}$. What is the probability of either of them becoming the girls prefect.

11. A roulette wheel has the numbers 1 to 40. What is the probability that the number which comes up will be either 15 or 36?

12. A card is drawn from a standard pack of 52 cards. What is the probability that it is either a Spade or Diamond?

13. A card is drawn from a standard pack of 52 cards. What is the probability that it is either a 3 or a picture card? (The picture cards are Kings, Queens and Jacks)

14. One scholarship is awarded each year to a student in a school. Kojo has probability $\frac{2}{5}$ of being awarded the scholarship next year, and Kofi's chance is $\frac{1}{3}$. What is the probability of either of them being awarded the scholarship?

Multiplication Law

Suppose we toss a fair coin two times; what is the probability that a head will turn up both in the first and second toss. The four possible outcomes are shown in the following table.

1st Toss	2nd Toss
H	H
H	T
T	H
T	T

Each of the outcomes is equally likely to occur. The two tosses of the coin will produce our specified pair of events, HH, in $\frac{1}{4}$ of the time. The probability of obtaining two heads in two tosses of a fair coin is therefore $\frac{1}{4}$. This result suggest that the probability of obtaining two heads in two tosses of a coin is obtained by multiplying the individual probabilities of the two events, i.e. $\frac{1}{2} \times \frac{1}{2} = \frac{1}{4}$. This example illustrates the multiplication law of probability.

The Multiplication Law

If two events are independent, then the probability of both events occurring is the product of the individual probabilities. For instance if A and B are two independent events then

$P(A \text{ and } B) = P(A) \times P(B)$

The symbol ∩ can be used for 'and'.

Example 3

(a) The probability that a man will be alive in 10 years is 0.6 and the probability that his wife will be alive in 10 years is 0.7. What is the probability that they will both be alive in 10 years?

Solution $P(both\ alive\ in\ 10\ years) = 0.6 \times 0.7 = 0.42$

(b) A box contains 6 red and 4 black balls. Find the probability that two balls drawn successively are both red if the balls are:

(i) replaced

Solution Since the first ball is replaced before the second draw, the two events are independent. The probability that both balls are red is $\frac{6}{10} \times \frac{6}{10} = \frac{9}{25}$.

(ii) not replaced

Solution Since the first ball is not replaced, the second draw is dependent on the first draw. The probability that both balls are red is $\frac{6}{10} \times \frac{5}{9} = \frac{1}{3}$

Exercise 22.2(b)

1. If A and B are independent events and if $P(A) = 0.3$ and $P(B) = 0.5$, find $P(A \cap B)$.

2. If A and B are independent events and if $P(A) = 0.6$ and $P(A \cap B) = 0.3$, find $P(B)$.

3. A and B are two events such that $P(A) = \frac{4}{21}$, $P(B) = \frac{7}{10}$ and $P(A \cap B) = \frac{1}{9}$. Show that A and B are independent.

4. Two fair dice are rolled. What is the probability that they will all be twos?

5. A marksman hits a target with probability $\frac{3}{5}$. Assuming independence for successive firings, find the probability of getting one miss followed by two hits.

6. If the probability that a student study Mathematics is $\frac{6}{7}$ and the probability he study Physics is $\frac{1}{5}$, and assuming independence, find the probability that a student chosen at random study both Mathematics and Physics.

7. Two letters are chosen from the word SOCCER. What is the probability that C's are chosen.

8. In a multiple choice exam, each question has 5 possible answers. A candidate answers three questions at random, find the probability that three answers are wrong.

9. A box contains 6 red and 12 white balls. If two balls are chosen, what is the probability that the first is red and the second is white?

10. A third of the students of a school like reading and a tenth like listening to music. If a student is picked at random find the probability that the student like reading and listening to music.

Non- exclusive Events

Two events are non- mutually exclusive if they have one or more outcomes in common. For example, in tossing a die, the event a multiple of 2 and the event of obtaining a multiple of 3 can occur together if a 6 is thrown.

If A and B are two non- mutually exclusive events then the probability of one or both of the two events happening is the sum of the probabilities of each event less the probability of both events happening together,

That is $P(A \cup B) = P(A) + P(B) - P(A \cap B)$, if $A \cap B \neq \emptyset$

When the two events are mutually exclusive, then $P(A \cap B) = 0$ and the law simplifies to $P(A \cup B) = P(A) + P(B)$

For example, the probability of throwing at least one six in a throw of two dice is
$P(at\ least\ one\ six) = \frac{1}{6} + \frac{1}{6} - \frac{1}{6} \times \frac{1}{6} = \frac{11}{36}$

Exercise 22.2(c)

1. A card is drawn from a standard pack of 52 cards. What is the probability that it is either a Spade or an Ace?

2. Two fair dice are rolled. What is the probability that either of them will be a 2?

3. The probability that fufu is on the menu for lunch is $\frac{2}{3}$, and the probability of beans is $\frac{1}{4}$. The probability of both is $\frac{1}{5}$. Find the probability that either fufu or beans will be on the menu.

4. The probability of a boy passing Mathematics and Science are 0.7 and 0.6 respectively. The probability of him passing both is 0.5. What is the probability that he will pass either subjects?

5. The probability that a day is wet is $\frac{1}{5}$, and that it is windy is $\frac{1}{3}$. The probability that it is both wet and windy is $\frac{1}{6}$. Find the probability that it is either wet or windy.

6. Esi is taking courses in both Mathematics and English. She estimates her probability of passing Mathematics is 0.3 and English is 0.7, and she estimates her probability of passing at least one of them is 0.8. What is her probability of passing both courses?

7. At a car repair shop, the manager has found that a car will require a tune-up with a probability of 0.5, a brake job with a probability of 0.2 and both with a probability of 0.3. What is the probability that a car requires either a tune-up or a brake job?

8. A factory needs two raw materials, A and B. The probability of not having an adequate supply of material A is 0.07, whereas the probability of not having an adequate supply of material B is 0.05. A study shows that the probability of a shortage of both A and B is 0.03. What is the probability of the factory being short of either material A or B?

22.3 Conditional Probability

Recall that whenever we roll a single die the sample space S is {1, 2, 3, 4, 5, 6}. Consider the experiment of rolling a single die. If we roll "a 4" the only outcome is 4, and the probability is $\frac{1}{6}$. However, if an even number is rolled, then the possible outcomes are 2, 4, 6, and the probability of rolling an even number is $\frac{3}{6}$. Since only three equally likely outcomes are now possible, the probability of rolling "a 4" is $\frac{1}{3}$. The ratio $\frac{1}{3}$ represents the conditional probability of rolling "a 4" given that an even number is rolled. If we let A be the event "a 4 is rolled" and let B be the event "an even number is rolled", then the conditional probability of A given B, denoted by P (A/B), is $\frac{1}{3}$.

Notice that in computing P (A/B), we form the ratio of the numbers of those entries in both A and B, with the numbers that are in B, i.e. $P(A/B) = \frac{n(A \cap B)}{n(B)}$, if $n(B) \neq 0$.

Now $P(A/B) = \frac{n(A \cap B)/n(S)}{n(B)/n(S)}$

$= \frac{P(A \cap B)}{P(B)}$

From this equation we get $P(A/B) = \frac{1/6}{3/6} = \frac{1}{3}$

Also from $P(A/B) = \frac{P(A \cap B)}{P(B)}$ we obtain

$P(A \cap B) = P(B) \cdot P(A/B)$

Note that if A and B are independent then P (A/B) = P (A).

Example 4

(a) A toy manufacturer knows that 15 % of the toy he produces are guns and 5 % of all toys he produces are gun which are defective. What is the probability that a gun selected at random is defective?

Solution   If A is the event that a toy is defective and B is the event that a toy is a gun.

$$P(A/B) = \frac{P(A \cap B)}{P(B)} = \frac{0.05}{0.15} = \frac{1}{3}$$

Hence the probability that any given gun is defective is $\frac{1}{3}$

(b) A pair of fair dice is thrown. If at least one of them shows a 5, what is the probability that the sum is 8?

Solution   The sample space contains 36 ordered pairs. If we let A be the event the sum is 8 and let B be the event at least one die shows a 5, then

$A = \{(2, 6), (3, 5), (4, 4), (5, 3), (6, 2)\}$

$B = \{(1,5), (2,5), (3,5), (4,5), (5,5), (6,5), (5,1), (5,2), (5,3), (5,4), (5,6)\}$

Hence $A \cap B = \{(3, 5), (5, 3)\}$

Thus $P(A/B) = \frac{P(A \cap B)}{P(B)} = \frac{2/36}{11/36} = \frac{2}{11}$

(c) A box contains 3 red balls and 2 black balls. Two balls are drawn at random, without replacement. What is the probability that both balls are red?

Solution   Let A be the event that the first ball drawn is red and B be the event that the second ball drawn is red.

Since there are 5 balls in the box, of which 3 are red, it follows that $P(A) = \frac{3}{5}$.

Because the ball is not replaced only 4 balls will be left in the box, of which 2 are red. So,

$$P(B/A) = \frac{2}{4} = \frac{1}{2}.$$

Hence $P(A \cap B) = P(A) \cdot P(B/A)$

$$= \frac{3}{5} \times \frac{1}{2}$$

$$= \frac{3}{10}$$

Exercise 22.3(a)

1. If the events A and B are such that $P(A) = 0.7$, $P(B) = 0.8$ and $P(A \cap B) = 0.6$, find P (A/B).

2. A pair of fair dice is thrown. If at least one of them shows a 2, what is the probability that the total is 7?

3. For a 3-child family find the probability of exactly 2 girls given that the first child is a girl.

4. A card is drawn at random from a standard pack of 52 cards. What is the probability that the card is a red ace given that an ace was picked?

5. A card is drawn at random from a standard pack of 52 cards. What is the probability that the card is a red ace given that a red card was picked?

6. A group of 500 teachers attending a conference, includes 80 Mathematics teachers and 240 female. There are 30 female Mathematics teachers. If a female is chosen at random, what is the probability that she is a Mathematics teacher?

7. Fifty percent of all cars manufactured by a company are deluxe models. Twenty percent of all models are models with defective steering. What is the probability that a car has defective steering given that it is a deluxe model?

8. A box contains 5 white and 4 blue balls. Two are drawn without replacement. Find the probability that both are white.

9. A box contains 4 red, 5 black and 6 white balls. Two balls are drawn without replacement. Find the probability that the first is red and the second is white.

10. If 2 cards are drawn from a standard pack of 52 cards without replacement, what is the probability that the second card is a queen?

11. Two students are chosen from 5 girls and 4 boys. What is the probability that they are both boys?

12. In a certain school, 25% of the students are girls and 75% are boys. Also 10% of the girls are left-handed and 5% of the boys are left-handed.

(a) What is the probability that a student chosen at random is a girl and left-handed?

(b) What is the probability that a student chosen at random is left-handed?

(c) What is the probability that a student is a girl given that the student is left-handed?

Probability Tree Diagrams

When we have successive experiments, a good way to illustrate the possible outcomes is by a tree diagram. Consider the example below.

A box contains 7 black balls and 5 red balls. If two balls are drawn without replacement, find the probability that one red ball is drawn.

First Draw	Second Draw	Outcome	Probability
$\frac{7}{12}$ B	$\frac{6}{11}$ B	BB	$\frac{7}{12} \times \frac{6}{11}$
	$\frac{5}{11}$ R	BR	$\frac{7}{12} \times \frac{5}{11}$
$\frac{5}{12}$ R	$\frac{7}{11}$ B	RB	$\frac{5}{12} \times \frac{7}{11}$
	$\frac{4}{11}$ R	RR	$\frac{5}{12} \times \frac{4}{11}$

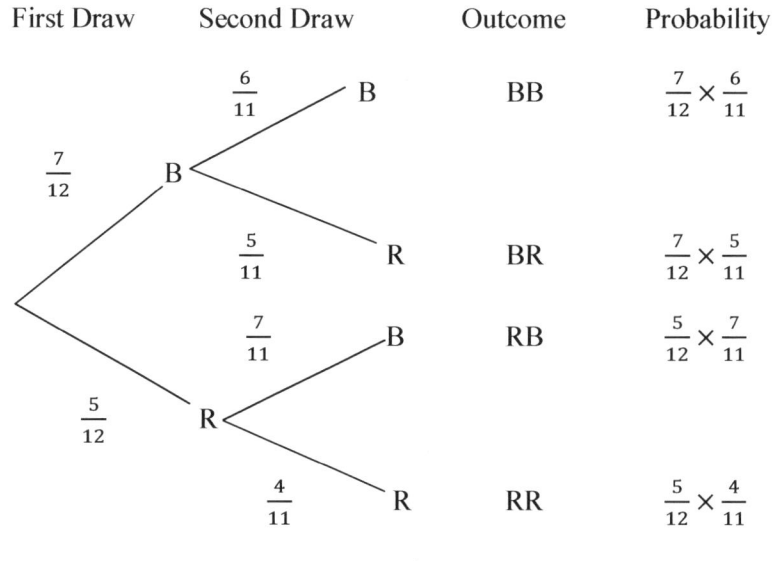

Figure 22.3

The first draw has two outcomes represented by the two initial branches as shown in Figure 22.3. The outcome is written at the end of the branch. The branch leading to B has probability $\frac{7}{12}$, written beside the branch. The branch leading to R has probability $\frac{5}{12}$. Since these are the only possible outcomes, the two probabilities add up to 1. Each of the initial branches has two other branches. The branch B to B is the conditional probability of drawing a black ball on the second draw after a black on the first draw, which is $\frac{6}{11}$. The branch from B to R is the conditional probability of drawing a red ball on the second draw after a black on the first draw, which is $\frac{5}{11}$. The remaining entries are obtained in a similar manner. The probability of drawing two balls is obtained by multiplying the branch probabilities. For example, the probability that the first ball drawn is black and the second ball drawn is red is $\frac{7}{12} \times \frac{5}{11} = \frac{35}{132}$.

The event of drawing one red ball occurs in two ways; drawing a red ball first and then a black ball or drawing a black ball first and then a red ball. Since the two events are mutually exclusive, the probability of drawing one red ball is sum of these two probabilities, i.e.,
$P(R \cap B) = P(R) \cdot P(B/R) + P(B) \cdot P(R/B)$

$$= \frac{5}{12} \times \frac{7}{11} + \frac{7}{12} \times \frac{5}{11} = \frac{35}{66}.$$

Notice that:

1. The sum of probabilities on adjacent branches of the tree is always 1.

2. The probability of each outcome is found by multiplying the probabilities along the branches.

3. The sum of the probabilities of all the possible outcomes is always 1.

Exercise 22.3(b)

1. A box contains 7 red and 3 white balls. If two balls are drawn without replacement, what is the probability that both balls are red?

2. A box contains 5 white and 3 black balls. If two balls are drawn at random one after the other without replacement, what is the probability that both balls are of the same colour?

3. A bag contains 5 red and 6 black balls. Two are drawn without replacement, what is the probability that they are different colours?

4. A box contains 4 white, 3 blue and 2 green balls. Two balls are drawn at random without replacement; use a tree diagram to find the probability that one white and one blue ball are drawn.

5. One bag contains 3 white balls and 4 blue balls and another contains 4 white balls and 5 blue balls. If one bag is selected at random and one ball is picked at random from that bag, what is the probability that it is white?

6. The probability that a man and his wife pass a particular exam are $\frac{2}{3}$ and $\frac{3}{4}$ respectively. Assuming independence, find the probability that at least one of them passes the exam.

7. A company has two plants to manufacture refrigerators. Plant I manufactures 70% of the refrigerators and Plant II manufactures 30%. At Plant I, 85 out of every 100 refrigerators are frost free. What is the probability that a customer obtains a frost free refrigerator if he buys a refrigerator from the company?

8. In a certain population of people 25% are female and 75% are male. Also, 10% of the female are left-handed and 5% of the male are left-handed. What is the probability that a person chosen at random is female and left-handed?

22.4 Using Counting Method

Many problems in probability require the ability to count the elements of a set. Example 5 shows how the counting principles is used to solve probability problems.

Example 5

(a) A bag contains 10 blue balls, 8 green balls and 7 red balls. Three balls are selected at random without replacement. Find the probability that they are all blue.

Solution The three balls can be selected in $^{10}C_3$ ways. Any three balls can be selected in $^{25}C_3$ ways. If E denotes the event of selecting three blue balls then

$$P(E) = \frac{^{10}C_3}{^{25}C_3} = \frac{6}{115} = 0.052$$

(b) A committee of 12 teachers, of which 5 are female, is formed to plan an anniversary programme for a school. If 3 members of the committee are chosen at random to form a subcommittee, what is the probability that:

(i) all 3 are female

Solution The number of elements in the sample space is equal to the number of combination of 12 teachers taken 3 at a time, i.e. $^{12}C_3 = 220$. Let E denote the event 3 female are selected. Then E can occur in $^5C_3$, i.e. 10 ways. Hence,

$$P(E) = \frac{^5C_3}{^{12}C_3} = \frac{10}{220} = 0.05$$

(ii) exactly 2 are female

Solution Let F denote the event 2 female are selected. To obtain 2 female when 3 teachers are chosen requires that we select 2 female from the 5 available female and 1 male from the 7 male. This can be done in $^5C_2 \cdot {^7C_1} = 10 \times 7 = 70$ ways. Hence,

$$P(F) = \frac{^5C_2 \cdot {^7C_1}}{^{12}C_3} = \frac{70}{220} = 0.32$$

(iii) at least 2 are female

Solution Let G denote the event at least 2 female. The event G is equivalent to either selecting exactly 2 or exactly 3 female. These events are mutually exclusive, therefore $P(G) = P(E) + P(F) = 0.05 + 0.32 = 0.37$

Exercise 22.4

1. What is the probability that a three-digit number has one or more repeated digits?

2. A fair coin is tossed 5 times. Find the probability that exactly 3 heads appear.

3. A person is dealt 4 cards from a standard pack of 52 cards. What is the probability that they are all hearts.

4. A person is dealt 3 cards from a standard pack of 52 cards. What is the probability that there were no king?

5. If the five letters in the word VOWEL are rearranged, what is the probability the word will begin with L?

6. If the eight letters in the word COMPUTER are rearranged, what is the probability the word will end in M and begin with U?

7. A bag contains 8 red, 5 green and 3 blue balls. Three are selected at random without replacement, find the probability that two are red and one is green.

8. A bag contains 6 blue, 5 green and 4 red balls. Three are selected at random without replacement. Find the probability that there is one of each colour.

9. A box contains 15 light bulbs of which 6 are defective. Three light bulbs are picked at random from the box, what is the probability that at least 2 are defective?

10. A box contains 12 light bulbs of which 5 are defective. Three light bulbs are picked at random from the box, what is the probability that exactly 2 are defective?

22.5 The Binomial Distribution

The binomial distribution is the probability distribution for the number of successes in a sequence of Bernoulli trials.

Bernoulli Trials

Repeated trials of an experiment are called Bernoulli trials if the following conditions are satisfied.

1. The same experiment is repeated several times.

2. Each trial results in one of two outcomes, one outcome is called success with probability, p, and the other failure with probability, q.

3. The trials are independent.

4. The probability of each outcome remains the same for each trial.

Notice that the probability of failure is equal to $q = 1 - p$.

A coin-tossing experiment is a common example of a Bernoulli trial. The experiment satisfies the following characteristic of the Bernoulli trail.

1. Each trail results in one of two outcomes (head or tail)

2. The probability of a particular outcome (say heads) remains the same from trail to trail.

3. The outcome of any trail is independent of the outcome of any other trail.

The Binomial Probability Distribution

A binomial experiment consists of a sequence of trials with probability of success p on each trial. The probability of r successes in n trials is given by

$$P(x=r) = {}^nC_r p^r q^{n-r}$$

where ${}^nC_r$ = number of ways of getting exactly r successes in n trails.

Example 6

(a) A fair die is thrown 10 times. Find the probability that there are 3 sixes.

Solution A success is throwing a six. The probability of success, p is $\frac{1}{6}$, so $q = 1 - \frac{1}{6} = \frac{5}{6}$. A trial is a single throw of a die hence the number of trials, n is 10.

Using the binomial probability formula with $r = 3$ yields

$$P(x=3) = {}^{10}C_3 {}^7\left(\tfrac{1}{6}\right)^3 \left(\tfrac{5}{6}\right)^7 = 0.1541$$

(b) Suppose that 5% of the items produced by a factory are defective. If 6 items are chosen at random, find the probability that at least 1 is defective.

Solution   A success is the choice of a defective item. The probability of success, p is 0.05, so $q = 0.95$. A trial is choosing a single item, hence the number of trials, n is 6.

The probability that at least one defective item is chosen can be obtained by adding each of the six individual probabilities. However, it would be easier to use the complement of the event "at least one six is chosen". Since the probabilities of two complementary events is 1 we get

$$P(x \geq 1) = 1 - P(x = 0)$$

$$= 1 - {}^6C_0 (0.05)^0 (0.95)^6$$

$$= 0.2649$$

Exercise 22.5

1. A fair coin is tossed 8 times. What is the probability of obtaining 1 head?

2. A fair coin is tossed 6 times. What is the probability of obtaining at least 4 tails?

3. A fair coin is tossed 8 times. What is the probability of obtaining 2 tails?

4. A fair die is tossed 6 times. What is the probability of obtaining 2 sixes?

5. A fair die is tossed 8 times. What is the probability of obtaining at least 1 six?

6. What is the probability of obtaining a total of 7 exactly twice in five tosses of a pair of fair dice?

7. A trail with probability $\frac{1}{3}$ of success is repeated independently six times. Let x be the number of successes. Find the probability that:

(a) $x = 2$ (b) $x > 0$ (c) $x \leq 4$

8. When an arrow is fired at a target, the probability of hitting a target is 0.1. If the arrow is thrown 20 times, and each throw is independent, find the probability of:

(a) exactly 3 hits

(b) at least one hit

9. A multiple choice exam contains 20 questions, with 5 possible answers for each question. A candidate answers the question at random. Find the probabilities that:

(a) he gets none right

(b) he gets 4 right

10. One in ten of articles manufactured by a company is defective. If 6 are bought, find the probability that two are defective.

11. Suppose that 5% of the items produced by a factory are defective. If 8 items are chosen at random, what is the probability that fewer than 3 are defective?

12. What is the probability that a family with exactly 6 children will have 3 boys and 3 girls?

13. Suppose that 60% of the voters intend to vote for a particular candidate. What is the probability that a survey polling 8 people reveals that 3 or fewer intend to vote for that candidate?

14. In a 15-item true-false examination, what is the probability that a student who guesses on each question will get at least 10 correct answers?

15. In a 20-item true-false examination, a student has 0.8 probability of correctly answering each question, what is the probability that this student will this student will answer exactly 12 questions correctly?

Review Exercise 22

1. Two coins are tossed simultaneously. Write down the sample space for the experiment.

2. An experiment consists of selecting a family from the set of families with three children. Observing whether each child is a boy (B) or a girl (G), write down the sample space for this experiment.

3. A blue die and a green die are thrown simultaneously. Write down the set representation of each of the following events;

A = the sum of the scores is divisible by 5

B = an odd score less than 3 is obtained on the blue die

4. Two coins are tossed simultaneously. Write down the set representation of the event, at least 1 head is tossed.

5. If the probability is 0.78 that a boy will be late for work at least once next week, what is the probability that he will not be late for work next week?

6. A bag contains 30 balls which are identical except for colour. 12 of the balls are red, 8 are blue and the rest write. A ball is picked at random, what is the probability that it is either blue or white?

7. In a survey of 600 voters, 215 said they will vote for the liberal party, 100 will vote for the conservative and the rest were undecided. If a voter is chosen at random, what is the probability that he or she is either undecided or will vote for the liberal party.

8. A card is selected randomly from a standard pack of 52 cards. What is the probability that a black card is drawn.

9. A card is drawn from a standard pack of 52 cards. What is the probability that it is either black or a Heart?

10. Two fair dice are rolled. What is the probability that either the first is a 2 or the second is a 3?

11. In a throw of two fair dice, what is the probability that the number on one die is double the number on the other?

12. A card is selected randomly from a pack of 52 cards. What is the probability that it is either a Heart or a 9?

13. If an integer is randomly selected from all positive 2-digit integers, find the probability that the integer chosen has:

(a) a "5" in the tens place

(b) at least one "5"

(c) no "5" in either place

14. In a box of 10 items, 2 are defective.

(a) If one item is chosen randomly from the box, what is the probability that it is not defective?

(b) If two parts are randomly chosen from the box, without replacement, what is the probability that both are defective?

15. $P(A \cup B) = 0.60$ and $P(A) = 0.20$.

(a) Find $P(B)$ given that the events A and B are mutually exclusive.

(b) Find $P(B)$ given that events A and B are independent.

16. The probabilities of a man and his wife passing a certain examination are 0.8 and 0.7 respectively. Find the probability that:

(a) both will pass the examination

(b) at least one of them will pass the examination

(c) neither of them will pass the examination

17. The probability that Kofi hits a target is $\frac{1}{4}$. The probability for Kojo and Kwame hitting the target are $\frac{1}{3}$ and $\frac{2}{5}$ respectively. If they all fire together, what is the probability that at least one shot hits the target?

18. A box contains 5 white balls and 3 black balls. If two balls are drawn at random one after the other without replacement, what is the probability that both balls are of the same colour.

19. One bag contains 4 white balls and 2 black balls and another contains 3 white balls and 5 black balls. If one ball is drawn from each bag, find the probability that the two balls are of different colours.

20. A bag contains 6 white and 4 red identical balls. If the balls are drawn at random, one after the other without replacement, what is the probability that:

(a) the first three balls are white?

(b) the first red ball is picked at the fifth draw?

(c) a red ball is picked at the second draw?

21. A bag contains 7 red and 2 blue, and a second bag contains 5 red balls and 1 blue balls. One of the bags is selected at random and a ball is drawn from it. Find the probability that the ball drawn is blue.

22. A bag contains 4 black and 7 white balls, and a second bag contains 3 black, 1 white and 4 yellow balls. One bag is selected at random and a ball is drawn, what is the probability that the ball drawn is black?

23. A box contains 40 items of which 5 are defective. If two items are selected at random, find the probability of selecting exactly one defective item, both with and without replacement.

24. A committee of 5 people including at least one woman is to be selected from a group of 5 men and 4 women. Calculate, correct to three decimal places, the probability that there will be fewer women than men on the committee.

25. A committee of 3 is formed from a panel of 5 men and 3 women. Find the probability that at least one woman is on the committee.

26. A committee of 5 members is to be formed from a teaching staff of 9 women and 3 men. What is the probability that the committee includes exactly two men?

27. A box contains five blocks numbered 1, 2, 3, 4 and 5. A boy picks a block and replaces it. Another boy then picks a block. What is the probability that the sum of numbers they picked is even?

28. A bag contains 6 blue balls, 5 green balls and 4 red balls. Three balls are selected at random without replacement. Find the probability that there is one of each colour.

29. If the probability of hitting a target is $\frac{1}{5}$ and 10 shots are fired independently, what is the probability of the target being hit at least twice?

30. In an examination 10% of the candidates passed. If 8 of the candidates are selected at random, calculate the probability that:

(a) exactly 5 failed

(b) at most 3 passed

31. A survey of a group of students showed that 2 out of every 10 students study science. If 5 students are randomly selected from the group, find correct to three decimal places the probability that:

(a) exactly 3 study science

(b) less than 4 study science

(c) at least 2 study science

32. The probability that items produced by a factory is defective is $\frac{2}{5}$. If 5 items are selected at random, find, correct to three decimal places, the probability that:

(a) exactly 2 items are defective

(b) none of the items is defective

(c) at least 1 item is defective

23 Statistics

23.1 Organization of Data

Statistics is the science of collecting, organising, analysing and interpreting data. Data can be summarized and represented in various forms including: tables, graphs and other diagrams.

Frequency Distribution

The following is a record of the ages of boys at a party:

```
12  16  15  12  15  16
13  18  15  16  16  13
15  17  14  13  18  17
14  15  16  16  17  14
15  14  16  17  14  16
```

Table 1

The data in Table 1 is presented in a frequency distribution shown in Table 2. The frequency distribution list the values arranged in order of magnitude (shown in the first column) and their corresponding frequencies.

Ages/ yrs	Tally	Frequency
12	//	2
13	///	3
14	////	5
15	//// /	6
16	//// ///	8
17	////	4
18	//	2

Table 2

The tally marks are used to determine the frequency with which each different value occurs. The tally is obtained by inspecting each of the values in turn and then putting 1's in the appropriate rows as shown in the second column of Table 2. Every fifth "/" is shown as an oblique line crossing the four previous "1's" to make counting easier.

The frequency distribution provides a quick summary of the data. Certain information available now becomes more evident. A look at the table clearly shows that 16 is the most frequently occurring age group and two boys were 12 years old.

When summarising large masses of data it is often useful to distribute the data into groups called class intervals and to determine the number of individuals belonging to each class interval, called the class frequency. Table 3 shows the frequency distribution of the height (mm) of a number of seedlings. Such a table is called a grouped frequency distribution or a grouped data.

Height (in mm)	Frequency
10 – 14	1
15 – 19	2
20 – 24	5
25 – 29	8
30 – 34	9
35 – 39	3
40 – 44	2

Table 3

Each class interval is identified by its "class limits", the highest and the lowest score in the interval. For example, the first class includes heights from 10 to 14 and is indicated as 10 – 14. The first number, 10 is called the lower class limit and the second number, 14 is called the upper class limit. The class interval 10 – 14 theoretically includes all measurements from 9.5 to 14.5. These numbers are called class boundaries or true class limits, the smaller number, 9.5 is called the lower class boundary and the larger number, 14.5 is called the upper class boundary. The class boundaries are obtained by dividing the sum of the upper limit of one class interval and the lower limit of the next higher class interval by 2. In practice, the lower class and upper class boundaries can be obtained by subtracting 0.5 from the lower class limit and adding 0.5 to the upper class limit respectively.

The Class Size (Class Width)

The class size of a class interval is defined as:

The class size = upper class boundary – lower class boundary

For example, the class size for the class interval 10 – 14 is 14.5 – 9.5 = 5.

The Class Midpoint (Class Mark)

The midpoint of a class interval is defined as:

$$Class\ midpoint = \frac{upper\ class\ limit + lower\ class\ limit}{2}$$

For example, the class midpoint for the class interval 10 – 14 is $\frac{10+14}{2} = 12$.

Notice that, it is impossible to see how the data is distributed within each class. We take the midpoint of each interval as a representative of each class.

We summarise the terms in the following diagram using the class interval 10 – 14 as our example.

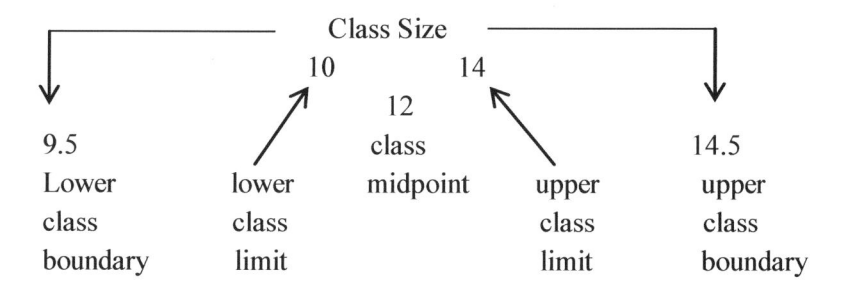

```
                    ┌──────── Class Size ────────┐
                         10        14
                            12
     9.5                   class                  14.5
     Lower       lower    midpoint    upper       upper
     class       class                class       class
     boundary    limit                 limit      boundary
```

Figure 23.1

Forming Frequency Distribution

Before a set of data can be converted into a grouped frequency distribution, we must decide how many class intervals to employ. There is no definite rule to help you make this decision but the following guide may be useful: the class should be few enough so that the resulting distribution reveals a group pattern, but not so few as to obscure a group pattern and cause great loss in the precision with which we can identify the individual score. As a rule of thumb, the number of classes should range from 5 to 12. Once the approximate number of class intervals has been chosen, we must determine the width of the class interval necessary to produce the desired number of intervals. We can roughly estimate the needed class intervals by dividing the range of the scores by the number of intervals wanted. In most situations, the division results in a fractional value. For example, if the largest and the smallest values of a frequency distribution are 89 and 11 respectively, and 10 groups are required then the

$$Class\ width = \frac{89-11}{10} = 7.8$$

Whole numbers are more convenient than fractions for widths of intervals. In this example, we have a choice of 8 or 10, but 10 is more convenient to use. Class widths of: 2, 3, 5 or 10 are often more convenient than other values.

Example 1

The marks of 30 candidates in an examination are given below. Construct a grouped frequency distribution.

15	20	42	24	22	39
48	29	33	35	38	27
25	36	26	21	16	44
32	19	53	34	32	30
40	33	46	28	31	34

Solution The largest number of the distribution is 53 and the smallest is 15. The range of the distribution is 53 – 15 = 38. If we decide to use ten intervals, the approximate class width will be 3.8. For convenience, we choose a class width of 5. Beginning the first interval at the lowest value 15, we form the class intervals, 15 – 19, 20 – 24, and so on. The classes must be chosen so that each score falls into one and only one class. There is no unique solution; Table 4 shows one possible frequency distribution.

Marks	Tally	Frequency
15 – 19	///	3
20 – 24	////	4
25 – 29	HH/	5
30 – 34	HH/ ///	8
35 – 39	////	4
40 – 44	///	3
45 – 49	//	2
50 – 54	/	1

Table 4

Exercise 23.1

1. The following scores were made on a 30-item test

```
16  18  18  16  17  18
17  19  15  20  16  15
18  16  19  17  17  17
15  17  20  15  19  16
17  16  17  18  16  17
```

(a) Form a frequency distribution table for the data

(b) Which item has the largest frequency?

2. The ages (in years) of 40 members of staff of a company are:

```
42  48  41  55  58  38  57  44
47  33  53  50  43  62  34  45
52  54  49  50  63  47  52  52
43  44  56  34  54  68  58  48
37  39  42  46  42  42  48  37
```

(a) Form a grouped frequency distribution, using a class interval of size 5 beginning with 31.

(b) Find the lower limit of the third class.

(c) Find the upper class limit of the fourth class.

(d) Find the midpoint of the second class.

3. The marks obtained by 50 students in an examination are:

55	54	74	93	81	62	76	62	26
68	54	88	74	87	18	95	45	34
43	73	15	32	53	92	44	61	51
55	42	66	82	44	75	44	33	48
31	65	66	62	63	25	72	52	28

(a) Form a grouped frequency distribution, using the class intervals, 10 – 19, 20 – 29, etc.

(b) Find the midpoint of the class interval having the highest frequency.

(c) Find the upper class boundary of the fourth class.

(d) Find the lower class boundary of the fifth class.

(e) Find the size of the third class.

4. The ages of a group of 40 retired high school teachers are:

71	73	77	64	66	79	72	74
76	66	67	61	76	75	77	78
74	71	70	71	63	79	72	69
71	69	68	70	75	67	74	68
73	69	70	73	72	69	70	67

(a) Find the range of the data.

(b) Form a frequency distribution for these data with seven classes, beginning with 60.

(c) Find the size of the class having the largest frequency

(d) Find midpoint of the fifth class.

23.2 Graphical Representation of Frequency Distribution

Data can be represented by various graphs. We will discuss only two of these graphs.

Histogram

The histogram is a method of representing a frequency distribution graphically. A histogram consists of a series of vertical rectangles (or bars) with width proportional to the class interval concerned and an area proportional to the frequency. However, if the distribution

Example 2

The heights in centimetres of plants were recorded and summarised in the table below. Plot a histogram of the distribution.

Height (cm)	130-134	135-139	140-144	145-149	150-154	155-159	160-164
Number of plants	1	3	8	10	6	5	2

Solution Draw two axes: the vertical and the horizontal axes. Mark equal intervals of length along the horizontal axis. Draw a bar over each interval, with height equal to the frequency of that class. The axes are labelled appropriately. The vertical axis is labelled frequency and the horizontal axis is labelled heights. The base is labelled with the class boundaries as shown in Figure 23.2. Another method is to use class midpoints. In this case, each class midpoint is marked at the middle of each bar. The histogram of the distribution is shown in Figure 23.2.

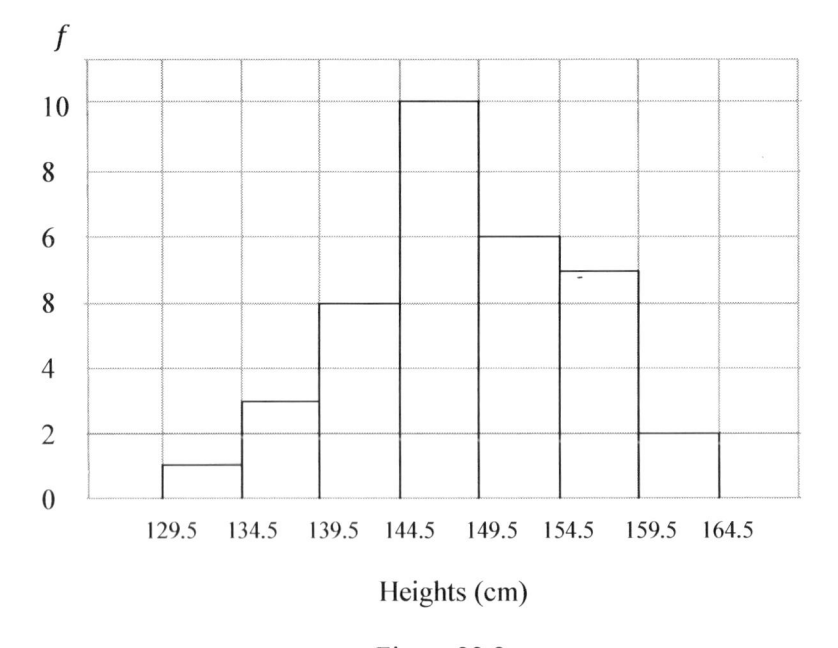

Figure 23.2

Recall that whenever the class intervals are unequal the area of the bars must be kept proportional to their frequencies by changing the height proportionately.

The distribution shown in Table 5 shows the ages in years of 80 workers at an end of year party of a company.

Ages (in yrs)	Frequency (f)	Class size (c)	Frequency density $=\frac{f}{c}$
16 – 20	6	5	1.2
21 – 30	20	10	2.0
31 – 40	27	10	2.7
41 – 50	13	10	1.3
51 – 70	14	20	0.7

Table 5

You may have noticed that the class intervals of the distribution are unequal. The most common class size is 10. We take this as the standard width. The first class interval is half the standard width, so its frequency must be multiplied by 2 to keep the areas equivalent. The last interval is twice the standard width, so its frequency must be halved to keep the areas equivalent. The frequency is stated for the standard class when the histogram is drawn.

Alternatively, the histogram can be drawn as shown in Figure 23.3. The rectangles of the histogram are drawn each with a width equal to the class interval size and height equal to the frequency density. In this case the label of the vertical axis should be frequency density.

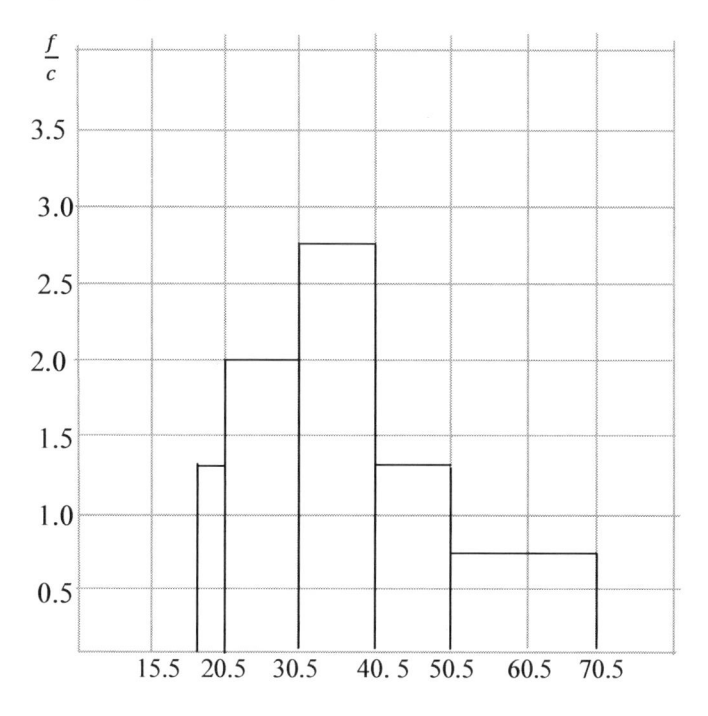

Figure 23.3

Exercise 23.2(a)

1. The masses in grams of 34 canned fish in a shop are:

41	56	44	48	50	47	49
47	40	44	49	51	43	46
48	46	59	54	42	60	47
29	72	32	46	50	43	38
60	45	21	68	33	55	

   (a) Form a frequency distribution using intervals 20 – 29, 30 – 39, etc.

   (b) Draw a histogram of the data.

2. The data given below are marks of students obtained in an examination:

81	83	87	74	76	89	82	84
86	76	77	71	86	85	87	88
84	81	80	81	73	89	82	79
81	79	78	80	85	77	84	78
83	79	80	83	82	79	80	77

   (a) Form a frequency distribution for these data, using class intervals, 70 – 72, 73 – 75, 76 – 78, etc.

   (b) Draw a histogram for the data.

3. The amount of money earned weekly by 50 factory workers is shown below:

80	86	82	75	80	91	85	76	82	78
83	71	81	83	87	78	87	85	84	85
77	84	79	88	72	81	78	82	77	75
81	74	88	80	84	85	81	73	90	86
74	82	84	77	83	82	79	85	79	80

   (a) Form a frequency distribution, using intervals 71 – 73, 74 – 76, 77 – 79, etc.

   (b) Draw a histogram for the data.

4. The table shows the ages of 40 people in a club.

Ages (in yrs)	1 - 20	21 - 30	31 - 40	41 - 50	51 - 60	61 – 80
Frequency	2	6	12	14	4	2

   Draw a histogram for the distribution

5. The table below shows the distribution of weights of students in a survey:

Weight(in kg)	1-10	11-15	16-20	21-25	26-30	31-40	41-60
Frequency	3	7	12	14	7	5	2

Draw a histogram for the data

Cumulative Frequency Distribution

Marks	Frequency	Marks less than	Cumulative frequency
49 – 49	5	49.5	5
50 – 59	7	59.5	12
60 – 69	10	69.5	22
70 – 79	13	79.5	35
80 – 89	9	89.5	44
90 – 99	6	99.5	50

Table 6

Table 6 shows a distribution of marks obtained by 50 students in a mathematics examination. Column 4 gives the value of the sum of all frequencies less than the upper class boundary of a given class. For example, the sum of all frequencies less than 69.5 is 5 + 7 + 10 = 22. Also, the sum of all frequencies less than 79.5 is (5 + 7 + 10 + 13) or (22 + 13) i.e. 35. The value obtained by adding all frequencies less than the upper class boundary of a specified class is called the cumulative frequency. The table giving values of cumulative frequency less than an upper class boundary is called a cumulative frequency distribution.

Cumulative Frequency Curve

The curve obtained by plotting the cumulative frequency against the upper class boundary, as shown in Figure 23.4, is called ogive or cumulative frequency curve.

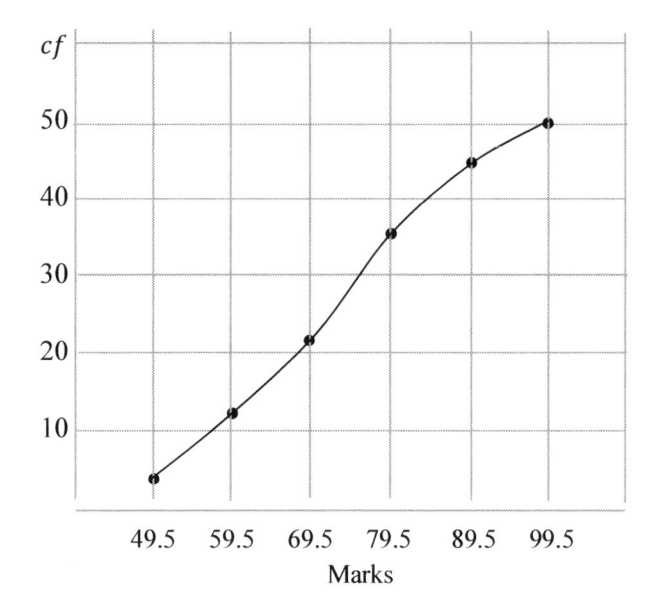

Figure 23.4

Exercise 23.2(b)

1. The heights (in centimetres) of a group of students are:

130	152	145	137	156	147	139	159	149	151
144	134	154	145	137	158	148	141	164	142
152	144	136	155	146	138	157	149	142	161
132	153	145	137	157	148	141	161	150	143
144	135	154	146	138	158	148	142	142	152

(a) Form a frequency distribution using intervals of size 5 centimetres starting with 130 – 134.

(b) Form a cumulative frequency distribution.

(c) Draw a cumulative frequency curve.

2. The marks obtained by a group of students in a mathematics examination are:

Marks	10-19	20-29	30-39	40-49	50-59	60-69	70-79	80-89	90-99
Frequency	2	4	10	11	10	6	4	2	1

(a) Form a cumulative frequency distribution

(b) Draw the cumulative curve for the data

3. The frequency distribution for the masses in grams of 50 toys manufactured in a factory is shown below.

Masses	71-73	74-76	77-79	80-82	83-85	86-88	89-91
Frequency	3	5	9	14	11	6	2

(a) Form a cumulative frequency distribution

(b) Draw the cumulative frequency curve

4. The frequency distribution of the ages in years of a group of 48 adults is:

Ages(in yrs)	20-29	30-39	40-49	50-59	60-69	70-79
Frequency	3	10	11	13	9	2

(a) Form a cumulative frequency distribution

(b) Draw a cumulative frequency curve

23.3 Measure of Central Location

We might have to answer questions such as the following. What is the average age of students in your class? or What is the average height of students in your class? Such questions ask for a single value that is representative of a given set of values. This representative value of a set of values is called a measure of central location. Three common measures of location, often called "average" for a set of discrete numerical values are arithmetic mean or simply the mean, the median and the mode.

The Arithmetic Mean

The arithmetic mean of n values is defined as the sum of all the n values divided by n, that is the number of values. For example, the arithmetic mean of the values 15, 18, 18, 24, 25 and 20 is:

$$\frac{15 + 18 + 18 + 24 + 25 + 20}{6} = 20$$

In general, the arithmetic mean, $\bar{x}$, of the set $x_1, x_2, x_3, \cdots, x_n$ is:

$$\bar{x} = \frac{x_1 + x_2 + x_3 \cdots + x_n}{n}$$

written briefly as $\bar{x} = \frac{\sum x}{n}$

where $\sum x = $ sum of all values

$n = $ number of values

The arithmetic mean of a frequency distribution is found by using the formula

$$\bar{x} = \frac{\sum fx}{\sum f}$$

where $fx = $ a value multiplied by its corresponding frequency

$\sum f = $ sum of all frequencies

For example, if 15, 18, 16 and 12 occur with frequencies 3, 2, 4 and 1 respectively, the arithmetic mean is:

$$\bar{x} = \frac{3(15) + 2(18) + 4(16) + 1(12)}{3 + 2 + 4 + 1}$$

$$= 15.7$$

Calculating Arithmetic Mean of a Frequency Distribution

Ungrouped Data

Example 3

The table shows the distribution of marks obtained by 30 students in a test. Find the mean mark of the distribution.

Marks	13	14	15	16	17	18	19	20
Frequency	1	3	8	5	5	2	4	2

Solution The work can be arranged as in Table 7. The third column shows the values of the product $f \times x$.

Marks (x)	Frequency (f)	fx
13	1	13
14	3	42
15	8	120
16	5	80
17	5	85
18	2	36
19	4	76
20	2	40
	$\sum f = 30$	$\sum fx = 492$

$$\text{Mean mark} = \frac{\sum fx}{\sum f}$$

$$= \frac{492}{30}$$

$$= 16.4$$

Grouped Data

To calculate the arithmetic mean of a grouped data we interpret x as the class midpoint.

Example 4

The table shows the ages of the members of a club. Find the mean age of the distribution.

Ages/yrs	11-15	16-20	21-25	26-30	31-35
Frequency	6	3	4	5	2

Solution The work may be arranged as in Table 8

Ages/yrs	Frequency (f)	Class midpoint (x)	fx
11 - 15	6	13	78
16 - 20	3	18	54
21 – 25	4	23	92
26 – 30	5	28	140
31 - 35	2	33	66
	$\sum f = 20$		$\sum fx = 430$

Table 8

$$\text{Mean age} = \frac{\sum fx}{\sum f}$$

$$= \frac{430}{20}$$

$$= 21.5$$

Using an Assumed Mean

If A is any guessed or assumed mean (which may be any real number) and if d denotes the deviation of each item of the data from the assumed mean ($d = X - A$) then the actual mean is:

$$\bar{x} = A + \frac{\sum d}{n}$$

For a frequency distribution $\bar{x} = A + \frac{\sum fd}{\sum f}$

Example 5

(a) Find the arithmetic mean for the numbers: 149, 150, 151, 153, 154, 155.

Solution Using 153 as an assumed mean we have

$$\sum d = -4 + (-3) + (-2) + 0 + 1 + 2 = -6$$

Thus, the arithmetic mean is:

$$\bar{x} = A + \frac{\sum d}{n}$$

$$= 153 + \left(\frac{-6}{6}\right)$$

$$= 152$$

(b) The frequency distribution shows the marks of students who took an examination in mathematics.

Marks	24	25	26	27	28	29	30
Frequency	2	3	5	7	8	3	2

Find the mean mark for the distribution.

Solution Table 9 illustrates the method used to calculate arithmetic mean using the assumed mean. We take 27 as the assumed mean.

Marks(X)	Frequency(f)	$d = X - A$	fd
24	2	-3	-6
25	3	-2	-6
26	5	-1	-5
27	7	0	0
28	8	1	8
29	3	2	6
30	2	3	6
	$\sum f = 30$		$\sum fd = 3$

Table 9

$$\text{Mean mark} = A + \frac{\sum fd}{\sum f}$$

$$= 27 + \left(\frac{3}{30}\right)$$

$$= 27.1$$

(c) The frequency distribution shows the heights in millimetre of 20 seedlings. Find the mean height of the distribution.

Heights(mm)	160-162	163-165	166-168	169-171	172-174
Frequency	3	5	6	4	2

Solution The work is shown in Table 10. We take 167 as the assumed mean.

Height (mm)	Frequency (f)	Class midpoint (x)	$d = X - A$	fd
160-162	3	161	-6	-18
163-165	5	164	-3	-15
166-168	6	167	0	0
169-171	4	170	3	12
172-174	2	173	6	12
	$\sum f = 20$			$\sum fd = -9$

Table 10

$$Mean\ heoght = A + \frac{\Sigma fd}{\Sigma f}$$

$$= 167 + \left(\frac{-9}{20}\right)$$

$$= 167 - 0.45$$

$$= 166.55$$

The mean height of the seedlings is 166.55 mm

The Mode

The modal value or mode of a set of values is the value that occurs most frequently. The mode is not always unique; a set of numbers can have more than one mode. If two values occur with the same frequency, the set is said to be "bimodal". A set of values has no mode if each data value has a frequency of 1 or if the frequency of all the data values are the same. For example:

1. The set of values: 2, 2, 5, 7, 9, 9, 2, 10, 2, 10, 11, 12 has one mode, 2.

2. The set of values: 3, 2, 4, 4, 3, 4, 7, 5, 5, 7, 7, 9 has two modes, 4 and 7

3. The set of values: 4, 5, 4, 8, 5, 8, 11, 11, 13, 12, 13, 12 has no mode.

Table 11 shows the ages of a number of 25 boys at a party.

Ages(yrs)	5	6	7	8	9	10
Number of boys	2	5	9	4	3	2

Table 11

The mode of a frequency distribution is the item with the highest frequency. In this example the modal age of the distribution is 7. It is important to note that 9 is not the mode of the distribution.

A class interval with the highest frequency is called the modal class interval.

Marks	1-5	6-10	11-15	16-20	21-25	26-30
Frequency	3	7	10	8	7	5

Table 12

Table 12 shows the marks obtained by 40 students in a test. The modal class for this frequency distribution is 11 – 15.

Determining Mode from Histograms

The mode is obtained by dividing the width of the highest rectangle in the histogram in proportion to the heights of the adjacent rectangles as illustrated in Figure 23.5. Note that the highest rectangle represents the modal class.

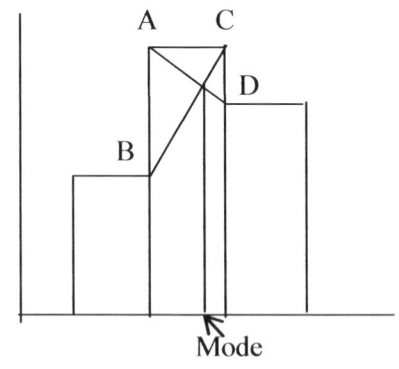

Figure 23.5

The modal class is the middle class. The value of the mode lies somewhere within this range and its value can be found by the construction shown. The value on the horizontal axis of the point of intersection of the diagonal lines AD and BC is taken as the mode of the distribution.

Example 6

Table 13 gives the distribution of ages of 38 members of a club. Draw a histogram for the distribution and use it to estimate the mode.

Ages(yrs)	10-19	20-29	30-39	40-49	50-59	60-69
Frequency	4	7	11	8	5	3

Table 13

Solution

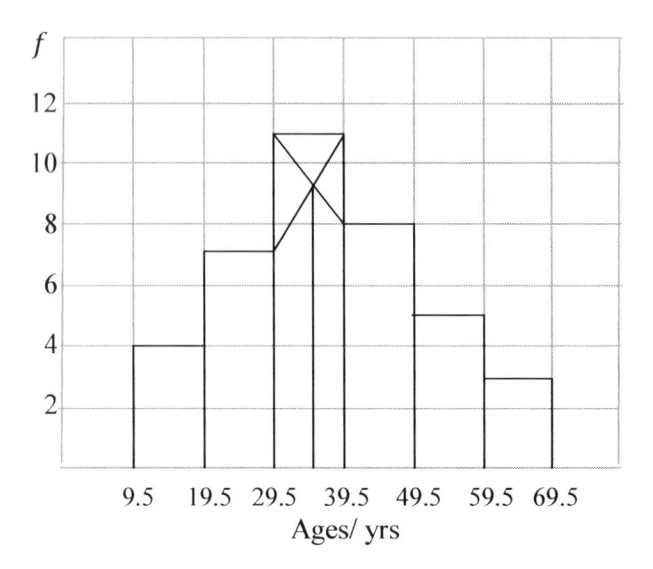

The point of intersection of the vertical line and the horizontal axis is taken as the mode of the distribution. In this case the mode of the distribution is 35.5.

Determining Mode from Frequency Distribution

From a frequency distribution the mode can be obtained from the following formula.

$$Mode = L + \left(\frac{\Delta_1}{\Delta_1 + \Delta_2}\right)c$$

where $L =$ lower class boundary of the modal class

$\Delta_1 =$ excess of modal frequency over frequency of next lower class

$\Delta_2 =$ excess of modal frequency over frequency of next higher class

$c =$ size of the modal class interval

Example 7

The table below gives the mass in kilograms of 50 ingots. Find the mode.

Marks(kg)	20-22	23-25	26-28	29-31	32-34	35-37	38-40
Frequency	3	7	16	10	8	5	1

Solution Recall that the class with largest frequency is called the modal class. In this case, the modal class is 26 – 28. The lower class boundary for this class is 25.5.

Here $\Delta_1 = 16 - 7 = 9$, $\Delta_2 = 16 - 10 = 6$ and $c = 28.5 - 25.5 = 3$, so the mode is:

$$Mode = 25.5 + \left(\frac{9}{9+6}\right) \times 3$$

$$= 25.5 + 1.8$$

$$= 27.3$$

The mode is 27.3 kg

The Median

The median of a set of numbers arranged in order is the middle value if the number of values is odd or the mean of the two middle values if the number of values is even.

For example, the median of the set of numbers 2, 3, 5, 6, 8 is 5.

The median of the set of numbers 11, 12, 13, 14, 15, 16, 17, 18 is the mean of the two middle numbers, i.e. the median is $\frac{14+15}{2} = 14.5$.

Median Position

If n numbers are arranged in order the median is the value in the $\frac{n+1}{2}$ th position. For example, the median of the set of numbers: 6, 7, 8, 9, 11, 12, 13, 14 and 15 is the value in the $\frac{9+1}{2}$ i.e 5 th position. For the set of numbers: 4, 5, 5, 7, 9, 11, 12 and 15 the median position is $\frac{8+1}{2}$ i.e. $4\frac{1}{2}$ th position. Since the $4\frac{1}{2}$ th position is mid-way between the 4 th and 5 th position, the mean of the numbers in the 4 th and 5 th position gives the median.

Calculating the Median from Frequency Distribution

The median may be determined from a frequency distribution. The two examples below show how the computation is done.

Ungrouped Data

Example 8

Find the median for the distribution below:

x	0	1	2	3	4	5
y	3	9	12	13	8	5

Solution The median can be calculated quickly if we add a row for the cumulative frequencies as shown in Table 14.

x	0	1	2	3	4	5
f	3	9	12	13	8	5
cf	3	12	24	37	45	50

Table 14

The median is the middle value of the set of data. If n is the total number of values the median is the $\frac{n+1}{2}$ th value. However, if the total frequency is large $\frac{1}{2}n$ th value is used. For this example the median value corresponds to the $\frac{50}{2}$, i.e., 25 th value. Going along the cumulative frequency row we find that 24 cases earned scores of 2 or below. The 25 th case is included among the 13 individuals who earned a score of 3. Thus the median is 3.

Grouped Data

For a grouped data the median is given by the following formula.

$$Median = L + \left(\frac{\frac{N}{2} - f}{f_m}\right)c$$

where L = lower class boundary of the median class

N = number of values in the data (i.e. total frequency)

f = sum of all the frequencies lower than the median class

f_m = frequency of the median class

c = the class size of the median class

The example below shows the method of determining the median of a grouped data.

Consider the distribution below:

Class interval	Frequency	Cumulative frequency
11 – 20	9	9
21 – 30	12	21
31 – 40	17	38
41 – 50	18	56
51 – 60	13	69
61 – 70	8	77
71 – 80	3	80

Table 15

Begin by locating the median class interval, that is the class that contains the middle value. The $\frac{80}{2} = 40$ th value lies in the class interval 41 – 50. So $L = 40.5$, $N = 80$, $f = 38$, $f_m = 18$ and $c = 10$. Substituting these values into the median formula we get

$$Median = 40.5 + \left(\frac{\frac{80}{2} - 38}{18}\right) \times 10$$

$$= 40.5 + \frac{2}{18} \times 10$$

$$= 41.6$$

The median is 41.6

Determining Median from Histogram

The frequency distribution below shows the marks obtained by 50 students in an examination. Draw a histogram for the distribution and determine the median mark.

Marks	21-30	31-40	41-50	51-60	61-70	71-80
Frequency	5	8	16	12	6	3
cf	5	13	29	41	47	50

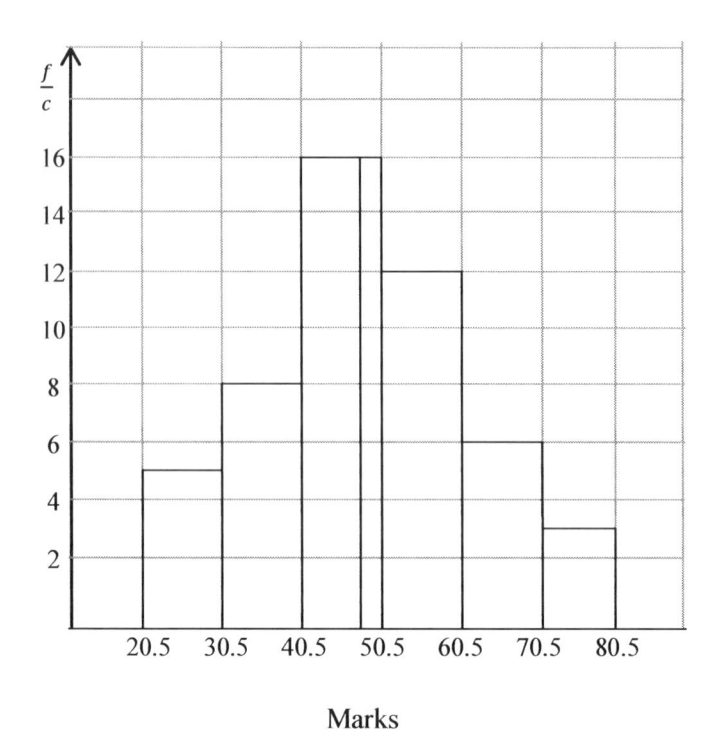

The vertical line inserted in the diagram divide the area of the histogram into two equal parts and the value on the horizontal axis that correspond to this line is the median. The total frequency is 50, hence the vertical line has 25 cases above it and 25 below it. The sum of the frequencies of the first two classes is 13. Thus to get the desired 25, we require (25 − 13), that is 12 of the 16 cases in the next class. Therefore the median class is in the third class, with boundaries 40.5 and 50.5 marks. The value of the median lies somewhere within this range and it is found by splitting the area of the third class so that $\frac{12}{16}$ of its width i.e. $\frac{12}{16} \times 10 = 7.5$, lies to the left of the vertical line. Thus the median value of the distribution is (40.5 + 7.5) i.e. 48 percent.

Exercise 23.3

1. Find the mean, median and mode of the given data:

(a) 31, 35, 53, 36

(b) 12, 11, 13, 14, 12, 15, 20, 21, 18

(c) 51, 61, 73, 80, 65

(d) 26, 31, 21, 29, 32, 26, 25, 28

2. The daily temperatures, in degrees Fahrenheit, in June were:

51, 52, 55, 55, 55, 58, 64, 64, 65 and 67

Find the mean, median and mode of the data.

3. 15, 16, 17, 18 and 19 occur with frequencies 4, 5, x, 3 and 2 respectively. Find x if the mean is 16.7.

4. The table below shows the distribution of weekly wages(GH¢) of 50 factory workers:

Weekly wages(GH¢)	45	46	47	48	49	50
Frequency	5	x	10	16	y	4

If the mean weekly wage of the distribution is 47.42, find the values of x and y.

5. The heights in centimetres of 100 people are shown in the table below. Calculate to the nearest centimetres the mean height.

Height(in cm)	150-156	157-163	164-170	171-177	178-184	185-191
Frequency	5	18	20	27	22	8

6. The table shows the distribution of length in centimetres of some pieces of rod:

Length(in cm)	14.5-15.5	16.5-17.5	18.5-19.5	20.5-21.5	22.5-23.5	24.5-25.5
Frequency	5	8	16	12	6	3

Find the mean length

7. Find the mean, median and mode for x given the frequency distribution below:

x	45	46	47	48	49	50	51	52	53	54	55
f	1	3	2	4	5	6	10	7	6	2	2

8. The table shows the distribution of marks obtained by students in a test:

Marks	25-29	30-34	35-39	40-44	45-49
Frequency	1	3	7	8	6

(a) Draw a histogram for the distribution

(b) Estimate the mode from your histogram

9. The following table shows the distribution of marks scored by 20 students in a mathematics test.

Marks	10-14	15-19	20-24	25-29	30-39
Frequency	3	4	7	2	4

Draw a histogram for the distribution and use it to estimate the mode of the distribution.

10. The table below shows the distribution of ages of people in a village.

Ages(in years)	11-30	31-40	41-50	51-60	61-70	71-90
Frequency	140	50	45	52	35	38

(a) Draw a histogram for the distribution

(b) Find the median age and indicate it on your histogram

11. The table below shows the distribution of the masses of canned food manufactured by a factory.

Mass(g)	1-20	21-40	41-60	61-80	81-100
Frequency	1	20	36	40	3

Draw a histogram and use it to estimate the median mass

12. The table below shows the distribution of marks obtained by a group of students in a test. Find the median and the mode.

Marks	18-20	21-23	24-26	27-29	30-32	33-35
Frequency	5	10	9	16	6	4

13. The table shows the ages (in years) of a group of people on a bus.

Ages(in years)	27-30	31-34	35-38	39-42	43-46	47-50
Frequency	3	5	9	11	7	5

Find the median and modal age.

14. The grouped frequency distribution table below gives the lengths, recorded to the nearest mm, of nails produced from a particular machine. Find the median mass.

Length(mm)	98-99	100-101	102-103	104-105	106-107	108-109
Frequency	3	7	13	14	8	5

15. The table below gives the distribution of the masses of 80 male students recorded to the nearest kilogram. Find the median mass and the mode.

Mass(kg)	50-54	55-59	60-64	65-69	70-74	75-79	80-84	85-89
Frequency	2	5	9	18	23	14	8	1

23.4 Quartiles, Deciles and Percentiles

Recall that the value which divides a set of data into two equal parts is called the median. You can separate a set of data into four equal parts as illustrated below.

For the set of numbers: 7, 8, 9, 10, 11, 12, 13, 14, 15, 16, 18.

9, 12 and 15 divide the set of data into four equal parts. These numbers are called the quartiles. In general, the values which divide a set of data into four equal parts are called quartiles, denoted by Q_1, Q_2 and Q_3 respectively. Q_1 is called the first quartile, Q_2 the second quartile and Q_3 the third quartile. Q_2 is the middle value and hence it is the same as the median. Q_1 and Q_3 are also called the lower and upper quartiles respectively.

It is also possible to divide a data into ten or hundred equal parts. The values which divide a data into ten or hundred equal parts are called deciles or percentiles respectively.

Evaluation of the first quartile and the third quartile from a grouped data may be obtained from the median formula by replacing $\frac{1}{2}n$ with $\frac{1}{4}n$ and $\frac{3}{4}n$ respectively.

Example 9

(a) Find the first and third quartiles for the set of values:

11, 12, 13, 14, 15, 16, 17, 18, 19, 20

Solution   First divide the data into two equal parts. Ignore the middle value if the number of values is odd. These are ten numbers so each half will have 5 numbers. The two halves are:

11, 12, 13, 14, 15 and 16, 17, 18, 19, 20

The median for the first half is 13, so the first quartile is 13, and the median of the upper half is 18, so the third quartile is 18.

(b) Find the first and third quartiles for the set of values:

24, 25, 27, 27, 28, 30, 31, 35, 37

Solution   Here the middle value, 28, is ignored. The two halves are:

24, 25, 27, 27 and 30, 31, 35, 37

The median of the lower half is $\frac{25+27}{2} = 26$

and the median of the upper half is $\frac{31+35}{2} = 33$

Hence, the first and third quartiles are 26 and 33 respectively.

Using Cumulative Frequency Curves

A cumulative frequency curve can be used to estimate the median, the first quartile and the third quartile.

Example 10

The frequency distribution shows the heights in centimetres of 40 students. Draw a cumulative frequency curve and use it to find the median, the first and third quartiles.

Heights (cm)	35-139	40-144	45-149	50-154	55-159	60-164	65-169
Frequency	3	5	9	12	5	4	2

Solution The cumulative frequency curve for the distribution is shown in Figure 23.7.

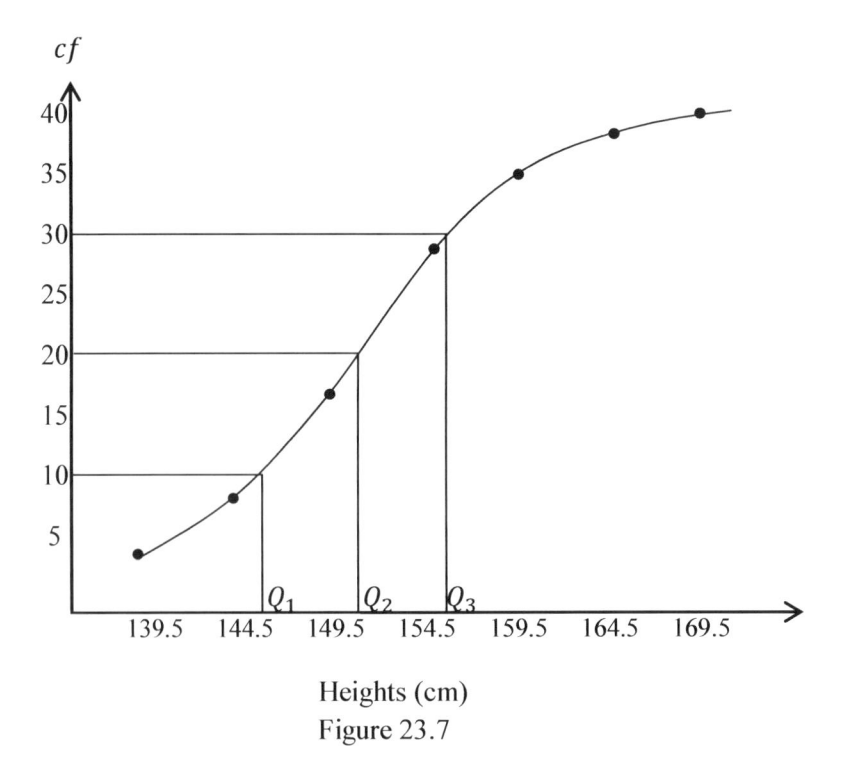

Heights (cm)
Figure 23.7

You can find the median from the curve by drawing a horizontal line to represent half of the total frequency. Here half of the total frequency is $\frac{1}{2} \times 40 = 20$. Draw a horizontal line from this point, i.e. 20 to the curve, and at the point of intersection a vertical line is drawn to the horizontal axis as shown in Figure 23.7. The median value is marked Q_2 in Figure 23.7, and is about 151.

You can estimate the first and third quartiles in a similar way. The first and third quartiles correspond to the $\frac{1}{4} \times 40 = 10$ th value and $\frac{3}{4} \times 40 = 30$ th value respectively. The first and third quartiles are marked Q_1 and Q_3 respectively on the horizontal axis. The first quartile and the third quartile are 146 and 155.5 respectively.

Exercise 23.4

1. Determine the median, first and third quartiles for the following data:

(a) 12, 13, 14, 15, 15, 17, 19, 21, 23, 24, 27

(b) 30, 27, 25, 24, 27, 37, 31, 27, 35

(c) 18, 54, 42, 23, 87, 34, 76, 45, 68

(d) 27, 37, 40, 28, 23, 30, 35, 24, 30, 32, 31, 28

2. Calculate the median and quartiles for the distribution shown in the table below.

x	60	61	62	63	64	65	66	67	68
f	1	2	3	9	8	7	5	3	2

3. Calculate the median and quartiles for the distribution shown in the table below.

x	13	14	15	16	17	18	19	20
f	2	9	15	20	30	12	8	4

4. The marks of some students who took an examination in mathematics are shown in the frequency distribution below:

Marks	10-29	30-39	40-49	50-59	60-69	70-79	80-89	90-99
Frequency	9	12	15	10	13	9	7	5

Calculate the median, the first and third quartiles for the distribution.

5. The table below gives the height of a group of children at a nursery.

Height(cm)	50-54	55-59	60-64	65-69	70-74	75-79	80-84
Frequency	3	7	10	5	8	2	1

Calculate the median, first and third quartiles for the distribution.

6. The frequency distribution given below shows the marks obtained by a group of students in mathematics examination:

Marks	25-29	30-34	35-39	40-44	45-49	50-54	55-59
Frequency	5	4	7	11	12	8	1

Draw the cumulative frequency curve for this data and use it to determine the median, the first and third quartiles.

7. The frequency distribution below shows the lengths of leaves of a certain plant in millimetres:

Length	20-24	25-29	30-34	35-39	40-44	45-49	50-54	55-59	60-64
Frequency	1	5	10	19	25	21	15	3	1

Draw the cumulative frequency curve for this data and use it to determine the median, the first and third quartiles.

8. The table shows the distribution of the masses of a group of people in a survey:

Mass(kg)	40-44	45-49	50-54	55-59	60-64	65-69	70-74	75-79
Frequency	4	6	13	54	58	50	13	2

Draw the cumulative frequency curve for the distribution and use it to determine the median, the first and third quartiles.

23.5 Measure of Dispersion

The data sets:

$$2, 6, 10, 14, 18$$

and $$8, 9, 10, 11, 12$$

have the same mean and median. The mean of each set is 10. Notice that the scores in the second set are more closely clustered around 10 than those in the first set. In addition to averages we need a statistical measure to indicate the extent to which scores are spread out. Such measures are called measures of dispersion. Examples include range, mean deviation, interquartile range, semi-interquartile range and standard deviation.

Range

The range is defined as the difference between the largest and smallest values. For example, the range of the data set: 4, 6, 8, 12, 14, 16, is $16 - 4 = 12$.

The range is a poor measure of dispersion since it depends on only two scores and tells us nothing about the rest of the scores.

The Mean Deviation

The mean deviation is the arithmetic mean of absolute differences of each value from the mean. The median deviation of the set of n number $x_1, x_2, x_3 \cdots x_n$ is given by

$$Mean\ deviation = \frac{\Sigma |x - \bar{x}|}{n}$$

where $\bar{x}$ is the arithmetic mean of the numbers.

If $x_1, x_2, x_3 \cdots x_n$ occur with frequencies $f_1, f_2, f_3 \cdots f_n$ respectively, the mean deviation is given by

$$Mean\ deviation = \frac{\Sigma f |x - \bar{x}|}{\Sigma f}$$

This form is useful for grouped data where x represnt class marks and f the corresponding class frequency.

Example 11

(a) Find the mean deviation of the set of numbers: 12, 13, 16, 18, 21

Solution $\quad Arithmetic\ mean = \frac{12+13+16+18+21}{5}$

$$= 16$$

$$Mean\ deviation = \frac{|12-16| + |13-16| + |16-16| + |18-16| + |21-16|}{5}$$

$$= \frac{|-4|+|-3|+|0|+|2|+|5|}{5}$$

$$= 2.8$$

(b) The following distribution shows the performance of 50 students in a test. Find the mean deviation of the distribution.

Marks	15-19	20-24	25-29	30-34	35-39
Frequency	5	10	17	11	7

Solution $\quad$ The work is arranged as in Table 16

Marks	f	Class marks (x)	fx	$\lvert x-\bar{x}\rvert$	$f\lvert x-\bar{x}\rvert$
15 – 19	5	17	85	10.5	52.5
20 – 24	10	22	220	5.5	55.0
25 – 29	17	27	459	0.5	8.5
30 – 34	11	32	352	4.5	49.5
35 – 39	7	37	259	9.5	66.5
	$\sum f = 50$		$\sum fx = 1375$		$\sum f\lvert x-\bar{x}\rvert = 232$

Table 16

$$\text{Mean} = \frac{\sum fx}{\sum f}$$

$$= \frac{1375}{50}$$

$$= 27.5$$

$$\text{Mean deviation} = \frac{\sum f\lvert x-\bar{x}\rvert}{\sum f}$$

$$= \frac{232}{50}$$

$$= 4.64$$

Interquartile range (IQR)

The interquartile range is the value obtained when the lower quartile is subtracted from the upper quartile.

$IQR = Q_3 - Q_1$

Example 12

Find the interquartile range of the set of values:

25, 26, 27, 30, 32, 33, 35, 38, 39, 40, 45

Solution The lower quartile and the upper quartile of the set of values are 27 and 39 respectively. Therefore the interquartile range is 39 – 27 = 12.

Semi-interquartile range (SIQR)

The semi-interquartile range is defined by $SIQR = \frac{Q_3 - Q_1}{2}$

Example 13

Find the semi-interquartile range of the set of values:

11, 32, 43, 45, 46, 47, 48, 49, 50, 57, 58, 60

Solution The lower quartile of the set of values is $\frac{43+45}{2} = 44$ and upper quartile is $\frac{50+57}{2} = 53.5$. the semi-interquartile range is $\frac{53.5-44}{2} = 4.75$.

Standard Deviation

The standard deviation for a data of n values is obtained as follows:

1. First calculate the arithmetic mean.

2. Find the difference between the mean and each value.

3. Square each of the differences.

4. Sum the squared values.

5. Divide the sum by n.

6. Finally take the nonnegative square root of the quotient.

In symbol, the standard deviation for the set: $x_1, x_2, x_3, \cdots x_n$ is:

$$s = \sqrt{\frac{\Sigma(x - \bar{x})^2}{n}}$$

where s is the standard deviation

$\bar{x}$ is the arithmetic mean

and n is the number of values

Example 14

Determine the standard deviation of the set of values: 6, 9, 9, 11, 15, 16

Solution The arithmetic mean $\bar{x} = \frac{\Sigma x}{n}$

$$\bar{x} = \frac{6+9+9+11+15+16}{6}$$

$$= 11$$

The rest of the solution is shown in Table 17

x	$x - 11$	$(x - 11)^2$
6	-5	25
9	-2	4
9	-2	4
11	0	0
15	4	16
16	5	25
		74

Table 17

$$s = \sqrt{\frac{\Sigma(x-\bar{x})^2}{n}}$$

$$= \sqrt{\frac{74}{6}}$$

$$= 3.5$$

Note that the standard deviation cannot be negative, and when two sets of data are compared, the one with the larger dispersion will have the larger standard deviation.

For frequency distribution

$$s = \sqrt{\frac{\Sigma f(x-\bar{x})^2}{\Sigma f}}$$

For a grouped data, x is the class midpoint value.

Example 15

The table below gives the distribution of the wages of 100 factory workers.

Wages(GH¢)	90-98	100-108	110-118	120-128	130-138	140-148
Number of workers	10	15	24	30	18	3

Calculate, to the nearest cedi

(a) the mean

(b) the standard deviation of the distribution

Solution The work may be arranged as in Table 18.

Wages	f	x	fx	$x - \bar{x}$	$(x - \bar{x})^2$	$f(x - \bar{x})^2$
90-98	10	94	940	-24	576	5760
100-108	15	104	1560	-14	196	2940
110-118	24	114	2736	-4	16	384
120-128	30	124	3720	6	36	1080
130-138	18	134	2412	16	256	4608
140-148	3	144	432	26	676	2028
	$\sum f = 100$		$\sum fx = 11{,}800$			$\sum f(x-\bar{x})^2 = 16800$

Table 18

$$\bar{x} = \frac{\sum fx}{\sum f}$$

$$= \frac{11800}{100}$$

$$= 118$$

The mean wage is 118

$$s = \sqrt{\frac{\sum f(x-\bar{x})^2}{\sum f}}$$

$$= \sqrt{\frac{16800}{100}}$$

$$= 12.96$$

The use of the two formulas above is sometimes fairly tedious. It may be more convenient to use the following formulas to compute the standard deviation of a distribution.

$$s = \sqrt{\frac{\sum d^2}{n} - \left(\frac{\sum d}{n}\right)^2} \qquad (1)$$

$$s = \sqrt{\frac{\sum fd^2}{\sum f} - \left(\frac{\sum fd}{\sum f}\right)^2} \qquad (2)$$

Example 16

Using the assumed mean find the standard deviation of the set of values:

28, 31, 32, 37, 40, 42, 45, 46, 48, 50

Solution We take any value around the centre of the set as the assumed mean. Here we take 40 as the assumed mean. The work is shown in Table 19.

x	d	d^2
28	-12	144
31	-9	81
32	-8	64
37	-3	9
40	0	0
42	2	4
45	5	25
46	6	36
48	8	64
50	10	100
	$\sum d = -1$	$\sum d^2 = 527$

Table 19

$$s = \sqrt{\frac{\sum d^2}{n} - \left(\frac{\sum d}{n}\right)^2}$$

$$= \sqrt{\frac{527}{10} - \left(\frac{-1}{10}\right)^2}$$

$$= \sqrt{52.69}$$

$$= 7.26$$

Example 17

The distribution gives the ages of 50 members of a club

Ages(in yrs)	11-20	21-30	31-40	41-50	51-60	61-70
Frequency	4	10	11	13	9	3

Calculate the standard deviation, correct to three significant figures.

Solution The work is shown in Table 20.

Ages	Frequency (f)	Class midpoint (x)	d	fd	d^2	fd^2
11-20	4	15.5	-30	-120	900	3600
21-30	10	25.5	-20	-200	400	4000
31-40	11	35.5	-10	-110	100	1100
41-50	13	45.5	0	0	0	0
51-60	9	55.5	10	90	100	900
61-70	3	65.5	20	60	400	1200
	$\sum f = 50$			$\sum fd = -280$		$\sum fd^2 = 10800$

Table 20

$$s = \sqrt{\frac{\Sigma fd^2}{\Sigma f} - \left(\frac{\Sigma fd}{\Sigma f}\right)^2}$$

$$= \sqrt{\frac{10800}{50} - \left(\frac{-280}{50}\right)^2}$$

$$= \sqrt{184.64}$$

$$= 13.5$$

Variance

The variance is calculated in the same way as the standard deviation, but without the final step of finding the square root.

The variance of the set of data $x_1, x_2, x_3, \cdots x_n$ is found from the formula:

$$s^2 = \frac{\Sigma(x - \bar{x})^2}{n}$$

This is equivalent to:

$$s^2 = \frac{\Sigma x^2}{n} - \bar{x}^2$$

Example 18

Find the variance of the set of values below: 11, 13, 15, 16, 19

Solution $Mean = \frac{11+13+15+16+19}{5} = 14.8$

It is easier to use the second formula to find the variance

$$s^2 = \frac{\Sigma x^2}{n} - \bar{x}^2$$

$$= \frac{11^2 + 13^2 + 15^2 + 16^2 + 19^2}{5} - 14.8^2$$

$$= 7.36$$

Exercise 23.5

1. Find the range of the following sets of values:

(a) 18, 20, 22, 23, 25, 27

(b) 31, 33, 28, 37, 29, 40

(c) 21, 16, 23, 17, 24, 18, 26, 30, 29

(d) 11, 26, 9, 15, 21, 17, 25, 9, 7, 8, 12

2. Find the interquartile range and semi-interquartile range for the following data:

(a) 30, 31, 34, 36, 37, 39, 40, 41, 43, 44, 45, 46, 50, 52, 55

(b) 25, 26, 27, 28, 29, 30, 31, 33, 34, 35, 37, 38, 39, 40, 42

(c) 21, 36, 19, 25, 31, 27, 35, 19, 17, 18, 22

(d) 31, 46, 29, 35, 41, 37, 45, 29, 27, 28, 32

3. Find the mean deviation of the sets:

(a) 3, 7, 9, 5

(b) 24, 16, 38, 41, 31

(c) 5, 3, 8, 4, 7, 6, 12, 4, 3, 2

(d) 12.5, 10.2, 10.6, 13.5, 11.2

4. Find the standard deviation of the following data:

(a) 21, 22, 24, 25, 28

(b) 27, 28, 29, 21, 22, 25, 27, 31, 35, 25

(c) 7, 9, 10, 10, 11, 12, 18

(d) 14, 15, 19, 19, 21, 24, 35

5. Find the variance of the following data:

(a) 52, 48, 60, 60

(b) 55, 65, 80, 80, 90

(c) 85, 75, 62, 78, 100

(d) 82, 72, 65, 65, 76

6. The table below gives the distribution of marks obtained by 100 candidates in an examination. Draw the cumulative frequency curve and use it to estimate the interquartile range

Marks(%)	0-9	10-19	20-29	30-39	40-49	50-59	60-69	70-79	80-89	90-99
Frequency	6	8	9	10	20	15	11	9	7	5

7. The number of days in a year taken off work by the employees of a factory are shown in the table below. Draw the cumulative frequency curve and use it to estimate the interquartile range of the distribution.

Number of days	25-29	30-34	35-39	40-44	45-49	50-54	55-59
Frequency	5	4	7	11	12	8	1

8. The following table shows the distribution of marks by 100 students in a chemistry examination. Draw the cumulative frequency curve and use it to estimate the semi-interquartile range.

Marks(%)	10-19	20-29	30-39	40-49	50-59	60-69	70-79	80-89	90-99
Frequency	5	7	12	19	23	14	10	9	1

9. For the distribution below find the mean deviation.

x	51	53	55	57	58	60	62
f	4	3	2	2	1	6	2

10. The table shows the frequency distribution of the ages (in years) of 50 people in a village. Find the mean deviation.

Ages(yrs)	10-14	15-19	20-24	25-29	30-34	35-39	40-44
Frequency	3	7	15	11	8	4	2

11. The heights of a collection of 80 plants are recorded in the table below. Find the mean deviation of the heights.

Heights(in cm)	0-4	5-9	10-14	15-19	20-24
Frequency	10	8	14	7	11

12. The distribution of values are given by the following frequency table. Find the variance and standard deviation of the values.

x	10	11	12	13	14
Frequency	2	6	3	2	4

13. The heights of a collection of 25 plants are recorded in the table below. Calculate the standard deviation for the distribution.

Height(mm)	0-3	4-7	8-11	12-15	16-19
Frequency	2	5	8	7	3

14. The ages of 40 members of a club are recorded in the table below. Calculate the standard deviation of the distribution.

Ages(in years)	10-16	17-23	24-30	31-37	38-44	45-53
Frequency	2	4	11	13	7	3

15. A group of 25 applicants for employment made the following scores on an aptitude test:

$$\begin{array}{ccccc}
59 & 57 & 42 & 47 & 55 \\
49 & 41 & 47 & 51 & 57 \\
60 & 61 & 54 & 60 & 44 \\
50 & 50 & 60 & 48 & 46 \\
55 & 55 & 42 & 59 & 48
\end{array}$$

Find the mean and standard deviation of these scores

16. The marks of 50 students in a test are given in the table. Using an assumed mean of 50, calculate the standard deviation of the distribution.

Marks	45	46	47	48	49	50	51	52	53	54	55
Frequency	1	3	2	4	5	8	10	7	6	2	2

17. The table below gives the distribution of masses (g) of canned tuna. Using an assumed mean of 14.0 g calculate the standard deviation of the distribution.

Masses(g)	12.5	13.0	13.5	14.0	14.5	15.0	15.5	16.0
Frequency	2	9	15	20	30	12	8	4

18. The net weight of some canned fruit juice in a shop are recorded in the table below. Using an assumed mean of 28.5 g calculate the standard deviation of the distribution.

Net Weight(g)	24-25	26-27	28-29	30-31	32-33	34-35
Frequency	2	9	25	9	4	1

19. The frequency distribution gives the lengths in centimetres of iron rods produced from a particular machine. Using an assumed mean of 17 centimetres calculate the standard deviation of the distribution.

Lengths(cm)	0-4	5-9	10-14	15-19	20-24	25-29
Frequency	3	23	33	31	21	9

20. The table shows the distribution of the masses of a group of students in a survey. Using an assumed mean of 57 kg calculate the standard deviation of the distribution.

Masses	45-49	50-54	55-59	60-64	65-69	70-74
Frequency	3	10	11	13	9	2

23.6 Correlation

Scatter Diagram

The test scores of seven students in Mathematics and Science are presented in Table 21.

Mathematics	82	53	71	58	62	81	78
Science	84	55	66	69	63	83	80

Table 21

It is difficult to see the relationship between the Mathematics and the Science test scores. Their relationship is easier to see if we plot these points on a graph, taking the Mathematics scores as x-coordinate of the point and the Science scores as y-coordinate. This gives the graph shown below. Such graphs are called scatter diagrams (or scatter graphs). Generally scatter diagrams help us determine whether one set of data is related to another.

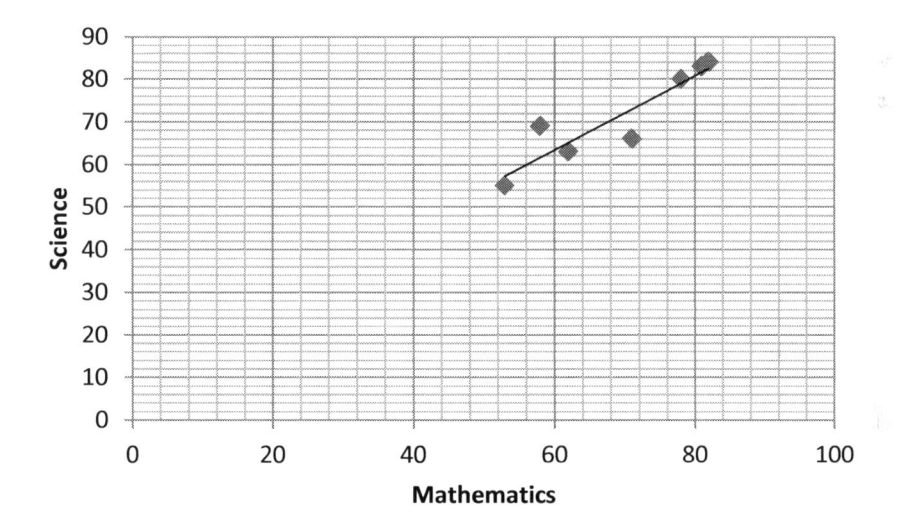

Figure 23.8

You can see from Figure 23.8 that a student who obtain high mark in Mathematics is likely to obtain high mark in Science.

Correlation

The word correlation is used to describe how closely the points of a scatter diagram cluster about a straight line. If the points on a scatter diagram are very nearly along a straight line, there is a high correlation between the variables.

Types of Correlation

The relationship between two sets of data can be described as positive, negative or no correlation.

Positive Correlation

(a) (b) (c)

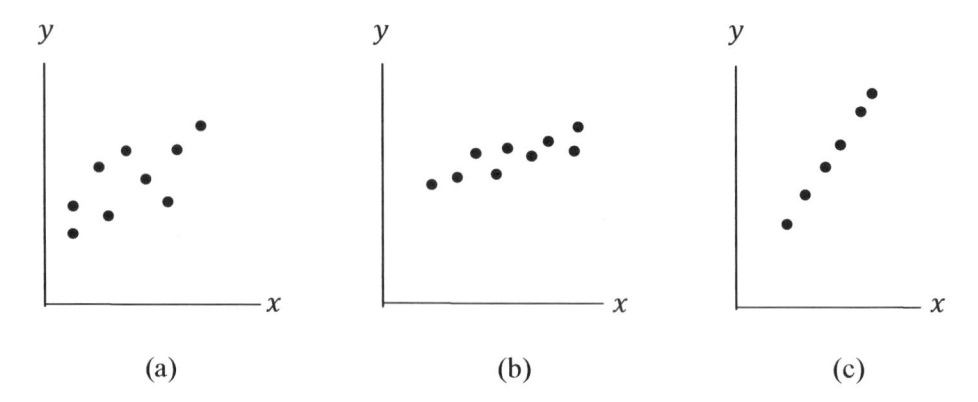

In each of the three cases, you can see that as x values increase, the y values tend to increase. We say that there is positive correlation between the data.

A pattern of points which tends to form a straight line suggests a high correlation between the set of data. The correlation is low in case (a) and high in case (b). In case (c) we have a perfect positive correlation.

Negative Correlation

(a) (b) (c)

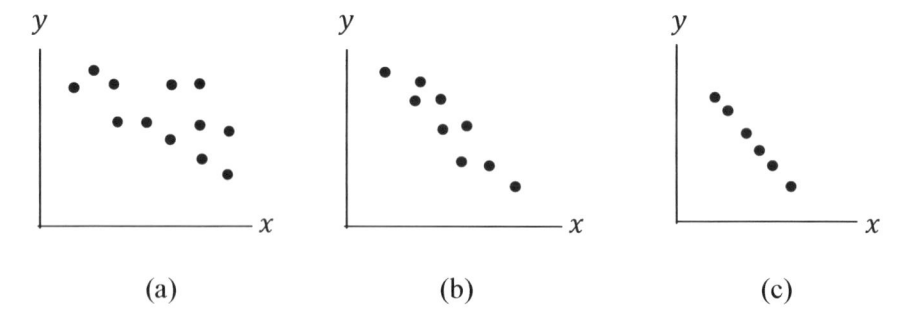

In each of the cases, you can see that as the x values increase, the y values tend to decrease. We say that there is negative correlation between the data. We have a low negative correlation in case (a) and high negative correlation in case (b). In case (c) we have a perfect negative correlation.

No Correlation

A set of data can give a scatter diagram as shown in the next page.

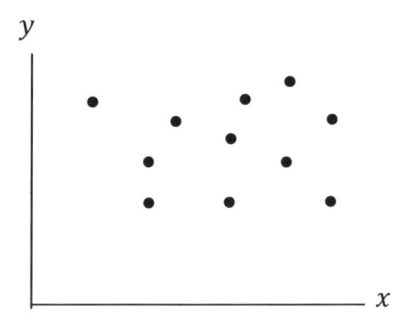

[scatter diagram with y and x axes showing randomly scattered points]

The points are scatted at random, and do not seem to cluster around a line. There is no obvious relationship between x and y. We say that there is no correlation between the data.

Line of Best Fit

Often the points plotted on a scatter diagram do not lie on a straight line unless there is a perfect correlation between the variables. However we can still draw a straight line which comes closer to fitting all the points. This line is called the line of best fit (or the line of regression). The line of best fit can be used to predict the approximate value of one variable when the value of another variable is given.

Drawing the Line of Best Fit

The position of the line of best fit can be estimated by eye. The line is positioned so that it passes through the point $(\bar{x}, \bar{y})$, where $\bar{x}$ is the mean of x values and $\bar{y}$ is the mean of y values. Also, the line must be drawn through as many points as possible and must be close to an equal number of points on each side.

When the independent variable is plotted on the x-axis and the dependent variable on the y-axis, the equation of the line of best fit is given by $y = ax + b$, where a is the gradient and b is the intercept on the y axis. In this case the line of best fit is called the regression line of y on x.

When y is made the independent variable and plotted on the horizontal axis and x the dependent variable and plotted on the vertical axis, the regression line is called the regression line of x on y.

Example 19

(a) The table gives marks scored by 8 students in a mathematics test and in a science test.

Mathematics(x)	58	67	72	64	52	83	76	78
Science(y)	55	60	78	69	45	80	77	86

(i) Draw a scatter diagram of the marks scored in the mathematics and science tests.

Solution Choose suitable scales for the horizontal and vertical axis. Plot the pairs of data points in the table, as shown in Figure 23.16.

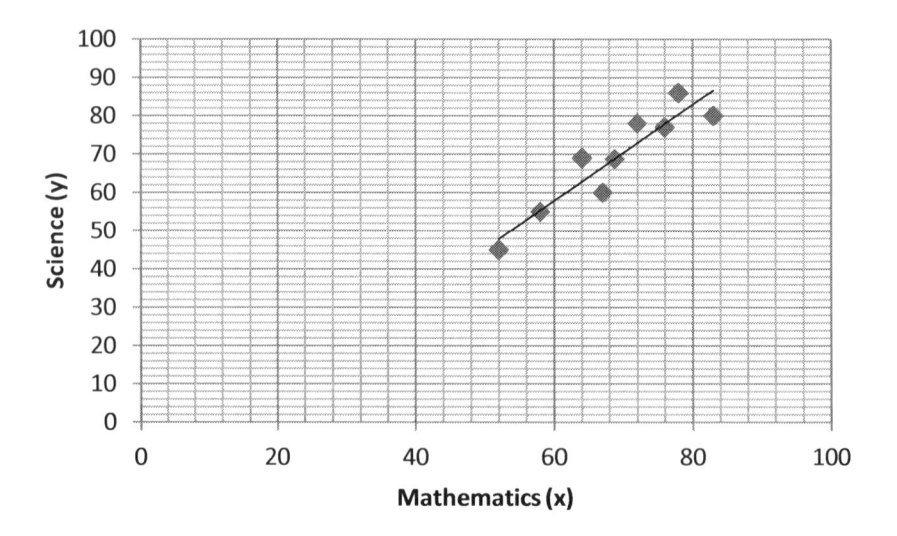

Figure 23.16

(ii) Draw the line of test fit.

Solution First find the mean, $\bar{x}$, of the mathematics scores and the mean, $\bar{y}$, of the science scores;

$$\bar{x} = \frac{58 + 67 + 72 + 64 + 52 + 83 + 76 + 78}{8} = 68.75$$

$$\bar{y} = \frac{55 + 60 + 78 + 69 + 45 + 80 + 77 + 86}{8} = 68.75$$

Next plot the point (68.75, 68.75) on the graph and then draw in the line of best fit to pass through it (see Figure 23.16).

(b) The table below shows the marks obtained by ten students in a test and the number of hours they spent studying for the test.

Marks(x)	4	5	6	5	8	7	10	8	10	8
Number of hours(y)	6	6	3	5	2	4	2	3	1	2

(i) Draw a scatter diagram for the data

Solution The scatter diagram is drawn in much the same way as described above.

(ii) Draw the line of best fit

Solution Find the value of $\bar{x}$ and the value of $\bar{y}$.

$$\bar{x} = \frac{4+5+6+5+8+7+10+8+10+8}{10} = 7.1$$

$$\bar{y} = \frac{6+6+3+5+2+4+2+3+1+2}{10} = 3.4$$

Plot the point (7.1, 3.4) on the graph and then draw a line to pass through it (see Figure 23.17).

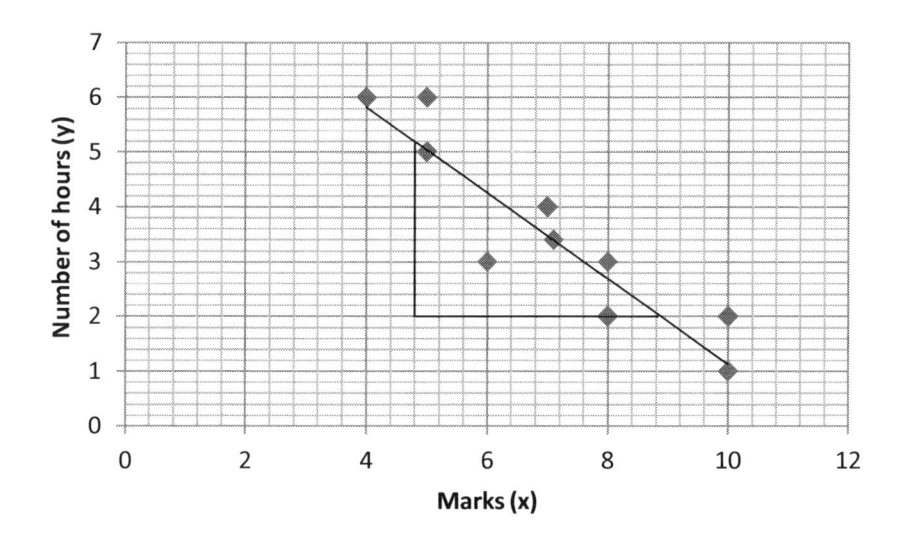

Figure 23.17

(iii) Find the equation of the line of best fit and use it to find the number of hours a student spent studying if he obtained 7 marks in the test.

Solution Choose two points on the line to find the gradient. Note; do not choose any of the data points. We choose the points (4.8, 5.2) and (8.8, 2).

The gradient of the line is given by $Gradient = \frac{5.2-2}{4.8-8.8} = -0.8$

Substitute the gradient and the coordinates of a point into $y = ax + b$, where a is the gradient. We substitute the coordinates of the point (7.1, 3.4).

So $3.4 = -0.8(7.1) + b$

$b = 9.08$

The equation of the line of best fit is $y = -0.80x + 9.08$

Notice that the value of b can be read from the graph if the line of best fit crosses the vertical axis.

You can find the number of hours a student studied by substituting 7 for x in the equation $y = -0.8x + 9.08$.

So $y = -0.80(7) + 9.08 = 3.48$

To the nearest unit, you can expect a student who studied 3 hours to obtain 7 marks.

Correlation Coefficient by Ranking

Spearman's Rank Coefficient

The Spearman's rank coefficient establishes whether there is any form of association between two variables when the variables are arranged in a ranked form. The Spearman's coefficient is given by

$$R = 1 - \frac{6\sum d^2}{n(n^2 - 1)}$$

where d = difference between the pairs of ranked values

n = number of pairs of rankings

Example 20

The marks of 7 students in two tests are as follows:

Mathematics(x)	34	30	58	24	47	12	68
Science(y)	47	33	72	58	88	44	73

Find the coefficient of rank correlation.

Solution The work is shown in Table 24. Rank the marks for Mathematics and the marks for Science as shown in the third and fourth columns. The fifth column shows the difference between each pair of ranks.

x	y	R_x	R_y	d	d^2
34	47	4	5	-1	1
30	33	5	7	-2	4
58	72	2	3	-1	1
24	58	6	4	2	4
47	88	3	1	2	4
12	44	7	6	1	1
68	73	1	2	-1	1
					$\sum d^2 = 16$

Table 24

$$R = 1 - \frac{6\sum d^2}{n(n^2-1)}$$

$$= 1 - \frac{6 \times 16}{7(5^2-1)}$$

$$= 0.71$$

The result indicate a high positive correction, so we say that there is a reasonable agreement between the student's performances in the two test.

Note:

1. R varies between + 1 and – 1.

2. If R = + 1 then there is perfect positive correlation between the two sets of ranks.

3. If R = - 1, then there is perfect negative correlation, or complete disagreement between the two sets of ranks.

Tied Ranking

When two or more observation are equal the average of the ranks that they would otherwise have is used. For example, if we rank the following scores:

77, 63, 77, 78, 72, 70, 68, 84, 65, 80

two observations are equal; 77 are tied for the 4 rank, so we assign $\frac{4+5}{2}$, i.e. 4.5 to each of them.

Example 21

The following marks were obtained by ten students in Mathematics and Physics test:

Mathematics(x)   42   61   54   64   73   80   68   47   79   68

Physics(y)   55   65   78   69   78   89   78   50   92   83

Find the Spearman's rank correlation coefficient, correct to two decimal places.

Solution   The work is shown in Table 25.

x	R_x	y	R_y	d	d^2
42	10	55	9	1	1
61	7	65	8	-1	1
54	8	78	5	3	9
64	6	69	7	-1	1
73	3	78	5	-2	4
80	1	89	2	-1	1
68	4.5	78	5	-0.5	0.25
47	9	50	10	-1	1
79	2	92	1	1	1
68	4.5	83	2	2.5	6.25
					$\sum d^2 = 25.50$

Table 25

$$R = 1 - \frac{6\sum d^2}{n(n^2-1)}$$

$$= 1 - \frac{6 \times 25.50}{10(10^2-1)}$$

$$= 0.85$$

Exercise 23.6

1. The table below shows the values of two variables x and y obtained from a survey:

x	21	22	24	25	26	28	29	30
y	91	83	63	64	53	39	25	19

(a) Draw a scatter diagram for the data.

(b) Draw the line of best fit of y on x.

(c) Find the equation of the line of best fit.

(d) Use your equation to find, correct to one decimal place, the value of y when $x = 27$.

2. The table below shows the marks obtained by seven students in a test in Mathematics and Physics

Mathematics(x)	13	17	14	19	21	26	18
Physics(y)	23	25	24	27	25	29	27

(a) Draw a scatter diagram of y against x.

(b) Draw the line of best fit and find the equation of the line.

(c) Estimate from your graph the possible mark obtained in Physics by a student who had 10 marks in Mathematics.

3. The table below shows the monthly salaries (in GH¢) of some factory workers and their years of experience.

Salaries (y)	122	123	138	133	124	122	128	126
Experience in years(x)	12	11	17	30	32	26	24	21

(a) Draw a scatter diagram for the data.

(b) Draw the line of best fit of y on x.

(c) Use your line of best fit to estimate the monthly salary of a worker with 13 years of experience.

4. The table below shows the height (in centimetres) and weight (in kilograms) of 12 boys in a class:

Weights(x)	43	35	40	35	37	42	50	28	33	38	41	38
Heights(y)	157	145	159	143	142	152	147	135	141	155	150	145

(a) Draw a scatter diagram of the data.

(b) Draw the line of best fit of y on x.

(c) Use your line of best fit to predict the value of a boy's height if his weight is 45 kilograms.

5. The table below shows the wages (in GH¢) of 8 people and the number of days in a mouth they worked part time in a factory:

Number of days (x)	10	11	12	13	14	15	16	17
Wages(in GH¢)(y)	25.00	37.50	42.00	57.00	64.00	73.50	82.50	96.00

(a) Draw a scatter diagram of the data.

(b) Draw the line of best fit of y on x and estimate its gradient.

(c) From your graph, estimate y when $x = 15.5$

6. A group of 8 science students are tested in Mathematics and Physics. Their rankings in the two were:

Students	A	B	C	D	E	F	G	H
Ranking Mathematics	2	7	6	1	4	3	5	8
Ranking Physics	3	6	4	2	5	1	8	7

(a) Calculate the Spearman's rank correlation coefficient.

(b) Interpret your result.

7. Test in Accounting and Mathematics were given to the following candidates and they were ranked as follows:

Candidates	A	B	C	D	E	F	G	H	I	J
Accounting	8	3	9	2	7	10	4	6	1	5
Mathematics	9	5	10	1	8	7	3	4	2	6

Find the Spearman's rank correlation coefficient.

8. The following marks were obtained by 8 candidates in History and English language.

Candidates	A	B	C	D	E	F	G	H
History Marks	53	44	63	78	38	48	64	51
English Marks	43	29	57	56	32	40	73	41

Rank these results and hence find Spearman's rank correlation coefficient between the two sets of results. Interpret your result.

9. Two judges were asked to rank ten girls participating in a beauty contest. The ranking are recorded in the following table:

Contestant	A	B	C	D	E	F	G	H	I	J
Judge 1	9	6	7	2	5	1	8	3	10	4
Judge 2	5	7	1	4	10	3	2	8	6	9

(a) Calculate the Spearman's rank correlation coefficient.

(b) Interpret your result.

10. The marks of 10 contestants in a beauty contest given by two judges are as follows:

Judge A	39	47	27	43	48	56	35	30	38	49
Judge B	44	66	43	65	69	76	64	59	46	56

Rank these results and calculate the Spearman's rank correlation coefficient between the two sets of results. Interpret your result.

11. The table below shows the heights of 12 boys and 12 girls in a school. Find the Spearman's rank correlation coefficient.

Heights of boys	55	53	57	54	58	52	60	56	58	57	59	61
Heights of girls	58	56	58	55	59	56	58	54	61	57	58	60

12. The table gives the scores of a group of eight students in Mathematics and Science. Compute Spearman's rank correlation coefficient for these data.

Mathematics	62	65	66	56	66	67	62	60
Science	55	74	68	64	67	73	65	71

Review Exercise 23

1. The masses, to the nearest kilogram, of 50 pupils are as shown below:

60	66	62	55	60	71	65	56	62	58
63	51	61	63	67	58	67	65	64	65
57	64	59	68	52	61	58	62	57	55
61	54	68	60	64	65	61	53	70	66
54	62	64	57	63	62	59	65	59	60

(a) Construct a frequency distribution using the class intervals, 51 – 53, 54 – 56, 57 – 59, etc.

(b) Draw a histogram for the distribution and use it to estimate the mode.

2. The table below shows the mark distribution in a test:

Marks	16-20	21-30	31-35	36-40	41-50
Frequency	5	7	10	16	12

Draw a histogram for the distribution.

3. The table below gives the distribution of the ages of a group of people at a party:

Ages(in years)	11-19	20-24	25-29	30-34	35-39
Frequency	1	3	7	8	6

(a) Draw a histogram for the distribution

(b) Use your histogram to estimate the mode.

4. The height of 50 students were measured to the nearest centimetre as follows:

Heights(cm)	135-144	145-149	150-154	155-159	160-169
Frequency	10	12	21	3	4

Draw a histogram of the distribution

5. The table shows the distribution of marks obtained by some students in an examination.

Marks	1-20	21-30	31-40	41-50	51-60	61-80
Frequency	10	32	18	20	12	8

(a) Draw a histogram for the distribution.

(b) Estimate the median mark from your histogram.

6. Find the mean, median and mode for each of the following sets of measurements:

(a) 3, 8, 10, 7, 5, 14, 2, 9, 8   (b) 26, 31, 21, 29, 32, 26, 25, 28

(c) 122, 123, 124, 126, 121, 129, 119, 126, 130, 127

7. The masses of ten toys in grams are:

   69, 48, 52, 60, 63, 55, 66, 52, 63, 66

Using an assumed mean of 63 grams, calculate the mean of the data.

8. The deviation of a set of values from 16 are: -5, -4, 4, -3, -1, 1, 0, 2, 3, -2. Find the mean value.

9. The table shows the distribution of marks obtained by some students in a test.

Marks	23	24	25	26	27	28	29
Frequency	3	7	8	5	10	5	2

Calculate the mean mark of the distribution.

10. The following table shows the ages of a group of retired public servants.

Ages(in years)	60	61	62	63	64	65	66	67
Frequency	3	5	8	11	13	5	3	2

Using an assumed mean of 63 years calculate the mean age.

11. The following distribution shows the ages of 50 members of a church:

Ages(in years)	1-10	11-20	21-30	31-40	41-50
Frequency	3	17	15	10	5

Find the mean age of the distribution.

12. The table below shows the weekly income, in GH¢, of 40 factory workers:

Income(in GH¢)	11-20	21-30	31-40	41-50	51-60	61-70
Frequency	11	14	7	4	3	1

Calculate the mean of the distribution.

13. The number of days in a year taken off by 40 workers of a factory are as shown in the following table.

No. of days	26-30	31-35	36-40	41-45	46-50	51-55
No. of workers	4	7	13	8	6	2

Using an assumed mean of 43 days calculate, correct to two significant figures, the mean of the distribution.

14. The table below shows the distribution of weights of students in a survey:

Mass(kg)	20-29	30-39	40-49	50-59	60-69	70-79	80-89
Frequency	5	9	12	7	10	4	3

Using an assumed mean of 54.5 kg calculate the mean mass.

15. The table below shows the distribution of the marks obtained by 50 students in a test marked out of 10.

Marks	3	4	5	6	7	8	9	10
Frequency	3	5	8	11	13	5	3	2

Find the median of the distribution.

16. The table below gives the distribution of masses (kg) of 40 youth at an interview.

Masses(kg)	50-54	55-59	60-64	65-69	70-74	75-79	80-94
Frequency	6	7	4	5	8	6	4

Calculate the median.

17. The table below shows the lengths of 50 rods measured to the nearest centimetre.

Length(cm)	20-24	25-29	30-34	35-39	40-44	45-49	50-54	55-59
Frequency	1	2	8	10	12	12	3	2

(a) Draw a cumulative frequency curve for the distribution.

(b) Use your curve to find:

   (i) the lower quartile

   (ii) the median

   (iii) the upper quartile

18. The table below shows the distribution of marks (in percentages) obtained by 100 students in a Mathematics examination:

Marks(%)	10-19	20-29	30-39	40-49	50-59	60-69	70-79	80-89	90-99
Frequency	5	7	12	19	23	14	10	9	1

(a) Draw the ogive for the distribution.

(b) Use your graph to find the interquartile range.

19. The table below shows the marks scored by 100 candidates in an examination.

Marks(%)	1-10	11-20	21-30	31-40	41-50	51-60	61-70	71-80	81-90	91-100
Frequency	4	6	10	20	30	12	8	6	3	1

(a) Draw the cumulative frequency curve for the distribution.

(b) Use your curve to estimate:

(i) the median

(ii) the number of students who passed if the pass mark for the examination is 42.

20. The table shows the distribution of marks obtained by candidates in an examination.

Marks	1-10	11-20	21-30	31-40	41-50	51-60	61-70	71-80	81-90	91-100
Frequency	5	7	13	25	52	45	30	14	6	3

(a) Draw a cumulative frequency curve for the distribution.

(b) Use your graph to estimate the semi-interquartile range.

21. The marks of 10 students in a test are: 59, 38, 41, 50, 53, 45, 56, 42, 53, 56

Using an assumed mean of 53 calculate the standard deviation of the data.

22. The deviation of a set of values from 12 are: 5, -4, 2, -3, 0, 3, -1, 4, 1, -2. Find the variance of the values.

23. The table shows the mean weight of a group of babies born on the same day at a hospital.

Mean weight(kg)	2.5	3.0	3.5	4.0	4.5	5.0	5.5	6.0
Frequency	2	9	15	20	30	12	8	4

Using an assumed mean of 4.0 kg calculate the standard deviation.

24. The distribution of marks of 40 students is shown in the following table.

Marks(%)	16-20	21-25	26-30	31-35	36-40	41-45
Frequency	4	7	13	8	6	2

Using an assumed mean of 35% calculate, correct to two significant figures, the variance of the distribution.

25. The table below shows the distribution of the weekly expenditure of a group of families.

Expenditure(GH¢)	18-22	23-27	28-32	33-37	38-42	43-47	48-52
No. of families	5	8	19	25	23	16	4

Using the assumed mean of GH¢ 35 calculate, for the distribution, the standard deviation.

26. The table below shows the amount (in GH¢) of money earned by 50 people working part-time in a factory.

Amount earned	10-14	15-19	20-24	25-29	30-34	35-39	40-44	45-49
Frequency	1	2	8	10	12	12	3	2

Using an assumed mean of GH¢ 27 calculate, for the distribution,

(a) the mean

(b) the standard deviation

27. The table below shows the values of two variables x and y obtained from a survey:

x	1	2	4	5	6	8	9	10
y	81	73	53	54	43	29	15	9

(a) Draw a scatter diagram for the data

(b) Draw the line of best fit of y on x

(c) Find the equation of your line of best fit and use it to find, correct to one decimal place, the value of y when $x = 7$.

28. The table below shows the length (in cm) and mass (in g) of 10 rods.

Length(cm) (x)	13	21	9	17	22	24	10	13	10	19
Masses(g) (y)	21	24	16	22	31	35	19	20	18	26

(a) Draw a scatter diagram for the data. (b) Draw the line of best fit of y on x.

(c) Use your line in (b) to find the length of a rod whose mass is 25 g.

29. In a test ten candidates obtained the following marks in Mathematics and English:

Candidates	A	B	C	D	E	F	G	H	I	J
Mathematics marks	61	73	74	38	48	54	45	88	63	58
English marks	51	42	31	67	83	66	53	36	50	39

Rank these results and hence find the Spearman's rank correlation coefficient.

30. Test in Physics and Mathematics were given to the following candidates and they were ranked as follows:

Candidates	A	B	C	D	E	F	G	H	I	J
Physics	8	3	9	2	7	10	4	6	1	5
Mathematics	9	5	10	1	8	7	3	4	2	6

Find the Spearman's correlation coefficient.

31. The table below shows marks obtained by ten students in English and Mathematics tests. Find the Spearman's rank correlation coefficient.

English	77	63	77	78	72	70	68	84	65	80
Mathematics	78	66	79	80	75	65	62	90	64	70

32. The following paired observations were obtained on two variables x and y. Calculate the Spearman's rank correlation coefficient.

x	22	37	36	38	42	58	58	37
y	53	68	42	49	51	65	51	49

33. The table below shows marks obtained by ten students in English and History. Compute Spearman's rank correlation coefficient.

English	29	33	26	27	39	35	33	29	36	22
History	24	32	31	27	33	28	24	20	34	21

Test covering Chapter 18 – 23

These tests are provided to help you in the process of reviewing the previous chapters. If you missed any answer go back and review the appropriate chapter section.

Take each test as you would take a test in class. Allow yourself 1 hour to take each test.

Test 5

1. Differentiate with respect to x:

(a) $(3x^2 + 5)^3$

(b) $5x^3(x - 3)^2$

2. Find the equation of the normal to the curve $y = x^2(x - 3)$ at the point where it cuts the x-axis.

3. Find the coordinates of the point at which the expression $y = x^3 - 6x^2 + 9x + 2$ is greatest.

4. A farmer erects a fence along three sides of rectangular pen; the fourth side is walled. Find the greatest area of the pen he can fence if the total length of the fence is to be 100 metres.

5. If the radius of a sphere is increased from 10 cm to 10.1 cm, what is the approximate increase in surface area?

6. An error of 3 % is made in measuring the radius of a sphere. Find the percentage error in volume.

7. Air is escaping from a spherical balloon at the rate of 18 cm^2 s^{-1}. If the radius is 4.5 centimetres, at what rate is it decreasing? Take $\pi = 3.142$

8. The distance s m of a particle moving in a straight line from a fixed point O at time t s, is given by $s = 24 + 10t - 20t^2$. Find the time when the particle is momentarily at rest.

9. Evaluate:

(a) $\int_0^1 (x^3 + 1)^2 \, dx$

(b) $\int_{-1}^2 \left(\frac{x^3 + 2x^2 - 3x}{x} \right) dx$

10. Find the area enclosed by the curve $y = x^2 + 3$ and the lines $x = -1$ and $x = 2$.

11. Find the volume of the solid generated when the area between the lines $x - 3y + 3 = 0, x = 0$ and $y = 2$ is rotated about the x-axis.

12. Find the area enclosed between the curves $y = 4x^2$ and $y = x^2 - 2x$.

13. If P (2, -1) is the turning point of the graph of $y = ax^2 + bx + 3$, find the values of a and b.

14. The points A (4, 7), B(p, q), C(-5, -80 and D (1, 4) are the vertices of a parallelogram ABCD find the values of p and q.

15. Calculate to the nearest degree the angle between $a = \begin{pmatrix} -2 \\ 5 \end{pmatrix}$ and $b = \begin{pmatrix} 1 \\ 1 \end{pmatrix}$.

Test 6

1. Find $a \cdot b$ if $a = 4i + 7j$ and $b = 3i - 5j$.

2. Given that $v = 3i + kj$ and $u = i - 2j$ are orthogonal, find k.

3. An object falls from the top of a cliff with a velocity of 25 m s^{-1} for 6 s. What is the height of the cliff? Take g = 9.8 m s^{-2}

4. Find the magnitude of the resultant of the forces $P = (25\sqrt{3}\ N, 030°)$ and $Q = (25\sqrt{3}\ N, 120°)$.

5. A force of 0.2 N acts on a mass of 10 g. Calculate its acceleration.

6. A particle of mass 400 g moving at a speed of 15 m s^{-1} collides with another particle moving in the opposite direction at 24 m s^{-1}. After the impact, both particles remain at rest. Find the mass of the second particle.

7. A mass of 4 kg is suspected from a string. Find the tension in the string.

8. Nine people are going to travel in two taxis. The larger has five seats and the smaller has four. In how many ways can the party be split up?

9. In how many ways can the letters of the word DOCUMENT be arranged in a row?

10. One bag contains 3 white and 4 black balls; another contains 4 white and 2 black balls. One ball is drawn for each bag, find the probability that both balls are white?

11. A bag contains 10 identical balls of which 6 are red and the rest blue. Two balls are drawn at random one after the other from the bag without replacement. What is the probability that they are of different colours?

12. The table below shows the distribution of marks scored by forty candidates in an examination.

Marks	0	1	2	3	4	5
Frequency	6	4	8	10	9	3

Find:

(a) the median of the distribution

(b) the interquartile range of the distribution

13. The distribution of a variable x was as follows:

Values of x	1	2	3	4	5
Frequency	6	7	3	8	7

Calculate the mean of x.

14. The following table shows the distribution of marks of students in an examination.

Marks	15-29	30-34	35-39	40-44	45-49
No. of students	5	6	10	5	4

Using an assumed mean of 37 calculate:

(a) the mean of the distribution

(b) the standard deviation of the distribution

15. The frequency distribution below shows the distribution of weekly incomes of 100 workers.

Weekly Incomes(GH¢)	25-29	30-39	40-44	45-54	55-59
No. of Workers	10	32	18	28	12

(a) Draw a histogram for the distribution

(b) Using your histogram find the median of the distribution.

Test 7

1. Differentiate with respect to x:

(a) $x^2(2x + 3)$

(b) $(x + 1)(3x - 2)$

2. Find the equation of the tangent to the curve $y = 3x^2 + 2$ at the point $x = 4$.

3. Find the coordinates of the point where the expression $y = x^3 - 5x^2 + 3x = 2$ has a minimum value.

4. A curve whose equation is $y = ax^2 + bx - 6$ passes through the point (2, 0) and its gradient at this point is 3. Find the values of a and b.

5. The radius of a sphere decreases from 5 cm to 4.99 cm. Find the approximate decrease in area of its surface.

6. One side of a rectangle is three times more than the other. If the perimeter increases by 2 %, what is the percentage increase in area?

7. The area of a circle is increasing at the rate of 3 cm^2 s^{-1}. Find the rate of change of the circumference when the radius is 2 cm.

8. A particle moves in a straight line so that after t s its distance from O is x cm, where $x = 9 - t + t^2$. Find its position when its speed is 15 cm s^{-1}.

9. A curve passes through the point (3, -1) and its gradient function is $2x + 5$. Find its equation.

10. Evaluate:

(a) $\int_1^2 3(t-1)^2 \, dx$

(b) $\int_{-1}^2 (4x^3 + 3x^2 + 2x) \, dx$

11. Find the area between the curve $\frac{4}{x^2}$, the x-axis and the ordinates $x = 1$ and $x = 4$.

12. The portion of the curve $xy = 4$ from $x = 2$ to $x = 4$ is rotated about the x-axis. Find the volume generated.

13. The value of the function $y = f(x)$ are given in the table below:

x	0	1	2	3	4	5	6
y	10	12	15	16	11	5	3

Use the trapezium rule to evaluate $\int_0^6 f(x) \, dx$.

14. Given the vector $p = \begin{pmatrix} 2 \\ 1 \end{pmatrix}$ and $q = \begin{pmatrix} 5 \\ 3 \end{pmatrix}$ evaluate $p \cdot q$

15. The vectors a and b are given by $a = \begin{pmatrix} 2 \\ 1 \end{pmatrix}$ and $b = \begin{pmatrix} 1 \\ 1 \end{pmatrix}$. Find, to the nearest degree, the angle between them.

Test 8

1. Given that $\mathbf{P} = (5 \text{ N}, 030°)$ express this in the form $a\mathbf{i} + b\mathbf{j}$ where a and b are scalars.

2. A ship which moves due East at 40 km h^{-1} encounters an ocean current of 5.6 km h^{-1} flowing in the direction S 60° E. At what speed will it now move?

3. A particle of mass 2 kg has an initial velocity $\boldsymbol{u} = \begin{pmatrix} 2 \\ 5 \end{pmatrix}$ m s^{-1} and final velocity $\boldsymbol{v} = \begin{pmatrix} -2 \\ 5 \end{pmatrix}$ m s^{-1}. Find the change in momentum.

4. A particle is acted upon by 2 coplanar forces **P** = (2 N, 060°) and **Q** = (3 N, 045°). If the particle is in equilibrium calculate the magnitude of the resultant force, correct to one decimal place.

5.

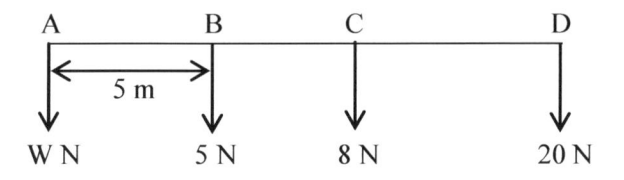

```
     A        B        C        D
     |←  5 m  →|
     ↓        ↓        ↓        ↓
    W N      5 N      8 N     20 N
```

NOT DRAWN TO SCALE

In the figure above find the value of the weight W if AD is a uniform rod of length 30 metres and C is the mid-point of AD.

6. Five boys and three girls join together to form a club. In how many ways can a committee consisting of three boys and two girls be chosen?

7. The probability that Kwame hits a target is $\frac{1}{3}$ and the probability that Yaw hits it is $\frac{1}{2}$. If Kwame and Yaw shoot at the target at the same time find the probability that exactly one of them will hit the target.

8. In how many ways can the letters of the word ACIDITY be arranged in a row?

9. A box contains four tickets numbered 1, 2, 3 and 4. Two tickets are drawn one after the other without replacement and their numbers added. Find the probability that the sum is prime.

10. The probabilities of Kofi and Kwesi scoring from penalties are 0.6 and 0.4 respectively. If they take a penalty kick each, what is the probability that exactly one goal will be scored?

11. The following distribution shows the ages, in years, of contestants at a beauty contest:

17, 16, 18, 21, 22, 19, 21, 18, 22, 24, 23

Determined the interquartile range using the above distribution.

12. The table shows the scores obtained by candidates in an examination:

Marks	0-9	10-19	20-29	30-39	40-49
Frequency	8	14	48	18	12

(a) Draw a cumulative frequency curve for the distribution

(b) Using your curve find:

　(i) the median

　(ii) the interquartile range

13. Find the semi-interquartile range of the following frequency distribution.

x	20-29	30-39	40-49	50-59	60-69	70-79
f	8	11	19	23	25	14

14. The table below gives the distribution of marks in a test:

Marks	0	1	2	3	4	5
Frequency	1	12	14	15	7	1

Find the standard deviation of the distribution.

15. The table shows the mid-term exam scores and final exam scores obtained by ten students in Mathematics.

Mid-Term Exam Scores(x)	71	79	84	76	62	93	88	91	68	77
Final Exams Scores(y)	80	85	88	81	75	90	87	96	82	83

(a) Draw a scatter diagram for the data

(b) Draw the line of best fit of y on x

(c) Find the equation of the line of best fit

(d) Use your equation to predict the final exam grade for a student who scored 87 on the mid-term exam.

Answers to Exercises

Chapter 1

Exercise 1.1

1. $2x + 6$ 2. $4x - 4y$ 3. $-3a + 6b$ 4. $6x + 3y$ 5. $-4y^2 + 8xy$
6. $3x^3 - 3x^2y$ 7. $6x + 2y$ 8. $-3x + 2y$ 9. $a^3 + ab$ 10. $11x + 7$
11. $-x + 18$ 12. $5x$ 13. $-4x^2 + 3xy$ 14. $-2y^2 + 2yz$ 15. $3y^3 - 8y^2 - 3y$

Exercise 1.2(a)

1. $2(x + 5)$ 2. $3(x + 2y)$ 3. $3(y + 5)$ 4. $5a(1 - 4b)$
5. $ax(3a + 2x)$ 6. $3xy(2y - x)$ 7. $3x^2(3x + 2y)$ 8. $9x^2y(x - 4y)$
9. $a^2b(a + b)$ 10. $ab(a - b)$ 11. $6p^2(p + 3)$ 12. $ax(x - 1)$

Exercise 1.2(b)

1. $(a + b)(x + y)$ 2. $(x - 2)(y + 3)$ 3. $(y + 1)(y^2 + 1)$ 4. $(x - 2y)(x - 3)$
5. $(b + 3y)(a - 2b)$ 6. $(xy - z)(x + y)$ 7. $(a - 2b)(3x - 1)$ 8. $(3b + y)(7a - x)$

Exercise 1.3(a)

1. $x^2 + 3x + 2$ 2. $x^2 - 8x + 12$ 3. $x^2 + 3x - 10$ 4. $2x^2 - x - 6$
5. $6x^2 + 13x + 6$ 6. $4x^2 - 16x + 15$ 7. $2x^2 - x - 45$ 8. $-3x + 11$

Exercise 1.3(b)

1. $x^2 + 6x + 9$ 2. $x^2 - 2x + 1$ 3. $x^2 + 10x + 25$ 4. $x^2 - 12x + 36$
5. $4x^2 + 12x + 9$ 6. $9x^2 - 30x + 25$ 7. $4x^2 - 20xy + 25y^2$ 8. $x^2 + 6xy + 9y^2$
9. $2x^2 + 2$ 10. $5x^2 - 2x + 10$ 11. $5x^2 - 10xy$ 12. $5x^2 - 8x + 14$

Exercise 1.3(c)

1. $x^2 - 4$ 2. $x^2 - 36$ 3. $16 - 9x^2$ 4. $9x^2 - 4$
5. $x^4 - y^4$ 6. $4x^4 - 9$ 7. $49x^2 - 25y^2$ 8. $a^4x^6 - 16y^4$

Exercise 1.4(a)

1. $(x+2)(x+3)$ 3. $(x+2)(x+5)$ 3. $(x-1)(x-5)$
4. $(x-3)(x-4)$ 5. $(x-2)(x+7)$ 6. $(x-1)(x+3)$
7. $(x-5)(x+4)$ 8. $(x-9)(x+2)$ 9. $(x-10)(x-2)$
10. $(x-8)(x+6)$ 11. $(x-3)(x-8)$ 12. $(x-4)(x+15)$
13. $(x-10)(x+8)$ 14. $(x-12)(x+2)$ 15. $(x-12)(x-5)$

Exercise 1.4(b)

1. $(2x+1)(x+3)$ 2. $(2x-3)(3x+1)$ 3. $(5x-1)(x+2)$
4. $(3x+2)(x+1)$ 5. $(4x+5)(x+1)$ 6. $(3x+1)(x-7)$
7. $(2x+3)(4x-3)$ 8. $(3x-1)(4x-1)$ 9. $(3x-4)(x-2)$
10. $(3+x)(5-6x)$ 11. $(3x-2)(x-1)$ 12. $(3x-2)(5x+3)$

Exercise 1.4(c)

1. $(1-x)(1+x)$ 2. $(2x-y)(2x+y)$ 3. $(x-4y)(x+4y)$
4. $(2x-3y)(2x+3y)$ 5. $3(x-1)(x+1)$ 6. $(3-2x)(3+2x)$
7. $(2x-5)(2x+5)$ 8. $(4x-9y)(4x+9y)$ 9. $(2x-7y)(2x+7y)$
10. $5(x-3y)(x+3y)$ 11. $(9x-10y^3)(9x+10y^3)$ 12. $3x^3(5x-3)(5x+3)$

Exercise 1.4(d)

1. $(x+6)^2$ 2. $(x-3)^2$ 3. $(x+5)^2$ 4. $(x-2)^2$
5. $(1-x)^2$ 6. $(2+3x)^2$ 7. $(2x+y)^2$ 8. $(3x-5y)^2$
9. $(x+7y)^2$ 10. $(6x-y)^2$ 11. $5x(3x+2)^2$ 12. $2y(5-2y)^2$

Exercise 1.5(a)

1. $\frac{3x}{4}$ 2. $-\frac{2x^2}{3y}$ 3. $\frac{3m^2}{4n}$ 4. $-\frac{2x^2z^2}{3y^2}$ 5. $\frac{a+b}{a}$ 6. $\frac{x-3}{x}$

7. $\frac{3x-1}{x}$ 8. $\frac{x+3}{y}$ 9. $\frac{x+3}{x-1}$ 10. $\frac{2x+3}{x+3}$ 11. $\frac{1}{3y-x}$ 12. $\frac{3x+2}{3x-2}$

Answers to Exercises 607

Exercise 1.5(b)

1. $\dfrac{4x^2}{5}$ 2. $2b$ 3. $\dfrac{2}{x+a}$ 4. $\dfrac{y}{x}$ 5. $\dfrac{3x(x+y)}{y}$ 6. $\dfrac{2(x+5)}{3(x+1)}$

7. $\dfrac{x(x+2)}{2y}$ 8. $\dfrac{2x}{3}$ 9. 1 10. $\dfrac{(x-1)(x+2)}{x+1}$ 11. $\dfrac{x-2y}{a}$ 12. $\dfrac{2y}{x}$

Exercise 1.5(c)

1. $\dfrac{y}{x}$ 2. $\dfrac{10x}{y}$ 3. $\dfrac{1}{y}$ 4. $\dfrac{x}{10a}$ 5. $\dfrac{1}{3(x+y)}$ 6. $\dfrac{a+2}{a-b}$

7. $\dfrac{x}{2x-1}$ 8. $\dfrac{2(x+3)}{3(a-b)}$ 9. $\dfrac{x}{y}$ 10. 1 11. 5 12. $\dfrac{2(x+1)}{x+3}$

Exercise 1.5(d)

1. $\dfrac{x-y}{2}$ 2. $\dfrac{x+2y}{3}$ 3. $x+3$ 4. $\dfrac{x+5}{x-5}$ 5. 2 6. $\dfrac{13x-32}{(x-3)(x-2)}$ 7. $\dfrac{3}{x+1}$

8. $\dfrac{5x-2}{x(x-2)}$ 9. $\dfrac{x^2+y^2}{(x-y)(x+y)}$ 10. $\dfrac{6y}{(y-1)(y+5)}$ 11. $\dfrac{x}{(x-2)(x-3)}$ 12. $\dfrac{3y^2-4y-10}{(y-5)(y-2)(y+3)}$

Exercise 1.5(e)

1. $\dfrac{2(y+1)}{y-1}$ 2. $\dfrac{x+3}{2(x-3)}$ 3. $\dfrac{x}{x-1}$ 4. $\dfrac{x-y}{y}$ 5. $\dfrac{2y-x}{xy}$ 6. $\dfrac{y}{x}$

7. $\dfrac{3x-1}{3-2x}$ 8. $\dfrac{3+x}{x-3}$ 9. $\dfrac{x-3y}{x+2y}$ 10. $\dfrac{1}{2y-x}$ 11. $\dfrac{x-3}{x-2}$ 12. $\dfrac{x-3}{x-5}$

Exercise 1.6(a)

1. 5 2. 2 3. 4 4. 4 5. 6 6. 3 7. 7 8. $3\tfrac{1}{2}$ 9. -4 10. 0 11. 13 12. 1

Exercise 1.6(b)

1. 7 2. -7 3. $1\tfrac{1}{2}$ 4. 6 5. -1 6. -5 7. -5 8. $9\tfrac{1}{2}$ 9. $2\tfrac{1}{4}$ 10. 9

Exercise 1.6(c)

1. -6 2. 10 3. $-1\tfrac{1}{3}$ 4. -10 5. $13\tfrac{1}{2}$ 6. $\tfrac{3}{8}$ 7. $-\tfrac{1}{3}$ 8. $\tfrac{3}{4}$ 9. 6
10. 7 11. 4 12. 10

Exercise 1.6(d)

1. $r = \sqrt[3]{\dfrac{3V}{4\pi}}$ 2. $v = \sqrt{\dfrac{2E}{m}}$ 3. $a = \dfrac{2A}{n} - b$ 4. $r = \sqrt{\dfrac{V}{\pi L}}$ 5. $g = \dfrac{2S}{t^2}$ 6. $v = \dfrac{fu}{u-f}$

7. $g = \dfrac{4\pi^2 l}{r^2}$ 8. $g = \dfrac{2(s-ut)}{t^2}$ 9. $s = \dfrac{v^2-u^2}{2a}$ 10. $R = \dfrac{100I}{PT}$ 11. $u = \dfrac{2s}{t} - v$

12. $v = \sqrt{\dfrac{2E}{m} + u^2}$

Exercise 1.6(e)

1. $x = 2, y = 1$
2. $x = 3, y = 2$
3. $x = 4, y = 1\frac{1}{2}$
4. $x = \frac{2}{3}, y = 3$
5. $x = 2, y = 2$
6. $x = 5, y = -5\frac{1}{2}$
7. $x = -1, y = 0$
8. $x = 1, y = 2$
9. $x = 3, y = 3$
10. $x = 3, y = 4$
11. $x = -2, y = 1\frac{1}{2}$
12. $x = 5, y = -3$
13. $x = 2\frac{1}{2}, y = 1$
14. $x = 3, y = 4$
15. $x = 3, y = 1\frac{1}{3}$
16. $x = 1, y = 2\frac{1}{2}$
17. $x = 4, y = -5$
18. $x = -2, y = -3$
19. $x = 2, y = -4$
20. $x = 2, y = 2$
21. $x = -3, y = 2$

Exercise 1.6(f)

1. $x = 1, y = 2$
2. $x = 3, y = 2$
3. $x = 3, y = 1$
4. $x = -1, y = -1$
5. $x = 2, y = 2$
6. $x = 0, y = 3$
7. $x = 0, y = -2$
8. $x = -1, y = 6$
9. $x = 7, y = 3$
10. $x = 4, y = 0$
11. $x = 2, y = -1$
12. $x = -3, y = 2$

Exercise 1.6(g)

1. $x = 1, y = 2, z = 3$
2. $x = 1, y = 2, z = -3$
3. $x = 2, y = 1, z = -1$
4. $x = 1, y = 2, z = 3$
5. $x = 0, y = 5, z = 3$
6. $x = 6, y = 4, z = -5$

Exercise 1.6(h)

1. $x = 2, y = 6; x = 4, y = 3$
2. $x = -3, y = -3\frac{1}{3}; x = 5, y = 2$
3. $x = 2, y = 2; x = 8, y = -11$
4. $x = 3, y = 1; x = 2\frac{5}{11}, y = 2\frac{7}{11}$
5. $x = -1, y = -4; x = 5, y = 8$
6. $x = 4, y = 1; x = -8, y = -3$
7. $x = 5, y = 3; x = -2\frac{2}{5}, y = -3\frac{4}{5}$
8. $x = 2, y = -1; x = 6, y = 5$
9. $x = 3, y = -3; x = \frac{3}{4}, y = 1\frac{1}{2}$
10. $x = 3, y = 4$
11. $x = 4, y = -1$
12. $x = -3, y = 1; x = -1, y = 3,$

Exercise 1.6(i)

1. – 2, - 5
2. 3, 4
3. – 5, 8
4. – 5, 2
5. – 5, - 3
6. – 3, 6
7. 4, 8
8. – 7, 3
9. 2, 5
10. – 4, 2
11. – 6, 7
12. 3

13. $-2, 6$ 14. $-5, -3$ 15. $-2, 3\frac{1}{3}$ 16. $1, 1\frac{1}{2}$ 17. $-\frac{1}{3}, \frac{1}{2}$ 18. $-\frac{1}{4}, 3$

19. $\frac{1}{4}, 0$ 20. $-5, 5$ 21. $-6, 2$ 22. $-1, 2\frac{1}{3}$ 23. $-\frac{1}{2}, 1\frac{1}{2}$ 24. $-5, 8$

25. $-4, 7$ 26. $-5, 6$ 27. $4, 6$ 28. $-2\frac{1}{2}, 1\frac{1}{2}$ 29. $-6, 2$ 30. $\frac{2}{3}, 2$

Review exercise 1

1(a) $14x + 7y$ (b) $-x + 10y$ (c) $2x^2 + 6xy$ (d) $-3x^3 + 3x^2y$

(e) $6x^3y^2 + 10x^2y^3$ (f) $-6a^3b + 9a^2b^2$

2(a) $5x + 15$ (b) $12a^2 - 5a$ (c) $38x - 9y$ (d) $8a - 7b$

(e) $11x$ (f) $3x^2 + 7xy - 4y^2$

3(a) $4y(2x + 3)$ (b) $3y(3x - z)$ (c) $3x(2y + 3z)$ (d) $a(a + b)$

(e) $7y(3y - 2)$ (f) $3d(3cd + 1)$

4(a) $(p - 3q)(2r + 5)$ (b) $(2x - 3y)(3x - 4)$ (c) $(3w - x)(2y + z)$

(d) $(a + b)(2m - 3n)$ (e) $(3c - 4d)(5a + b)$ (f) $(x + 3)(x - 3)^2$

5(a) $6x^2 + 5x - 21$ (b) $12x^2 - 29x + 15$ (c) $2x^2 + 9x + 4$

(d) $x^2 - 16$ (e) $4x^2 - y^2$ (f) $9x^2 - 16$ (g) $x^2 + 4xy + 4y^2$

(h) $9a^2 - 12ab + 4b^2$ (i) $4x^2 - 12x + 9$

6. (a) $(x + 3)(x - 7)$ (b) $(x - 2)(x - 5)$ (c) $(x + 3)(x + 10)$

(d) $(x - 5)(x - 6)$ (e) $(x - 4)^2$ (f) $(x - 5)^2$ (g) $3(x + 2y)(x - 2y)$

(h) $(4x^3 + 9y^2)(4x^3 - 9y^2)$ (i) $3a(2a + 3b^2)(2a - 3b^2)$

7. (a) $(x + 1)(2x + 5)$ (b) $(y + 1)(5y - 7)$ (c) $(x + 2)(2x - 7)$

(d) $(y + 2)(2y + 3)$ (e) $(a - b)(3a + 5b)$ (f) $(1 + 3x)(3 - 5x)$

(g) $(x + y)(3x + y)$ (h) $x(x - 4)(x + 6)$ (i) $y(y - 5)^2$

8. (a) $5y(5y - 6x)$ (b) $x(x + 4y)$ (c) $8(2 - a)(4a - 1)$

(d) $5x(x + 2)$ (e) $(5 - a)(5a - 1)$ (f) $5a(a - 2b)$

(g) $5(x - y)(x + y)$ (h) $24x(x + 4)$ (i) $15(4x - 3)$

9. (a) $\frac{y}{2x}$ (b) $\frac{2}{x+a}$ (c) $\frac{ab}{2}$ (d) $\frac{2(a+b)}{a-b}$ (e) $\frac{(x-1)(x+2)}{x+1}$ (f) $\frac{3x+1}{x+4}$

10. (a) $\dfrac{9c^2d}{4ab}$ (b) $\dfrac{3}{2xz}$ (c) $\dfrac{7b(a-1)}{2a-7b}$ (d) 5 (e) $\dfrac{2(x+1)(x-2)}{(x+3)(x+4)}$ (f) $\dfrac{y(y+3)}{(y-2)(y-3)}$

11. (a) $\dfrac{13x-32}{(x-2)9x-3)}$ (b) $\dfrac{x^2+y^2}{(x+y)(x-y)}$ (c) $\dfrac{2x^2-2x+41}{(x+4)(x-5)}$ (d) $\dfrac{2(x-5)}{(x-2)(x+4)}$

(e) $\dfrac{a^2+3}{(a-3)(a-2)(a+1)}$ (f) $\dfrac{2}{(y-4)(y-6)}$

12. (a) $\dfrac{-x(y+x)}{y}$ (b) 1 (c) $\dfrac{x-y}{x+y}$ (d) $\dfrac{x-3}{x-1}$ (e) $\dfrac{a-2}{a}$ (f) x

13(a) 11 (b) 4 (c) 7 (d) 11 (e) 1 (f) -5

14. (a) $t = \dfrac{v-u}{a}$ (b) $r = \sqrt{\dfrac{s}{4\pi}}$ (c) $l = \dfrac{2s}{n} - a$ (d) $C = \dfrac{5}{9}(F-32)$

(e) $h = \sqrt{L^2 - 4r^2}$ (f) $n = \dfrac{Rl}{E-lr}$

15(a) $x = 2, y = 3$ (b) $x = 8, y = 7$ (c) $x = 2, y = 7$

(d) $x = 12, y = 18$ (e) $a = 4, b = 3$ (f) $x = 3, y = 4$

16. (a) $x = 2, y = 3, z = 4$ (b) $x = 2, y = -1, z = 1$

(c) $x = 1, y = 3, z = -2$ (d) $x = -3, y = \dfrac{1}{2}, z = -1$

17(a) 3, 7 (b) -5, -2 (c) -1, $1\dfrac{1}{2}$ (d) $\dfrac{3}{5}$, 2 (e) -3, $-\dfrac{2}{3}$

(f) -2, $3\dfrac{1}{2}$ (g) -4, 3

18 (a) $x = -1, y = 3; x = -3, y = 1$ (b) $x = 0, y = 2; x = 2, y = 0$

(c) $x = -1, y = -4; x = 3, y = 0$ (d) $x = 1, y = 0; x = -\dfrac{1}{2}, y = 1\dfrac{1}{2}$

(e) $x = -2, y = 4; x = 4, y = -2$ (f) $x = 1, y = 2; x = -3, y = 0$

Chapter 2

Exercise 2:1

1.(a) statement (b) statement (c) not (d) statement

(e) statement (f) not (g) statement (h) not

2. (a) compound ~ (b) compound ∨ (c) compound ↔ (d) simple

(e) compound → (f) simple (g) compound ∧ (h) simple

3. (a) conjunction (b) disjunction (c) negation (d) biconditional

Answers to Exercises 611

(e) conditional (f) disjunction (g) biconditional (h) negation

4. (a) Every man has a car (b) Some shops are open on Sunday

(c) Some babies do not walk after two years (d) Every dog barks

(e) Someone wants to buy my phone (f) Some of the men like music

(g) Every person who works hard own a house

5. (a) $p \rightarrow q$ (b) $p \vee q$ (c) $p \wedge q$ (d) $p \rightarrow q$ (e) $p \leftrightarrow q$

(f) $p \vee q$ (g) $p \leftrightarrow q$ (h) $p \rightarrow q$

6. (a) Ofori went home or Kwebena studied French

(b) Ofori went home and Kwebena studied French

(c) Ofori do not go home

(d) If Ofori went home then Kwebena studied french

(e) Ofori went home if and only if Kwebena studied French

7. (a) He did not work hard or he did not have two cars

(b) If he did not work hard then he did not have two cars

(c) He did not have two cars if and only if he did not work hard

(d) He work hard or he did not have two cars

(e) He did not work hard and he has two cars

8. (a) $(p \wedge \sim q) \leftrightarrow \sim p$ biconditional (b) $(p \vee q) \rightarrow r$ conditional

(c) $p \leftrightarrow (p \rightarrow \sim r)$ biconditional (d) $q \rightarrow (\sim r \vee p)$ conditional

(e) $(p \wedge \sim q) \rightarrow (\sim p \wedge q)$ conditional (f) $p \rightarrow (q \vee \sim r)$ conditional

9. (a) $(p \rightarrow q) \vee r$ disjunction (b) $p \rightarrow (q \vee r)$ conditional

(c) $(\sim p \rightarrow \sim q) \wedge r$ conjunction (d) $(p \vee q) \rightarrow r$ conditional

(e) $(p \leftrightarrow \sim q) \vee r$ disjunction (f) $(p \rightarrow q) \leftrightarrow \sim p$ biconditional

Exercise 2:2

1. (a)

p	q	P ∨ ~q
T	T	T
T	F	T
F	T	F
F	F	T

(b)

p	q	~p ∨ ~q
T	T	F
T	F	T
F	T	T
F	F	T

(c)

p	q	~p ∧ ~q
T	T	F
T	F	F
F	T	F
F	F	T

(d)

p	q	~p ∧ q
T	T	F
T	F	F
F	T	T
F	F	F

(e)

p	q	(p ∨ q) → p
T	T	T
T	F	T
F	T	F
F	F	T

(f)

p	q	(p ∨ ~q) ∧ ~p
T	T	F
T	F	F
F	T	F
F	F	T

(g)

p	q	~(~p ∨ ~q)
T	T	T
T	F	F
F	T	F
F	F	F

(h)

p	q	(p → q) ↔ (q → p)
T	T	T
T	F	F
F	T	F
F	F	T

(i)

p	q	$(p \vee \sim q) \wedge (q \wedge \sim p)$
T	T	F
T	F	F
F	T	F
F	F	F

(j)

p	q	$(\sim p \vee q) \leftrightarrow \sim(p \wedge \sim q)$
T	T	T
T	F	T
F	T	T
F	F	T

(k)

p	q	$(\sim p \rightarrow q) \leftrightarrow (p \vee q)$
T	T	T
T	F	T
F	T	T
F	F	T

(m)

p	q	$(p \wedge \sim q) \vee (q \wedge \sim p)$
T	T	F
T	F	T
F	T	T
F	F	F

(n)

p	q	r	$(p \wedge q) \vee (p \wedge r)$
T	T	T	T
T	T	F	T
T	F	T	T
T	F	F	F
F	T	T	F
F	T	F	F
F	F	T	F
F	F	F	F

(o)

p	q	r	$(p \wedge \sim q) \vee r$
T	T	T	T
T	T	F	F
T	F	T	T
T	F	F	T
F	T	T	T
F	T	F	F
F	F	T	T
F	F	F	F

(p)

p	q	r	$(\sim p \vee q) \wedge \sim r$
T	T	T	F
T	T	F	T
T	F	T	F
T	F	F	F
F	T	T	F
F	T	F	T
F	F	T	F
F	F	F	T

2. (a) Converse: If the car has air conditioning, then the car is new

Inverse: If car is not new, then the car has no air conditioning

Contrapositive: If the car has no air conditioning, then the car is not new

(b) Converse: If I am not able to pay the bill, then the electric bill is too high

Inverse: If the electric bill is not too high, then I am able to pay the bill

Contrapositive: If I am able to pay the bill, then the electric bill is not too high

(c) Converse: If we go for a walk, then the light went off

Inverse: If the light does not go off, then we will not go for a walk

Contrapositive: If we do not go for a walk, then the light does not go off

(d) Converse: If I will not take the job, then the pay is not good

Inverse: If the pay is good, then I will take the job

Contrapositive: If I do take the job, then the pay is good

(e) Converse: If the library is not open, then today is a holiday

Inverse: If today is not a holiday, then the library is open

Contrapositive: If the library is open, then today is not a holiday

3. (a) Converse: If two angles of a triangle are equal, then the triangle is isosceles. True

Inverse: If the triangle is not isosceles, then two angles are not equal. False

Contrapositive: If two angles of a triangle are not equal, then the triangle is not isosceles. True

(b) Converse: If two lines are not parallel, then the two lines intersect in at least one point. True

Inverse: If two lines do not intersect in at least one point, then the two lines are parallel. True

Contrapositive: If two lines are parallel, then the two lines do not intersect in at least one point. True

(c) Converse: If the quadrilateral is a parallelogram, then the quadrilateral is a rectangle. False

Inverse: If the quadrilateral is not a rectangle, then the quadrilateral is not a parallelogram. False

Contrapositive: If the quadrilateral is not a parallelogram, then the quadrilateral is not rectangle. True

(d) Converse: If the polygon is a quadrilateral, then the sum of the interior angles of the polygon is 360°. True

Inverse: If the sum of the interior angles of a polygon is not 360°, then the polygon is not a quadrilateral. True

Contrapositive: If the polygon is not a quadrilateral, then the sum of the interior angles of the polygon is not 360°. True

(e) Converse: If the quadrilateral is a rectangle, then the quadrilateral is a square. False

Inverse: If the quadrilateral is not a square, then the quadrilateral is not a rectangle. False

Contrapositive: If the quadrilateral is not a rectangle, then the quadrilateral is not a square. True

(f) Converse: If 2 divides the natural number, then 2 divides the unit digit of the natural number. True

Inverse: If 2 does not divide the natural number, then 2 does not divide the unit digit of the natural number. True

4. (a) converse: $q \rightarrow \sim p$ (b) converse: $\sim q \rightarrow \sim p$

Inverse: $\sim(\sim p) \rightarrow \sim q$ Inverse: $\sim(\sim p) \rightarrow \sim(\sim q)$

Contrapositive: $\sim q \rightarrow \sim(\sim p)$ Contrapositive: $\sim(\sim q) \rightarrow \sim(\sim p)$

(c) converse: $\sim q \rightarrow p$ (d) converse: $r \rightarrow (p \wedge q)$

inverse: $\sim p \rightarrow \sim(\sim q)$ Inverse: $\sim(p \wedge q) \rightarrow \sim r$

Contrapositive: $\sim(\sim q) \rightarrow \sim p$ Contrapositive: $\sim r \rightarrow \sim(p \wedge q)$

(e) converse: $\sim r \rightarrow (p \wedge \sim q)$ (f) converse: $(\sim q \vee r) \rightarrow (p \vee q)$

Inverse: $\sim(p \wedge \sim q) \rightarrow \sim(\sim r)$ Inverse: $\sim(p \vee q) \rightarrow \sim(\sim q \vee r)$

Contrapositive: $\sim(\sim r) \rightarrow \sim(p \wedge \sim q)$ Contrapositive: $\sim(\sim q \vee r) \rightarrow \sim(p \vee q)$

Exercise 2:3

1. (a) Equivalent   (b) not equivalent   (c) equivalent   (d) equivalent

(e) not equivalent   (f) equivalent   (g) not equivalent   (h) equivalent

Exercise 2:4

1. (a) valid (b) valid (c) not valid (d) valid (e) valid (f) valid

2. (a) not valid (b) valid (c) valid (d) valid

Review exercise 2

1. (a) statement (b) not (c) statement (d) statement (e) not

2. (a) compound (b) simple (c) compound (d) compound (e) simple (f) compound

3. (a) I will not study hard (b) Esi is not talented (c) I will not wash my car (d) Afua will not do her home work

4. (a)

p	q	~q	q ∨ ~q	P ∧ (q ∨ ~q)
T	T	F	T	T
T	F	T	T	T
F	T	F	T	F
F	F	T	T	F

(b)

p	q	p ∨ q	~q	p ∧ ~q	(p ∨ q) ∧ (p ∧ ~q)
T	T	T	F	F	F
T	F	T	T	T	T
F	T	T	F	F	F
F	F	F	T	F	F

(c)

p	q	(p → q) ∧ ~q	[(p → q) ∧ ~q] → ~p
T	T	F	T
T	F	F	T
F	T	F	T
F	F	T	T

(d)

p	q	p ∧ q	~p ∧ ~q	(p ∧ q) ∨ (~p ∧ ~q)
T	T	T	F	T
T	F	F	F	F
F	T	F	F	F
F	F	F	T	T

5. (a) p ∨ q   (b) p ∧ q   (c) p ↔ q   (d) p → q

6. (a) Converse: If it is raining then it is cloudy

Contrapositive: If it is not raining then it is not cloudy

Inverse: If it is not raining then it is not cloudy

(b) Converse: If I miss the program then I walked home

Contrapositive: If I did not miss the program then I did not walk home

Inverse: If I did not walk home then I did not miss the program

(c) Converse: Esi will be expelled if Esi does not obey school rules

Contrapositive: Esi will not be expelled if she obeys school rules

Inverse: If Esi obeys school rules, she will not be expelled

(d) Converse: If I do not go fishing then I will go to school

Contrapositive: If I go fishing then I will not go to school

Inverse: If I do not go school then I will go finishing

7. (a) Equivalent   (b) not equivalent   (c) Equivalent   (d) Equivalent

(e) not equivalent   (f) equivalent

8. (a) valid   (b) valid   (c) not valid

Chapter 3

Exercise 3:1

1. (a) true   (b) true   (c) false   (d) false   (e) true   (f) false

2. (a) {11}   (b) {April, May, June, September, November}

(c) {4, 6, 8}   (d) {-2}

3. (a) $\subset$ (b) $\not\subset$ (c) $\in$ (d) $\supset$ (e) $\notin$

4. (a) P = {2, 4, 6, 8}   (b) Q = {3, 5, 7}   (c) R = {2, 3, 5, 7}

5. (a) 4   (b) 3   (c) 7   6. $\emptyset$, {a}, {b}, {c}, {a, b}, {a, c}, {b, c}, {a, b, c}

7. 16   8. 7   9 (a) infinite   (b) finite   (c) infinite   (d) finite

10 (a) {2, 4, 5}   (b) $\emptyset$ (c) {1, 3}   (d) {1, 2, 3, 4, 5}

11(a) $A' = \{1, 3, 6\}$ (b) $B' = \{2, 4, 5\}$ (c) $(A')' = \{2, 4, 5\}$ (d) $(B')' = \{1, 3, 6\}$

12 (a) not empty   (b) empty   (c) not empty   (d) empty

Exercise 3.2

1. (a) $A \cup B = \{-3, -2, -1, 0, 3, 4, 5\}$ (b) $B \cap C = \{-1\}$
(c) $(A \cup B) \cup C = \{-3, -2, -1, 0, 2, 3, 4, 5, 6\}$ (d) $A \cap (B \cap C) = \{-1\}$

2. (a) $\{2, 3\}$ (b) $\{2, 3\}$ (c) $\{1, 2, 3, 4, 5, 6\}$ (d) $\{1, 2, 3, 4, 5, 6$
(e) $\{2, 3, 4, 5\}$ (f) $\{2, 3, 4, 5$ (g) $\{1, 2, 3, 4, 5, 6\}$ (h) $\{1, 2, 3, 4, 5, 6\}$

3. (a) $\emptyset$ (b) $\{1, 2, 3, 4, 5, 7\}$ 4. (a) $\{d\}$ (b) $\{d\}$ (c) $\{a, b, d, e\}$ (d) $\{a, b, d, e\}$

5. (a) $\{x: -3 \le x < 7\}$ (b) $\{x: 2 < x \le 4\}$ 6. (a) $\{x: x > -1\}$ (b) $\{x: 4 < x < 6\}$

7. (a) $\{x: -4 \le x \le 3\}$ (b) $\{x: -1 < x < 1\}$

8. (a) $\{x: x < 5\}$ (b) $\{x: x - 3 < x < 4\}$ 9. $\{4, 5, 10\}$ 10. (a) $\{3, 6\}$ (b) $\{3, 6\}$

11. (a) $\{1, 3\}$ (b) $\{2, 4, 5, 6, 7, 8, 9, 10, 11, 12, 13, 14, 15\}$
(c) $\{2, 4, 6, 7, 8, 9, 10, 11, 12, 13, 14\}$ (d) $\{1, 2, 3, 4, 5, 6, 12, 15\}$
(e) $\{5, 7, 8, 9, 10, 11, 13, 14, 14, 15\}$ (f) $\{2, 4, 5, 6, 7, 8, 9, 10, 11, 12, 13, 14, 15\}$

12. (a) $\{1, 2, 3, 4, 6, 12\}$ (b) $\{1, 2, 4, 5, 10, 20\}$ (c) $\{1, 2, 3, 4, 5, 6, 10, 12, 20\}$
(d) $\{1, 2, 4\}$

Exercise 3.3

1. 35   2. 380

3. (a) $\{p\}$ (b) $\{w, x\}$ (c) $\{q, s, t\}$ (d) $\{p, q, r\}$ (e) $\{p, t, w\}$ (f) $\{w\}$

4. (a) 72 (b) 28 (c) 90 (d) 85 (e) 203 (f) 80 (g) 65

5. (a) 90, (b) 14   6. (a) 7 (b) 6 (c) 17   7. (a) 6 (b) 34

Review exercise 3

1. (a) $\in$ (b) $\notin$ (c) $\not\subset$ (d) $\supset$ (e) $\subset$ (f) $\in$

2. (a) $\{3, 8\}$ (b) $\{3, 4, 5, 6, 7, 8\}$ (c) $\{1, 2, 5, 7, 9\}$ (d) $\{1, 2, 4, 6, 9\}$

3. (a) $\{5, 6\}$ (b) $\{2, 3, 4, 5, 6, 7, 8, 10\}$ (c) $\{1, 7, 8, 9, 10\}$
(d) $\{1, 2, 3, 4, 9\}$

4. (a) $\{5\}$ (b) $\{1, 2, 3, 4, 5, 6, 7\}$ (c) $\{1, 2, 3, 4\}$ (d) $\{6, 7\}$
(e) $\{5, 6, 7\}$ (f) $\{1, 2, 3, 4\}$

5 (a) {3, 6, 8, 9}　　(b) {6}　　(c) {2, 3, 6, 7}

6 (a) {1, 2, 4, 5, 7}　　(b) {1, 3, 5}　　(c) {1, 2, 4, 5, 7}

7 (a) {2, 5, 8}　(b) {1, 3, 4, 6, 7}　(c) {6}　　$A \cup B' = (A' \cap B)'$

8 (a) $\{x: x > -2\}$　(b) $\{x: -1 < x < 4\}$

9 (a) $\{x: -3 < x < 7\}$　(b) $x: -2 \leq x < 5\}$　　10. 12　　11 (a) 25　(b) 16

Chapter 4

Exercise 4.1

1. $2\sqrt{5}$　　2. $2\sqrt{7}$　　3. $3\sqrt{3}$　　4. $2\sqrt{10}$　　5. $3\sqrt{5}$　　6. $4\sqrt{3}$　　7. $5\sqrt{2}$

8. $3\sqrt{6}$　　9. $2\sqrt{15}$　　10. $6\sqrt{2}$　　11. $5\sqrt{3}$　　12. $4\sqrt{5}$　　13. $4\sqrt{6}$　　14. $7\sqrt{2}$

15. $7\sqrt{3}$　　16. $5\sqrt{6}$　　17. $9\sqrt{2}$　　18. $6\sqrt{5}$　　19. $8\sqrt{3}$　　20. $7\sqrt{5}$　　21. $12\sqrt{2}$

22. $8\sqrt{5}$　　23. $9\sqrt{5}$　　24. $14\sqrt{3}$　　25. $11\sqrt{5}$　　26. $15\sqrt{3}$　　27. $11\sqrt{6}$

28. $6\sqrt{15}$　　29. $11\sqrt{7}$　　30. $12\sqrt{5}$　　31. $13\sqrt{2}$　　32. $9\sqrt{7}$

Exercise 4.2

1. $7\sqrt{3}$　　2. $8\sqrt{2}$　　3. $7\sqrt{5}$　　4. $2\sqrt{7}$　　5. $-3\sqrt{3}$　　6. $-2\sqrt{5}$　　7. $6\sqrt{2}$

8. $6\sqrt{3}$　　9. $4\sqrt{5}$　　10. $-3\sqrt{7}$　　11. $4\sqrt{3}$　　12. $7\sqrt{2}$　　13. $2\sqrt{3}$　　14. $2\sqrt{5}$

15. $16\sqrt{2}$　　16. $7\sqrt{2}$　　17. $\sqrt{5}$　　18. $3\sqrt{2}$　　19. 0　　20. $-\sqrt{5}$　　21. $4\sqrt{7}$

22. $15\sqrt{2}$　　23. 0　　24. $15\sqrt{2}$　　25. $\sqrt{5}$　　26. $7\sqrt{3}$　　27. $-\sqrt{2}$　　28. $3\sqrt{2}$

29. $-4\sqrt{5}$　　30. $5\sqrt{6}$　　31. $-9\sqrt{3}$　　32. $-6\sqrt{2}$

Exercise 4.3

1. $2\sqrt{3}$　　2. $5\sqrt{3}$　　3. $3\sqrt{6}$　　4. $10\sqrt{15}$　　5. $6\sqrt{10}$　　6. $15\sqrt{6}$　　7. $6\sqrt{15}$

8. 6　　9. $4\sqrt{5}$　　10. $4\sqrt{10}$　　11. $9\sqrt{5}$　　12. $6\sqrt{6}$　　13. $10\sqrt{15}$　　14. $35\sqrt{6}$

15. $4\sqrt{30}$　　16. $2\sqrt{15}$　　17. $12\sqrt{5}$　　18. $30\sqrt{2}$　　19. $18\sqrt{5}$　　20. $180\sqrt{3}$

Exercise 4.4

1. $\frac{2\sqrt{3}}{3}$　　2. $2\sqrt{5}$　　3. $2\sqrt{2}$　　4. $3\sqrt{3}$　　5. $\frac{\sqrt{30}}{2}$　　6. $4\sqrt{3}$　　7. $\frac{3\sqrt{10}}{2}$　　8. $3\sqrt{3}$

9. $2\sqrt{7}$ 10. $\frac{3\sqrt{2}}{8}$ 11. $\sqrt{2}$ 12. $\sqrt{10}$ 13. $\frac{3\sqrt{2}}{2}$ 14. $\frac{\sqrt{15}}{3}$ 15. $\frac{\sqrt{6}}{2}$ 16. $\frac{2}{3}$

17. $\frac{\sqrt{6}}{2}$ 18. $\frac{2\sqrt{15}}{5}$ 19. $\frac{\sqrt{5}}{2}$ 20. $\frac{2\sqrt{3}}{3}$

Exercise 4.5

1. $\sqrt{15} - 3$ 2. $2\sqrt{15} - 9$ 3. $\sqrt{15} + 3$ 4. $2\sqrt{21} + 21$ 5. $6 + \sqrt{30}$

6. $2\sqrt{3} - 2$ 7. $30 - 10\sqrt{6}$ 8. $4\sqrt{6} + 12$ 9. $7\sqrt{3} + 13$ 10. $7 - 3\sqrt{5}$

11. $6\sqrt{6} - 6\sqrt{2} - 2\sqrt{3} + 6$ 12. $2\sqrt{6} - 2$ 13. $-3\sqrt{6} + 3\sqrt{2} - 2\sqrt{3} + 2$

14. $\sqrt{10} - \sqrt{3}$ 15. $5\sqrt{5} + \sqrt{15}$ 16. $4\sqrt{5} + 5\sqrt{2} - 2\sqrt{10} - 8$ 17. 1

18. -18 19. -15 20. 4 21. 3 22. -20 23. $4\sqrt{3} + 7$

24. $14 - 6\sqrt{5}$ 25. $2\sqrt{6} + 5$ 26. $11 - 4\sqrt{6}$ 27. $12\sqrt{2} + 22$

28. $30 - 12\sqrt{6}$ 29. $6 - 12\sqrt{3}$ 30. $4\sqrt{6} + 11$

Exercise 4.6

1. $\sqrt{2} - 1$ 2. $\sqrt{3} + 1$ 3. $\sqrt{3} + 2$ 4. $6 - 2\sqrt{7}$ 5. $\sqrt{3} + \sqrt{2}$

6. $2 - \frac{\sqrt{6}}{2}$ 7. $\frac{5\sqrt{7}}{6} + \frac{7}{6}$ 8. $\frac{12\sqrt{2}}{11} + \frac{2\sqrt{6}}{11}$ 9. $7 - 4\sqrt{3}$

10. $\frac{6\sqrt{2}}{7} + \frac{11}{7}$ 11. $\sqrt{3} + 2$ 12. $\frac{17}{7} - \frac{4\sqrt{15}}{7}$ 13. $17 - 12\sqrt{2}$

14. $-\sqrt{15} - 4$ 15. $\frac{1}{5} - \frac{2\sqrt{14}}{5}$ 16. $-4\sqrt{3} - 7$ 17. $-\frac{5\sqrt{5}}{2} - \frac{11}{2}$

18. $\frac{2\sqrt{6}}{3} + \sqrt{3}$ 19. $\frac{13\sqrt{15}}{30} + \frac{3}{2}$ 20. $\frac{\sqrt{21}}{5} + \frac{1}{5}$ 21. $-\frac{14\sqrt{3}}{3} + 6\sqrt{2}$

Exercise 4.7

1. 3 2. 3 3. -5 4. 3 5. $\frac{1}{2}$ 6. 7 7. 3 8. 3 9. -4, -2 10. 2, 5 11. 10

12. 4 13. 9 14. $\frac{9}{16}$ 15. 7 16. 7 17. 2 18. 3 19. 7 20. $1\frac{2}{2}$, $1\frac{3}{4}$

Review exercise 4

1. (a) $-5\sqrt{6}$ (b) $4\sqrt{2}$ (c) $-6\sqrt{5}$ (d) $-2\sqrt{2}$ (e) $\sqrt{2}$ (f) $4\sqrt{2}$

2(a) $6\sqrt{3}$ (b) $4\sqrt{5}$ (c) $4\sqrt{7}$ (d) $6\sqrt{15}$ (e) $25\sqrt{6}$ (f) $6\sqrt{15}$

(g) $18\sqrt{6}$ (h) $6\sqrt{6}$ (i) $60\sqrt{5}$

3. (a) $\frac{\sqrt{2}}{10}$ (b) $\frac{4\sqrt{2}}{3}$ (c) $\frac{4\sqrt{2}}{9}$ (d) $\frac{3\sqrt{6}}{8}$ (e) $\frac{\sqrt{3}}{3}$ (f) $\frac{2\sqrt{3}+3}{3}$

4. (a) $7\sqrt{6} + 23$ (b) $3\sqrt{14} + 4$ (c) $2\sqrt{6} + 5$ (d) $79 - 20\sqrt{3}$

(e) $2 - 3\sqrt{6}$ (f) $-3\sqrt{35} - 53$

5(a) $\frac{3\sqrt{2}+\sqrt{5}}{13}$ (b) $-9\sqrt{2} - 12$ (c) $\frac{5\sqrt{3}+8}{11}$ (d) $2\sqrt{6} + 5$ (e) $\frac{3\sqrt{6}-5\sqrt{2}}{4}$

(f) $\frac{5\sqrt{6}-6}{19}$

6. (a) 3   (b) 1, 3   (c) 26   (d) 15

Chapter 5

Exercise 5:1

1. (a) 3   (b) 12   (c) -64   (d) -16   (e) 19   (f) $1 - \sqrt{3}$ (g) 9

(h) 53   (i) -13   (j) 127

2. -4, 2   3. 8, -2   4. -7, 1   5. 1   6. $1\frac{1}{2}$ 7. 5   8. $b = \frac{3}{2}a$

9. (i) -2   (ii) -5   10. 3

Exercise 5:2

1. commutative   2. Not commutative   4. associative   5. Not associative

6. Not closed   7. $\frac{1}{4}$ 8. $\frac{a}{a-3}$ 9. $\frac{4}{x}$ 10. $\frac{-a}{1+2a}$ 11. 1, -6   12. -4, 2

Review exercise 5

1. (a)

*	0	1	2	3
0	0	1	2	3
1	1	1	1	1
2	2	1	0	-1
3	3	1	-1	-3

(b) not closed   (c) 0   (d) 2   2. (a) 3   (b) none   (c) 4

4. (a) 0   (b) $\frac{x}{2x-1}$ $\frac{1}{2}$ 5. (b) $\frac{2a}{a-2}$ (c) 2   6. (a) $\frac{x}{x-1}$ (b) $\frac{3}{4}$

7. $-2 - \sqrt{5}$ 8. (a) -1, 3   (b) $1 + \frac{1}{6}\sqrt{3}$

9. (a) (i) 36   (ii) 30   (b) 1, -6   10. (a) 7   (b) -1

Chapter 6

Exercise 6.1

1. (b), (d)

2. (a) Domain = {-1, 0, 1}, Range = {2, 3}   (b) Domain = {-3, -2}, Range = {0, 2}
(c) Domain = {-1, 0, 1}, Range = {0}   (d) Domain = {-2, -1, 0}, Range = {3, 4}

3. (a), (c)

4. (a) $\{x: x \in R\}$ (b) $\{x: x \geq 0\}$ (c) $\{x: x \in R\}$ (d) $\{x: x \in R\}$
(e) $\{x: x \leq 0\}$ (f) $\{x: x \in R\}$

5. (a) 10   (b) 5   (c) 13   (d) 26   6. (a) −5   (b) 0   (c) 4   (d) 3

7. (a) $a^2 + 2a + 1$ (b) $4a$ (c) 4   (d) $x^2 + 2xy + y^2$

8. 2   9. −5, 3   10. 4

11. (a) $\{x: x \in R\}$ (b) $\{x: x \in R\}$ (c) $\{x: x \in R, x \neq 5\}$ (d) $\{x: x \in R, x \neq -3, x \neq 2\}$

12. (a) $\{x: x \leq 7\}$ (b) $\{x: x \geq 6\}$ (c) $\{x: -5 \leq x \leq 5\}$

Exercise 6.2

1. (a) one-to-one   (b) not   (c) one-to-one   (d) not

3. (a) not   (b) one-to-one   (c) not   (d) one-to-one

Exercise 6.3

1. (b), (c), (e), (f)

2. (a) $\{y: y \in R\}$ (b) $\{y: y \in R, y \geq -4\}$ (c) $\{y: y \in R, y \leq 3\}$
(d) $\{y: y \in R, y \geq -2\}$ (e) $\{y: y \in R, y > 0\}$ (f) $\{y: y \in R, y > 0\}$

5. (a) $f^{-1}(x) = \frac{x+8}{3}$ (b) $g^{-1}(x) = \sqrt{4-x}$ (c) $h^{-1}(x) = \frac{5-x}{2x}, x \neq 0$

(d) $f^{-1}(x) = \frac{4x+3}{x-2}, x \neq 2$

6. 5   7. $\frac{2}{5}$ 8. 1   9. −2

10. (a) $\{y: y \in R\}$ (b) $\{y: y \in R, y \geq 2\}$ (c) $\{y: y \in R, y \leq 8\}$ (d) $\{y: y \in R, y \geq 0\}$

Exercise 6.4

1. (a) $5x - 1$ (b) $x - 7$ (c) $6x^2 + x - 12$ (d) $\frac{3x-4}{2x+3}$

2. (a) $2x^2 + 4x + 3$ (b) $2x^2 + 2x - 3$ (c) $2x^3 + 9x^2 + 9x$ (d) x

3. (a) $x^2 + x - 6$ (b) $12 + x - x^2$ (c) $x^3 + 3x^2 - 9x - 27$ (d) $\frac{1}{x-3}$

4. (a) $\frac{2x+1}{3x+2}$ (b) $\frac{2x-7}{3x+2}$ (c) $\frac{8x-12}{(3x+2)^2}$ (d) $\frac{2x-3}{4}$

5. (a) $3x^2 + 4x - 8$ (b) $1 + 9x - 2x^2$

6. (a) $4x^5 - 12x^4 - 15x^3 + 50x^2$ (b) $\frac{x+2}{x^2}$

Exercise 6.5

1. (a) $25x + 18$ (b) $9x - 8$ (c) $15x - 7$ (d) $15x + 7$

2. (a) $2x^2 + 1$ (b) $4x^2 + 12x + 8$ (c) $4x + 9$ (d) $x^4 - 2x^2 + 1$

3. (a) $9x^2 - 24x + 16$ (b) $9x^2$ (c) $9x^2 + 12x + 2$ (d) $3x^2$

4. (a) $\frac{3x-12}{x+4}, x \neq -4$ (b) $\frac{6x-9}{2x+1}, x \neq -\frac{1}{2}$

5. (a) x (b) x (c) x (d) x (e) $\frac{2x-4}{3}$ (f) $\frac{2x-14}{3}$

6. (a) $12x^2 + 36x + 25$ (b) 1 (c) $6x^2 - 1$ (d) 23

7. (a) 26 (b) $9x^2 - 6x + 1$ (c) 196 (d) 11

8. (a) $-9\frac{1}{2}$ (b) 3 (c) $-\frac{3}{22}$ (d) $3\frac{5}{6}$ 9. (a) 9 (b) 48 (c) 5 (d) 63

10. $g(x) = x + 3$ 11. (a) 2 (b) 3 12. (a) $-2, 1$ (b) $-5, 5$ 13. $-1, 6$

Review exercise 6

1. (a) Domain of $f = \{x : x \in R, x \leq -5 \text{ or } x \geq 5\}$

Domain of $g = \{x : x \in R, x \neq \frac{2}{3}\}$ (b) $\{y : y \in R, y \geq 25\}$ (c) one-to-one

2. (a) $g^{-1}(x) = \frac{5x+4}{3-x}, x \neq 3$ (b) $\{y : y \in R, y \neq -5\}$

3. (a) Domain of $f = \{x : x \in R, x \neq 4\}$ Domain of $g = \{x : x \in R, x \neq -\frac{1}{2}\}$

(b) $f^{-1}(x) = \frac{4x+2}{x-3}, x \neq 3$ $gof^{-1}(x) = \frac{19x-1}{9x+1}$

4. (a) $\frac{4x+2}{x-1}$ (b) $\frac{9x-3}{x-3}$ 5. (a) one-to-one   (b) $h^{-1}(x) = \frac{2x+1}{x-1}, x \neq 1$

6. (a) $\dfrac{2x^2+3}{x^2+2}$ (b) $1\dfrac{2}{3}$ 7. $1\dfrac{1}{4}$ 8. (a) $h^{-1}(x) = \dfrac{2}{x-1}, x \neq 1$ (b) $\dfrac{7}{9}$

9. $p = 3, q = 2$ 10. $p = 2\dfrac{1}{5}, q = -\dfrac{2}{5}$

Chapter 7

Exercise 7.1

1. (a) 13 (b) 5 (c) 8 (d) $\sqrt{13}$ (e) $\sqrt{13}$ (f) $2\sqrt{2}$

2. $AB = 3\sqrt{2}$, $AC = 2\sqrt{17}$, $BC = 2\sqrt{5}$, right-angled triangle

3. isosceles triangle 4. 3 5. –4, 2 6. 4

Exercise 7.2

1. (a) (3, 3) (b) (5, 6) (c) $\left(3, -\dfrac{1}{2}\right)$ (d) (4, -1) (e) $\left(-2\dfrac{1}{2}, -1\dfrac{1}{2}\right)$ (f) (2, 2}

2.(a) (5, 3); $\left(-1\dfrac{1}{2}, -1\dfrac{1}{2}\right)$ $\left(\dfrac{1}{2}, -\dfrac{1}{2}\right)$ 3. (4, 8)   4. $x = 9, y = 5$

Exercise 7.3

1. (a) $\dfrac{7}{4}$ (b) –2 (c) $\dfrac{1}{11}$ (d) $\dfrac{3}{2}$ (e) $-\dfrac{4}{5}$ (f) $-\dfrac{5}{6}$

4. 14   5. 5   6. 2   7. 1   8. $a = 2, b = 1$

Exercise 7.4(a)

1. (a) $\dfrac{3}{2}$, (0, 2)   (b) $\dfrac{3}{4}$, $\left(0, 2\dfrac{1}{2}\right)$ (c) –3, $\left(0, 2\dfrac{1}{2}\right)$ (d) $\dfrac{4}{5}$, $\left(0, -2\dfrac{3}{5}\right)$

(e) $-\dfrac{4}{3}$, $\left(0, \dfrac{2}{3}\right)$ (f) $-\dfrac{1}{2}$, $\left(0, 1\dfrac{1}{2}\right)$

2. (a) $y = 3x + 4$ (b) $y = -2x + 5$ (c) $y = 4x$ (d) $y = -\dfrac{3}{4}x - 4$

(e) $y = \dfrac{2}{3}x + 1$ (f) $y = -\dfrac{3}{2}x - 3$

3. (a) –1   (b) 4   (c) 5   (d) 8   (e) –5   (f) –4

Exercise 7.4(b)

1. (a) $y = 3x - 2$ (b) $y = -3x - 5$ (c) $y = -4x + 5$ (d) $y = 2x - 9$

(e) $y = -7x - 12$ (f) $y = 4x - 5$

2. (a) $y = x + 1$ (b) $y = 2x + 7$ (c) $y = -x + 4$ (d) $y = 3$ (e) $y = 3x - 17$

(f) $y = -2x + 3$ (g) $y = 3x + 10$ (h) $y = -3x + 17$ (i) $y = 2x + 8$

Answers to Exercises 625

Exercise 7.4(c)

1. (a) $3x + 2y - 24 = 0$ (b) $5x - 3y + 9 = 0$ (c) $2x - 5y - 18 = 0$

(d) $3x + 2y + 19 = 0$ (e) $4x - 3y + 1 = 0$ (f) $3x + 5y - 13 = 0$

2. (a) $3x + 2y + 2 = 0$ (b) $4x + 3y - 6 = 0$ (c) $3x + 4y + 22 = 0$

(d) $5x + 4y - 32 = 0$ (e) $3x - 4y = 0$ (f) $2x - 3y - 7 = 0$

3. (a) $4x + 3y - 12 = 0$ (b) $2x - 3y + 6 = 0$ (c) $3x + 2y + 6 = 0$

(d) $3x + 2y - 1 = 0$ (e) $3x - 4y + 1 = 0$

Exercise 7.5(a)

1. (a) parallel (b) not parallel (c) parallel (d) parallel

2. (a) parallel (b) not parallel (c) parallel (d) parallel

3. (a) $4x - 3y - 3 = 0$ (b) $5x - 2y + 13 = 0$ (c) $2x + 3y = 0$

(d) $3x - 2y + 4 = 0$ (e) $x = -2$ (f) $y = 3$

4. 11   5. 3   6. $6x + 4y - 29 = 0$

Exercise 7.5(b)

1. (a) $-\frac{1}{3}$ (b) $\frac{1}{2}$ (c) -2 (d) 3 (e) $-\frac{4}{3}$ (f) $\frac{2}{3}$

2. (a) perpendicular (b) perpendicular (c) not perpendicular (d) perpendicular

3. (a) perpendicular (b) perpendicular (c) not perpendicular (d) not perpendicular

4. (a) $3x - 5y + 3 = 0$ (b) $y = 3x + 11$ (c) $2x - 3y + 12 = 0$

(d) $3x + 2y - 1 = 0$ (e) $2x - 3y + 4 = 0$ (f) $3x + 2y - 12 = 0$

5. $x + y + 1 = 0$ 6. -1 7. $2\frac{1}{4}$ 8. $3x + 2y - 12 = 0$ 9. 6.5

Exercise 7.6

1. (a) (2, 1) (b) (3, 2) (c) (3, 3) (d) $\left(5, -5\frac{1}{2}\right)$ (e) (2, - 4) (f) (2, 2)

(g) (- 1, 1) (h) (3, - 3)

2. (-9, 3)   3. (- 3, 5)   4. $P\left(1\frac{1}{3}, 0\right), Q(0, -2), 1\frac{1}{3}$ 5. (1, 3)   6. (1, 10)

7. P(-1, 0), $Q\left(0, 1\frac{1}{3}\right), \frac{2}{3}$ 8. P(2, 0) Q(0, - 3), 3

Exercise 7.7

1. (a) 2 (b) $\frac{7}{13}\sqrt{13}$ (c) 13 (d) 3 (e) 1 (f) b 2 (g) $2\sqrt{5}$ (h) 2

2. (a) $\frac{3}{2}\sqrt{2}$ (b) 6   3. (a) (-5, 4) (b) 19.5 (c) 3.42   4. $\frac{4}{5}$ 5. 3

Exercise 7.8

1. (a) $\frac{15}{16}$ (b) $\frac{5}{31}$ (c) $\frac{17}{6}$ (d) $\frac{1}{7}$

2. (a) 18.4° (b) 63.4° (c) 16.3° (d) 64.7°

Review exercise 7

1. (0, 11)   2. (5, 0)   6. $\left(1\frac{1}{2}, -1\frac{1}{2}\right), \left(-\frac{1}{2}, -4\frac{1}{2}\right), \left(-3\frac{1}{2}, \frac{1}{2}\right), \left(-1\frac{1}{2}, 3\frac{1}{2}\right)$

8. (5, 3), (1, -1), (9, -3)   10. $y = 3x - 6$ 11. $x - 3y + 6 = 0$

12. (a) $y = x + 1$, $x - 2y + 5 = 0$ $2x - 3y + 4 = 0$

(b) $3x - 5y + 9 = 0$ $y = 4$, $3x - 4y + 5 = 0$

13. (-8, -12), (16, 12)   14. $7x - 8y + 31 = 0$

15. $3x + 4y + 1 = 0$ $4x - 3y - 7 = 0$ 16. (a) $S\left(-4, 2\frac{1}{2}\right)$ T (- 3, 1) (b) 9.79

17. (a) 2 (b) $\sqrt{13}$ (c) 6 (d) $4\sqrt{5}$ 18. (a) 18.4° (b) 7.1° (c) 17.7° (d) 59°

Chapter 8

Exercise 8.2

2. $x = 5, y = 0$ 3. 75   4. (a) 3 (b) 12

Exercise 8.3

1. 25 short sleeve shirt, 15 long sleeve shirt   2. 6 AM radio, 18 FM radio

3. 20 of first product, 20 of second product   4. GH¢110

5. 21 m³ of mix A, 28 m³ of mix B

6. 9 of model A truck, 6 of model B truck   GH¢342,000

Review exercise 8

2. -12   3. $-9\frac{3}{4}$, $2\frac{7}{9}$ 4. 4, -1

5. GH¢1,200 in bonds, GH¢1,200 in treasury bills

6. 0 standard model, 40 deluxe model

7. 3 cups of drink X, 2 cups of drink Y

8. 0 of GH¢1500 model, 250 of GH¢2000 model

Chapter 9

Exercise 9.1(a)

1. (a) 36 (b) 49 (c) 16 (d) 64 (e) $\frac{9}{4}$ (f) $\frac{25}{4}$ (g) $\frac{1}{4}$ (h) $\frac{1}{16}$ (i) $\frac{1}{36}$

2. (a) $\pm\sqrt{38} - 6$ (b) -2, 4 (c) $\pm 2\sqrt{3} - 5$ (d) $\frac{\pm\sqrt{74}-3}{2}$ (e) $\frac{\pm 3\sqrt{13}-3}{2}$

(f) $\frac{\pm\sqrt{37}+7}{2}$ (g) $\frac{\pm\sqrt{17}-1}{4}$ (h) $\frac{\pm\sqrt{73}+1}{6}$ (i) $\frac{\pm\sqrt{21}+3}{6}$ (j) $\frac{\pm\sqrt{5}-2}{2}$ (k) $\frac{\pm\sqrt{37}+1}{3}$

(l) $\frac{\pm\sqrt{22}+1}{7}$

Exercise 9.1(b)

1. – 13, 5   2. – 2, 7   3. – 1.45, 3.45   4. 0.46, 6.54   5. – 10, 3   6. $-1, \frac{1}{5}$

7. 12   8. 0.18, 2.82   9. $-1, \frac{1}{3}$ 10. – 0.28, 1.78   11. – 1.46, 1.71

12. – 1.06, 0.95   13. – 0.89, 0.23   14. – 3.7, 2.7   15. – 1.9, 2.9

16. – 0.18, 1.85   17. – 0.15, 1.65   18. – 1.45, 3.45   19. – 0.69, 2.17

20. – 3.65, 1.65   21. – 3.19, 0.52   22. – 0.61, 4.11

Exercise 9.2

1. (a) real roots   (b) imaginary roots   (c) equal roots   (d) imaginary roots

(e) real roots

2. (a) 64, two   (b) 25, two   (c) 0, one   (d) 0, one   (e) 37, two

(f) -39, no real solution   (g) 25, two   (h) -63, no real solution

3. 1, 9   4. 4   5. $-\frac{1}{3}, 1$ 6. $k \leq 2$ 7. $k > -\frac{1}{4}$ 8. $k = 2$

9. $k < 1 \text{ or } k > 5$ 10. $k > 2$

Exercise 9.3

1. (a) 3, 2   (b) -7, 10   (c) 4, -21   (d) $\frac{1}{2}, -\frac{3}{2}$ (e) $-\frac{13}{5}, -\frac{6}{5}$ (f) 4, -12

(g) 5, 1   (h) $\frac{5}{2}, -\frac{3}{2}$ (i) $-\frac{1}{3}, \frac{4}{3}$

2. (a) $x^2 - 4x + 3 = 0$ (b) $x^2 + 6x + 5 = 0$ (c) $x^2 - 2x - 3 = 0$
   (d) $3x^2 - 2x = 0$ (e) $9x^2 + 3x - 2 = 0$ (f) $x^2 - 5 = 0$

3. (a) $\frac{(\alpha+\beta)^2 - 2\alpha\beta}{(\alpha\beta)^2}$ (b) $\frac{(\alpha+\beta)^2 - 2\alpha\beta}{\alpha\beta}$ (c) $\alpha\beta(\alpha + \beta)$ (d) $\sqrt{(\alpha+\beta)^2 - 4\alpha\beta}$
   (e) $\alpha\beta[(\alpha+\beta)^2 - 2\alpha\beta]$ (f) $(\alpha+\beta)[(\alpha+\beta)^2 - 3\alpha\beta]$

4. (a) -1   (b) $\frac{4}{5}$ (c) $\frac{11}{2}$ (d) 5   (e) $-\frac{2}{5}$ (f) $-\frac{7}{2}$

5. (a) $3x^2 - 16x + 22 = 0$ (b) $9x^2 - 4x + 4 = 0$ (c) $27x^2 - 24x + 8 = 0$
   (d) $9x^2 + 4x + 6 = 0$ (e) $9x^2 - 36x + 38 = 0$ (f) $9x^2 + 8 = 0$

6. $3x^2 - 4x = 0$ 7. -12   8. 5

Exercise 9.4

1. (a) $2\frac{1}{4}$ (b) 6   (c) $12\frac{1}{4}$ (d) $6\frac{1}{8}$ (e) $3\frac{1}{3}$ (f) 4

2. (a) $4\frac{1}{4}$ (b) -10 (c) 2   (d) $-5\frac{1}{8}$ (e) 1   (f) $\frac{1}{2}$

3(a) min. Value = 1, $x = 3$ (b) max. value = $6\frac{1}{4}$, $x = -\frac{1}{2}$
 (c) max. value = 3, $x = -1$ (d) min. Value = 1, $x = -2$
 (e) min. Value = 4, $x = -1$ (f) max. value = 5, $x = 2$

Exercise 9.6(a)

1. $-4 < x < 3$ 2. $x < -5$ or $x > 2$ 3. $x < -6$ or $x > 2$ 4. $-5 < x < 3$
5. $x \leq 2$ or $x \geq 3$ 6. $-5 \leq x \leq -2$ 7. $-2 \leq x \leq 1\frac{1}{2}$ 8. $-\frac{2}{3} \leq x \leq 4$
9. $-1 < x < \frac{3}{4}$ 10. $x \leq -8$ or $x \geq 3$

Exercise 9.6(b)

1. $x < -1$ or $x > 4$ 2. $-2 < x < 4$ 3. $-3 \leq x \leq 4$ 4. $x \leq -4$ or $x \geq -3$
5. $-2 < x < 5$ 6. $x < -9$ or $x > 3$ 7. $-3 \leq x \leq 4$ 8. $x < -3$ or $x > 6$
9. $-\frac{3}{2} < x < -1$ 10. $x \leq -\frac{5}{3}$ or $x \geq -1$ 11. $-1 \leq x \leq \frac{1}{2}$ 12. $x \leq -\frac{2}{3}$ or $x \geq 3$

Review exercise 9

1 (a) -4, -2 (b) $-1\frac{1}{2}, 2$ (c) -1.19, 4.19 (d) -1.55, 0.22

2 (a) -7, 3 (b) $-\frac{2}{3}, 2$ (c) -1.93, 4.19 (d) $1, -2\frac{1}{2}$

3 (a) -4, -2 (b) -2, 10 (c) -2.32, 4.32 (d) $-2\frac{1}{2}, 1$ (e) 0.28, 2.39

(f) -0.48, 1.68 (g) 3.59, 6.41 (h) -1.43, 2.10 (i) 0.58, 2.42 (j) -1.66, 0.91

4 (a) $x \leq -5$ or $x \geq 1$ (b) $-1 < x < 7$ (c) $-5 \leq x \leq 2$ (d) $2 < x < 6$

(e) $-2 \leq x \leq 1\frac{1}{3}$ (f) $x \leq -2\frac{1}{2}$ or $x \geq 4$

5. $-3 \leq x \leq 2$ 6. $x < -2$ or $x > -1$ 7. $k \leq 1$ or $k \geq 5$ 8. 2 9. 0, 12

10. 2 11. $2 < k < 6$ 12. $k > \frac{1}{3}$ 13. $k > 3$

14 (a) $\frac{37}{4}$ (b) $49x^2 - 37x + 4 = 0$

15 (a) $\frac{4}{9}$ (b) $3x^2 + 2x + 1 = 0$ 16. $2x^2 - 5 = 0$

17 (a) min. value $= -6\frac{1}{4}$, $x = \frac{1}{2}$ (b) max. value $= 20\frac{1}{4}$, $x = \frac{1}{2}$

18. min. Value $= \frac{1}{2}$, $x = 1\frac{1}{2}$ 19. max. value $= 10\frac{1}{8}$, $x = -1\frac{1}{4}$

Chapter 10

Exercise 10.1

1. (a) -13 (b) 2 (c) 0 (d) 4 (e) 0 (f) 0 (g) 1 (h) 7

2. (a) $x^2 + 3x + 6$; 7 (b) $x^2 - x + 2$; 2 (c) $x^2 + 2x - 3$; 10

(d) $2x^2 + x - 4$; 10 (e) $3x^2 + x + 6$; 4 (f) $3x^2 - 6x + 6$; -10

(g) $x^2 - 3x + 4$; 0 (h) $x^2 + 2x + 2$; -10

Exercise 10.2

1. (a) 2 (b) 0 (c) 31 (d) -16 (e) 4 (f) 4

2. (a) $(x-3)(x-1)(x+1)$ (b) $(x-5)(x-1)(x+1)$

(c) $(x-2)(x+1)(x+3)$ (d) $(2x-3)(x+1)(x+2)$

(e) $(2x-1)(x-2)(x+1)$ (f) $(x-3)(2x+1)(x+3)$

3. (a) $(x-2)(x^2+3)$ (b) $(x+1)(x+2)(x+3)$
(c) $(x-2)(x+3)(x+4)$ (d) $(x-2)(2x+1)(x+3)$
(e) $(x-2)(x-1)(5x+7)$ (f) $(2x-1)(x+2)(x+3)$

4. 4 5. −4 6. $-\frac{1}{2}, 1$ 7. $p=1, q=-3$ 8. $a=2, b=-10$

9. $a=-5, b=-3$ 10. $a=5, b=-2$ 11. $p=1, q=2$ 12. $a=1, b=-3$

Review exercise 10

1. (a) x^2+x+2 (b) x^2-1 (c) $2x^2-x+3$ (d) x^2-1

2. (a) 18 (b) 0 (c) 0 (d) 2

3. (a) $(x-4)(x-2)(x+1)$ (b) $(x-2)(3x-5)(x+1)$
(c) $(x-2)(2x-1)(x+2)$ (d) $(x-2)(x-1)(x+3)$

5. $(x-1); (2x-3)$

6. (a) $(x-3)(x-1)(x+2); -2, 1, 3$ (b) $(x-3)(x-2)(x+1); -1, 2, 3$
(c) $(x-3)(x-2)(x-1); 1, 2, 3$ (d) $(2x-1)(x-2)(x+2); -2, \frac{1}{2}, 2$
(e) $(2x-1)(x^2+1); \frac{1}{2}$ (f) $(x-5)(x-3)(x-1)(x+1); -1, 1, 3,$

7. $a=-7, b=-2$ 8. $p=-2,-1,,\frac{1}{2},1$

9. (a) $a=3, b=-4$ (b) $(x+2), (x+3)$ (c) 30

10. (a) $a=-2, b=-5$ (b) −2, 1, 3 11. (a) $\frac{2}{3}, 2$ (b) 6

12. (a) $a=2, b=-1, c=-6; -1\frac{1}{2}, 2$ 13. $a=2, b=-3$

Chapter 11

Exercise 11.1(a)

1. $-\frac{1}{2}$ 2. -4 3. 1 4. -2, 2 5. $-1, \frac{1}{2}$ 6. -3, 3

Exercise 11.1(b)

1. $\frac{3}{2}$. 2. 2, 3 3. -4, 2 4. -3 5. -2, 2 6. $-\frac{3}{2}, 2$

Answers to Exercises

Exercise 11.1(c)

1. $\{y: y \in R,\ y \neq 0\}$ 2. $\{y: y \in R, y \neq 0\}$ 3. $\{y: y \in R, y \neq 3\}$
4. $\{y: y \in R, y \neq 3\}$ 5. $\{y: y \in R, 0 \leq y \leq 1]$ 6. $\{y: y \in R, y \leq 0\ or\ y \geq 1\}$

Exercise 11.2

1. $\dfrac{2}{x+1} + \dfrac{3}{x+2}$ 2. $\dfrac{-2}{x-1} + \dfrac{3}{x-2}$ 3. $\dfrac{4}{3(x-4)} - \dfrac{1}{3(x-1)}$ 4. $\dfrac{-3}{2(x+1)} + \dfrac{3}{2(x-1)}$

5. $\dfrac{1}{2(x-2)} - \dfrac{1}{2(x+2)}$ 6. $\dfrac{4}{3(x+2)} + \dfrac{5}{3(x-1)}$ 7. $\dfrac{-7}{x-2} + \dfrac{9}{x-3}$ 8. $\dfrac{-1}{5(x-2)} - \dfrac{9}{5(x+3)}$

9. $\dfrac{1}{x-1} - \dfrac{x+1}{x^2+1}$ 10. $\dfrac{-x}{x^2+1} + \dfrac{1}{x-2}$ 11. $\dfrac{1}{2(x-1)} - \dfrac{1}{2(x+1)}$ 12. $\dfrac{-7}{4(x+3)} - \dfrac{1}{4(x-1)}$

13. $\dfrac{1}{2(x-1)} + \dfrac{1}{2(x+1)}$ 14. $\dfrac{1}{x+1} + \dfrac{2}{(x+1)^2}$ 15. $\dfrac{1}{x+2} - \dfrac{3}{(x+2)^2}$

Review exercise 11

1. -1, 1   2. $\{y: y \in R, y \neq -2\}$ 3. -2, -1, 2

4.(a) $\dfrac{-3}{x-2} + \dfrac{4}{x-5}$ (b) $\dfrac{2}{3x+4} + \dfrac{1}{x-1}$ (c) $\dfrac{2}{x+4} + \dfrac{3}{x-7}$ (d) $2 + \dfrac{5}{x+3} + \dfrac{3}{x-4}$

5. $A = 3,\ B = 2$

6. (a) $\dfrac{8}{x+5} - \dfrac{1}{x-2}$ (b) $\dfrac{-2}{x-1} + \dfrac{5}{x-3}$ (c) $1 + \dfrac{1}{x-2} - \dfrac{1}{x+2}$ (d) $2 - \dfrac{3}{x-2} + \dfrac{1}{x+1}$

(e) $\dfrac{2}{1+x} + \dfrac{3}{(1+x)^2}$ (f) $\dfrac{1}{2(x-1)} + \dfrac{x+1}{2(x^2+1)}$

7.(a) $\{x: x \in R, x \neq -2, -1, 2\}$ (b) -2, 3   (c) $A = 2, B = -1,\ C = -2$

8. (a) $\{x: x \in R, x \neq \pm 1, -2\}$ (b) $-1, \tfrac{1}{2}$ (c) $A = \tfrac{5}{3},\ B = \tfrac{1}{3},\ C = \tfrac{1}{3}$

Chapter 12

Exercise 12.1(a)

1. (b), (c)   2. (a) $x^2 + y^2 - 6x - 4y - 3 = 0$ (b) $x^2 + y^2 + 4x - 6y - 12 = 0$

(c) $x^2 + y^2 + 10y + 21 = 0$ (d) $x^2 + y^2 - 6x + 8 = 0$

(e) $x^2 + y^2 + 6x - 8y = 0$ (f) $x^2 + y^2 - 10x - 8y + 32 = 0$

3. (a) (3, -1), 2   (b) (3, -4), 6   (c) (1, 3), 4   (d) (2, -1), 2

(e) (-1, 0), 4   (f) (0, -4), 5

4. (a) $\left(\frac{3}{4}, -\frac{1}{2}\right)$, $\frac{1}{4}\sqrt{5}$ (b) $\left(-1, \frac{1}{2}\right)$, $\frac{1}{6}\sqrt{21}$ (c) $\left(-1, \frac{1}{2}\right)$, 2

(d) $\left(\frac{2}{3}, 1\right)$, 2   (e) $\left(-2, 1\frac{1}{2}\right)$, $\sqrt{5}$

5. $x^2 + y^2 - 2x - 4y - 3 = 0$ 6. $x^2 + y^2 + 4x - 6y + 3 = 0$

7. $x^2 + y^2 - 4x - 6y - 13 = 0$ 8. $x^2 + y^2 - 2x + 5y - 8 = 0$

9. $x^2 + y^2 + 8x - 8y + 16 = 0$ 10. $x^2 + y^2 - 2x - 6y - 3 = 0$

11. $x^2 + y^2 + 6x - 4y + 4 = 0$ 12. $x^2 + y^2 - 10x - 4y + 4 = 0$

Exercise 12.1(b)

1.(a) $3x + 4y - 23 = 0$ (b) $y = 4x - 17$ (c) $3x - 2y - 25 = 0$

(d) $x + 4y - 11 = 0$ (e) $9x + 4y + 5 = 0$

2. (a) $3x - 4y - 11 = 0$ (b) $x - 4y - 7 = 0$ (c) $3x + 2y - 8 = 0$

(d) $y = 2x - 5$ 3. $x + 3y + 5 = 0$

4. $5x + 7y - 29 = 0$, $5x - 7y + 4 = 0$ 5. $x^2 + y^2 + 6x + 8y = 0$

6. $x^2 + y^2 - 8x - 2y = 0$ 7. $y = 2x + 6$ 8. $y = x + 1$, $y = -x + 5$

9. (1, 2), (-1, 0)   10. $x = -2$, $y = 2$, $y = -x$

Exercise 12.1(c)

1. (a) $\sqrt{14}$ (b) $\sqrt{15}$ (c) $2\sqrt{7}$ (d) $\sqrt{29}$ 2. $4\sqrt{5}$ 3. $3\sqrt{5}$

Exercise 12.2

1. $x^2 + y^2 - 4x - 2y - 4 = 0$ 2. $8x + 2y + 5 = 0$ 3. $16x - 2y - 41 = 0$

4. $8x + 7 = 0$ 5. $y^2 = 6x + 9$ 6. $x^2 = 4y$ 7. $3x^2 + 3y^2 - 6x - 9 = 0$

8. $3x^2 + 3y^2 - 38x + 36y + 159 = 0$ 9. $x^2 - 3y^2 + 48 = 0$

10. $x^2 + y^2 - 2x - 3y = 0$

11. $x^2 + y^2 - 2x - 6y - 13 = 0$, Circle centre (1, 3), radius $\sqrt{23}$

12. Circle centre $\left(2\frac{1}{2}, 1\frac{1}{2}\right)$, radius $\frac{5}{2}\sqrt{2}$

Exercise 12.3

1. (a) $(1, 0)$ (b) $\left(-1\frac{1}{2}, 0\right)$ (c) $\left(2\frac{1}{2}, 0\right)$ (d) $\left(-1\frac{1}{4}, 0\right)$

2. (a) $(0, -1)$ (b) $\left(0, 1\frac{1}{2}\right)$ (c) $(0, -3)$ (d) $\left(0, 1\frac{1}{4}\right)$

3. (a) $x = -2$ (b) $x = 1$ (c) $x = -3$ (d) $x = \frac{3}{4}$

4. (a) $y = 2$ (b) $y = -1$ (c) $y = 2$ (d) $y = -\frac{3}{4}$

5. (a) $y^2 = 12x$ (b) $y^2 = -8x$ (c) $y^2 = 28x$ (d) $y^2 = -16x$

6. (a) $x^2 = 4y$ (b) $x^2 = 8y$ (c) $x^2 = -12y$ (d) $x^2 = -8y$

7. (a) $(1, -2), (-1, -2), x = 3$ (b) $(-3, 2), \left(-3, 1\frac{3}{4}\right), y = \frac{9}{4}$

(c) $\left(2, \frac{1}{4}\right), \left(2, 1\frac{1}{4}\right), y = -\frac{3}{4}$ (d) $(8, -1), (9, -1)\ x = 7$

(e) $(-3, -2), (-3, -4)\ y = 0$ (f) $\left(-1\frac{1}{2}, 3\right), (0, 3),\ x = -3$

8. (a) $(x + 2)^2 = -4(y + 1)$ (b) $(x - 3)^2 = 8(y - 3)$

(c) $(y - 4)^2 = -4(x + 2)$ (d) $(y - 3)^2 = 8(x - 4)$

9. (a) $(y - 2)^2 = -8(x - 4)$ (b) $(x + 1)^2 = -8(y - 3)$

(c) $x^2 = 8(y - 3)$ (d) $(y + 1)^2 = 4(x - 3)$

10. (a) $x + 2y - 3 = 0,\ y = 2x + 9$ (b) $x + y + 2 = 0,\ y = x - 6$

(c) $2x + y + 2 = 0,\ x - 2y - 9 = 0$ (d) $y = x - 2, -x + 6$

(e) $4x + 3y - 12 = 0,\ 3x - 4y - 34 = 0$ (f) $2x + 3y + 9 = 0,\ 3x - 2y + 33 = 0$

11. (a) $x + 2y + 4 = 0,\ x - 2y + 4 = 0,\ (-4, 0)$

(b) $y = 2x + 1,\ x + 3y + 18 = 0,\ (-3, -5)$

(c) $3x - 2y - 4 = 0,\ x + y - 3 = 0,\ (2, 1)$

(d) $x - 2y + 3 = 0,\ 3x + 2y + 1 = 0,\ (-1, 1)$

12. (a) $x - 4y + 8 = 0,\ 2x + 2y + 1 = 0$ (b) $x + 2y - 3 = 0,\ 3x - 2y - 1 = 0$

(c) $y = 2x + 1,\ y = x + 2$ (d) $3x - 2y - 4 = 0,\ y = -x + 3$

Review exercise 12

1. (a) $x^2 + y^2 - 8x - 6y - 75 = 0$ (b) $x^2 + y^2 + 4x - 10y - 7 = 0$

(c) $x^2 + y^2 - 10x + 6y - 47 = 0$ (d) $4x^2 + 4y^2 + 40x + 16y - 91 = 0$

2. (a) $x^2 + y^2 + 4x - 2y + 1 = 0$ (b) $x^2 + y^2 - 6x - 4y - 4 = 0$

(c) $x^2 + y^2 - 10x - 6y + 9 = 0$ (d) $x^2 + y^2 + 6x + 10y + 26 = 0$

3. (a) (2, 1); 2 (b) (-3, 2); 4 (c) (1, -3); 5 (d) (7, -4); 3

4. $\left(-\frac{2}{3}, \frac{1}{3}\right)$, 1 5. $x^2 + y^2 - 8x - 10y + 21 = 0$ 6. 4

7. $(-1, 2)$; $x^2 + y^2 - 6x - 10y + 9 = 0$ 8. (2, -2) 9. $5\sqrt{2}$

10. $3x - 4y + 18 = 0$; $13\frac{1}{2}$ 11. $x^2 + y^2 - 8x - 2y = 0$

12. (a) (0, -3); (2, 1) (b) $3x - 4y - 12 = 0$ (c) $\sqrt{89}$

13. (a) (-1, 2) (b) $x^2 + y^2 + 2x - 4y - 20 = 0$ (c) $(0, 2 + 2\sqrt{6})$; $(0, 2 - 2\sqrt{6})$

14. $\left(2\frac{2}{3}, -4\right)$; $\frac{2}{3}\sqrt{13}$ 15. $8x^2 + 8y^2 - 2x + 40y + 31 = 0$

16. $\left(\frac{1}{2}, 2\right)$, $\frac{1}{2}\sqrt{13}$ 17. $x^2 + y^2 - 2x - 6 = 0$

18. (a) $x^2 = -3y$ (b) $y^2 = 12x$ (c) $x^2 = 4y$ (d) $y^2 = -16x$

19. (a) $\left(0, \frac{1}{2}\right)$; $y = -\frac{1}{2}$ (b) $\left(\frac{3}{4}, 0\right)$; $x = -\frac{3}{4}$

(c) $\left(0, -2\frac{1}{2}\right)$; $y = \frac{5}{2}$ (d) $\left(\frac{1}{4}, 0\right)$; $x = -\frac{1}{4}$

20. (a) $y = -x + 2$; $y = x + 6$ (b) $4x + 3y - 4 = 0$; $9x - 12y - 34 = 0$

(c) $x + 2y = 0$; $2x - 2y + 21 = 0$ (d) $2x - 2y + 15 = 0$; $2x + 2y - 15 = 0$

21. $\left(\frac{1}{3}, 2\right)$ 22. $x + 2y - 18 = 0$; $\left(-9, 13\frac{1}{2}\right)$

23. $x + y + 3 = 0$; $x - y + 3 = 0$, $(-2, 1)$

24. $5x - 2y + 1 = 0$; $x + 2y + 5 = 0$

Chapter 13

Exercise 13.1

1. (a) $1 + 12x + 60x^2 + 160x^3 + 240x^4 + 192x^5 + 64x^6$

(b) $x^4 - 8x^3y + 24x^2y^2 - 32xy^3 + 16y^4$

(c) $1 - 5x + 10x^2 - 10x^3 + 5x^4 - x^5$

(d) $16x^4 + 96x^3y + 216x^2y^2 + 216xy^3 + 81y^4$

(e) $64x^3 - 48x^2 + 12x - 1$

(f) $81x^4 + 108x^3y + 54x^2y^2 + 12xy^3 + y^4$

(g) $1 - 12x + 54x^2 - 108x^3 + 81x^4$

(h) $64a^6 + 192a^5x + 240a^4x^2 + 160a^3x^2 + 60a^2x^4 + 12ax^5 + a^6$

(i) $16a^4 - 96a^3x + 216a^2x^2 - 216ax^3 + 81x^4$

2. $1 - 5x + 10x^2 - 10x^3 + 5x^4 - x^5$; 0.904

3. $1 + 4x + 6x^2 + 4x^3 + x^4$; 1.0406

4. $1 - 10x + 40x^2 - 80x^3 + 80x^4 - 32x^5$; 0.904

Exercise 13.2

1. (a) 84 (b) 1 (c) 10 (d) 15 (e) 35 (f) 252 (g) 220 (h) 792

2. (a) $240x^2$ (b) $-22680x^4$ (c) $-\frac{63}{8}x^5$ (d) $4032x^4$ (e) $55427328x^4y^8$

(f) $-4320x^3$ (g) $7920x^4y^8$ (h) $3328x^3$ (i) $1180980x^2y^3$ (j) $\frac{945}{16}x^2$

3. (a) $1 - 10x + 45x^2 - 120x^3$ (b) $1 + 18x + 144x^2 + 672x^3$

(c) $1024 - 5120x + 11520x^2 - 15360x^3$ (d) $1 + 3x + \frac{15}{4}x^2 + \frac{5}{2}x^3$

(e) $2187 - 3402x + 2268x^2 - 840x^3$ (f) $128 + 224x + 168x^2 + 70x^3$

4. (a) 1.127 (b) 0.817 (c) 535.506 (d) 3856.887

5. 240 6. $40x$ 7. $70x^4y^{-4}$ 8. $1 - 20x + 180x^2 - 960x^3$; 0.8171

9. $1 + \frac{6}{5}x + \frac{3}{5}x^2 + \frac{4}{25}x^3$; 1.062

10. (a) $1 + 7x + 14x^2$ (b) $1 - 12x + 78x^2$ (c) $1 - 6x + 14x^2 - 14x^3$

(d) $1 + 9x + 30x^2 + 40x^3$

Exercise 13.3

1.(a) $1 + 2x - 3x^2 + 4x^3$ (b) $1 - x + 2x^2 - \frac{14}{3}x^3$ (c) $1 - \frac{1}{3}x - \frac{2}{27}x^2 - \frac{5}{81}x^3$

(d) $1 + x - \frac{1}{2}x^2 + \frac{1}{2}x^3$ (e) $1 + x + \frac{3}{4}x^2 + \frac{1}{2}x^3$ (f) $1 - \frac{2}{3}x + \frac{5}{9}x^2 - \frac{40}{81}x^3$

(g) $1 - 3x + 9x^2 - 27x^3$ (h) $1 - \frac{1}{2}x + \frac{3}{8}x^2 - \frac{5}{16}x^3$ (i) $1 - x + \frac{3}{4}x^3 - \frac{1}{2}x^3$

(j) $1 - x + \frac{3}{2}x^2 - \frac{5}{2}x^3$ (k) $\sqrt{2}\left\{1 - \frac{1}{4}x - \frac{1}{32}x^2 - \frac{1}{128}x^3\right\}$

(j) $\frac{1}{3} - \frac{1}{9}x + \frac{1}{27}x^2 - \frac{1}{81}x^3$

2. (a) 1.0050 (b) 0.9426 (c) 0.999 (d) 0.9899

3. $1 + \frac{1}{3}x - \frac{1}{9}x^2 + \frac{5}{27}x^3$; 4.017   4. $1 + \frac{1}{2}x - \frac{1}{8}x^2 + \frac{1}{16}x^3$; 4.123

Review exercise 13

1. (a) $x^3 + 6x^2 + 12x^3 + 8$ (b) $x^5 + 15x^4 + 90x^3 + 270x^2 + 405x + 243$

(c) $x^4 - 4x^3y + 6x^2y^2 - 4xy^3 + y^4$

(d) $32x^5 - 80x^4 + 80x^3 - 40x^2 + 10x - 1$

(e) $64x^6 + 192x^5y + 240x^4y^2 + 160x^3y^3 + 60x^2y^4 + 12xy^5 + y^6$

(f) $1 - 12x + 60x^2 - 160x^3 + 240x^4 - 192x^5 + 64x^6$

(g) $x^3 + 12x^2 + 48x + 64$

(h) $32x^5 - 80x^4y + 80x^3y^2 - 40x^2y^3 + 20xy^4 - y^5$

2. (a) $512 + 2304x + 4608x^2 + 5376x^3$

(b) $2187 - 5103x + 5103x^2 - 2835x^3$ (c) $1 + 24x + 264x^2 + 1760x^3$

(d) $1 - 24x + 252x^2 - 1512x^3$ (e) $1 + \frac{7}{2}x + \frac{21}{4}x^2 + \frac{35}{8}x^3$

(f) $1 - 9x + \frac{135}{4}x^2 - \frac{135}{2}x^3$

3. (a) $-120x^7y^3$ (b) $5940x^3y^9$ (c) $5103x^5y^2$ (d) $-129024a^4b^5$

4. (a) – 120 (b) 1365 (c) – 1760 (d) 120 (e) 54 (f) – 5940

5. (a) 240 (b) $\frac{7}{16}$ (c) 11520

Answers to Exercises 637

6. $x^4 + 4x^3y + 6x^2y^2 + 4xy^3 + y^4$; 15.682

7. $1 + 7x + 21x^2 + 35x^3 + 35x^4$; 0.97919

8. $1 - \frac{8}{3}x + \frac{28}{9}x^2 - \frac{56}{27}x^3 + \frac{70}{81}x^4$; 0.97625

9. $1 - 12x + 60x^2 - 160x^3 + 240x^4 - 192x^5 + 64x^6$; 0.690

10. $\pm\frac{1}{4}$ 11. ± 3

12. (a) $192 + 80x - 3x^2 - \frac{15}{2}x^3$ (b) $1 - \frac{23}{2}x + 54x^2 - 130x^3$

(c) $256 - 512x - 256x^2 + 1536x^3$ (d) $64 + 64x - 80x^2 - 128x^3$

13. $1 + \frac{1}{3}x - \frac{1}{9}x^2 + \frac{5}{81}x^3$; 10.0332   14. $1 + x - \frac{1}{2}x^2 + \frac{1}{3}x^3$; 1.732

15. $1 + \frac{1}{4}x + \frac{3}{32}x^2 + \frac{5}{128}x^3$; 3.1623

Chapter 14

Exercise 14.1(a)

1. (a) 6 (b) 5 (c) 3 (d) 8 (e) 8 (f) 16 (g) 0 (h) $\frac{1}{5}$ (i) $\frac{1}{5}$ (j) $\frac{1}{32}$

(k) $\frac{1}{8}$ (l) $\frac{1}{27}$ (m) 8 (n) 9 (o) 8

2. (a) $\frac{9}{16}$ (b) $\frac{8}{27}$ (c) $\frac{36}{25}$ (d) $\frac{4}{9}$ (e) $\frac{27}{64}$ (f) $\frac{9}{4}$ (g) 1 (h) $\frac{10}{3}$

3. (a) $4\frac{1}{2}$ (b) 10 (c) 3 (d) 9 (e) 15 (f) 1 (g) 16 (h) 2 (i) 5

4. (a) $\frac{x^2}{y^4}$ (b) 1 (c) $\frac{2yz^3}{x^2}$ (d) $\frac{x^3}{y^2z}$ (e) $\frac{y}{x}$ (f) x^4 (g) $\frac{1}{x}$ (h) yz

5. (a) $x + x^2$ (b) $x^2 - x$ (c) $\frac{2x+4}{2x+3}$ (d) $\frac{5x+3}{(x+1)^2}$ (e) $\frac{1}{\sqrt{1-x}}$ (f) $\frac{3+2x}{(1+x)^2}$

Exercise 14.1(b)

1. (a) 2 (b) $1\frac{1}{2}$ (c) $\frac{5}{7}$ (d) 2 (e) -2 (f) $2\frac{1}{3}$ (g) 5 (h) 243

(i) 1 (j) -3 (k) -1 (l) $\frac{1}{9}$

2. (a) 7 (b) $3\frac{1}{2}$ (c) $\frac{7}{8}$ (d) $4\frac{2}{3}$

3. (a) 1, 2 (b) 1, 3 (c) 0, 1 (d) 1, 2 (e) $-1, 1$ (f) 3 (g) $-1, 2$ (h) 1, 3

4. (a) $x = 1, y = 2$ (b) $x = 2\frac{1}{2}, y = 4\frac{1}{2}$ (c) $x = 5, y = 0$ (d) $x = 1, y = 2$

(e) $x = -1, y = 2$ (f) $x = \frac{1}{2}, y = 1\frac{1}{2}$

Exercise 14.2(a)

1(a) $\log_5 125 = 3$ (b) $\log_2 4 = 2$ (c) $\log_{32} 2 = \frac{1}{5}$ (d) $\log_e 1 = 0$

(e) $\log_{27} \frac{1}{3} = -\frac{1}{3}$ (f) $\log_{10} 1000 = 3$

2. (a) $8 = 2^3$ (b $64 = 4^3$) (c) $9 = 9^1$ (d) $\frac{1}{16} = 4^{-2}$ (e) $10000 = 10^4$

(f) $\frac{1}{27} = 9^{-3/2}$

3. (a) 5 (b) 5 (c) 3 (d) -7 (e) 3 (f) -2 (g) $\frac{1}{2}$ (h) 6

(i) -2 (j) -18 (k) -3 (l) -2

Exercise 14.2(b)

1. $3\log_a x + 2\log_a y$ 2. $2\log_a x - \log_a y$ 3. $2\log x + \frac{1}{3}\log y$ 4. $-5\log_2 x$

5. $3\ln x + \frac{1}{2}\ln y$ 6. $2\log x + 3\log y - 2\log z$ 7. $2\log_5 x + \log_5 y - \frac{1}{3}\log_5 z$

8. $4\log_{10} x + \frac{1}{3}\log_{10} y - 3\log_{10} z$ 9. $\frac{1}{4}\log x + \frac{1}{2}\log y - \frac{1}{2}\log z$

10. $\frac{1}{3}\log x + \frac{2}{3}\log y - \frac{4}{3}\log z$ 11. $2\log x + 3\log y - \frac{5}{2}\log z$

12. $\log_5 10 + \frac{2}{3}\log_5 x - \log_5 y - \frac{4}{3}\log_5 z$

Exercise 14.2(c)

1. $\log_2 x^2 y^3$ 2. $\log_3 \frac{x}{y^2}$ 3. $\log \frac{x^3}{y}$ 4. $\log_5 x \cdot \sqrt[3]{y^2}$ 5. $\log \frac{\sqrt[3]{x^2}}{y^2}$

6. $\ln \left(\frac{x^3 y^2}{z}\right)$ 7. $\log_3 \left(\frac{x^2}{y^2 z}\right)$ 8. $\log \left(\frac{\sqrt{x}}{z^3 \cdot \sqrt[3]{y^2}}\right)$ 9. $\log \left(\frac{x^3 z^6}{y^9}\right)$

10. $\log_2 \frac{\sqrt{x} \cdot \sqrt[4]{y^3}}{z}$ 11. $\log \frac{1000 x^2}{\sqrt[4]{y}}$ 12. $\log \left(\frac{10 \sqrt[3]{x^2}}{y^3}\right)$

Exercise 14.2(d)

1. 5 2. 7 3. 5 4. 3 5. $\frac{1}{2}$ 6. 3 7. 4 8. -3 9. $-\frac{1}{3}$

10. $\frac{1}{3}$ 11. -3 12. 2 13. $\frac{3}{2}$ 14. $\frac{3}{2}$ 15. 2

Answers to Exercises 639

Exercise 14.2(e)

1. 4.70   2. 1.585   3. 1.431   4. 1.893   5. 1.277   6. 2.708   7. 1.667

8. 0.9464   9. 1.285   10. 1.723

Exercise 14.2(f)

1. $\frac{5}{6}$ 2. $\frac{1}{2}$ 3. $\frac{2}{5}$ 4. -5 5. -3 6. $\frac{3}{7}$ 7. $\frac{1}{9}$ 8. -6 9. 1.292

10. 2.161   11. 0.6309   12. 0.6701   13. 1.453   14. 2.341   15. 3.819

Exercise 14.3

1. (a) 8   (b) 5   (c) 2   (d) $2\frac{1}{2}$ (e) 2   (f) 8   (g) $\frac{2}{3}$ (h) 125   (i) 7   (j) 13

(k) 8   (l) 20   (m) 9   (n) 4   (o) 2   (p) 3   (q) $\frac{1}{3}$, 9   (r) 2

2. (a) 1   (b) 2   (c) $1\frac{1}{2}$ (d) 6   (e) $2\frac{3}{8}$ (f) $\frac{12}{19}$ (g) $3\frac{3}{4}$ (h) 6

Exercise 14.4

1. $a = 5.6, n = 2$ 2. $a = 3, b = 2$ 3. $a = 15, n = -2, x = 27$

4. $y = 4.2x^{1.6}$ 5. $a = 1.5, b = 8, x = 18$ 6. $a = -1.2, b = 7$

7. $a = 3, b = 2.4$ 8. $a = 0.45, b = 1.2$ 9. $a = 4.8, b = -0.5$

Review exercise 14

1. (a) $\frac{1}{9}$ (b) 8   (c) $\frac{1}{8}$ (d) 25   (e) $\frac{4}{9}$ (f) $\frac{27}{8}$ (g) 4   (h) $\frac{9}{25}$

2. (a) $\frac{y^4 z^8}{x^2}$ (b) $\frac{1}{xy^5}$ (c) $\frac{x}{y^3 z^4}$

3. (a) $2x + 3x^2$ (b) $\frac{2+x}{(1+x)^2}$ (c) $\frac{1+x-x^2}{\sqrt{1-x^2}}$ (d) $\frac{11x+4}{(2x+1)^2}$

4. (a) 10   (b) $\frac{2}{3}$ (c) 2   (d) 0   (e) 2   (f) -3

5(a) $2logx + logy - 3logz$ (b) $5logx + 3logy - 2logz$

(c) $4 + 2log_2 x + 3log_2 y - 4log_2 z$

(d) $2lnx + 3lny - 4lnz$ (e) $\frac{5}{2}log_5 x - 4log_5 y - \frac{7}{2}log_5 z$

(f) $2logx + \frac{2}{3}logy - \frac{4}{3}logz$

640 Further Mathematics

6 (a) $\log x^3 y^5$ (b) $\log \dfrac{x^2}{y^3}$ (c) $\log \dfrac{x^5 y^3}{z^2}$ (d) $\log \sqrt[5]{\dfrac{x^3}{y^4}}$ (e) $\log \dfrac{\sqrt[3]{x^2 y^4}}{z^3}$

(f) $\ln x^2 \cdot \sqrt[4]{y}$ (g) $\ln \dfrac{\sqrt{x} \cdot z^2}{y^3}$ (h) $\log_2 \left(\dfrac{x^3}{256 y^2}\right)$

7 (a) 3 (b) 6 (c) −2 (d) 1 (e) −3 (f) $\dfrac{1}{3}$ (g) $\dfrac{4}{3}$ (h) 3 (i) $\dfrac{2}{3}$

8 (a) $\dfrac{7}{5}$ (b) $\dfrac{3}{2}$ (c) $\dfrac{3}{4}$ (d) $\dfrac{3}{2}$

9 (a) 4.106 (b) 1.292 (c) 1.585 (d) 1.608 (e) 0.7337 (f) 0.8549

10 (a) 9 (b) 3 (c) 5 (d) 36 (e) $\dfrac{3}{4}$ (f) 2

11 (a) $1\dfrac{3}{4}$ (b) 6 12 (a) 2 (b) $\dfrac{1}{2}$ (c) 2

13. 4.1625 14. $\dfrac{1}{5}$, 25 15 (a) 1.1761 (b) 0.2552

16 (a) $x = 4,\ y = \dfrac{1}{2}$ (b) $x = 9,\ y = 3$

17 (a) 3, 5 (b) 4, 6 18 $a = b^2$ 19 (a) 6 (b) $-1\dfrac{1}{2}$

20 (a) $a = 3,\ b = 1.5$ (b) 18.6 21. $a = 1.6,\ n = 2$

22 (a) $a = 0.5,\ b = 4.2$ (b) 17.0

Chapter 15

Exercise 15.1(a)

1.(a) 3 (b) 4 (c) -6 (d) -12 (e) $\dfrac{1}{2}$ (f) $\dfrac{3}{4}$

2. (a) linear, 2 (b) not linear (c) linear, -2 (d) not linear (e) linear, 4

(f) linear, $\dfrac{3}{2}$

3. (a) 16, 19 (b) 16, 21 (c) 0, -2 (d) 9, $\dfrac{23}{2}$ (e) 1, $\dfrac{7}{4}$ (f) $\dfrac{13}{4}$, 3

Exercise 15.1(b)

1.(a) 114 (b) -60 (c) $26\dfrac{1}{2}$ (d) 2 (e) -2

2. (a) 23 (b) 13 (c) 13 (d) 41 (e) 31

3. 3, 4   4. -5   5. 5, 50   6. 3, 5, 7

7. 5, 7, 9   8. GH¢1.65   9. GH¢300   10. 12

Answers to Exercises 641

Exercise 15.1(c)

1. (a) 2550 (b) 648 (c) 2091 (d) 966 (e) 448

2. (a) 456 (b) -40 (c) 115 (d) -536 (e) 111

3. 1683 4. 3850 5. $\frac{4}{5}$ 6. 2 7. 8, 4, 540 8. 24 9. 52, 12

10. 59250 11. GH¢116.25 12. 1560 13. 12 14. 126

15. 210 16. 246,000 17. 480 18. 79.5 m

Exercise 15.2(a)

1. (a) 2 (b) 3 (c) $\frac{1}{10}$ (d) $-\frac{1}{3}$ (e) $\frac{1}{5}$ (f) $-\frac{1}{3}$

2. (a) exponential, $\frac{1}{2}$ (b) not exponential (c) exponential, 2 (d) exponential, $-\frac{1}{3}$

(e) not exponential (f) exponential, $-\frac{2}{3}$

3. (a) 32, 64 (b) $\frac{3}{4}, \frac{3}{8}$ (c) -40, 80 (d) $-\frac{1}{8}, \frac{1}{16}$ (e) 192, 768

(f) 0.008, 0.0016

Exercise 15.2(b)

1. (a) 486 (b) $\frac{1}{8}$ (c) $\frac{2}{27}$ (d) 1024 (e) $\frac{1}{8}$

2. (a) 9 (b) 8 (c) 9 (d) 8 (e) 10

3. 6, 2 4. -3, 3, $\frac{1}{9}$ 5. 2 6. 12, 27

7. 18225 8. 7430 9. 1934.84 10. 13,500

Exercise 15.2(c)

1. (a) 510 (b) 13.48 (c) 6.248 (d) 5.25 (e) 3.996 (f) -255

2. (a) $\frac{1}{5}$; $-1, -\frac{1}{5}, -\frac{1}{25}, -\frac{1}{125}$ $-\frac{5}{4}\left[1 - \left(\frac{1}{2}\right)^n\right]$ (b) 3; $\frac{1}{9}, \frac{1}{3}, 1, 3$ $\frac{1}{18}[3^n - 1]$

(c) 2; $\frac{1}{2}, 1, 2, 4$ $\frac{1}{2}[2^n - 1]$ (d) $\frac{3}{2}$; $\frac{1}{2}, \frac{3}{4}, \frac{9}{8}, \frac{27}{16}$ $\left[\left(\frac{3}{2}\right)^n - 1\right]$

3. 96, 765 4. -3, 2; 258, 93 5. 16, 211 6. 7 7. 9 8. 1257.79

9. 99197.86 10. 1628.89, 628.89 11. 72 12. 3493

Exercise 15.2(d)

1. (a) 27 (b) 24 (c) $\frac{40}{3}$ (d) $\frac{3}{2}$ (e) $\frac{2}{3}$ (f) $\frac{10}{9}$ (g) 10 (h) $\frac{5}{9}$

2. (a) $\frac{2}{3}$ (b) $\frac{19}{30}$ (c) $\frac{3}{11}$ (d) $\frac{2}{11}$ (e) $5\frac{2}{9}$ (f) $3\frac{8}{11}$

3. $\frac{1}{2}$ 4. $\frac{1}{3}, \frac{2}{3}$; 18, 9

Exercise 15.3

1. (a) 1, 2, 3, 7, 11 (b) 1, 0, -1, -4, -17 (c) -1, 1, 5, 13, 29 (d) 1, 2, 6, 15, 31

(e) 1, 0, 2, 1, 3 (f) 1, 4, 14, 45, 139

2. (a) 2, 5, 8, 11, 14 $3n - 1$ (b) 3, 1, -1, -4, -6 $5 - 2n$

(c) 3, 8, 13, 18, 23 $5n - 2$ (d) 1, 2, 4, 8, 16 2^{n-1}

3. 5, 9, 13; 860   4. $a = 1$ $b = 2$ $u_0 = 1$ $u_4 = 9$

Review exercise 15

1. (a) linear, 4 (b) linear, -3 (c) exponential, -2 (d) linear, 2

(e) exponential, $\frac{1}{2}$ (f) exponential, 3 (g) linear, 10 (h) exponential, $\frac{1}{3}$

2. (a) 45 (b) 384 (c) $-\frac{1}{8}$ (d) 68 (e) -20 (f) $\frac{4}{27}$ (g) 72 (h) $\frac{3}{125}$

3. (a) 10 (b) 10 (c) 16 (d) 15 (e) 6 (f) 12 (g) 5 (h) 10

4. (a) -14,762 (b) 522 (c) 26.53 (d) 1850 (e) -4095 (f) 13,120

(g) 12200 (h) 66

5. (a) $13\frac{1}{2}$ (b) 48 (c) $\frac{5}{9}$ (d) 3

6. (a) $\frac{5}{9}$ (b) $\frac{7}{9}$ (c) $\frac{14}{15}$ (d) $\frac{31}{60}$ (e) $\frac{9}{11}$ (f) $\frac{54}{111}$

7. (a) -2, -1, 1, 4, 8 (b) -1, 1, 5, 11, 27 (c) -2, -5, -13, -36, -104

(d) 1, 2, 3, 15, 31

8. 10, 3   9. 7   10. $4n - 2$ 11. (a) $6n + 3$ (b) 6   12. 15

13. (a) $81\left(\frac{1}{3}\right)^n$ (b) $\frac{81}{2}\left[1 - \left(\frac{1}{3}\right)^n\right]$ (c) $\frac{81}{2}$ 14. (a) $\frac{1}{2}$ (b) 8 (c) $16\left[1 - \left(\frac{1}{2}\right)^n\right]$

Answers to Exercises   643

15. 7   16. (a) 2, 6, 18, 54   (b) 7   17. $\frac{7}{4}$

18. (a) $\frac{7}{4}$, $\frac{37}{16}$, $\frac{175}{64}$ (b) $4\left[1-\left(\frac{3}{4}\right)^n\right]$ (c) 4

19. (a) (i) 9, 13, 17   (ii) $1+4n$ (b) $n(3+2n)$, 324

20. $\frac{1}{2}n(2n+8)$, 8   21. GH¢25.42   22. GH¢168,000   23. 182

24. (a) 5.25 m   (b) 79.5 m   25. 7   26. GH¢39,150.47   27. GH¢150,935

28. GH¢15,938.48   29. GH¢7096   30. 18.4°

Chapter 16

Exercise 16.1

1. (a) 70°   (b) 20°   (c) 80°   (d) 40°   (e) 36°   (f) 60°   (g) 10°

(h) 23°   (i) 87°   (j) 41°   (k) 40°   (l) 10°

2. (a) 3rd (b) 3rd (c) 4th (d) 2nd (e) 3rd (f) 1st (g) 1st

(h) 4th (i) 2nd (j) 3rd (k) 1st (l) 2nd

3. (a) $-\frac{\sqrt{3}}{2}$ (b) $-\sqrt{3}$ (c) $-\frac{1}{2}$ (d) 1   (e) $\frac{1}{\sqrt{2}}$ (f) 0   (g) $\frac{\sqrt{3}}{2}$

(h) $-\frac{\sqrt{3}}{2}$ (i) 1   (j) $-\sqrt{3}$ (k)) $\frac{1}{\sqrt{2}}$ (l) 1   (m) $-\frac{1}{2}$ (n) 1   (o) 0

4. (a) max 3, 180°; min -3, 540°   (b) max 2, 180° ; min -2, 0°

(c) max 5, 0° ; min 1, 180°   (d) max 3, 90°; min -1, 30°

(e) max 2, , 90°; min -2, 270°   (f) max $-\frac{1}{2}$, 0°; min $-\frac{3}{2}$, 180°

(g) max 1, , 90°; min $\frac{1}{5}$, 270°   (h) max 1, , 0° ; min -3, 180°

(i) max 2, 90°; min $\frac{2}{7}$, 270°

Exercise 16:2(a)

1. (a) $\frac{1}{6}\pi$ (b) $\frac{1}{3}\pi$ (c) $\frac{1}{2}\pi$ (d) $\frac{2}{3}\pi$ (e) $\frac{5}{4}\pi$ (f) $-\frac{7}{4}\pi$ (g) $\frac{5}{4}\pi$

(h) $\frac{3}{2}\pi$ (i) $\frac{9}{4}\pi$ (j) $-\frac{4}{3}\pi$ (k) $-\frac{11}{6}\pi$ (l) $\frac{5}{3}\pi$

2. (a) 36°   (b) 135°   (c) 120°   (d) 540°   (e) -225°   (f) 240°   (g) -252°

(h) -300° (i) 80° (j) 18° (k) -720° (l) 450°

3. (a) 0 (b) $\frac{\sqrt{2}}{2}$ (c) $\frac{\sqrt{3}}{3}$ (d) $\frac{2\sqrt{3}}{3}$ (e) –1 (f) –1 (g) $-\sqrt{3}$ (h) -1

(i) $-\frac{1}{2}$ (l) 1 (k) $\frac{-\sqrt{3}}{3}$ (l) $\frac{2\sqrt{3}}{3}$ (m) –1 (m) $-\frac{2}{2}$ (o) $\sqrt{3}$

Exercise 16:2(b)

1. (a) 10.8 cm, 81 cm^2 (b) 20 m, 80 m^2 (c) 7.2 m, 21.6 m

(d) 25.1 cm, 150.8 cm^2 (e) 94.2 cm, 942.5 cm^2 (f) 39.3m, 294.5 m^2

2. (a) 1.2 rad. (b) 2.5 rad. (c) 2.1 rad. (d) 0.8 rad.

3. 4 rad.   4. 6 cm   5. 6 cm   6. 7 cm   7. 6.75 cm^2 8. 19.4 cm

9. 45.4 cm   10. 57.4 cm   11. 416.2 cm   12. 37-07 m^2 13. 7.8 cm

14. (a) 93.2 cm, 676.7 cm^2 15. 2 : 25

16. (a) 20.9 cm (b) 43.3 cm^2 (c) 61.4 cm^2 17. 3.26 cm^2 18. 45.3 cm^2

Exercise 16.3(a)

1. $\cos\theta$ 2. $\csc\theta$ 3. $\tan^2\theta$ 4. $\sec^2\theta$ 5. $1 + \cot\theta$ 6. $\csc\theta + 1$

7. $\tan^2\theta$ 8. $\tan\theta$ 9. 1   10. 2   11. $\csc\theta$ 12. $2\sin\theta\cos\theta$

Exercise 16.4

1. (a) 60°, 300°   (b) 45°, 225°   (c) 45°, 135°, 225°, 315°

(d) 60°, 120°, 240°, 300°   (e) 60°, 120°, 240°, 300°   (f) 270°,

(g) 0°, 120°, 300°, 360°   (h) 45°, 71.6°, 225°, 251.6°   (i) 60°, 300°

(j) 45°, 135°, 225°, 315°

2. (a) $\frac{1}{6}\pi, \frac{5}{6}\pi$ (b) $\frac{1}{4}\pi, \frac{7}{4}\pi$ (c) $\frac{3}{2}\pi$ (d) $\frac{3}{4}\pi, \frac{7}{4}\pi$ (e) $\frac{1}{3}\pi, \frac{2}{3}\pi, \frac{4}{3}\pi, \frac{5}{3}\pi$

(f) $\frac{1}{3}\pi, \frac{2}{3}\pi, \frac{4}{3}\pi, \frac{5}{3}\pi$ (g) $0, \frac{3}{4}\pi, \pi, \frac{7}{4}\pi, 3\pi$ (h) $\frac{3\pi}{4}, \frac{7\pi}{4}$

3. (a) ±120°   (b) – 135°, 45°   (c) - 60°, 120°   (d) 0°, 120°   (e) 30°, 120°

(f) ±60°   (g) ±45°, ±135°   (h) ±120°   (i) ±60°, ±120°   (j) - 90°, 0°, 180°

(k) - 135°, 0°, 45°, 180°   (l) 0°, ±120°   (m) – 150, - 90°, -30°   (n) - 90°

(o) – 63.4°, -18.4°, 116.6°, 161.6°   (p) - 90°, 30°, 150°

Answers to Exercises   645

Exercise 16.5

1. (a) 3 (b) $105°, 255°$ 2. $52°$ 3. $40°$ 4 (a) 3 (b) $11°$

5 (a) 2.23   (b) $87°, 145°$ 6 (b) $6°, 106°$ 7 (b) $12°, 62°$

8 (a) max. 2.23, min. -2.23   (b) $-\frac{1}{5}\pi, \frac{1}{2}\pi$ 9. $12°, 62°$ 10. $24°, 204°$

Exercise 16.6

1. (a) $\frac{1}{4}\sqrt{2}(\sqrt{3}+1)$ (b) $\frac{1}{4}\sqrt{2}(\sqrt{3}+1)$ (c) $\frac{1}{2}(1+\sqrt{3})^2$ (d) $-\frac{1}{2}(1+\sqrt{3})^2$

(e) $\frac{1}{4}\sqrt{2}(\sqrt{3}-1)$ (f) $\frac{1}{2}(\sqrt{3}-1)^2$ (g) $\frac{1}{4}\sqrt{2}(1-\sqrt{3})$ (h) $-\frac{1}{4}\sqrt{2}(1-\sqrt{3})$

(i) $\frac{1}{2}(1-\sqrt{3})^2$ 2. (a) $\cos x$ (b) $\cos x$ (c) $-\tan x$

3. (a) $\frac{1}{2}$ (b) $\frac{1}{2}$ (c) $\frac{1}{2}\sqrt{2}$ (d) $-\frac{1}{2}\sqrt{2}$ (e) $\frac{1}{3}\sqrt{3}$

4. (a) $\cos(x-60°)$ $\sin(x+30°)$ (b) $\sin(x-45°)$ (c) $\tan(x+60°)$

(d) $-\frac{1}{2}\sqrt{2}$

5. (a) $-\frac{16}{65}$ (b) $-\frac{63}{65}$ (c) $\frac{16}{63}$ 6. (a) $\frac{63}{65}$ (b) $\frac{16}{65}$ (c) $\frac{33}{56}$

7. (a) $-\frac{16}{65}$ (b) $\frac{56}{35}$ (c) $-\frac{63}{65}$

Exercise 16.7

1. (a) $\frac{1}{2}$ (b) $\frac{1}{3}\sqrt{3}$ (c) $\frac{1}{2}\sqrt{2}$ (d) $-\frac{1}{2}\sqrt{2}$ (e) $-\frac{1}{2}\sqrt{3}$ (f) $\frac{1}{3}\sqrt{3}$

2. (a) $\frac{24}{25}, -\frac{7}{25}$ (b) $\frac{336}{625}, \frac{527}{625}$ (c) $\frac{120}{169}, -\frac{119}{169}$

3. (a) $270°$ (b) $0°, 41.4°, 180°, 318.6°$ (c) $30°, 150°, 270°$

(d) $60°, 109.5°, 250.5°, 300°$ (e) $26.6°, 116.6°, 206.6°, 296.6°$

(f) $35.3°, 144.7°, 215.3°, 324.7°$

Exercise 16.8

1. (a) $2\cos(\theta+60°)$ (b) $\sqrt{10}\cos(\theta-18.4°)$ (c) $5\sin(\theta-53.1°)$

(d) $\sqrt{13}\sin(\theta+56.3°)$

2. (a) max $\sqrt{2}, 45°$; min $-\sqrt{2}, 225°$ (b) max $\sqrt{13}, 33.7°$; min $-\sqrt{13}, 213.7°$

(c) max 5, 143.1°; min -5, 323.1°   (d) max 2, 30°; min -2, 210°

3. (a) 119.6°, 346.7°   (b) 162.4°, 310.2°   (c) 90°, 330°   (d) 130.8°

(e) 45°   (f) 53.2°, 360°

Exercise 16.9(a)

1. $C = 80°$ $a = 6.2$ $b = 4.6$ 2. $B = 75°$ $a = 2.3$ $b = 4.5$

3. $A = 65°$ $b = 6.4$ $c = 8.5$ 4. $C = 58°$ $b = 23.9$ $c = 21.5$

5. $C = 30°$ $a = 6.4$ $b = 5.7$

Exercise 16.9(b)

1. $B = 38°$ $C = 62°$ $c = 7.2$ 2. $B = 25.7°$ $C = 110.3°$ $c = 10.8$

3. $A = 77.7°$ $C = 41.1°$ $a = 8.5$ 4. $B = 31.1°$ $C = 111.9°$ $c = 21.6$

5. $A = 59.1°$ $B = 67.7°$ $a = 17.3$

Exercise 16.9(c)

1. $A = 36.4°$ $B = 56.6°$ $c = 53.8$ 2. $A = 36.3°$ $B = 65.7°$ $c = 82.6$

3. $A = 5.1°$ $C = 144.9°$ $b = 11.3$ 4. $B = 70.9°$ $C = 49.1°$ $a = 4.6$

5. $A = 32.8°$ $B = 27.2°$ $c = 20.8$

Exercise 16.9(d)

1. $A = 54.9°$ $B = 84.2°$ $C = 40.9°$ 2. $A = 42.8°$ $B = 60.9°$ $C = 76.3°$

3. $A = 38.9°$ $B = 40.1°$ $C = 101°$ 4. $A = 127.8°$ $B = 23.6°$ $C = 28.6°$

5. $A = 48.4°$ $B = 53.3°$ $C = 78.3°$

Exercise 16.10

1. (a) 55.8°   (b) 3.3 m   2. 4 m   3. 225 m, 086.8°   4. 6 m, 5.6 m

5. 8.2 km   6. (a) 122.6 m   (b) 40.7°   44.5°   94.8°   7. 331.6°, 154.2

8. 193 km   9. (a) 36.3 m   (b) 15.7 m   10. N 51.8° E

Review exercise 16

1. (a) 15°   (b) 50°   (c) 38°   (d) 80°   2. (a) 2nd   (b) 2nd   (c) 4th   (d) 4th

3. (a) -1   (b) $-\frac{\sqrt{3}}{2}$ (c) $-\frac{\sqrt{3}}{2}$ (d) $\frac{1}{2}$

4. (a) max 2, 0°; min -2, 360° (b) max 2, 270°; min -2, 90°

(c) max 5, 270°; min 1, 90° (d) max 3, 60°; min -1, 0°

5. (a) $\frac{1}{6}\pi$ (b) $-\frac{1}{4}\pi$ (c) $-\frac{5}{6}\pi$ (d) $\frac{11}{6}\pi$

6. (a) 60° (b) 225° (c) −315° (d) 900°

7. (a) 27 cm, 121.5 cm² (b) 37.5 m, 281.25 m² (c) 20.04 cm, 85.15 cm²

8. (a) $\frac{3}{5}$ rad (b) 2.5 rad (c) 2.4 rad (d) 0.2 rad

9. 3 cm 10. 1.9 rad 11. 24π rad

12. $\frac{1}{22}$ s 13. (a) sin θ (b) cot² θ (c) 2 (d) 1

15. (a) 0°, 135°, 180°, 315°, 360° (b) 39.2°, 140.8°, 219.2°, 320.8°

(c) 90°, 189.6°, 350.5° (d) 60°, 109.5°, 250.5°, 300°

(e) 60°, 120°, 240°, 300° (f) 45°, 225°

16. (a) ±180°, ±120° (b) ±180° (c) ±180°, ±60° (d) ±60°

(e) ±120°, ±60° (f) ±135°, ±45° 18. 1.6

19. (a) $\frac{\sqrt{2}}{4}(\sqrt{3}+1)$ (b) $\frac{\sqrt{2}}{4}(1-\sqrt{3})$ (c) $\frac{1}{2}(\sqrt{3}+1)^2$ (d) $-\frac{\sqrt{2}}{4}(1+\sqrt{3})$

(e) $\frac{\sqrt{2}}{4}(\sqrt{3}+1)$ (f) 1

20. (a) $\frac{\sqrt{3}}{2}$ (b) $\frac{1}{2}$ (c) 1 21. (a) $\frac{63}{65}$ (b) $\frac{56}{65}$ (c) $\frac{33}{56}$

22. (a) $-\frac{16}{65}$ (b) $-\frac{33}{65}$ (c) $-\frac{56}{33}$

23. (a) $\frac{\sqrt{3}}{2}$ (b) $\frac{\sqrt{2}}{2}$ (c) $-\sqrt{3}$ (d) $\frac{\sqrt{3}}{2}$

24. (a) 90°, 210°, 270°, 330° (b) 90°, 194.5°, 345.5° (c) 120°, 240°

26. (a) max $\sqrt{13}$, 33.7°; min $-\sqrt{13}$, 213.7° (b) max 5, 126.9°; min -5, 306.9°

27. (a) 0°, 120° (b) 17.6°, 229.8° (c) 240.5°

28. (a) $B = 71°$ $a = 24.3$ $c = 17.7$ (b) $P = 51°$ $p = 6.3$ $q = 6.1$

(c) $F = 58°$ $f = 21.5$ $e = 23.8$

29. (a) $A = 36.4°$ $B = 56.6°$ $c = 53.8$ (b) $B = 39.4°$ $C = 62.6°$ $a = 77$

(c) $A = 63°$ $C = 63°$ $b = 5.7$

30. (a) $A = 130.6°$ $B = 22.3°$ $C = 27.1°$ (b) $A = 38.9°$ $B = 40.1°$ $C = 101°$

(c) $A = 86.1°$ $B = 56.3°$ $C = 37.6°$

31. 3.08 cm   32. 9.47 km   33. 11.5 km   34. 153.3 m   35. 13.2 km h^{-1}

Chapter 17

Exercise 17.1(a)

1. (a) $x = 3, y = 1, z = 2$ (b) $x = 2, y = 3$

2. (a) $\begin{bmatrix} 11 & 7 \\ 2 & 5 \end{bmatrix}$ (b) $\begin{bmatrix} 6 & 1 \\ 9 & 1 \end{bmatrix}$ (c) $\begin{bmatrix} 5 & 5 \\ 1 & 1 \end{bmatrix}$ (d) $\begin{bmatrix} 3 & 5 \\ 1 & 4 \end{bmatrix}$ (e) $\begin{bmatrix} -2 & 1 & -1 \\ 2 & 3 & -3 \end{bmatrix}$

(f) $\begin{bmatrix} 1 & -1 & 1 \\ 2 & 4 & -2 \end{bmatrix}$ (g) $\begin{bmatrix} -1 & -1 \\ 5 & 2 \\ 1 & -1 \end{bmatrix}$ (h) $\begin{bmatrix} 2 & 2 & -4 \\ 2 & 4 & 2 \\ -1 & 1 & -1 \end{bmatrix}$

3. $a = 1, b = -1, c = -5, d = 3$ 4. $w = -1, x = -2, y = 5, z = 3$

5. $x = 2, y = 1$ 6. $x = 1, y = 1$

7. (a) $\begin{bmatrix} 2 & -3 \\ -1 & 4 \end{bmatrix}$ (b) $\begin{bmatrix} 3 & 2 \\ -5 & 1 \end{bmatrix}$ (c) $\begin{bmatrix} -6 & 0 \\ 0 & 6 \end{bmatrix}$ (d) $\begin{bmatrix} -3 & 2 & 0 \\ 1 & -5 & 2 \end{bmatrix}$

(e) $\begin{bmatrix} 4 & 0 \\ -3 & -1 \\ 2 & 1 \end{bmatrix}$ (f) $\begin{bmatrix} -2 & 5 & -3 \\ 3 & 0 & 2 \\ -4 & 3 & -2 \end{bmatrix}$

8. (a) $\begin{bmatrix} 1 & 1 \\ 0 & -1 \end{bmatrix}$ (b) $\begin{bmatrix} 3 & -6 \\ -2 & 3 \end{bmatrix}$ (c) $\begin{bmatrix} 7 & -4 \\ -3 & 4 \end{bmatrix}$ (d) $\begin{bmatrix} 0 & 4 \\ -3 & 2 \end{bmatrix}$

(e) $\begin{bmatrix} 1 & 8 & -3 \\ 5 & -6 & -5 \end{bmatrix}$ (f) $\begin{bmatrix} 8 & 4 & 4 \\ 1 & -1 & 1 \end{bmatrix}$ (g) $\begin{bmatrix} 11 & 1 \\ -10 & 1 \\ 3 & -5 \end{bmatrix}$ (h) $\begin{bmatrix} 5 & 1 & -1 \\ -4 & 2 & 2 \\ 7 & -12 & 3 \end{bmatrix}$

9. (a) $\begin{bmatrix} 3 & 6 \\ -6 & 3 \end{bmatrix}$ (b) $\begin{bmatrix} 6 & -2 \\ -2 & 4 \end{bmatrix}$ (c) $\begin{bmatrix} -4 & 0 \\ 0 & 4 \end{bmatrix}$ (d) $\begin{bmatrix} 1 & -3 \\ -2 & 0 \end{bmatrix}$ (e) $\begin{bmatrix} -4 & 0 \\ 8 & -6 \end{bmatrix}$

(f) $\begin{bmatrix} -6 & 3 \\ 9 & -3 \end{bmatrix}$ (g) $\begin{bmatrix} -4 & 2 & 6 \\ 0 & -6 & 4 \end{bmatrix}$ (h) $\begin{bmatrix} 0 & 3 \\ -3 & 6 \\ -6 & 9 \end{bmatrix}$ (i) $\begin{bmatrix} -1 & 0 & 2 \\ 1 & -3 & 1 \\ 0 & 1 & -2 \end{bmatrix}$

10. (a) $\begin{bmatrix} -8 & 5 \\ 5 & 5 \end{bmatrix}$ (b) $\begin{bmatrix} 4 & -11 \\ 1 & -4 \end{bmatrix}$ (c) $\begin{bmatrix} 9 & -12 \\ -4 & 4 \end{bmatrix}$ (d) $\begin{bmatrix} 2 & -5 \\ -1 & 12 \end{bmatrix}$

Exercise 17.1(b)

1. (a) $\begin{bmatrix} 11 & 0 \\ 12 & -3 \end{bmatrix}$ (b) $\begin{bmatrix} -7 & -2 \\ -2 & -8 \end{bmatrix}$ (c) $\begin{bmatrix} 6 & -14 \\ 10 & -16 \end{bmatrix}$ (d) $\begin{bmatrix} 28 & 6 \\ 5 & 0 \end{bmatrix}$ (e) $\begin{bmatrix} 8 \\ 16 \end{bmatrix}$

Answers to Exercises 649

(f) $\begin{bmatrix} -1 \\ -1 \end{bmatrix}$ (g) $\begin{bmatrix} -14 \\ 3 \end{bmatrix}$ (h) $\begin{bmatrix} -1 \\ 3 \end{bmatrix}$ (i) $\begin{bmatrix} -10 & -4 \\ -7 & -4 \end{bmatrix}$ (j) $\begin{bmatrix} 8 & -4 \\ 8 & -2 \\ -3 & 1 \end{bmatrix}$

(k) $\begin{bmatrix} 1 & -7 \\ 4 & -2 \\ 2 & -8 \end{bmatrix}$ (l) $\begin{bmatrix} 11 & -7 & 5 \\ 11 & 0 & 1 \\ 6 & 3 & 0 \end{bmatrix}$

2. (a) $\begin{bmatrix} 0 & 0 \\ 0 & 0 \end{bmatrix}$ (b) $\begin{bmatrix} 0 & 0 \\ 0 & 0 \end{bmatrix}$

3. (a) $\begin{bmatrix} -2 & -3 \\ 9 & 1 \end{bmatrix}, \begin{bmatrix} -11 & -4 \\ 12 & -7 \end{bmatrix}$ (b) $\begin{bmatrix} 7 & 2 \\ -4 & -1 \end{bmatrix}, \begin{bmatrix} -19 & -5 \\ 10 & 3 \end{bmatrix}$ (c) $\begin{bmatrix} 1 & 0 \\ 0 & 1 \end{bmatrix}, \begin{bmatrix} 1 & 0 \\ 2 & -1 \end{bmatrix}$

4. $x = 2, y = -5, z = -11$ 5. $x = 3, y = -2, z = 7$

6. $a = -2, b = 1, c = -1, d = 2$ 7. $a = 2, b = -3, c = 3, d = -2$

Exercise 17.1(c)

2. (a) -1 (b) 2   (c) 2   (d) -1 (e) 1   (f) -2

3. (a) -7 (b) -14 (c) 98   (d) 98   (e) 98

4. (a) -5 (b) $-1, 1$ (c) 1, 6

5. (a) inverse   (b) no inverse   (c) no inverse   (d) inverse   (e) inverse

6. (a) $\begin{bmatrix} 0 & \frac{1}{2} \\ \frac{1}{3} & \frac{1}{3} \end{bmatrix}$ (b) $\begin{bmatrix} \frac{3}{2} & \frac{-5}{2} \\ -1 & 2 \end{bmatrix}$ (c) $\begin{bmatrix} \frac{-3}{2} & -\frac{1}{2} \\ 2 & 1 \end{bmatrix}$ (d) $\begin{bmatrix} \frac{-5}{2} & 2 \\ \frac{3}{2} & -1 \end{bmatrix}$

(e) $\begin{bmatrix} \frac{1}{5} & \frac{1}{5} \\ 0 & \frac{1}{2} \end{bmatrix}$ (f) $\begin{bmatrix} \frac{1}{4} & -\frac{1}{4} \\ \frac{1}{8} & \frac{3}{8} \end{bmatrix}$

Exercise 17.2

1. (a) $x = -1, y = 2$ (b) $x = 1, y = 0$ (c) $x = 2, y = 3$

(d) $x = -2, y = 2$ (e) $x = 3, y = -2$ (f) $x = 2, y = 3$

2. 12, 10   3. 300, 200   4. £500, £300   5. 200 ml, 300 ml

6. 120 km h^{-1}, 60 km h^{-1}

Exercise 17.3

1. (a) $\begin{pmatrix} 3 & 1 \\ 2 & -3 \end{pmatrix}$ (b) $\begin{pmatrix} 2 & -3 \\ 1 & 2 \end{pmatrix}$ (c) $\begin{pmatrix} -3 & 2 \\ 4 & 5 \end{pmatrix}$ (d) $\begin{pmatrix} -1 & 4 \\ 1 & -3 \end{pmatrix}$

2. (a) $\begin{pmatrix} 4 & -1 \\ -3 & 2 \end{pmatrix}$ (b) $\begin{pmatrix} 2 & -1 \\ 3 & 1 \end{pmatrix}$ (c) $\begin{pmatrix} 5 & -1 \\ 2 & 3 \end{pmatrix}$ (d) $\begin{pmatrix} -2 & 4 \\ -4 & 1 \end{pmatrix}$

3. (a) $\begin{pmatrix} -1 & 0 \\ 0 & -1 \end{pmatrix}$ (b) $\begin{pmatrix} -1 & 0 \\ 0 & 1 \end{pmatrix}$ (c) $\begin{pmatrix} 1+a & a \\ b & 1+b \end{pmatrix}$ (d) $\begin{pmatrix} 0 & 1 \\ 1 & 0 \end{pmatrix}$

4. (a) Anticlockwise rotation about the origin through an angle 270°

(b) Enlargement from the origin with scale factor -3

(c) Reflection in the line $y = -x$

5. (a) (0, -1) (5, 8) (6, 9) (b) (7, 3) (4, 1) (9, 3) (c) (0, 0) (10, -15) (12, -18)

6. (-7, 7)

7. $\begin{pmatrix} 0 & -1 \\ -1 & 0 \end{pmatrix}$ Reflection in the line $y = -x$ $\begin{pmatrix} 0 & 1 \\ 1 & 0 \end{pmatrix}$ Reflection in the line $y = x$

$\begin{pmatrix} -2 & 0 \\ 0 & -2 \end{pmatrix}$ Enlargement from the origin with scale -2

8. (a) $\begin{pmatrix} 0 & 1 \\ -1 & 0 \end{pmatrix}$ (b) $\begin{pmatrix} 0 & 2 \\ -2 & 0 \end{pmatrix}$ (c) $\begin{pmatrix} 0 & 1 \\ -1 & 0 \end{pmatrix}$ (d) $\begin{pmatrix} 0 & 2 \\ 2 & 0 \end{pmatrix}$

9. $x = 3$ 10. $5x - 4y - 12 = 0$ 11. $\left(\frac{1}{2}, \frac{3}{2}\right)$ 12. $\left(-\frac{3}{2}, \frac{5\sqrt{3}}{6}\right)$

Review exercise 17

1. (a) $\begin{bmatrix} 4 & 1 \\ 4 & 6 \end{bmatrix}$ (b) $\begin{bmatrix} 5 & 2 \\ 8 & 0 \end{bmatrix}$ (c) $\begin{bmatrix} -4 & -5 \\ -12 & 14 \end{bmatrix}$ (d) $\begin{bmatrix} 3 & 5 \\ 4 & 2 \end{bmatrix}$ (e) $\begin{bmatrix} -5 & 4 \\ -8 & 0 \end{bmatrix}$

(f) $\begin{bmatrix} -6 & 13 \\ -4 & -14 \end{bmatrix}$ (g) $\begin{bmatrix} 0 & -5 \\ -4 & 10 \end{bmatrix}$ (h) $\begin{bmatrix} 6 & 15 \\ 13 & -6 \end{bmatrix}$ (i) $\begin{bmatrix} -23 & -6 \\ -40 & 8 \end{bmatrix}$

(j) $\begin{bmatrix} -6 & 6 \\ -12 & 6 \end{bmatrix}$ (k) $\begin{bmatrix} -6 & 3 \\ -12 & 6 \end{bmatrix}$ (l) $\begin{bmatrix} 6 & -3 \\ 12 & -6 \end{bmatrix}$

2. (a) $\begin{bmatrix} 12 & 10 \\ 10 & -8 \end{bmatrix}$ (b) $\begin{bmatrix} 4 \\ 2 \end{bmatrix}$ (c) $\begin{bmatrix} 18 & -8 \\ 2 & 2 \end{bmatrix}$ (d) $\begin{bmatrix} -2 \\ 7 \end{bmatrix}$ (e) $\begin{bmatrix} 4 & 2 \\ -1 & 5 \end{bmatrix}$

(f) $\begin{bmatrix} 0 & 4 \\ -1 & 1 \end{bmatrix}$ (g) $\begin{bmatrix} 22 & 12 \\ 12 & 4 \end{bmatrix}$ (h) $\begin{bmatrix} -2 & 14 \\ -4 & 0 \end{bmatrix}$ (i) $\begin{bmatrix} 16 \\ -4 \end{bmatrix}$

3. (a) $\begin{bmatrix} 9 & -1 \\ 0 & 4 \end{bmatrix}$ (b) $\begin{bmatrix} -27 & 7 \\ 0 & 8 \end{bmatrix}$

4. (a) $\begin{bmatrix} 10 & -6 \\ 9 & 5 \end{bmatrix}$ (b) $\begin{bmatrix} -3 & 4 \\ -4 & -3 \end{bmatrix}$ (c) $\begin{bmatrix} 22 & 14 \\ 19 & 13 \end{bmatrix}$ (d) $\begin{bmatrix} 0 & 2 \\ -25 & 11 \end{bmatrix}$

5. $a = 3, b = 1, c = 1, d = -2$ 6. $x = -5, y = 9$ 7. $x = 4, y = 5$

8. (a) 2 (b) 7 (c) 28 (d) 14 (e) 2 (f) – 6

9. (a) – 5 (b) – 1 , 1 (c) 1, 6 (d) $-1\frac{1}{2}, 4$

10. (a) $\begin{bmatrix} 1 & -1 \\ -1 & 2 \end{bmatrix}$ (b) $\begin{bmatrix} 1 & 1 \\ 2 & 3 \end{bmatrix}$ (c) $\begin{bmatrix} 1 & -\frac{5}{2} \\ -1 & 3 \end{bmatrix}$ (d) $\begin{bmatrix} -1 & -\frac{1}{2} \\ -3 & -\frac{1}{2} \end{bmatrix}$

(e) $\begin{bmatrix} 2 & 1 \\ -\frac{3}{2} & -\frac{1}{2} \end{bmatrix}$ (f) $\begin{bmatrix} -3 & 2 \\ 2 & -1 \end{bmatrix}$

11. (a) $\begin{bmatrix} 0 & 2 \\ -2 & 4 \end{bmatrix}$ (b) $\begin{bmatrix} 5 & -2 \\ 2 & 1 \end{bmatrix}$ 12. $\begin{bmatrix} 0.18 & -0.14 \\ -0.07 & 0.11 \end{bmatrix}$

13. (a) $\begin{bmatrix} \frac{3}{5} & \frac{2}{5} \\ \frac{2}{15} & -\frac{1}{15} \end{bmatrix}$ (b) $\begin{bmatrix} \frac{1}{31} & -\frac{4}{31} \\ \frac{6}{31} & \frac{7}{31} \end{bmatrix}$ 14. $\begin{bmatrix} -2 & 1 \\ \frac{3}{2} & -\frac{1}{2} \end{bmatrix}$ 15. $\begin{bmatrix} \frac{3}{7} & -\frac{1}{7} \\ \frac{1}{7} & \frac{2}{7} \end{bmatrix}, \begin{bmatrix} 3 & 5 \\ -5 & 8 \end{bmatrix}$

16. (a) $x = 3, y = -4$ (b) $x = 3, y = 2$ (c) $x = -1, y = 2$

17. (a) $x = 3, y = 2$ (b) $x = -1\frac{2}{3}, y = 1$

18. (a) $x = -2, y = 3$ (b) $x = 2, y = 1$

19. $(-y, x)$ Clockwise rotation about the origin through 90° 20. $\begin{pmatrix} 2 & -3 \\ 4 & -1 \end{pmatrix}$

Chapter 18

Exercise 18.1

1. (a) 9 (b) 6 (c) 3 (d) -5 (e) 6 (f) 3 (g) 6 (h) 4 (i) 3

2. (a) $\frac{3}{8}$ (b) 6 (c) -3 (d) 8 (e) 3 (f) $\frac{4}{3}$ (g) 0 (h) 3 (i) 4 (j) $\frac{2}{3}$

(k) 3 (l) 6

3. (a) $\frac{1}{2}$ (b) $\frac{2}{5}$ (c) 0 (d) -2 (e) 0 (f) -2 (g) 1 (h) $-\frac{1}{2}$ (i) $\frac{3}{2}$

4. (a) 2 (b) $6a + 3h$ (c) 2 (d) $4a + 2h$ (e) $3a^2$ (f) $-\frac{1}{a^2}$

Exercise 18.2(a)

1. 1   2. 5   3. $6x$ 4. $-4x$ 5. 0   6. $-x$ 7. $2x^2$ 8. $-6x$ 9. $-6x^2$

10. $6x$ 11. $2 - 6x$ 12. $2x + 3$

Exercise 18.2(b)

1. -1 2. 0   3. $4x^3$ 4. $6x^5$ 5. $3x^{-4}$ 6. $-x^{-2}$ 7. $-8x^{-9}$

8. $-\frac{3}{4}x^{-7/4}$ 9. $-\frac{1}{2}x^{-1/2}$ 10. $-\frac{3}{2}x^{-5/2}$ 11. $\frac{2}{3}x^{-5/3}$ 12. $-\frac{1}{3}x^{-2/3}$

Exercise 18.2(c)

1. -2 2. $12x^3$ 3. $-10x^4$ 4. $2x^7$ 5. $4x^5$ 6. $2x^{-6}$ 7. $-6x^{-4}$

8. $12x^{-4}$ 9. $-6x^{-3}$ 10. $x^{-2/3}$ 11. $-3x^{1/2}$ 12. $3x^{-1/2}$

13. $4x^{-5/3}$ 14. $-6x^{-7/4}$ 15. $-6x^{-5/3}$

Exercise 18.2(d)

1. $6x^2 + 12x^3$ 2. $12x^4 - 2$ 3. $-4t^3 + 5t^4$ 4. $2t^{-2}$ 5. $12t^2 + 3$

6. $20x^3 - 15x^4$ 7. $3x^{-4}$ 8. $6x^2 + 6x^{-3}$ 9. $12t^4 - 4$ 10. $4t - 3$

11. $12x + 5$ 12. $t^2 + t^{-3}$ 13. $x^{-2/3} + x^{-3/2}$ 14. $-3t^{-7/4} + t^2$

15. $4x^{-5/3}$ 16. $-\frac{1}{x^4}$ 17. $3 + \frac{1}{2\sqrt{x^3}} - \frac{1}{x^2}$ 18. $-\frac{10}{x^3} + \frac{3}{2\sqrt{x^5}}$ 19. $2t - \frac{12}{t^4}$

20. $6 - 4x$ 21. $6x - 12$ 22. $1 - \frac{4}{3}x$ 23. $3 + \frac{1}{x^2}$ 24. $-\frac{5}{t^2} + \frac{6}{t^3}$

Exercise 18.2(e)

1. $20x^4 - 9x^2$ 2. $4x + 5$ 3. $18x^2 + 6x$ 4. $12x - 1$ 5. $8x^3 - 9x^2$

6. $6 - 2x - 6x^2$ 7. $3\sqrt{x} + \frac{3}{2\sqrt{x}}$ 8. $2x - 8x^3$ 9. $\frac{2}{3\sqrt[3]{x^2}} + \frac{4\sqrt[3]{x}}{3}$

10. $-\frac{6}{x^2} + \frac{18}{x^3}$ 11. $7x^6 + 4x^3 - 3x^2$ 12. $18x^2 - 30x + 4$

Exercise 18.2(f)

1. $\frac{1}{(x+1)^2}$ 2. $\frac{2}{(x-1)^2}$ 3. $-\frac{2x}{(x^2-1)^2}$ 4. $\frac{-12x}{(3x^2-1)^2}$ 5. $\frac{-3x^2-6}{(x^2-2)^2}$

6. $\frac{6x^2-42x+10}{(2x-7)^2}$ 7. $\frac{-1}{\sqrt{x}(1+\sqrt{x})^2}$ 8. $\frac{4x}{(1-x^2)^2}$ 9. $\frac{x^2+2x}{(x+1)^2}$ 10. $\frac{-2x}{(x^2-1)^2}$

11. $\frac{2x^2+6x}{(2x+3)^2}$ 12. $\frac{-1}{2\sqrt{x}(\sqrt{x}-1)}$

Exercise 18.2(g)

1. $3(x-5)^2$ 2. $15(3x-2)^4$ 3. $24x(1-2x^2)^5$ 4. $8x(x^2+1)^3$

5. $21x^2(x^3-2)$ 6. $-6(2x+1)^{-4}$ 7. $8(3-2x)^{-5}$ 8. $-8(4x+3)^{-3}$

9. $4x(1-2x^2)^{-2}$ 10. $x(x^2-1)^{-1/2}$ 11. $-2(3x+2)^{-5/3}$ 12. $3x(2x^2-2)^{-1/4}$

13. $\frac{-4(3x+1)}{(3x^2+2x)^3}$ 14. $\frac{1}{3\sqrt[3]{(1-x)^4}}$ 15. $\frac{-x}{\sqrt{(x^2-1)^3}}$ 16. $\frac{-3x^2}{2\sqrt{(2+x^3)^3}}$

Exercise 18.2(h)

1. $4(x+3)^3$ 2. $9(3x-2)^2$ 3. $20(2x^2+3)^4$ 4. $x(x^2+1)^{-1/2}$

5. $-2x^2(2x^3-1)^{-4/3}$ 6. $-4x(3+x^2)^{-3}$ 7. $-6x(1-x^2)^2$

Answers to Exercises 653

8. $-18x(3x^2-1)^{-4}$ 9. $-12x(1-x^2)^5$ 10. $\dfrac{-8x}{(x^2-5)^5}$ 11. $\dfrac{-2x}{\sqrt{(2x^2+3)^3}}$

12. $\dfrac{-2x^2+3}{3\sqrt[3]{(x^3+3x)^4}}$

Exercise 18.2(i)

1. $\dfrac{1}{3\sqrt[3]{x^2}}$ 2. $\dfrac{1}{\sqrt{2x}}$ 3. $6x^5$ 4. $\dfrac{-2}{3\sqrt[3]{x^5}}$ 5. $\dfrac{-1}{2\sqrt{3x^3}}$ 6. $\dfrac{3\sqrt{x}}{2}$

Exercise 18.2(j)

1. t 2. $\dfrac{1}{t}$ 3. $\dfrac{2}{3}t$ 4. $\dfrac{-t}{t+1}$ 5. $\dfrac{t}{(2t+1)^2}$ 6. $\dfrac{2t-1}{-3(1-t)^2}$ 7. $2t-t^2$

8. $\dfrac{9(2-t)^2}{4(t+3)^2}$ 9. $\dfrac{2t}{1+t^2}$ 10. -3

Exercise 18.2(k)

1. $2(x-3)(5x-8)(2x+1)^2$ 2. $(x-3)(5x-11)(x-1)^2$

3. $6(3x+2)(2x+1)$ 4. $2(x-1)(3x^2-x+2)$ 5. $\sqrt{\dfrac{(x+1)(2x-1)}{x-1}}$

6. $\dfrac{2x^2+x-1}{\sqrt{x^2-1}}$ 7. $\dfrac{x(x^2+2)}{\sqrt{(x^2+1)^3}}$ 8. $\dfrac{3x^2-4x-3}{(x^2+1)^2}$ 9. $\dfrac{4(1-x)}{(x-2)^3}$ 10. $\dfrac{-1}{\sqrt{(1-x)(x+1)}}$

11. $\dfrac{\sqrt{x}(3-x)}{2(x-1)^2}$ 12. $\dfrac{3(x+1)(1-x)}{(x^2-x+1)^2}$

Exercise 18.2(l)

1. (a) 6   (b) 0   (c) $6x-6$ (d) $12x^2+2$ (e) 6   (f) $2-6x$

2. (a) $20x^{-6}$ (b) $24x^{-5}$ (c) $6x^{-3}$ (d) $12x^{-5}$ (e) $12x^{-4}$

(f) $-48x^{-5}$ (g) $\dfrac{6}{x^5}-\dfrac{6}{x^4}+\dfrac{2}{x^3}$

3. (a) $\dfrac{-1}{4\sqrt{x^3}}$ (b) $\dfrac{-2}{9\sqrt[3]{x^5}}$ (c) $\dfrac{3}{4\sqrt{x}}$ (d) $\dfrac{-3}{16\sqrt[4]{x^5}}$ (e) $\dfrac{2}{9\sqrt[3]{x^5}}$ (f) $\dfrac{3}{4\sqrt{x^5}}$

(g) $\dfrac{4}{9\sqrt[3]{x^7}}-\dfrac{3}{2\sqrt{x^5}}+\dfrac{6}{x^3}$

4. (a) $6x(5x^3+2)$ (b) $\dfrac{-1}{2\sqrt{x^3}}$ (c) $24(2x-3)$ (d) $2(3x+4)$

(e) $12(x-3)(2x-3)$ (f) $2(3x+4)$

5. (a) $\dfrac{-2}{(x+1)^3}$ (b) $\dfrac{4}{(1-x)^3}$ (c) $\dfrac{-3x}{\sqrt{(1+x^2)^5}}$ (d) $\dfrac{3}{4\sqrt{(1+x)^5}}$ (e) $\dfrac{4(1-3x^2)}{(1-x^2)^3}$

(f) $\frac{4+2x-3x^2}{2\sqrt{(x-1)^5}}$

6. $\frac{-1}{6t^4}$ 7. $\frac{1}{6at}$ 8. $\frac{-3(t^2+1)}{16(t^2-1)^3}$

Exercise 18.2(m)

1. $\frac{x}{y}$ 2. $\frac{1}{y}$ 3. $\frac{2y-x}{y-2x}$ 4. $\frac{x+y}{y-1}$ 5. $\frac{-3y}{2x}$ 6. $\frac{x^2}{y^2}$ 7. $\frac{3x-2y}{2x}$ 8. $\frac{2x+3y}{2y-3x}$

9. $\frac{x+4}{y-3}$ 10. $\frac{2+2x+2y^2}{3-2x-2y}$

Exercise 18.3

1. (a) $-2\sin x$ (b) $5\cos 5x$ (c) $-6\sin 3x$ (d) $-3\cos 6x$ (e) $-\sin(3x-2)$

(f) $2\cos(2x+3)$ (g) $2x\cos x^2$ (h) $-3\sin\frac{1}{2}x$ (i) $6x\sec^2 3x^2$

(j) $2\sec^2(2x-5)$ (k) $\frac{2}{3}x\sec^2\frac{1}{3}x^2$ (l) $2x\sec^2(x^2+1)$

2. (a) $-4\cos 2x \sin 2x$ (b) $6\sin x\cos x$ (c) $4\tan 2x\sec^2 2x$ (d) $6\sin 3x\cos 3x$

3. (a) $\cos x - x\sin x$ (b) $2x\cos 2x + \sin 2x$ (c) $x^2\cos x + 2x\sin x$

(d) $x^3\sec^2 x + 3x^2\tan x$

4. (a) $x\cos\frac{1}{2}x + 2\sin\frac{1}{2}x$ (b) $-2x\sin 2x + \cos 2x$ (c) $-x^2\sin x + 2x\cos x$

(d) $2x^2\sec^2 2x + 2x\tan 2x$ (e) $-\sin^2 x + \cos^2 x$ (f) $\frac{x\cos x - \sin x}{x^2}$

(g) $\frac{-2x\sin 2x - \cos 2x}{x^2}$ (h) $\cos x\sec^2 x - \tan x\sin x$

5. (a) $2\cos x + 3$ (b) $2\sin 2x$ (c) $2x + 3\sec^2 x$ (d) $-2\sin x$

(e) $2\cos x + 2x$ (f) $6x^2 + 3\sin x$ (g) $4x + 5\sec^2 x$ (h) $6\cos 2x + \sin x$

(i) $x\sec^2 x + \tan x - 6x$ (j) $x^2\cos x + 2x\sin x$ (k) $x^2\sec^2 x + 2x\tan x - \frac{1}{x^2}$

(l) $x^2\cos x + 2x\sin x + \frac{2}{x^3}$

Exercise 18.4

1. (a) 3   (b) -8 (c) 0   (d) 9   (e) 4   (f) 0   (g) -9 (h) $-\frac{5}{2}$

2. (a) $-\frac{2}{3}$ (b) $\frac{1}{3}$ (c) $-\frac{1}{4}$ (d) -3 (e) $-\frac{7}{4}$ (f) -1

3. (a) $y = -4x + 1$ (b) $y = 9x - 23$ (c) $y = -4x + 9$ (d) $y = x + 7$

(e) $y = x$ (f) $y = x$

4. (a) $x + 5y - 21 = 0$ (b) $x - 8y + 42 = 0$ (c) $y = 6$ (d) $x + 5y - 6 = 0$

(e) $x + 23y + 71 = 0$ (f) $x + 12y - 13 = 0$

5. (a) $y = -4x + 14$; $x - 4y + 5 = 0$ (b) $8x + 7y - 22 = 0$; $7x - 8y + 9 = 0$

(c) $x + 4y - 14 = 0$; $y = 4x - 5$ (d) $4x + 3y - 5 = 0$; $3x - 4y - 10 = 0$

6. $\left(\frac{1}{2}, -\frac{9}{4}\right)$ 7. $\left(-\frac{2}{3}, -\frac{44}{27}\right)$; $(2, -4)$ 8. $\left(\frac{2}{3}, \frac{26}{9}\right)$

Exercise 18.5(a)

1. $(2, -1)$; min. point   2. $(1, 6)$; max. point

3. $(0, 0)$; min. point $\left(1\frac{1}{3}, 1\frac{5}{27}\right)$; max. point   4. $(0, 0)$; min. point $(2, 4)$; max. point

5. $(1, -9)$; min. point $(-1, -5)$; max. point   6. $\left(1\frac{1}{3}, -1\frac{5}{27}\right)$ min. point $(0, 0)$; max. point

7. $(1, -1)$; min. point $(-2, 26)$; max. point   8. $\left(1\frac{1}{3}, -4\frac{19}{27}\right)$; min. point $(-1, 8)$ max. point

9. $\left(-5, -58\frac{1}{3}\right)$; min. point $(3, 27)$; max. point   10. $\left(3, 4\frac{1}{2}\right)$; min. point $\left(1\frac{1}{3}, 1\frac{5}{27}\right)$; max. point

11. $(0, 0)$ min. point $\left(1\frac{1}{3}, 1\frac{5}{27}\right)$; max. point   12. $(3, -9)$; min. point $\left(\frac{2}{3}, 1\frac{5}{27}\right)$; max. point

Exercise 18.5(b)

1. (a) -9 min. value (b) $\frac{1}{16}$; max. value (c) 3 min. value   (d) 14 max. value

(e) $-4\frac{19}{27}$ min. value; 8 max. value   (f) 2 min. value; -1 max. value

(g) -14 min. value; $-1\frac{8}{27}$ max. value   (h) -3 max. value

2. (a) $\left(-\frac{4}{5}, -3\frac{1}{5}\right)$; max. point   (b) $(1, 5)$ min. point $(-1, 9)$ max. point

(c) $(-1, -2)$; min. point $(1, 2)$ max. point   (d) $(3, 8)$; min. point $(1, 12)$ min. point

(e) $(2, 0)$; min. point $(0, 4)$ max. point   (f) $\left(-\frac{2}{3}, -1\frac{7}{9}\right)$; min. point $\left(\frac{2}{3}, 1\frac{7}{9}\right)$; max. point

(g) $(0, 0)$; min. point $\left(1\frac{1}{3}, 1\frac{5}{27}\right)$; max. point   (h) $\left(\frac{1}{2}, 4\right)$ min. point $\left(-\frac{1}{2}, -4\right)$; max. point

(i) $\left(\frac{1}{2}, 1\frac{1}{2}\right)$; min. point (j) $(-2, -3)$; max. point

.6

1. 2.2 cm   2. 0.68 m, 1.37 m   3. 1.47 cm   4. $\frac{9}{4}$ 5. 4   6. 64 cm^2

7. 5000 cm^3 50 cm, 100 cm   8. 60 cm 30 cm, 15 cm   9. 1 cm   10. 2 cm

11. $A = x^2 + \frac{16000}{x}$, 20   12. $ 5200, 20 cm, 20 cm, 10 cm

Exercise 18.8

1. 2 cm   2. 3.84 cm^2 3. 50.7π   4. 6 %   5. $2\frac{1}{2}$%   6. 40π cm   7. 3%

8. 4%, 6%   9. 1.8π cm^3 10. $1\frac{1}{3}$%

11. (a) 2.325   (b) 3.011   (c) 2.000125   (d) 4.15   (e) 3.00185

12. (a) 3.074   (b) 12.125   (c) 2.05   (d) 25.04   (e) 10.0233

Exercise 18.9

1. 16 cm^2 s^{-1} 2. 4.8π cm^2 s^{-1}, 7.2π cm^3 s^{-1} 3. $\frac{3}{5\pi}$ cm s^{-1} 4. $\frac{2}{45\pi}$ cm s^{-1}

5. $\frac{1}{8}$ cm s^{-1}, 6 cm^2 s^{-1} 6. 100π cm^3 s^{-1} 7. $\frac{1}{20\pi}$ ms^{-1} 8. $\frac{1}{2\pi}$ m^3 s^{-1}

9. $\frac{1}{20\pi}$ cm^3 s^{-1} 10. $\frac{3}{10\pi}$ cm s^{-1}

Exercise 18.10

1. 31 m, 36 m s^{-1}, 30 m s^{-2} 2. 14 m, 15 m s^{-1}, 12 m s^{-2} 3. 11 m, 32 m

4. 24 m s^{-1}, 42 m s^{-2} 5. -1 s, 2 s   6. $1\frac{2}{5}$ m s^{-1} 7. 12 m s^{-2}

8. -7 m, 5 m, 8 m s^{-2} 9. 3 s, 44.1 m   10. 1 s, 3 s, 4 m

11. 0 s, 2 s, 3 s; -2 m s^{-1}, 9 m s^{-1}, 6 m s^{-1}; -10 m s^{-2}, 2 m s^{-2}, 8 m s^{-2}

12. 3 m s^{-1}, -2 m s^{-2} ; $10\frac{2}{3}$ m   13. 0 m s^{-1}, 30 m s^{-2}; $31\frac{1}{4}$ m   14. 27 m, -5 m

15. 7:30 am, 9 am

Exercise 18.11

1. (a) $\frac{1}{3}x^3 + c$ (b) $\frac{1}{5}x^5 + c$ (c) $\frac{1}{10}x^{10} + c$ (d) $\frac{1}{12}x^{12} + c$ (e) $5x + c$ (f) $\frac{1}{3}x + c$

2. (a) $3x^2 + c$ (b) $\frac{2}{3}x^6 + c$ (c) $2x^4 + c$ (d) $\frac{1}{4}x^3 + x$ (e) $\frac{1}{4}x^5 + c$ (f) $\frac{1}{12}x^8 + c$

3. (a) $-\frac{1}{2}x^{-2} + c$ (b) $-x^{-1} + c$ (c) $-\frac{1}{4}x^{-4} + c$ (d) $-\frac{1}{3x^3} + c$

(e) $-\frac{1}{6x^6}+c$ (f) $-\frac{1}{6x^5}+c$

4. (a) $\frac{3}{4}x^{4/3}+c$ (b) $\frac{4}{7}x^{7/4}+c$ (c) $\frac{2}{5}x^{5/2}+c$ (d) $\frac{5}{6}x^{6/5}+c$

(e) $\frac{3}{5}x^{5/3}+c$ (f) $\frac{4}{9}x^{9/4}+c$

5. (a) $\frac{4}{3}x^{3/4}+c$ (b) $3x^{1/3}+c$ (c) $-3x^{-1/3}+c$ (d) $4x^{1/4}+$

(e) $\frac{-2}{\sqrt{x}}+c$ (f) $4\sqrt[4]{x}$

6. (a) $\frac{2}{3}x^3+\frac{1}{4}x^4+c$ (b) $\frac{1}{2}x^4-x^5+c$ (c) x^3-x^2+x+c

(d) $3x-x^2+\frac{4}{3}x^3+c$ (e) $x-x^2+2x^3+c$ (f) $x^5-x^3+x^2+c$

7. (a) $\frac{1}{4}x^3-x^2+c$ (b) x^4+x^3+c (c) $\frac{2}{3}x^3+\frac{3}{5}x^5+c$ (d) $\frac{1}{6}x^6-x^3+c$

(e) $x^4-\frac{2}{5}x^5+c$ (f) $\frac{3}{5}x^5+x^3+c$

8. (a) $\frac{1}{3}x^3-x^2-3x+c$ (b) $\frac{2}{3}x^3+\frac{7}{2}x^2+3x+c$ (c) $9x-6x^2+\frac{4}{3}x^3+c$

(d) $\frac{1}{4}x^4-\frac{4}{3}x^3+2x^2+c$ (e) $3x^3+6x^2+4x+c$ (f) $\frac{1}{7}x^7-\frac{2}{3}x^6+\frac{4}{5}x^5+c$

9. (a) $\frac{1}{2}x^2-\frac{1}{x}+c$ (b) $5x-\frac{3}{2}x^2+c$ (c) $\frac{-1}{2x^2}-\frac{2}{3x^3}+c$ (d) $\frac{-3}{x}-4x^2+c$

(e) $\frac{-4}{x}-3x+x^2+c$ (f) $\frac{-1}{x}-x-\frac{2}{3}x^3+c$

10. (a) $\frac{-2}{\sqrt{x}}+\frac{5}{x}+c$ (b) $2\sqrt{x}+\frac{3}{5}\sqrt[3]{x^5}+c$ (c) $\frac{3}{10}\sqrt[3]{x^{10}}-\frac{12}{7}\sqrt[3]{x^7}+3\sqrt[3]{x^4}+c$

(d) $\frac{2}{3}\sqrt{x^3}+\frac{3}{2}x^2-2x+c$ (e) $\frac{1}{2}x^2-\frac{12}{7}\sqrt[3]{x^{11}}+\frac{3}{5}\sqrt[3]{x^5}+c$

(f) $\frac{2}{5}\sqrt{x^5}+\frac{3}{2}x^2-\frac{2}{3}\sqrt{x^3}-3x+c$

11. $y=x^2+1$ 12. $y=2x^2-3x+3$ 13. $y=x^3+\frac{1}{2}x^2-\frac{1}{2}$

14. $y=x^3+x^2-9$ 15. $y=2x^2-\frac{1}{4}x^4-4$ 16. $y=\frac{1}{4}x^4+\frac{1}{2}x^2+\frac{5}{3}$

Exercise 18.12

1. (a) $1\frac{1}{2}$ (b) $8\frac{2}{3}$ (c) 60   (d) 27   (e) $6\frac{1}{5}$ (d) 1

2. (a) $5\frac{1}{3}$ (b) $5\frac{1}{3}$ (c) 44   (d) $2\frac{1}{4}$ (e) 6   (f) $-1\frac{1}{3}$

3. (a) $10\frac{5}{6}$ (b) $43\frac{1}{2}$ (c) $45\frac{3}{7}$ (d) 10   (e) 3   (f) $1\frac{11}{15}$

658 Further Mathematics

4. (a) $1\frac{1}{2}$ (b) 4 (c) $\frac{1}{3}$ (d) $\frac{3}{8}$ (e) $\frac{1}{9}$ (f) $4\frac{2}{3}$

5. (a) $4\frac{2}{3}$ (b) 2 (c) -2 (d) -3 (e) $10\frac{2}{5}$ (f) $21\frac{1}{3}$

Exercise 18.13

1. $\frac{1}{12}(3x+2)^4 + c$ 2. $\frac{-1}{2(2x-1)} + c$ 3. $\frac{3}{4}\sqrt[3]{(x-2)^4} + c$ 4. $\frac{1}{6}(x^2+3)^3 + c$

5. $\frac{1}{10}(x^2+1)^5 + c$ 6. $\frac{1}{30}(2x^3+3)^5 + c$ 7. $-\frac{1}{3}\sqrt{(1-x^2)^3} + c$

8. $\frac{2}{9}\sqrt{(x^3-2)^3} + c$ 9. $-\frac{1}{6}\sqrt{(2-x^4)^3} + c$ 10. $\frac{2}{9}\sqrt{(x^3+4)^2} + c$

11. $-\frac{1}{3}\sqrt{(3-x^2)^3} + c$ 12. $\frac{1}{9}\sqrt{(2x^2-3)^3} + c$ 13. $-\frac{1}{6(2x-5)^3} + c$

14. $\frac{1}{2}\sqrt[3]{(3x+1)^2} + c$ 15. $\frac{2}{3}\sqrt{(1+3x^2)} + c$ 16. $-\frac{1}{2(x^2-1)^3} + c$

17. $\frac{1}{5}$ 18. $12\frac{2}{3}$ 19. $\frac{1}{3}$ 20. $-\frac{3}{5}$ 21. $\frac{1}{4}$ 22. $\frac{3}{32}$

Exercise 18.14

1. (a) $\frac{1}{2}sin2x + c$ (b) $-\frac{1}{3}cos3x + c$ (c) $\frac{1}{2}tan4x + c$ (d) $-\frac{3}{2}cos2x + c$

(e) $\frac{1}{2}cos6x + c$ (f) $\frac{3}{2}cos4x + c$ (g) $-\frac{9}{4}cos\frac{1}{3}x + c$ (h) $2tan\frac{1}{2}x + c$

2. (a) $\frac{1}{2}sin(2x+1) + c$ (b) $\frac{3}{2}cos(3-2x) + c$ (c) $\frac{1}{5}sin(5x+2) + c$

(d) $\frac{1}{3}tan(3x-2) + c$ (e) $2\cos\left(3-\frac{1}{2}x\right) + c$ (f) $4\tan\left(\frac{3}{2}x+2\right) + c$

Exercise 18.15

1. 1.12 2. 25.66 3. 1.40 4. 3.25 5. 266 6. 0.403 7. 92.1 8. 44.68

Exercise 18.16(a)

1. (a) $37\frac{1}{2}$ (b) $8\frac{2}{3}$ (c) $7\frac{1}{2}$ (d) 12 (e) $2\frac{2}{3}$ (f) $\frac{7}{8}$

2. (a) $6\frac{1}{3}$ (b) 4 (c) $1\frac{5}{12}$ (d) $6\frac{1}{4}$

3. $18\frac{5}{6}$ 4. $1\frac{1}{3}$ 5. $1\frac{1}{3}$ 6. $8\frac{5}{6}$ 7. 8 8. $2\frac{2}{3}$ 9. 12

10. (a) 9 (b) 12 (c) $2\frac{1}{3}$ (d) 80

Exercise 18.16(b)

1. (a) $57\frac{1}{6}$ (b) 36 (c) $75\frac{1}{6}$ (d) $1\frac{1}{3}$ (e) $\frac{1}{6}$ (f) $20\frac{5}{6}$ (g) $20\frac{5}{6}$ (h) 16

2. (a) 9 (b) $2\frac{2}{3}$ (c) $\frac{1}{6}$ (d) $\frac{5}{12}$

Exercise 18.17

1. (a) π (b) 192π (c) $\frac{28}{15}\pi$ (d) $\frac{1}{6}\pi$ (e) $\frac{9}{2}\pi$ (f) $\frac{1}{2}\pi$ (g) $\frac{77}{3}\pi$ (h) $\frac{56}{15}\pi$

2. (a) $\frac{9}{2}\pi$ (b) $\frac{3}{5}\pi$ (c) 18π (d) $\frac{32}{5}\pi$ (e) $\frac{15}{2}\pi$ (f) $96\frac{3}{5}\pi$

3. $\frac{1}{3}\pi r^2 h$ 4. $\frac{4}{3}\pi r^3$ 5. $\frac{1}{3}\pi$ 6. $\frac{96}{5}\pi$

Exercise 18.18

1. $s = t^3 + 2t^2 + 2$ 2. $v = 3t^2 - 8t + 6$ $s = t^3 - 4t^2 + 6t + 4$

3. 90 m   4. 42 m s⁻¹, 30 m   5. 14 m s⁻¹ 12 m   6. 192 m   7. 32 m s⁻¹, 156 m

8. 144 m   9. 4 s   10. 102.4 km   11. 39.6 m s⁻¹, 59.6 m   12. $31\frac{1}{4}$ m

Review exercise 18

1. (a) 1 (b) -1 (c) $\frac{1}{2}$ (d) -2 (e) $-\frac{1}{3}$ (f) 3

2. (a) $6x + 2$ (b) $2 - 8x$ (c) $4x^3 + 1$ (d) $4x^3 - 2x$ (e) $-\frac{1}{\sqrt{x^3}} + \frac{3}{x^2}$

(f) $16x^4 + \frac{16}{x^5} - \frac{1}{\sqrt[3]{x^4}}$

3. (a) $2x - 2$ (b) $-10x^4 + 9x^2 - 8x$ (c) $-24x + 16x^3$ (d) $3 - \frac{1}{x^2}$

(e) $\frac{1}{2\sqrt{x}} + \frac{4}{\sqrt{x^3}}$ (f) $-\frac{1}{6\sqrt[6]{x^7}} - \frac{1}{4\sqrt[4]{x^5}} + \frac{3}{2\sqrt{x^3}}$

4. (a) $\frac{3}{2(3x-7)^{1/2}}$ (b) $10x(2 + x^2)^4$ (c) $\frac{-2x}{(1-2x^2)^{1/2}}$ (d) $\frac{1-3x}{(1-2x)^{1/2}}$

(e) $\frac{4x-1}{(x^2-1)^{3/2}}$ (f) $-\frac{1}{2\sqrt{x(x-1)^3}}$

5. (a) $2\cos 2x$ (b) $-5\sin 5x$ (c) $6\cos 3x$ (d) $3\sin 6x$ (e) $3\cos(3x - 2)$

(f) $3\cos\frac{1}{2}x$ (g) $-2\sin(2x + 3)$ (h) $3\sec^2(3x + 5)$ (i) $-2x\sin x^2$

(j) $2x\sec^2(x^2 - 1)$ (k) $2x\cos(x^2 + 3)$ (l) $6x\sec^2 3x^2$

6. (a) $3t$ (b) $2t(1-t)^2$ (c) $\frac{2t+3t^2}{1-4t}$ (d) $\frac{2t(1-t)^2}{(1-t^2)^2}$

7. (a) $\frac{x}{y}$ (b) $\frac{2x+3}{1-2y}$ (c) $\frac{2y-x}{y-2x}$ (d) $\frac{2x-9x^2y}{3x^3+4}$

8. (a) 4 (b) 12 (c) $-12x$ (d) $12x^2 - 24x$ (e) $\frac{-3}{4\sqrt{x^3}} + 4$ (f) $6(x+1)$

9. 32 10. $y = 2 + x^2 - x^3$ 11. (a) 2 (b) $\frac{1}{2}$ (c) $-\frac{3}{4}$ (d) 3 (e) $-\frac{4}{3}$ (f) 4

12. (a) $7x + y + 1 = 0$ $x - 7y + 43 = 0$ (b) $y = 5x - 2$ $x + 5y - 16 = 0$

(c) $y = 4x + 8$ $x + 4y + 2 = 0$ (d) $x + 2y - 6 = 0$ $y = 2x - 2$

13. $y = -5x + 10$ 14. $y = -8x - 2$ $y = x + 16$ 15. $y = 9x - 16$ (-4, -52)

16. (a) (-1, -3) (b) (2, -16), (-2, 16) (c) $\left(-\frac{2}{3}, 1\frac{13}{27}\right)$, (2, -8)

(d) $\left(\frac{1}{3}, -\frac{5}{27}\right)$, (-1, 1)

19. $\sqrt[3]{\frac{2000}{\pi}}$ 19. 20 20. 144 cm^2 21. 7 22. 60 m, £240 23. 2.7 cm^3 24. 4 %

25. $2\frac{1}{2}\%$ 26. 25 cm^{2}s^{-1} 27. 1.24π cm^{2}s^{-1} 28. $\frac{1}{45\pi}$ cm s^{-1}

29. (a) $y = x^2 + 1$ (b) $y = x^3 + \frac{1}{2}x^2 - \frac{1}{2}$ (c) $y = \frac{1}{3}x^3 + \frac{1}{2}x^2 + x + \frac{7}{6}$

(d) $y = \frac{1}{4}x^4 + \frac{1}{2}x^2 - \frac{3}{4}$

30. (a) $x^2 + c$ (b) $x^3 + x^2 + x + c$ (c) $5x + \frac{1}{3}x^3 + \frac{1}{2}x^4 + c$

(d) $\frac{3}{4}x^4 + \frac{7}{2}x^2 + c$ (e) $\frac{1}{4}x^4 - x^3 + x^2 - 6x + c$ (f) $x^4 - \frac{2}{5}x^5 + c$

31. (a) $\frac{1}{3}sin3x + c$ (b) $-\frac{1}{2}cos2x + c$ (c) $2tan2x + c$ (d) $\frac{1}{2}\cos(2x+1) + c$

(e) $\frac{1}{4}\tan(4x - 3)$ (f) $8\sin\left(\frac{1}{2}x + 3\right)$

32. (a) 33 (b) $\frac{3}{10}$ (c) $14\frac{2}{3}$ (d) $43\frac{1}{2}$ (e) $41\frac{3}{7}$ (f) $10\frac{2}{5}$

33. (a) $\frac{1}{6}(2x+3)^3 + c$ (b) $-\frac{1}{3}(3x-2)^{-1} + c$ (c) $\frac{1}{3}\sqrt{(2x+1)^3} + c$

(d) $-\frac{1}{4}\sqrt{(2-3x)^4} + c$ (e) $\frac{1}{4}(x^3 - 1)^2 + c$ (f) $\frac{1}{12}(3x^2 - 4)^2 + c$

(g) $\frac{1}{6}\sqrt{(2x^2 - 3)^3} + c$ (h) $\frac{2}{9}\sqrt{(x^3 + 2)^3} + c$ (i) $-\frac{1}{2(x^2-3)^2} + c$

34. (a) $-6\frac{3}{4}$ (b) $12\frac{1}{2}$ (c) -3 (d) $-\frac{1}{3}$ (e) $\frac{7}{200}$ (f) $-\frac{3}{16}$

35. 2.055   36. 10.69   37. 316   38. (a) $8\frac{2}{3}$ (b) $\frac{19}{20}$ (c) 6   (d) $\frac{1}{4}$

39. (a) π (b) $\frac{9}{2}\pi$ (c) $\frac{9}{10}\pi$ (d) $\frac{62}{3}\pi$ (e) 156π (f) 12π

40. $s = t^3 + 2t^2 + 2$ 41. $v = 3t^2 - 8t + 6$; $s = t^3 - 4t^2 + 6t + 4$

42. (a) $v = \frac{3}{2}t^2 - 4t + 6$; $s = \frac{1}{2}t^3 - 2t^2 + 6t + 5$

(b) $v = -10t + 13$; $s = -5t^2 + 13t - 2$

43. $\frac{4}{3}$ m ; 1 m s^{-1} 44. (a) $a = 1 - \frac{1}{6}t$ $s = \frac{1}{2}t^2 - \frac{1}{36}t^3 + 6$ (b) 6 s; 18 s

Chapter 19

Exercise 19.1(a)

1. $\begin{pmatrix}4.0\\6.9\end{pmatrix}$ 2. $\begin{pmatrix}10.4\\-6.0\end{pmatrix}$ 3. $\begin{pmatrix}-5.7\\-8.2\end{pmatrix}$ 4. $\begin{pmatrix}-1.0\\5.9\end{pmatrix}$

Exercise 19.1(b)

1. (a) $\begin{pmatrix}3\\-1\end{pmatrix}$ (b) $\begin{pmatrix}-1\\2\end{pmatrix}$ (c) $\begin{pmatrix}0\\1\end{pmatrix}$ (d) $\begin{pmatrix}-1\\-1\end{pmatrix}$ (e) $\begin{pmatrix}-6\\7\end{pmatrix}$ (f) $\begin{pmatrix}3\\-2\end{pmatrix}$

2. (a) $x = 5$, $y = 5$ (b) $x = -3$, $y = 2$ (c) $x = 1$, $y = 2$ (d) $x = -3$, $y = 4$

3. $x = -3$, $y = 4$

Exercise 19.1(c)

1. (a) $\begin{pmatrix}-1\\-3\end{pmatrix}$ (b) $\begin{pmatrix}-3\\4\end{pmatrix}$ (c) $\begin{pmatrix}2\\-5\end{pmatrix}$ (d) $\begin{pmatrix}2\\0\end{pmatrix}$

2. (a) $\begin{pmatrix}-2\\-3\end{pmatrix}$ (b) $\begin{pmatrix}2\\-4\end{pmatrix}$ (c) $\begin{pmatrix}-6\\3\end{pmatrix}$ (d) $\begin{pmatrix}4\\-1\end{pmatrix}$

Exercise 19.1(d)

1. (a) $\begin{pmatrix}2\\3\end{pmatrix}$ (b) $\begin{pmatrix}7\\-1\end{pmatrix}$ (c) $\begin{pmatrix}0\\4\end{pmatrix}$ (d) $\begin{pmatrix}-2\\-9\end{pmatrix}$ (e) $\begin{pmatrix}-3\\1\end{pmatrix}$ (f) $\begin{pmatrix}6\\5\end{pmatrix}$

2. (a) $\begin{pmatrix}1\\-1\end{pmatrix}$ (b) $\begin{pmatrix}2\\4\end{pmatrix}$ (c) $\begin{pmatrix}-1\\-5\end{pmatrix}$

Exercise 19.1(e)

1. (a) $\begin{pmatrix}2\\3\end{pmatrix}$ (b) $\begin{pmatrix}7\\-1\end{pmatrix}$ (c) $\begin{pmatrix}0\\4\end{pmatrix}$ (d) $\begin{pmatrix}-2\\-9\end{pmatrix}$ (e) $\begin{pmatrix}-3\\1\end{pmatrix}$ (f) $\begin{pmatrix}6\\5\end{pmatrix}$

2. (a) $\begin{pmatrix}1\\-1\end{pmatrix}$ (b) $\begin{pmatrix}2\\4\end{pmatrix}$ (c) $\begin{pmatrix}-1\\-5\end{pmatrix}$

Exercise 19.1(f)

1. Parallel   2. Not parallel   3. Not parallel   4. Parallel

Exercise 19.1(g)

1. (a) 5   (b) $2\sqrt{5}$ (c) $\sqrt{34}$ (d) 13

2. (a) $\sqrt{85}$ (b) $\sqrt{73}$ (c) $5\sqrt{10}$ (d) $\sqrt{145}$

Exercise 19.1(h)

1. (a) $\begin{pmatrix} -4/5 \\ 3/5 \end{pmatrix}$ (b) $\begin{pmatrix} 15/17 \\ 8/17 \end{pmatrix}$ (c) $\begin{pmatrix} -12/13 \\ -5/13 \end{pmatrix}$ (d) $\begin{pmatrix} 7/25 \\ -24/25 \end{pmatrix}$

2. (a) $\frac{3\sqrt{10}}{10}i - \frac{\sqrt{10}}{10}j$ (b) $-\frac{\sqrt{2}}{2}i + \frac{\sqrt{2}}{2}j$ (c) $\frac{2\sqrt{5}}{5}i + \frac{\sqrt{5}}{2}j$ (d) i

Exercise 19.1(i)

1. (a) $4i + 3j$ (b) $3i - 2j$ (c) $-5i - 2j$ (d) $5j$ (e) $-7i$

2. (a) $-i - j$ (b) $-3i - 7j$ (c) $12i + 5j$ (d) $-13i + 12j$

3. (a) $\sqrt{10}$ (b) 5   (c) 12   (d) 17

4. (a) $\frac{\sqrt{2}}{2}i + \frac{\sqrt{2}}{2}j$ (b) $-\frac{3\sqrt{13}}{13}i + \frac{2\sqrt{13}}{13}j$ (c) $\frac{2\sqrt{5}}{5}i - \frac{\sqrt{5}}{5}j$ (d) $-\frac{\sqrt{2}}{10}i - \frac{7\sqrt{2}}{10}j$

Exercise 19.1(j)

1. (a) $\begin{pmatrix} 4 \\ 2 \end{pmatrix}$ (b) $\begin{pmatrix} -5 \\ 3 \end{pmatrix}$ (c) $\begin{pmatrix} 0 \\ 4 \end{pmatrix}$ (d) $\begin{pmatrix} -3 \\ -2 \end{pmatrix}$ (e) $\begin{pmatrix} -5 \\ 0 \end{pmatrix}$

2. (a) $3i + 2j$ (b) $i + 3j$ (c) $-2i$ (d) $4i - 5j$ (e) $2j$

3. (a) $\begin{pmatrix} 2 \\ 1 \end{pmatrix}$ (b) $\begin{pmatrix} 2 \\ -3 \end{pmatrix}$ (c) $\begin{pmatrix} -1 \\ -5 \end{pmatrix}$ (d) $\begin{pmatrix} -4 \\ 1 \end{pmatrix}$

4. $x = 5, y = 3$ 5. (6, -2)   6. (1, -7)   7. $m = 1, n = 2$ 8. $3i + 4j$

9. $17i - 2j$ 10. $\begin{pmatrix} 10 \\ -1 \end{pmatrix}$

Exercise 19.1(k)

2. (a) $\overrightarrow{EF} = \frac{1}{2}\overrightarrow{AC}$ (b) $\overrightarrow{HG} = \frac{1}{2}\overrightarrow{AC}$ 8. $\frac{1}{3}(a + c)$

Exercise 19.2

1. (a) $\left(-\frac{4}{5}, \frac{7}{5}\right)$ (b) $\left(0, \frac{5}{3}\right)$ (c) $\left(\frac{7}{4}, -\frac{3}{8}\right)$ 2. (a) (-5, 13)   (b) $\left(5, -\frac{1}{3}\right)$ (c) $\left(-1, \frac{23}{3}\right)$

3. (a) $\overrightarrow{OQ} = \frac{1}{2}b$ (b) $\overrightarrow{AQ} = \frac{1}{3}(b + a)$ (c) $\overrightarrow{OB} = a + b$ (d) $\overrightarrow{OP} = \frac{1}{3}(a + b)$

(e) $X = P$

Exercise 19.3(a)

1. (a) 10 (b) 9 (c) -8 (d) -3 (e) 17 (f) 0

2. (a) perpendicular (b) not perpendicular (c) perpendicular (d) perpendicular

(e) not perpendicular (f) perpendicular

3. (a) 47.7° (b) 75.7° (c) 82.9° (d) 78.7° (e) 75.0° (f) 70.3°

Exercise 19.3(b)

1. (a) 23 (b) -11 (c) 18   2. (a) -7 (b) 10 (c) 1   4. – 6   5. – 1

6. (a) 31.3° (b) 130.2° (c) 55.3°   7. (a) 157.6° (b) 10.3° (c) 78.7°

10. (a) $\overrightarrow{AC} = a + b$ (b) $\overrightarrow{BD} = b - a$ (c) $|a| = |b|$

12. (a) $\binom{3}{4}$ (b) $\binom{5}{0}$ (c) 15, 5, 5 (d) 53.1°

13. (a) $\overrightarrow{BA} = \binom{4}{-4}$ $\overrightarrow{BC} = \binom{5}{1}$ (b) 16, $4\sqrt{2}$, $\sqrt{26}$ (c) 56.3°

14. (a) $a = b + c$ (b) 0

15. $AB = 3\sqrt{2}, AC = 5\sqrt{2}, BC = 2\sqrt{17}; \widehat{A} = 90°, \widehat{B} = 59.0°, \widehat{C} = 31.0°$

16. (a) (5, 1) (b) (2, 4) (c) $\binom{2}{4}$ (d) 71.6°

Review exercise 19

1. (a) $i + 8j$ (b) $7i – 15j$ (c) $13i$ 2. $x = -1, y = 3$ 3. $m = 2, n = 3$

4. $p = -1, q = -15$ 5. $m = 3, n = -2$ 6. (a) $\sqrt{41}$ (b) $\sqrt{221}$ (c) $\sqrt{65}$

7. (a) 5, 53.1°   (b) 13, 67.4°   (c) $\sqrt{13}$, 56.3°   (d) $\sqrt{41}$, 51.3°

8. (a) $-\frac{3}{5}i + \frac{4}{5}j$ (b) $\frac{3\sqrt{10}}{10}i - \frac{\sqrt{10}}{10}j$ (c) $\frac{2\sqrt{5}}{5}i + \frac{\sqrt{5}}{5}j$ (d) $-\frac{5}{13}i + \frac{12}{13}j$

9. $\overrightarrow{BA} = \binom{-4}{4}$ $\overrightarrow{BC} = \binom{1}{5}$

10. $\overrightarrow{AB} = -I + 3J$ $\overrightarrow{BC} = -3i + 3j$ $\overrightarrow{CA} = 4i - 6j$

11. $- 2i - 5j$ 12. $6i + 3j$ 13. $- 2i + 17j$ 14. $i + 6j$

15. $m = 1$, $n = 2$ 16. 5, 36.9°   19. $x = 1, y = -9$ 20. (a) $\frac{1}{3}a$

21. $\overrightarrow{AC} = c - a$ $\overrightarrow{AY} = \frac{1}{2}(c - a)$ $\overrightarrow{OX} = \frac{1}{2}b$ $\overrightarrow{OY} = \frac{1}{2}b$

X and Y are the same point

22. (a) 0   (b) 4   (c) -7   24. 53.1°   25. 24, 26.6°   27. 109.4°

28. 45°   29. 56.3°   30. (b) 106.3°

Chapter 20

Exercise 20.1(a)

1. 2.5 m s^{-2}, 100 m   2. 732 m, 12 s   3. 40 m s^{-1}, 4 s   4. 2 m s^{-2}, 27 m s^{-1}

5. 2.5 m s^{-2}, 180 m   6. 10 s   7. $\frac{5}{9}$ m s^{-2}, $\frac{20}{3}$ m s^{-1} 8. $\frac{80}{7}$ s   9. $\sqrt{12}$ s

10. $\frac{2}{5}$ m s^{-2}, 5 s   11. 10 m s^{-2}, 3 s   12. 3 m s^{-2}, 26 m s^{-1}

Exercise 20.1(b)

1. 40 m s^{-1} 2. $\frac{5}{3}$ m s^{-2} 3. 10 s, 1.5 m s^{-2} 4. (a) 5 s   (b) 12 m s^{-2} (c) 18 m s^{-2}

5. (a) 7.5 km   (b) 45 km h^{-1} (c) 1.2 m s^{-2} 6. (a) $\frac{4}{3}$ min.   (b) $\frac{3}{4}$ m s^{-2}

Exercise 20.1(c)

1. 5 m, 20 m, 125 m   2. 180 m, 380 m   3. 1300 m, 26 s   4. 4 s, 40 m s^{-1}

5. 4 s, 50 m s^{-1} 6. 10 s   7. (a) 45 m   (b) 6 s   (c) 7 s   8. 6.25 m, 5 s, 35 m s^{-1}

9. 5.5 s, 123.75 m   10. 451.2 m

Exercise 20.2

1. (a) (30.5 km h^{-1}, 155°)   (b) 30.5 km   2. 085°, 602.1 km

3. 237.5 km h^{-1}, 127°   4. 152°, 3.47 hr   5. 107°, 3.97 hr   6. 049°, 227 km h^{-1}

7. 037°, 40 s   8. 139°, 97 km   9. 9.4 ms^{-1}, 58°   10. 5 ms^{-1}, 76.5°

11. (a) (13 ms^{-1}, 067°)   (b) 7.5 s   (c) 37.5 m   12. (a) 207°   (b) 2.44 hours

Exercise 20.3(a)

1. 60 N   2. 1.5 kg   3. 4 m s^{-2} 4. 2.5 N   5. 30 N   6. 4000 N   7. 900 N

8. 180 N   9. 25,000 N   10. 150,000 N   11. 2000 N   12. 0.7 m s^{-2}

13. 4 m s^{-2} 14. 20.8 g   15. 92 kg   16. 2.5 m s^{-2} 17. 62.5 kg

18. $2\mathbf{i} + \frac{3}{2}\mathbf{j}$ 19. 2 m s^{-2} 20. 150 N

Exercise 20.3(b)

1. (a) 8.4 N, 36°   (b) 6 N, 35°   (c) 9 N, 56°   (d) 18 N, 74°   (e) 41 N, 28°

2. 38 N, 30°   3. 54 N   4. 10 N   5. 1.5 N

Exercise 20.3(c)

1. (a) 10.2 N, 073.9°   (b) 6.1 N, 055.3°   (c) 9 N, 033.7°   (d) 18 N, 043.9°

(e) 37 N, 059.5°

2. 3.8 N   3. 5 N   4. 6.25 N, 76°   5. 1.7 N

Exercise 20.3(d)

1. (a) 22.8 N   (b) 2.1 N   (c) 7 N

2. (a) 24.8 N, 136.5°   (b) 5 N, 353.1°   (c) 13.2 N, 193.1°   (d) 3.4 N, 024.4°

3. (a) 9.5 N   (b) 1.2 m s^{-2} 4. (a) 11.7 N   070°   (b) 1.95 m s^{-2}

Exercise 20.3(e)

1. 65 N, 67.4°   2. 20.7 N   3. 60 N   4. 41 N, 32 N   5. 32 N, 38 N

6. 56.5°, 33.5°   7. 40.6 N, 58.6°

Exercise 20.3(f)

1. 25 N   2. 0.56   3. 30 N   4. 0.9 m   5. 0.2   6. 112.5 N   7. 166.4 N

8. 30 N, 52 N   9. 274.8 N, 589.2 N   10. (a) 171 N   (b) 231 N   11. 0.56 m s^{-2}

12. 5.8 N, 1.6 N, 0.31   13. 8.3 N, 21.1 N   14. 1.07 m s^{-2} 15. 0.09

16. (a) 7.3 N   (b) 39.7 N   17. 37.3 s, 653 m   18. 3.46 s, 96.4 m

Exercise 20.3(g)

1. 1.2 m   2. 7.5 kg   3. 22.5 kg   4. 20 kg   5. 75 N   6. 2 cm from hinge

7. 1.6 m from one end   8. 8.33 kg, 155 N   9. 3 kg   10. 13.7 N, 221.5 N

Exercise 20.4

1. 2.1 kg m s^{-1} 2. 60 N   3. 5 m s^{-1} 4. 8 m s^{-1} 5. 8 N s   6. 1.44 m s^{-1}

7. 8.25 m s^{-1} 8. 1.2 m s^{-1} 9. 1600 m s^{-1} 2400 N 10. 0.98 m s^{-1}

11. 27.2 m s^{-1} 12. 225 kg 13. 750 g 14. 12 m s^{-1}

15. (a) 5.75 m s^{-1} (b) 1.75 m s^{-1} 16. 11.2 N s 17. $\frac{1}{3}i + \frac{4}{3}j$

18. $2i + 14j$ 19. 5.21 m s^{-1} 20. 6.1 m s^{-1}

Review exercise 20

1. 1.5 m s^{-2}, 31.25 m 2. 10 m s^{-1}, 30 m s^{-1} 3. 300 m, 10 s 4. 20 m s^{-1}, 2 s

5. 1.25 m s^{-2}, 10 m s^{-1} 6. 180 m 7. 6 s 8. 5 s 9. 30 m s^{-1} 10. 6 s

11. 6120 m, 0.225 m s^{-2} 12. 50 m s^{-1} 13. 2 m s^{-2} 14. 40 s

15. $\frac{1}{5}$ h, 187.5 km h^{-2} 16. 130 knot 17. 154.9 km h^{-1}, 138.9°

18. 164.9 km h^{-1}, 303.9° 19. 10 N 20. 5 s 21. 15 m s^{-1}

22. $\frac{1}{5}$ m s^{-2}, 125,000 kg 23. 1.5 m s^{-2} 24. 7.27 N, 090.8° 25. 8.1 N, 141.7°

26. 2.78 N, 289° 27. 4.74 N, 130.4° 28. $\frac{5}{2}I - \frac{3}{2}J$ 29. 206 N

30. 124 N, 14° 31. 2.6 N, 1.5 N 32. 38 N, 32 N 33. 7.25 m s^{-1}

34. (a) 7.5 m s^{-1} (b) 0.15 m s^{-1} 35. 3.75 m s^{-1} 36. 13 m s^{-1}, 022.6°

37. 2.8 m s^{-2} 38. 41 N 39. 170 N, 380 N 40. 35 kg, 470 N

Chapter 21

Exercise 21.1(a)

1. 20 2. 18 3. 864 4. 1320 5. 360 6. 5040

7. 36 8. 544,320 9. 20,000 10. 128 11. 60 12. 50

Exercise 21.1(b)

1. 720 2. 720 3. 40,320 4. 6,840 5. 90 6. 5,040 7. 43,680

8. 5040 9. 120 10. 151,250 11. 720 12. 13,800 13. 648

14. 24,360 15. 20,160 16. 6,720 17. (a) 60 (b) 125 18. 729 19. 50

20. 45 21. 72 22. 12 23. 120 24. 240 25. 36 26. 60 27. 420, 120

28. 4,989,600 29. 181,440 30. 27,720

Exercise 21.2

1. 56   2. 126   3. 210   4. 2300   5. 495   6. 60   7. 60   8. 6435

9. 4845   10. 36   11. 220   12. 35   13. 35   14. 120   15. 650   16. 20

17. 1260   18. 1800   19. 150   20. 110   21. 140   22. 40   23. 165

24. (a) 1716   (b) 140   (c) 1008   25. (a) 1435   (b) 2835   (c) 1218

Review exercise 21

1. 120   2. 720   3. 110   4. 210   5. 6720   6. 24   7. 182   8. 288

9. 210   10. 336, 56   11. 10   12. (a) 27, 216   (b) 90,000

13. (a) 300   (b) 144   14. 240   15. (a) 360   (b) 96

16. (a) 3003   (b) 1260   (c) 861   17. (a) 924   (b) 28   (c) 672

Chapter 22

Exercise 22.1

1. {HH, HT, TH, TT}, $\frac{1}{2}$ 2. $\frac{3}{4}$ 3. $\frac{1}{3}$ 4. $\frac{1}{3}$ 5. $\frac{2}{5}$ 6. $\frac{1}{2}$ 7. $\frac{1}{2}$ 8. $\frac{2}{3}$ 9. $\frac{3}{10}$

10. $\frac{7}{15}$ 11. $\frac{1}{10}$ 12. $\frac{3}{7}$ 13. $\frac{1}{3}$ 14. $\frac{7}{10}$ 15. 0.3   16. $\frac{7}{9}$ 17. $\frac{2}{5}$ 18. $\frac{12}{13}$

19. $\frac{5}{1000}$ 20. 0.2

Exercise 22.2(a)

1. 0.55   2. 0.85   3. 0.70   4. 0.40   6. $\frac{2}{13}$.   7. $\frac{8}{15}$ 8. $\frac{2}{5}$ 9. $\frac{7}{12}$ 10. $\frac{1}{6}$

11. $\frac{1}{20}$ 12. $\frac{1}{2}$ 13. $\frac{4}{13}$ 14. $\frac{11}{15}$

Exercise 22.2(b)

1. 0.15   2. 0.5   4. $\frac{1}{36}$ 5. $\frac{18}{125}$ 6. $\frac{6}{35}$ 7. $\frac{1}{15}$ 8. $\frac{64}{125}$ 9. $\frac{2}{9}$ 10. $\frac{1}{30}$

Exercise 22.2(c)

1. $\frac{4}{13}$ 2. $\frac{11}{36}$ 3. $\frac{43}{60}$ 4. 0.8   5. $\frac{11}{30}$ 6. 0.2   7. 0.4   8. 0.09

Exercise 22.3(a)

1. $\frac{3}{4}$ 2. $\frac{1}{3}$ 3. $\frac{1}{2}$ 4. $\frac{1}{2}$ 5. $\frac{1}{13}$ 6. $\frac{3}{8}$ 7. $\frac{2}{5}$ 8. $\frac{5}{18}$ 9. $\frac{4}{35}$ 10. $\frac{1}{13}$ 11. $\frac{1}{6}$

12. (a) 0.025   (b) 0.0625   (c) 0.4

Exercise 22.3(b)

1. $\frac{7}{15}$ 2. $\frac{13}{28}$ 3. $\frac{6}{11}$ 4. $\frac{1}{3}$ 5. $\frac{55}{126}$ 6. $\frac{5}{12}$ 7. $\frac{79}{100}$ 8. $\frac{1}{40}$

Exercise 22.4

1. 0.28   2. $\frac{5}{16}$ 3. 0.0026   4. 0.783   5. $\frac{1}{5}$ 6. $\frac{1}{56}$ 7. 0.25   8. 0.264

9. 0.348   10. 0.318

Exercise 22.5

1. 0.03125   2. 0.4219   3. 0.1445   4. 0.2344   5. 0.995   6. 0.1608

7. (a) 0.3292   (b) 0.9122   (c) 0.9822   8. (a) 0.1901   (b) 0.8784

9. (a) 0.0115   (b) 0.218   10. 0.7984   11. 0.9942   12. 0.3125   13. 0.1736

14. 0.1509   15. 0.02216

Review exercise 22

1. {HH, HT, TH, TT}   2. {BBB, BBG, BGB, BGG, GBB, GBG, GGB, GGG}

3. (a) A = {(1,4),(2,3),(3,2),(4,1),(4,6),(5,5),(6,4)}

(b) B = {(1,1),(2,1),(3,1),(4,1),(5,1),(6,1)}

4. {HH, HT, TH}   5. 0.22   6. $\frac{3}{5}$ 7. $\frac{5}{6}$ 8. $\frac{1}{2}$ 9. $\frac{3}{4}$ 10. $\frac{11}{36}$ 11. $\frac{1}{6}$

12. $\frac{4}{13}$ 13. (a) $\frac{1}{9}$ (b) $\frac{1}{5}$ (c) $\frac{4}{5}$ 14. (a) $\frac{4}{5}$ (b) $\frac{1}{45}$ 15. (a) 0.40   (b) 0.50

16. (a) 0.56   (b) 0.94   (c) 0.06   17. $\frac{7}{10}$ 18. $\frac{13}{28}$ 19. $\frac{13}{24}$

20. (a) $\frac{1}{6}$ (b) $\frac{1}{21}$ (c) $\frac{4}{15}$ 21. $\frac{7}{36}$ 22. $\frac{65}{176}$ 23. $\frac{7}{32}, \frac{35}{156}$ 24. 0.6349

25. 0.8214   26. 0.03182   27. $\frac{13}{25}$ 28. 0.2637   29. 0.6242

30. (a) 0.0331   (b) 0.5645   31. (a) 0.0512   (b) 0.6656   (c) 0.5904

32. (a) 0.3456   (b) 0.0778   (c) 0.9222

Chapter 23

Exercise 23.1

1 (a)

x	f
15	4
16	7
17	9
18	5
19	3
20	2

(b) 17

2(a)

Ages(yrs)	f
31 – 35	3
35 – 40	4
41 – 45	10
46 – 50	9
51 – 55	7
56 – 60	4
61 – 65	2
66 - 70	1

(b) 41   (c) 50   (d) 38

3. (a)

Marks	f
10 – 19	2
20 – 29	4
30 – 39	5
40 – 49	7
50 – 59	8
60 – 69	9
70 – 79	6
80 – 89	5
90 - 99	4

(b) 64.5   (c) 49.5

(d) 49.5   (e) 10

4(a) 18

Ages	f
60 – 62	1
63 – 65	2
66 – 68	7
69 – 71	12
72 – 74	9
75 – 77	6
78 - 80	3

(c) 3   (d) 7

Exercise 23.2(a)

1. (a)

Mass(g)	f
20 – 29	2
30 – 39	3
40 – 49	18
50 – 59	7
60 – 69	4
70 - 79	1

2(a)

Marks	f
70 – 72	1
73 – 75	2
76 – 78	7
79 – 81	12
82 – 84	9
85 – 87	6
88 - 90	3

3. (a)

Amount earned	f
71 – 73	3
74 – 76	5
77 – 79	9
80 – 82	13
83 – 85	12
86 – 88	6
89 - 91	2

Exercise 23.2(b)

1. (a)

Heights	f	cf
130 – 134	3	3
135 – 139	8	11
140 – 144	10	21
145 – 149	11	32
150 – 154	8	40
155 - 159	7	47
160 - 164	3	50

2.

Marks	f	cf
10 – 19	2	2
20 – 29	4	6
30 – 39	10	16
40 – 49	11	27
50 – 59	10	37
60 – 69	6	43
70 – 79	4	47
80 – 89	2	49
90 - 99	1	50

3.

Masses	f	cf
71 – 73	3	3
74 – 76	5	8
77 – 79	9	17
80 – 82	14	31
83 – 85	11	42
86 – 88	6	48
89 - 99	2	50

4.

Ages	f	cf
20 – 29	3	3
30 – 39	10	13
40 – 49	11	24
50 – 59	13	37
60 – 69	9	46
70 - 79	2	48

Exercise 23.3

1. (a) 38.75, 35.5, no mode   (b) 15.11, 14, 12   (c) 66, 65, no mode

(d) 27.25, 27, 26

2. 58.6, 56.5, 55   3. 6   4. $x = 9$, $y = 6$ 5. 171.69   6. 18

7. 50.5, 51, 51   8. 41.5   9. 21.5   10. 38.5   11. 57

12. 26.69, 27.74   13. 39.59, 39.83   14. 103.79   15. 70.8, 71.29

Answers to Exercises 671

Exercise 23.4

1. (a) 17, 14, 23 (b) 27, 26, 33 (c) 45, 28.5, 72 (d) 30, 27.5, 33.5

2. 64, 10, 65 3. 17, 15, 17 4. 53.5, 38.7, 70.6 5. 63.5, 58.8, 70.8

6. 43.1, 36.6, 48.3 7. 42.5, 36.9, 48.1 8. 61.5, 57, 66

Exercise 23.5

1. (a) 9 (b) 9 (c) 14 (d) 19   2. (a) 10, 5 (b) 10, 5 (c) 12, 6 (d) 12, 6

3. (a) 2 (b) 8 (c) 2.28 (d) 1.12   4. (a) 2.45 (b) 3.92 (c) 3.21 (d) 6.52

5. (a) 27 (b) 154 (c) 155.6 (d) 42.8   6. 34.4   7. 11.6   8. 13.0   9. 3.5

10. 1.23   11. 0.6   12. 1.88, 1.37   13. 4.47   14. 8.6   15. 6.7   16. 2.34

17. 0.82   18. 2.0   19. 3.84   20. 6.45

Exercise 23.6

1. (c) $y = -8x + 259$ (d) 43   2. (b) $y = 0.42x + 18.12$ (c) 22

3. 126   4. 153   5. (b) 9.75   (c) 78.6   6. 0.74   7. 0.85   8. 0.90

9. -0.90   10. 0.66   11. 0.74   12. 0.5

Review exercise 23

1.

Masses(kg)	f
51 – 53	3
54 – 56	5
57 – 59	9
60 – 62	13
63 – 65	12
66 – 68	6
69 - 71	2

(b) 60

3. 29.8   5. 34.9   6. (a) 7.3, 8, 8   (b) 27.25, 27, 26   (c) 124.7, 125, 126

7. 59.4   8. 15.5   9. 25.875   10. 63.26   11. 24.9   12. 29.75   13. 39.375

14. 50.9   15. 6   16. 67.5   17. (b) (i) 35   (ii) 41   (iii) 46.5   18. 26

19. (b) (i) 43.5   (ii) 55   20. 10.5   21. 6.93   22. 8.25   23. 0.79   24. 6.61

25. 7.35   26. (a) 30.8   (b) 7.59   27. (c) $y = -8x + 89$, 33.0   28. 18

29. $y = 10 + 0.43x$ 30. -0.85 31. 0.85 32. 0.67 33. 0.23 34. 0.63

Answers to test

Test 1

1. (a) -15 (b) $-3, 2\frac{2}{3}$ 2. (a) $9(x^2 + 2y^3)(x^2 - 2y^3)$ (b) $(x + 2y)(4x - 5)$

3. (a) $\frac{x+5}{3x-1}$ (b) 0   4. (a) $-4\sqrt{2}$ (b) $6 - 5\sqrt{2}$ 5. $x = -10, y = 8\frac{2}{3}$

6. 0   7. $g^{-1}: x \to \frac{3x+1}{x+2}, x \neq -2$ 8. $4x^2 + 1$ 9. $\{x: x \in R, x \neq \pm 1\}$

10. $(1, 2)$ 11. $\frac{11}{3}, \frac{2}{3}$ 12. $\frac{65}{4}$ 13. 1   14. 2, 6   15. $\{x: -\frac{3}{2} \leq x \leq 1\}$

16. $2x^2 + 9x - 5$ 17. $a = 1, b = 2$ 18. 3, 6   19. $\frac{3}{5(x-3)} + \frac{7}{59x+2)}$ 20. (a)

Test 2

1. 3   2. (a) $\frac{a+2}{2a}$ (b) $\frac{x+2}{x-2}$ 3. $x = -2, y = 5; x = 5, y = -2$

4. (a) $\frac{5}{3}(\sqrt{5} + \sqrt{2})$ (b) $\frac{6\sqrt{2}+11}{7}$ 5. $\frac{x}{1-x}$ 6. 6   7. $x + 1$ 8. (b), (c)

9. $3x - 2y - 12 = 0$ 10. $y = -x + 7$ 11. $\frac{20}{9}$ 12. $\frac{121}{8}$

13. $\{k: k \leq -2, k \geq 6\}$ 14. $\{x: -5 < x < 3\}$ 15. C   16. $a = 1, b = 2$

17. $a = 1, b = -3$ 18. -3, 4   19. $x(x - 3)$ 20. $x + \frac{1}{2(x+1)} + \frac{1}{2(x-1)}$

Test 3

1. $(-4, 5), 10$ 2. 4   3. $(-3, 2), (1, 4)$ 4. 1.05101   5. $560a^6b^4$

6. $1 - \frac{1}{3}x + \frac{2}{9}x^2 - \frac{14}{81}x^3 + \frac{35}{243}x^4$ 7. 1.00995   8. 12   9. 0, 1.89   11. $6p$

12. $x = 1, y = -1; x = \frac{25}{4}, y = \frac{5}{2}$ 13. 39, 210   14. 15.9   15. 3   16. 676

17. $\frac{24}{25}$ 18. $\frac{12}{13}$ 19. $\frac{3}{4}$ 20. $30°, 150°; \frac{\sqrt{3}}{2}, -\frac{\sqrt{3}}{2}$ 21. $63.4°, 71.6°, 243.4°, 251.6°$

22. $x = 3, y = 2$ 23. $\begin{pmatrix} 5 & 0 \\ 7 & 3 \end{pmatrix}$ 24. $\begin{pmatrix} -5 & 3 \\ 2 & -2 \end{pmatrix}$ 25. $(8, 2)$

Test 4

1. $x^2 + y^2 - 2x - 4y = 0$ 2. $y = 2x - 7$ 3. $(6, 0), (0, 4), 12$

4. $1 - 12x + 60x^2 + 160x^3 + 240x^4 + 192x^5 + 64x^6$; 0.886

5. 240   6. $1 + 4x + 10x^2 + 20x^3 + 35x^4$ 7. 1.00503   8. $9/x^{5/3}$ 9. 8

10. 2   11. $x = \frac{5}{2}$ $y = 250$ 12. 3, 7   13. $\frac{3}{2}$ 14. 2   15. $\frac{1}{2}$, 1 $511\frac{1}{2}$ 16. 7

17. 1432 m   18. 5   19. $\frac{56}{65}$ 20. 35.8°, 125.8°, 215.8°, 305.8°

22. $\begin{pmatrix} -2 & -2 \\ 3 & -5 \end{pmatrix}$ 23. $\begin{pmatrix} 0 & 0 \\ 0 & 0 \end{pmatrix}$ 24. $x = -2$ $y = 3$ 25. (1, 2)

Test 5

1. (a) $18x(3x^2 + 5)^2$ (b) $5x^2(x - 3)(5x - 9)$ 2. $x + 9y - 3 = 0$ 3. (1, 6)

4. 1250 m²   5. 8π 6. 9%   7. 0.32 cm s⁻¹   8. $\frac{1}{4}$ s   9. (a) $1\frac{9}{14}$ (b) -3

10. 12   11. 5π 12. $\frac{4}{27}$ 13. $a = 1, b = -4$ 14. $p = -2, q = -5$ 15. 67°

Test 6

1. -23   2. $\frac{3}{2}$ 3. 326.4 m   4. 75 N   5. 20 m s⁻²   6. 250 g   7. 40 N   8. 126

9. 81   10. $\frac{2}{7}$ 11. $\frac{8}{15}$ 12. (a) 3   (b) 3   13. 3.2   14. (a) 36.5   (b) 6.2   15. 43.9

Test 7

1. (a) $6x(x + 1)$ (b) $6x + 1$ 2. $y = 24x - 46$ 3. (3, -7)   4. $a = 0, b = 3$

5. 0.4π 6. 4%   7. $\frac{3}{2}$ cm s⁻¹   8. 65 cm   9. $y = x^2 + 5x - 25$ 10. (a) 2   (b) 27

11. 3   12. 4π 13. 32.75   14. 13   15. 18.4°

Test 8

1. $2.5i + 4.33j$ 2. 19.7 km h⁻¹   3. 8 kg m s⁻¹   4. 3.97   5. 16.7 N   6. 30

7. $\frac{1}{2}$ 8. 2520   9. $\frac{1}{4}$ 10. 0.52   11. 4   12. (b) (i) 25.5   (ii) 11   13. 11.2

14. 1.11   15. (c) $y = 0.62x + 36$ (d) 90

Made in the USA
Lexington, KY
17 June 2018